Discrete Probability Mo

Model	pdf $f(x)$	Support $\mathcal{X}$	Mean μ	Variance σ^2
Equilikely(a, b)	$\dfrac{1}{b - a + 1}$	$x = a, a + 1, \ldots, b$	$\dfrac{a + b}{2}$	$\dfrac{(b - a + 1)^2 - 1}{12}$
Bernoulli(p)	$p^x (1 - p)^{1-x}$	$x = 0, 1$	p	$p(1 - p)$
Geometric(p)	$p^x (1 - p)$	$x = 0, 1, 2, \ldots$	$\dfrac{p}{1 - p}$	$\dfrac{p}{(1 - p)^2}$
Pascal(n, p)	$\dbinom{n + x - 1}{x} p^x (1 - p)^n$	$x = 0, 1, 2, \ldots$	$\dfrac{np}{1 - p}$	$\dfrac{np}{(1 - p)^2}$
Binomial(n, p)	$\dbinom{n}{x} p^x (1 - p)^{n-x}$	$x = 0, 1, \ldots, n$	np	$np(1 - p)$
Poisson(μ)	$\dfrac{e^{-\mu} \mu^x}{x!}$	$x = 0, 1, 2, \ldots$	μ	μ

Continuous Probability Models

Model	pdf $f(x)$	Support $\mathcal{X}$	Mean μ	Variance σ^2
Uniform(a, b)	$\dfrac{1}{b - a}$	$a < x < b$	$\dfrac{a + b}{2}$	$\dfrac{(b - a)^2}{12}$
Exponential(μ)	$\dfrac{1}{\mu} e^{-x/\mu}$	$x > 0$	μ	μ^2
Erlang(n, b)	$\dfrac{1}{b(n - 1)!} (x/b)^{n-1} e^{-x/b}$	$x > 0$	nb	nb^2
Normal$(0, 1)$	$\dfrac{1}{\sqrt{2\pi}} e^{-x^2/2}$	$-\infty < x < \infty$	0	1
Normal(μ, σ)	$\dfrac{1}{\sigma\sqrt{2\pi}} e^{-(x-\mu)^2/2\sigma^2}$	$-\infty < x < \infty$	μ	σ^2
Lognormal(a, b)	$\dfrac{1}{bx\sqrt{2\pi}} e^{-(\ln(x)-a)^2/2b^2}$	$x > 0$	$e^{a+b^2/2}$	$e^{2a+b^2}(e^{b^2} - 1)$
Chisquare(n)	$\dfrac{1}{2\Gamma(n/2)} (x/2)^{n/2-1} e^{-x/2}$	$x > 0$	n	$2n$
Student(n)	$\dfrac{(1 + x^2/n)^{-(n+1)/2}}{\sqrt{n}\, B(1/2, n/2)}$	$-\infty < x < \infty$	0	$\dfrac{n}{n - 2}$

Discrete-Event Simulation: A First Course

Discrete-Event Simulation: A First Course

Lawrence M. Leemis
Professor of Mathematics
The College of William & Mary
Williamsburg, VA

Stephen K. Park
Late Professor of Computer Science
The College of William & Mary
Williamsburg, VA

PEARSON
Prentice
Hall

Upper Saddle River, New Jersey 07458

Library of Congress Cataloging-in-Publication Data

Leemis, Lawrence M.
  Discrete-event simulation : a first course / Lawrence Leemis, Steve Park.
    p. cm.
  Includes bibliographical references and index.
  ISBN 0-13-142917-5
    1. Discrete-time systems—Textbooks. 2. Computer simulation—Textbooks. 3.
  Mathematical models—Textbooks. 4. Random variables—Textbooks. I. Park, Stephen
  Keith. II. Title.

  QA402.L44 2006
  003′.3—dc22   2005054983

Vice President and Editorial Director, ECS: *Marcia J. Horton*
Senior Editor: *Holly Stark*
Associate Editor: *Alice Dworkin*
Editorial Assistant: *Richard Virginia*
Executive Managing Editor: *Vince O'Brien*
Managing Editor: *David A. George*
Production Editor: *Scott Disanno*
Director of Creative Services: *Paul Belfanti*
Art Director: *Jayne Conte*
Cover Designer: *Bruce Kenselaar*
Art Editor: *Greg Dulles*
Manufacturing Manager: *Alexis Heydt-Long*
Manufacturing Buyer: *Lisa McDowell*

© 2006 Pearson Education, Inc.
Pearson Prentice Hall
Pearson Education, Inc.
Upper Saddle River, NJ 07458

Printed in the United States of America

10 9 8 7 6 5 4 3 2 1

ISBN: 0-13-142917-5

Pearson Education Ltd., *London*
Pearson Education Australia Pty. Ltd., *Sydney*
Pearson Education Singapore, Pte. Ltd.
Pearson Education North Asia Ltd., *Hong Kong*
Pearson Education Canada, Inc., *Toronto*
Pearson Educación de Mexico, S.A. de C.V.
Pearson Education—Japan, *Tokyo*
Pearson Education Malaysia, Pte. Ltd.
Pearson Education, Inc., *Upper Saddle River, New Jersey*

In memory of:
Rosa

To those who remain:
Leone, Karin, John, Ashley, Andrew, Kristopher, Kelly, Brittany

and to
my simulation students.
SKP

To my family:
Jill, Lindsey, Mark, Logan

To my original family:
Arthur, Ivah, Richard, Cindy, Nancy

and to
the great "I am" whose reality we simulate.
LML

Contents

Preface

This book presents an introduction to computational and mathematical techniques for *modeling*, *simulating*, and *analyzing* the performance of various systems by using simulation. For the most part, the system models studied are as follows: *stochastic* (at least some of the system-state variables are random); *dynamic* (the time evolution of the system-state variables is important); and *discrete-event* (significant changes in system-state variables are associated with events that occur at discrete time instances only). Therefore, the book represents an introduction to what is commonly known as *discrete-event simulation*. There is also a significant, but secondary, emphasis on *Monte Carlo simulation* and its relation to *static* stochastic systems. Deterministic systems, static or dynamic, and stochastic dynamic systems that evolve continuously in time are not considered in any significant way.

Discrete-event simulation is a multidisciplinary activity studied and applied by students of applied mathematics, computer science, industrial engineering, management science, operations research, statistics, and various hybrid versions of these disciplines found in schools of engineering, business, management, and economics. As it is presented in this book, discrete-event simulation is a computational science—a mix of theory and experimentation, with a computer as the primary piece of laboratory equipment. In other words, discrete-event simulation is a form of computer-aided model-building and problem-solving. The goal is insight: a better understanding of how systems operate and respond to change.

PREREQUISITES

In terms of formal academic background, we presume the reader has taken the undergraduate equivalent of the first several courses in a conventional computer-science program, two calculus courses, and a course in probability or statistics. In more detail, and in decreasing order of importance, these prerequisites are as follows.

Computer Science: Readers should be able to program in a contemporary high-level programming language—for example C, C++, Java, Pascal, or Ada—and have a working knowledge of algorithm complexity. Because the development of most discrete-event simulation programs necessarily involves an application of queues and event lists, some familiarity with dynamic data structures is prerequisite, as is the ability to program in a language that naturally supports such objects. By design, the computer-science prerequisite is strong. We firmly believe that the best way to learn about discrete-event simulation is by hands-on model-building. We consistently advocate a structured approach wherein a model is constructed at three levels—conceptual, specification, and computational. At the computational level, the model is built as a computer program; we believe that this construction is best done with a standard and widely available general-purpose high-level programming language, by the use of tools (editor, compiler, debugger, etc.) already familiar.*

Calculus: Readers should be able to do single-variable differential and integral calculus. Although still relatively strong, the calculus prerequisite is as weak as possible—for example, we have generally avoided the use of multivariate calculus—however, single-variable integration and differentiation is used as appropriate in the discussion of, for example, continuous random variables. In addition, we freely use the analogous, but computationally more intuitive, discrete mathematics of summation and differencing in the discussion of discrete random variables. By design, we maintain a balance between continuous and discrete stochastic models, generally using the more easily understood, but less common, discrete (noncalculus) techniques to provide motivation for the corresponding continuous (calculus) techniques.

Probability: Readers should have a working knowledge of probability, including random variables, expected values, and conditioning. Some knowledge of statistics is also desirable, but not necessary. Those statistical tools most useful in discrete-event simulation are developed as needed. Because of the organization of the material, our classroom experience has been that students with strength only in the computer science and calculus prerequisites can use this book to develop a valid intuition about things stochastic. In this way, the reader can learn about discrete-event simulation and, if necessary, also establish the basis for a later formal study of probability and statistics. That study is important for serious students of discrete-event simulation, because without the appropriate background, a student is unlikely ever to be proficient at modeling and analyzing the performance of stochastic systems.

ORGANIZATION AND STYLE

The book has 10 chapters, organized into 41 sections. The shortest path through the text could *exclude* the 15 optional Sections 2.4, 2.5, 4.4, 5.3, 6.4, 6.5, 7.4, 7.5, 7.6, 8.5, 9.3, and 10.1–10.4. All the 26 remaining core sections are consistent with a 75-minute classroom lecture; together, they define a traditional one-semester, three-credit-hour course.† Generally, the optional sections in the first nine chapters are also consistent with a 75-minute presentation and so can be used in a classroom setting as supplemental lectures. Each section in the tenth chapter provides relatively detailed specifications for a variety of discrete-event simulation projects designed to integrate much of the core material. In addition, there are seven appendices that provide background or reference material.

---

*The alternative to using a general-purpose high-level programming language is to use a (proprietary, generally unfamiliar, and potentially expensive) special-purpose simulation language. In some applications, this can be a superior alternative, particularly if the simulation language is already familiar (and paid for); see Chapter 1 and Appendix A for more discussion of this trade-off.

†Because of its multidisciplinary nature, there is not universal agreement on what constitutes the academic core of discrete-event simulation. It is clear, however, that the core is large, sufficiently so that we have not attempted to achieve comprehensive coverage. Instead, the core sections in the first nine chapters provide a self-contained, although limited, first course in discrete-event simulation.

In a traditional one-semester, three-credit-hour course, there might not be time to cover more than the 26 core sections. In a four-credit-hour course, there will be time to cover the core material and some of the optional sections (or appendices), and, if appropriate, to structure the course around the projects in the tenth chapter as a culminating activity. Similarly, some optional sections can be covered in a three-credit-hour course, provided that student background is sufficient to warrant not devoting classroom time to some of the core sections.

The book is organized in a way consistent with a dual philosophy: (i) begin to model, simulate, and analyze simple-but-representative systems as soon as possible; (ii) whenever possible, encourage the experimental exploration and self-discovery of theoretical results before their formal presentation. As an example of (i), detailed trace-driven computational models of a single-server queue and a simple inventory system are developed in Chapter 1, then used to motivate the need for the random-number generator developed in Chapter 2. The random-number generator is used to convert the two trace-driven models into stochastic models that can be used to study both transient and steady-state system performance in Chapter 3. Similarly, as an example of (ii), an experimental investigation of sampling uncertainty and interval estimation is motivated in Chapters 2 and 3. A formal treatment of this topic is presented in Chapter 8.

We have tried to achieve a writing style that emphasizes *concepts* and *insight* without sacrificing rigor. Generally, formalism and proofs are not emphasized. When appropriate, however, definitions and theorems (most with proofs) are provided, particularly if their omission could create a sense of ambiguity that might impede a reader's ability to understand concepts and develop insights.

SOFTWARE

Software is an integral part of the book. We provide this software as source code for several reasons. Because a computer program is the logical product of the three-level approach to model-building we advocate, an introductory discrete-event simulation book based on this philosophy would be deficient if a representative sampling of such programs were not presented. Moreover, many important exercises in the book are based on the idea of extending a system model at the computational level; these exercises are conditioned on access to the source code. The software consists of many complete discrete-event programs and a variety of libraries for random-number generation, random-variate generation, statistical data analysis, priority-queue access, and event-list processing.

The software has been translated from its original development in Turbo Pascal to ANSI C with *units* converted to C libraries. Although experienced C programmers will no doubt recognize the Pascal heritage, the result is readable, structured, portable, and reasonably efficient ANSI C source code.* A C-like syntax is also used on the pseudocode that is used to describe the algorithms developed in the text.

EXERCISES

There are exercises associated with each chapter and some appendices—about 400 in all. They are designed to reinforce and extend the material and to encourage computational experimentation.

---

*All the programs and libraries compile successfully, without warnings, under the GNU C compiler gcc with the -ansi -Wall switches set. Alternatively, the GNU C++ compiler g++ can be used instead.

Some exercises are routine, others are more advanced. Some of the advanced exercises are sufficiently challenging and comprehensive to merit consideration as (out-of-class) exam questions or projects. Serious readers are encouraged to work a representative sample of the routine exercises and, time permitting, a large portion of the advanced exercises.

Consistent with the computational philosophy of the book, a significant number of exercises require some computer programming. If required, the amount of programming is usually small for the routine exercises and more significant for the advanced exercises. For some of the advanced exercises, the amount of programming could be significant. In most cases in which programming is required, the reader is aided by access to source code for the programs and to related software tools that the book provides.

Our purpose is to give an introductory, intuitive development of algorithms and methods used in Monte Carlo and discrete-event simulation modeling. More comprehensive treatments are given in the textbooks referenced throughout the text.

ANCILLARIES

The ANSI C programs associated with the text have been converted to Java. A solutions manual and overhead slides are available for instructors.

CAVEATS

Although the material presented in this text provides a foundation for understanding discrete-event simulation, we recommend that students take additional courses that view discrete-event simulation from other perspectives. One such perspective is a course in modeling real-world systems by using a discrete-event simulation package that includes animation. Such a course presents challenges that are not encountered in this text: interacting with a client, collecting real-world data, building stochastic input models from that data, learning the syntax of a discrete-event package, verifying and validating a discrete-event simulation model, animating a discrete-event simulation model, analyzing the output, and presenting the results to the client. A second alternative perspective emphasizes the probabilistic and statistical aspects of simulation and requires more probability/statistics background than is required for this book. Many of the simulation textbooks listed as references in this text provide the appropriate perspective for these types of companion courses.

ACKNOWLEDGMENTS

We have worked diligently to make this book as readable and as free of errors as possible. We have been helped in this by student feedback and by the comments of several associates. Our thanks to all for your time and effort. We would like to acknowledge the contribution of Don Knuth, whose TEX makes the typesetting of technical material so rewarding, and of Michael Wichura, whose PICTEX macros give TEX the ability to do graphics. Particular thanks go to former students Mousumi Mitra and Rajeeb Hazra, who TEX-set preliminary versions of some material; to Tim Seltzer, who PICTEX-ed preliminary versions of some figures; to Dave Geyer, who converted a significant amount of

Pascal software into C; and to Rachel Siegfried, who proofread much of the manuscript. Special thanks go to our colleagues and their students who have class-tested this text and provided us lists of typos and suggestions: David Nicol at William & Mary, Tracy Camp at The University of Alabama, Dan Chrisman at Radford University, Rahul Simha at William & Mary, Evgenia Smirni at William & Mary, Barry Lawson at the University of Richmond, Andy Miner at Iowa State University, Ben Coleman at Moravian College, Ed Williams at the University of Michigan–Dearborn, and Mike Overstreet at Old Dominion University. Thanks also to Barry Lawson, Andy Miner, and Ben Coleman, who have prepared PowerPoint slides to accompany the text.

We appreciate the help, comments, and advice on various parts of this text and associated software from Sigrún Andradóttir, Kerry Connell, Matt Duggan, Jason Estes, Diane Evans, Andy Glen, James Henrikson, Elise Hewett, Whit Irwin, Charles and Dana Johnson, Rex Kincaid, Pierre L'Ecuyer, Lindsey Leemis, Chris Leonetti, David Lutzer, Jeff Mallozzi, Nathan and Rachel Moore, Lisa Moya, Bob Noonan, Steve Roberts, Eric and Kristen Rozier, Jes Sloan, Jim Swain, Virginia Torczon, Michael Trosset, Ed Walsh, Ed Williams, Marianna Williamson, Zheng Zhang, and several anonymous reviewers. Bruce Schmeiser, Barry Nelson, and Michael Taaffe provided valuable guidance on the framework introduced in Appendix G. Thanks to Barry Lawson and Nathan and Rachel Moore for contributing exercise solutions to the manual that is being edited by Matt Duggan and Ed Walsh. The authors are grateful to Evgenia Smirni and her students Richard Dutton and Jun Wang for converting the C code to Java. A special word of thanks goes to Barry Lawson for his generous help with TEXand PICTEX, with setting up *make* files, and with proofreading the text. We are both very grateful to our wives, Rosa and Jill, for their unwavering support for this writing project. The authors also thank The College of William & Mary for some teaching relief needed to complete this text.

When Steve Park passed away on April 16, 2001, the computer science, simulation, and academic communities lost a talented teacher, writer, researcher, and leader. He is still deeply missed by family, friends, and colleagues. Steve served as the chair of the Computer Science Department at William & Mary from 1991 to 2000 and laid the foundation for the department it has become. The lecture notes that Steve developed for the department's primary simulation course (Computer Science 426) have provided the basis for this text. A favorite quote (attributed to R. Coveyou) he used in that simulation course was, "The generation of random numbers is too important to be left to chance." Steve felt the same way about teaching. He combined his creative research ability with a very organized and disciplined writing style, which you will see throughout the text. Any mathematical or grammatical errors that remain are my fault and I would be grateful if you would bring them to my attention. I hope you enjoy the book.

LARRY LEEMIS

1

Models

The modeling approach in this book is based on the use of a general-purpose programming language for model implementation at the computational level. The alternative approach is to use a special-purpose simulation language; for a survey of several such languages, see Appendix A.

This chapter presents an introduction to discrete-event simulation, with an emphasis on model-building. The focus in Section 1.1 is on the multiple steps required to construct a discrete-event simulation model. By design, the discussion in this section is at a high level of generality, with few details. In contrast, two specific discrete-event system models are presented in Sections 1.2 and 1.3, with a significant amount of detail. A single-server queue model is presented in Section 1.2; a simple inventory system model is presented in Section 1.3. Both of these models are of fundamental importance because they serve as a basis for a significant amount of material in later chapters.

Although the material in this chapter also can be found in other modeling and simulation texts, there is a relatively novel emphasis on model-building at the conceptual, specification, and computational levels. Moreover, in Sections 1.2 and 1.3, there is a significant amount of notation, terminology, and computational philosophy, which extends to subsequent chapters. For these reasons, this chapter is important to any reader of the book, even those already familiar with the rudiments of discrete-event simulation.

1.1 INTRODUCTION

This book is the basis for a first course on *discrete-event simulation* — that is, the book provides an introduction to computational and mathematical techniques for *modeling*, *simulating*, and *analyzing* the performance of discrete-event stochastic systems. By definition, the nature of discrete-event simulation is that one does not actually experiment with or modify an actual system.

Instead, one develops and then works with a discrete-event simulation *model*. Consistent with that observation, the emphasis in this first chapter is on model-building.

1.1.1 Model Characterization

Briefly, a discrete-event simulation model is both stochastic and dynamic with the special discrete-event property that the system-state variables change value at discrete times only (see Definition 1.1.1). But what does that mean?

A system model is *deterministic* or *stochastic*. A deterministic system model has no stochastic (random) components. For example, provided that the conveyor belt and machine never fail, a model of a constant-velocity conveyor belt feeding parts to a machine with a constant service time is deterministic. At some level of detail, however, all systems have some stochastic components: Machines fail, people are not robots, service requests occur at random, and so on. An attractive feature of discrete-event simulation is that stochastic components can be accommodated, usually without a dramatic increase in the complexity of the system model at the computational level.

A system model is *static* or *dynamic*. A static system model is one in which time is not a significant variable. For example, if three million people play the state lottery this week, what is the probability that there will be at least one winner? A simulation program written to answer this question should be based on a static model; when during the week these three million people place their bets is not significant. If, however, we are interested in the probability of no winners in the next four weeks, then this model needs to be dynamic. That is, experience has revealed that each week that there are no winners, the number of players in the following week increases (because the pot grows). When this happens, a dynamic system model must be used, because the probability of at least one winner will increase as the number of players increases. The passage of time always plays a significant role in dynamic models.

A dynamic system model is *continuous* or *discrete*. Most of the traditional dynamic systems studied in classical mechanics have state variables that evolve continuously. A particle moving in a gravitational field, an oscillating pendulum, or a block sliding on an inclined plane are examples. In each of these cases, the motion is characterized by one or more differential equations, which model the continuous time evolution of the system. In contrast, the kinds of queuing, machine

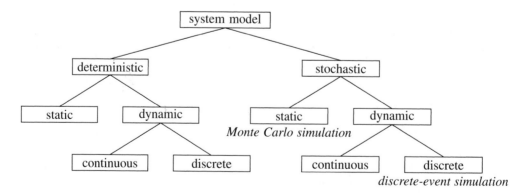

Figure 1.1.1   System model taxonomy.

repair, and inventory systems studied in this book are discrete, because the state of the system is a *piecewise constant* function of time. For example, the number of jobs in a queuing system is a natural state variable that changes value only at those discrete times when a job arrives (to be served) or departs (after being served).

The characterization of a system model can be summarized by a tree diagram that starts at the system-model root and steps left or right at each of the three levels, as illustrated in Figure 1.1.1.

As summarized by Definition 1.1.1, the system model characterized by the rightmost branch of this tree is of primary interest in this book.

Definition 1.1.1   A *discrete-event simulation model* is defined by three attributes:

- *stochastic*—at least some of the system-state variables are random;
- *dynamic*—the time evolution of the system-state variables is important;
- *discrete-event*—significant changes in the system-state variables are associated with events that occur at discrete time instances only.

One of the other five branches of the system-model tree is of significant, but secondary, interest in this book. A *Monte Carlo simulation model* (Fishman, 2006) is stochastic and static—at least some of the system-state variables are random, but the time evolution (if any) of the system-state variables is not important. Accordingly, the issue of whether time flows continuously or discretely is not relevant.

Because of space constraints, the remaining four branches of the system-model tree are not considered—that is, there is no material about deterministic systems, static or dynamic, or about stochastic dynamic systems that evolve continuously in time.

1.1.2 Model Development

It is naive to think that the process of developing a discrete-event simulation model can be reduced to a simple sequential algorithm. As an instructional device, however, it is useful to consider two algorithms that outline, at a high level, how to develop a discrete-event simulation model (Algorithm 1.1.1) and then conduct a discrete-event simulation study (Algorithm 1.1.2).

Algorithm 1.1.1   If done well, a typical discrete-event simulation model will be developed in a way consistent with the following six steps. Steps (2) through (6) are typically iterated, perhaps many times, until a valid computational model, i.e., a computer program, has been developed.

(1) Determine the *goals* and *objectives* of the analysis once a system of interest has been identified. These goals and objectives are often phrased as simple Boolean decisions (e.g., should an additional queuing network service node be added?) or numeric decisions (e.g., how many parallel servers are necessary to provide satisfactory performance in a multiple-server queuing system). Without specific goals and objectives, the remaining steps lack meaning.

(2) Build a *conceptual* model of the system that is based on (1). What are the *state variables*, how are they interrelated, and to what extent are they dynamic? How comprehensive should the model be? Which state variables are important; which have an effect so negligible that they can be ignored? This is an intellectually challenging but rewarding activity, one that should not be avoided just because it is hard to do.

(3) Convert the conceptual model into a *specification* model. If this step is done well, the remaining steps are made much easier. If, instead, this step is done poorly (or not at all), the remaining steps are probably a waste of time. This step typically involves collecting and statistically analyzing data to provide the input models that drive the simulation. In the absence of such data, the input models must be constructed in an ad-hoc manner by using stochastic models believed to be representative. This step also includes the development of equations and algorithms required to simulate the system.

(4) Turn the specification model into a *computational* model, a computer program. At this point, a fundamental choice must be made—to use a general-purpose programming language, or to use a special-purpose simulation language. For some, this is a religious issue not subject to rational debate.

(5) *Verify* the model. As with all computer programs, the computational model should be consistent with the specification model—did we implement the computational model correctly? This verification step is not the same as the next step.

(6) *Validate* the model. Is the computational model consistent with the system being analyzed—did we build the right model? Because the purpose of simulation is insight, some (including the authors) would argue that the *act* of developing the discrete-event simulation model—steps (2), (3), and (4)—is frequently as important as the tangible *product*. However, given the blind faith many people place in any computer-generated output, the validity of a discrete-event simulation model is always fundamentally important. One popular nonstatistical, Turing-like technique for model validation is to place actual system output alongside similarly formatted output from the computational model. This output is then examined by an expert familiar with the system. Model validation is indicated if the expert is not able to determine which is the model output and which is the real thing. Interactive computer graphics (animation) can be very valuable during the verification and validation steps.

Example 1.1.1

This *machine-shop* model helps illustrate the six steps in Algorithm 1.1.1.

A new machine shop has 150 identical machines; each operates continuously, eight hours per day, 250 days per year, until failure. Each machine operates independently of all the others. As machines fail, they are repaired, in the order in which they fail, by a service technician. As soon as a failed machine is repaired, it is put back into operation. Each machine produces a net income of $20 per hour of operation. All service technicians are hired at once, for two years, at the beginning of the two-year period, with an annual salary expense of $52,000 each. Because of vacations, each service technician works only 230 eight-hour days per year. By agreement, vacations are coordinated to maximize the number of service technicians on duty each day. How many service technicians should be hired?

(1) The objective seems clear—to find that number of service technicians that maximizes the profit. One extreme solution is to hire one technician for each machine; this produces a huge service-technician overhead but maximizes income by minimizing the amount of machine down-time. The other extreme solution is to hire just one technician; this minimizes overhead at the potential expense of large down-times and associated loss of income. In this case, neither extreme is close to optimal for typical failure and repair times.

(2) A reasonable conceptual model for this system can be expressed in terms of the state of each machine (failed or operational) and each service technician (busy or idle). These state variables provide a high-level description of the system at any time.

(3) To develop a specification model, more information is needed. Machine failures are random events; what is known (or can be assumed) about the time between failures for these machines? The time to repair a machine is also random; what, for example, is the distribution of the repair time? In addition, to develop the associated specification model, one must devise some systematic method to simulate the time evolution of the system-state variables.

(4) The computational model will likely include a simulation-clock data structure to keep track of the current simulation time, a queue of failed machines, and a queue of available service technicians. Also, to characterize the performance of the system, there will be statistics-gathering data structures and associated procedures. The primary statistic of interest here is the total profit associated with the machine shop, given as a function of the number of technicians.

(5) The computational model must be verified, usually by extensive testing. Verification is a software-engineering activity made easier if the model is developed in a contemporary programming environment.

(6) The validation step is used to see whether the verified computational model is a reasonable approximation of the machine shop. If the machine shop is already operational, the basis for comparison is clear. If, however, the machine shop is not yet operational, validation is based primarily on *consistency checks*. If the number of technicians is increased, does the time-averaged number of failed machines go down; if the average service time is increased, does the time-averaged number of failed machines go up?

System Diagrams.   Particularly at the conceptual level, the process of model development can be facilitated by drawing system diagrams. Indeed, when asked to explain a system, our experience is that, instinctively, many people begin by drawing a system diagram. For example, consider this system diagram of the machine-shop model in Example 1.1.1.

The box at the top of Figure 1.1.2 represents the pool of machines. The composite object at the bottom of the figure represents the four service technicians and an associated single queue.

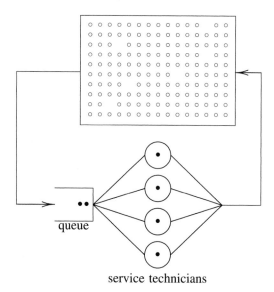

Figure 1.1.2   Machine-shop system diagram.

Operational machines are denoted with a ◦ and broken machines with a •. Conceptually, as machines break they change their state from operational (◦) to broken (•) and move along the arc on the left from the box at the top of the figure to the queue at the bottom of the figure. From the queue, a broken machine begins to be repaired as a service technician becomes available. As each broken machine is repaired, its state is changed to operational and the machine moves along the arc on the right, back to the pool of operational machines.*

As time evolves, there is a continual counterclockwise circulation of machines: from the pool at the top of Figure 1.1.2, to the service technicians at the bottom of the figure, and then back again. At the "snapshot" instant illustrated, there are six broken machines; four of these are being repaired, and the other two are waiting in the queue for a service technician to become available.

In general, the application of Algorithm 1.1.1 should be guided by the following observations.

- Throughout the development process, the operative principle should always be to make every discrete-event simulation model as simple as feasible, but never simpler. The goal is to capture only the relevant characteristics of the system. The dual temptations of (*i*) ignoring relevant characteristics and (*ii*) including characteristics that are extraneous to the goals of the model should be avoided.
- The actual development of a complex discrete-event simulation model will not be as sequential as Algorithm 1.1.1 suggests, particularly if the development is a team activity, in which case some steps will surely be worked in parallel. The different characteristics of each step should always be kept clearly in mind—to avoid, for example, the natural temptation to merge steps (5) and (6).

---

*The movement of the machines to the servers is conceptual, as is the queue. In practice, the servers would move to the machines and there would not be a physical queue of broken machines.

- There is an unfortunate tendency on the part of many to largely skip over steps (1), (2), and (3), jumping rapidly to step (4). Skipping these first three steps is an approach to discrete-event simulation virtually certain to produce large, inefficient, unstructured computational models that cannot be validated. Discrete-event simulation models should *not* be developed by those who like to think a little and then program a lot.

1.1.3 Simulation Studies

Algorithm 1.1.2 Following the successful application of Algorithm 1.1.1, use of the resulting computational model (computer program) involves the following steps.

(7) Design the simulation experiments. This is not as easy as it may seem. If there are a significant number of system parameters, each with several possible values of interest, then the combinatoric possibilities to be studied make this step a real challenge.

(8) Make production runs. The runs should be made systematically, with the value of all initial conditions and input parameters recorded along with the corresponding statistical output.

(9) Analyze the simulation results. The analysis of the simulation output is statistical in nature, because discrete-event simulation models have stochastic (random) components. The most common statistical-analysis tools (means, standard deviations, percentiles, histograms, correlations, etc.) will be developed in later chapters.

(10) Make decisions. The results of step (9) will be used in reaching decisions that result in actions taken. If so, the extent to which the computational model correctly predicted the outcome of these actions is always of great interest, particularly if the model is to be further refined in the future.

(11) Document the results. If you really did gain insight, summarize it in terms of specific observations and conjectures. If not, why did you fail? Good documentation facilitates the development (or avoidance) of subsequent similar system models.

Example 1.1.2
As a continuation of Example 1.1.1, consider the application of Algorithm 1.1.2 to a verified and validated machine-shop model.

(7) The objective of the model is to find the optimal number of service technicians to hire to maximize profit, so the number of technicians is the primary system parameter to be varied from one simulation run to the next. Other issues also contribute to the design of the simulation experiments. What are the initial conditions for the model (e.g., are all machines initially operational)? For a fixed number of service technicians, how many replications are required to reduce the natural sampling variability in the output statistics to an acceptable level?

(8) If many production runs are made, management of the output results becomes an issue. A discrete-event simulation study can produce *a lot* of output files, which will consume large amounts of disk space if not properly managed. Avoid the temptation to archive "raw data" (e.g., a detailed time history of simulated

machine failures). If this kind of data is needed in the future, it can always be reproduced. Indeed, the ability to reproduce previous results *exactly* is an important feature that distinguishes discrete-event simulation from other, more traditional, experimental sciences.

(9) The statistical analysis of simulation output often is more difficult than classical statistical analysis, where observations are assumed to be *independent*. In particular, time-sequenced simulation-generated observations are often correlated with one another, making the analysis of such data a challenge. If the current number of failed machines is observed each hour, for example, consecutive observations will be found to be significantly positively correlated (since consecutive observations tend to both be above or below the mean number of failed machines). A statistical analysis of these observations based on the (false) assumption of independence could produce erroneous conclusions.

(10) For this example, a graphical display of profit versus the number of service technicians yields both the optimal number of technicians and a measure of how sensitive the profit is to variations about this optimal number. In this way, a policy decision can be made. Provided that this decision does not violate any external constraints, such as labor-union rules, the policy should be implemented.

(11) Documentation of the machine-shop model would include a system diagram, explanations of assumptions made about machine failure rates and service repair rates, a description of the specification model, software for the computational model, tables and figures of output, and a description of the output analysis.

Insight.   An important benefit of developing and using a discrete-event simulation model is that valuable insight is acquired. As conceptual models are formulated, computational models are developed, and output data are analyzed, subtle system features and component interactions might be discovered that otherwise would not have been noticed. The systematic application of Algorithms 1.1.1 and 1.1.2 can result in better actions taken, ones due to insight gained by an increased understanding of how the system operates.

1.1.4 Programming Languages

There is a continuing debate in discrete-event simulation over whether to use a general-purpose programming language or a (special-purpose) simulation-programming language. For example, two standard discrete-event simulation textbooks provide the following contradictory advice. Bratley, Fox, and Schrage (1987, page 219) state "... for any important large-scale real application, we would write the programs in a standard general-purpose language, and avoid all the simulation languages we know." In contrast, Law and Kelton (2000, page 204) state "... we believe, in general, that a modeler would be prudent to give serious consideration to the use of a simulation package."

General-purpose languages are more flexible and familiar; simulation languages allow modelers to build computational models quickly. There is no easy way to resolve this debate in general. However, for the specific purpose of this book—learning the principles and techniques of discrete-event simulation—the debate is easier to resolve. Our approach is facilitated by using a familiar, general-purpose programming language, a philosophy that has dictated the style and content of this book.

General-Purpose Languages.   Because discrete-event simulation is a specific instance of scientific computing, any *general-purpose* programming language suitable for scientific computing is similarly suitable for discrete-event simulation. Therefore, a history of the use of general-purpose programming languages in discrete-event simulation is really a history of general-purpose programming languages in scientific computing. Although this history is extensive, we will try to summarize it in a few paragraphs.

For many years, FORTRAN was the primary general-purpose programming language used in discrete-event simulation. In retrospect, this was natural and appropriate; there was no well-accepted alternative. By the early 80's, things had begun to change dramatically. Several general-purpose programming languages created in the 70's, primarily C and Pascal, were as good as or superior to FORTRAN in most respects, and they began to gain acceptance in many applications, including in discrete-event simulation, where FORTRAN had once been dominant. Because of its structure and its relative simplicity, Pascal became the de facto first programming language in many computer science departments; because of its flexibility and power, the use of C became common among professional programmers.

Personal computers became popular in the early 80's and were followed soon thereafter by increasingly more powerful workstations. Concurrently with this development, it became clear that networked workstations or, to a lesser extent, stand-alone personal computers were ideal discrete-event simulation engines. The popularity of workstation networks then helped to guarantee that C would become the general-purpose language of choice for discrete-event simulation: The usual workstation network was Unix-based, and, in that environment, C was the natural general-purpose programming language of choice. The use of C in discrete-event simulation became widespread by the early 90's, when C became standardized and when C++, an object-oriented extension of C, gained popularity.

In addition to C, C++, FORTRAN, and Pascal, other general-purpose programming languages are used occasionally in discrete-event simulation. Of these, Ada, Java, and Visual BASIC are probably the most common. This diversity is not surprising; every general-purpose programming language has its advocates, some quite vocal, and, no matter what the language, there is likely to be an advocate to argue that it is ideal for discrete-event simulation. We leave that debate for another forum, however, confident that our use of ANSI C in this book is appropriate.

Simulation Languages.   Simulation languages have built-in features that provide many of the tools needed to write a discrete-event simulation program. Because of this, simulation languages support rapid prototyping and have the potential to decrease programming time significantly. Moreover, animation is a particularly important feature now built into most of these simulation languages. This capability is important, because animation can increase the acceptance of discrete-event simulation as a legitimate problem-solving technique. By using animation, dynamic graphical images can be created that enhance verification, validation, and the development of insight. The most popular discrete-event simulation languages historically are GPSS, SIMAN, SLAM, and SIMSCRIPT. Because of our emphasis in this book on the use of general-purpose languages, any additional discussion of simulation languages is deferred to Appendix A.

Because it is not discussed in Appendix A, for historical reasons it is appropriate here to mention the simulation language Simula. This language was developed in the 60's as an object-oriented ALGOL extension. Despite its object orientation and several other novel (for the time) features, it never achieved much popularity, except in Europe. Still, like other premature-but-good ideas, the impact of Simula has proven to be profound, especially by its serving as the inspiration for the creation of C++.

1.1.5 Organization and Terminology

We conclude this first section with some brief comments about the organization of the book and the sometimes ambiguous use of the words *simulation*, *simulate*, and *model*.

Organization.   The material in this book could have been organized in several ways. Perhaps the most natural sequence would be to follow, in order, the steps in Algorithms 1.1.1 and 1.1.2, devoting a chapter to each step. However, that sequence is not followed. Instead, the material is organized in a manner consistent with the experimental nature of discrete-event simulation. That is, we begin to model, simulate, and analyze simple-but-representative systems as soon as possible (indeed, in the next section). Whenever possible, new concepts are first introduced in an informal way that encourages experimental self-discovery, with a more formal treatment of the concepts deferred to later chapters. This organization has proven to be successful in the classroom.

Terminology.   The words "model" and "simulation" (or "simulate") are commonly used inter-changeably in the discrete-event simulation literature, both as a noun and as a verb. For peda-gogical reasons, this word interchangeability is unfortunate because, as indicated previously, a "model" (the noun) exists at three levels of abstraction: conceptual, specification, and computa-tional. At the computational level, a system model is a computer program; this computer program is what most people mean when they talk about a system simulation. In this context, a simulation and a computational system model are equivalent. It is uncommon, however, to use the noun "simulation" as a synonym for the system model at either the conceptual or specification level. Similarly, "to model" (the verb) implies activity at three levels, but "to simulate" is usually a computational activity only.

When appropriate, we will try to be careful with these words, generally using "simulation" or "simulate" in reference to a computational activity only. This is consistent with common usage of the word *simulation* to characterize not only the computational model (computer program) but also the computational process of using the discrete-event simulation model to generate output statistical data and thereby analyze system performance. In those cases when there is no real need to be fussy about terminology, we will yield to tradition and use the word simulation or simulate even though the word model might be more appropriate.

1.1.6 Exercises

1.1.1   There are six leaf nodes in the system model tree in Figure 1.1.1. For each leaf node, describe a specific example of a corresponding physical system.

1.1.2   The distinction between model *verification* and model *validation* is not always clear in practice. Generally, in the sense of Algorithm 1.1.1, the ultimate objective is a valid discrete-event simulation model. If you were told that "this discrete-event simulation model has been verified but it is not known whether the model is valid," how would you interpret that statement?

1.1.3   The *state* of a system is important, but difficult to define in a general context. (*a*) Locate at least five contemporary textbooks that discuss system modeling and, for each, research and comment on the extent to which the technical term "state" is defined. If possible, avoid example-based definitions or definitions based on a specific system. (*b*) How would you define the state of a system?

1.1.4   (*a*) Use an Internet search engine to identify at least ten different simulation languages that support discrete-event simulation. (Note that the '-' in discrete-event is not a universal convention.) Provide a URL, phone number, or mailing address for each and, if it is a commercial product, a price. (*b*) If you tried multiple search engines, which produced the most meaningful hits?

1.2 A SINGLE-SERVER QUEUE

In this section, we will construct a *trace-driven* discrete-event simulation model (i.e., a model driven by external data) of a single-server service node. We begin the construction at the conceptual level.

1.2.1 Conceptual Model

Definition 1.2.1   A *single-server service node* consists of a *server* plus its *queue*, as illustrated in Figure 1.2.1.*

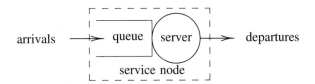

Figure 1.2.1   Single-server service-node system diagram.

Jobs (customers) arrive at the service node at random points in time, seeking service. When service is provided, the service time involved is also random. At the completion of service, jobs depart. The service node operates as follows: As each (new) job arrives, if the server is busy, then the job enters the queue, else the job immediately enters service; as each (old) job departs, if the queue is empty, then the server becomes idle, else a job is selected from the queue to immediately

_____

*The term "service node" is used in anticipation of extending this model, in later chapters, to a *network* of service nodes.

enter service. At any time, the state of the server will be either *busy* or *idle* and the state of the queue will be either *empty* or *not empty*. If the server is idle, the queue must be empty; if the queue is not empty then the server must be busy.

Example 1.2.1
If there is just one service technician, the machine-shop model presented in Examples 1.1.1 and 1.1.2 is a single-server service-node model. That is, the "jobs" are the machines to be repaired and the "server" is the service technician. (In this case, whether the jobs move to the server or the server moves to the jobs is not an important distinction because the repair time is the primary source of delay.)

Definition 1.2.2 Control of the queue is determined by the *queue discipline*—the algorithm used when a job is selected from the queue to enter service. The standard algorithms are the following:

- FIFO—first in, first out (the traditional computer-science queue data structure);
- LIFO—last in, first out (the traditional computer-science stack data structure);
- SIRO—service in random order;
- Priority—typically, shortest job first (SJF) or equivalently, in job-shop terminology, shortest processing time (SPT) first.

The maximum possible number of jobs in the service node is the *capacity*. The capacity can be either *finite* or *infinite*. If the capacity is finite, then jobs that arrive and find the service node full will be *rejected* (unable to enter the service node).

Certainly the most common queue discipline is FIFO (also known as FCFS—first come, first served). If the queue discipline is FIFO, then the order of arrival to the service node and the order of departure from the service node are the same; there is no passing. In particular, upon arrival, a job will enter the queue if and only if the *previous* job has not yet departed the service node. This is an important observation that can be used to simplify the simulation of a FIFO single-server service node. If the queue discipline is not FIFO then, for at least some jobs, the order of departure will differ from the order of arrival. In this book, the *default* assumptions are that the queue discipline is FIFO and that the service-node capacity is infinite, unless otherwise specified. Discrete-event simulation allows these assumptions to be altered easily, for more realistic modeling.

There are two important additional default assumptions implicit in Definition 1.2.1. First, service is *nonpreemptive*—once initiated, service on a job will be continued until completion. Hence, a job in service cannot be preempted by another job arriving later. (Preemption is commonly used with priority queue disciplines to prevent a job with a large service time requirement from producing excessive delays for small jobs arriving soon after service on the large job has begun.) Second, service is *conservative*—the server will never remain idle if there is at least one job in the service node. If the queue discipline is not FIFO *and* if the next arrival time is known in advance then, even though one or more jobs are in the service node, it could be desirable for a nonconservative server to remain idle until the next job arrives. This is particularly true in nonpreemptive job-scheduling applications, if a job in the service node has a much larger service requirement than the next job scheduled to arrive.

1.2.2 Specification Model

The following variables, illustrated in Figure 1.2.2, provide the basis for moving from a conceptual model to a specification model. At their arrival to the service node, jobs are indexed by $i = 1, 2, 3, \ldots$. For each job, there are six associated time variables.

- The *arrival time* of job i is a_i.
- The *delay* of job i in the queue is $d_i \geq 0$.
- The time that job i begins service is $b_i = a_i + d_i$.
- The *service time* of job i is $s_i > 0$.
- The *wait* of job i in the service node (queue and service) is $w_i = d_i + s_i$.
- The time that job i completes service (the *departure time*) is $c_i = a_i + w_i$.

The term "wait" can be confusing; w_i represents the total time job i spends in the service node, *not* just the time spent in the queue. The time spent in the queue (if any) is the delay d_i. In many computer science applications, the term *response time* is used. To some authors this means *wait*, to others it means *delay*. Because of this ambiguity, we will generally avoid using the term "response time" and instead will consistently use the terminology specified previously. Similarly, we avoid the use of the common terms *sojourn time*, *flow time*, and *system time*, in place of *wait*.

Arrivals.   As a convention, if the service-node capacity is finite, then rejected jobs (if any) are not indexed—that is, although rejected jobs may be counted for statistical purposes (for example, to estimate the probability of rejection), the index $i = 1, 2, 3, \ldots$ is restricted to only those jobs that actually enter the service node.

Rather than specify the *arrival* times $a_1, a_2, \ldots$ explicitly, in some discrete-event simulation applications it is preferable to specify the *interarrival* times $r_1, r_2, \ldots$, thereby defining the arrival times implicitly, as shown in Figure 1.2.3 and defined in Definition 1.2.3.

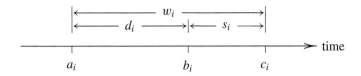

Figure 1.2.2   Six variables associated with job *i*.

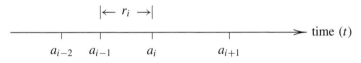

Figure 1.2.3   Relationship between arrival and interarrival times.

> **Definition 1.2.3** The *interarrival time* between jobs $i - 1$ and i is $r_i = a_i - a_{i-1}$. That is, $a_i = a_{i-1} + r_i$ and so (by induction), with $a_0 = 0$ the arrival times are*
>
> $$a_i = r_1 + r_2 + \cdots + r_i \qquad i = 1, 2, 3, \ldots.$$
>
> (We assume that $r_i > 0$ for all i, thereby eliminating the possibility of *bulk* arrivals. That is, jobs are assumed to arrive one at a time.)

Algorithmic Question. The following algorithmic question is fundamental. Given a knowledge of the arrival times a_1, a_2, ... (or, equivalently, the interarrival times r_1, r_2, ...), the associated service times s_1, s_2, ..., and the queue discipline, how can the delay times d_1, d_2, ... be computed?

As discussed in later chapters, for some queue disciplines this question is more difficult to answer than for others. If the queue discipline is FIFO, however, then the answer is particularly simple: As is demonstrated next, if the queue discipline is FIFO, then there is a simple algorithm for computing d_i (also b_i, w_i, and c_i) for all i.

Two Cases. If the queue discipline is FIFO, then the delay d_i of job $i = 1, 2, 3, \ldots$ is determined by when the job's arrival time a_i occurs relative to the departure time c_{i-1} of the previous job. There are two cases to consider.

- **Case I.** If $a_i < c_{i-1}$ (i.e., if job i arrives before job $i - 1$ departs), then, as illustrated in Figure 1.2.4, job i will experience the delay $d_i = c_{i-1} - a_i$. Job $i - 1$'s history is displayed above the time axis and job i's history is displayed below the time axis in Figure 1.2.4.
- **Case II.** If, instead, $a_i \geq c_{i-1}$ (i.e., if job i arrives after, or just as, job $i - 1$ departs), then, as illustrated in Figure 1.2.5, job i will experience no delay, and so $d_i = 0$.

Algorithm. The key point in the associated algorithm development is that, if the queue discipline is FIFO, then the truth of the expression $a_i < c_{i-1}$ determines whether job i will experience a delay. The computation of the delays, based on this logic, is summarized by Algorithm 1.2.1. This algorithm, like all those presented in this book, is written in a C-like pseudocode that is easily translated into other general-purpose programming languages.

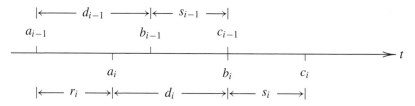

Figure 1.2.4 Job i arrives before job $i - 1$ departs.

---

*All arrival times are referenced to the virtual arrival time a_0. Unless explicitly stated otherwise, in this chapter and elsewhere we assume that elapsed time is measured in such a way that $a_0 = 0$.

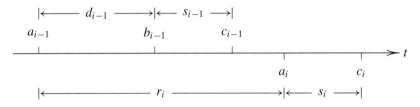

Figure 1.2.5   Job i arrives after job $i - 1$ departs.

Although it is not an explicit part of Algorithm 1.2.1, an equation can be written for the delay that depends on the interarrival and service times only:

$$c_{i-1} - a_i = (a_{i-1} + d_{i-1} + s_{i-1}) - a_i$$

$$= d_{i-1} + s_{i-1} - (a_i - a_{i-1})$$

$$= d_{i-1} + s_{i-1} - r_i.$$

If $d_0 = s_0 = 0$ then $d_1, d_2, d_3, \ldots$ are defined by the nonlinear equation

$$d_i = \max\{0, d_{i-1} + s_{i-1} - r_i\} \qquad i = 1, 2, 3, \ldots.$$

This equation is commonly used in theoretical studies to analyze the stochastic behavior of a FIFO service node.

Algorithm 1.2.1   If the arrival times $a_1, a_2, \ldots$ and service times $s_1, s_2, \ldots$ are known and if the server is initially idle, then this algorithm computes the delays $d_1, d_2, \ldots$ in a single-server FIFO service node with infinite capacity.

```
c₀ = 0.0;                                    // assumes that a₀ = 0.0
i = 0;
while ( more jobs to process ) {
    i++;
    aᵢ = GetArrival();
    if (aᵢ < cᵢ₋₁)
        dᵢ = cᵢ₋₁ - aᵢ;                     // calculate delay for job i
    else
        dᵢ = 0.0;                            // job i has no delay
    sᵢ = GetService();
    cᵢ = aᵢ + dᵢ + sᵢ;                       // calculate departure time for job i
}
n = i;
return d₁, d₂, ..., dₙ;
```

The GetArrival and GetService procedures read the next arrival and service time from a file. (An algorithm that does not rely on the FIFO assumption is presented in Chapter 5.)

Example 1.2.2

If Algorithm 1.2.1 is used to process $n = 10$ jobs according to the input indicated below (for simplicity, the a_i's and s_i's are integer time units—e.g., seconds or minutes) then the output is the sequence of delays, calculated as follows:

i	1	2	3	4	5	6	7	8	9	10
read from file a_i	15	47	71	111	123	152	166	226	310	320
from algorithm d_i	0	11	23	17	35	44	70	41	0	26
read from file s_i	43	36	34	30	38	40	31	29	36	30

For future reference, note that the last job arrived at time $a_n = 320$ and departed at time $c_n = a_n + d_n + s_n = 320 + 26 + 30 = 376$.

As will be discussed in more detail later in this section, it is a straightforward programming exercise to produce a computational model of a single-server FIFO service node with infinite capacity by using Algorithm 1.2.1. The ANSI C program `ssq1` is an example. Three features of this program are noteworthy: (i) Because of its reliance on previously recorded arrival- and service-time data read from an external file, `ssq1` is called a *trace-driven* discrete-event simulation program. (ii) Because the queue discipline is FIFO, program `ssq1` does not need to use a queue data structure. (iii) Rather than produce a sequence of delays as output, program `ssq1` computes four averages instead: those of the interarrival time, the service time, the delay, and the wait. These four job-averaged statistics and three corresponding time-averaged statistics are discussed next.

1.2.3 Output Statistics

One basic issue that must be resolved when constructing a discrete-event simulation model is the question of what statistics should be generated. The purpose of simulation is insight, and we gain insight about the performance of a system by looking at meaningful statistics. Of course, a decision about what statistics are most meaningful is dependent upon your perspective. For example, from a job's (customer's) perspective, the most important statistic might be the average delay or the 95th percentile of the delay—in either case, the smaller the better. On the other hand, particularly if the server is an expensive resource whose justification is based on an anticipated heavy workload, from management's perspective, the server's utilization (the proportion of busy time, see Definition 1.2.7) is most important—the larger the better.

Job-Averaged Statistics

> **Definition 1.2.4** For the first n jobs, the *average interarrival time* and the *average service time* are, respectively,[*]
>
> $$\bar{r} = \frac{1}{n} \sum_{i=1}^{n} r_i = \frac{a_n}{n} \qquad \text{and} \qquad \bar{s} = \frac{1}{n} \sum_{i=1}^{n} s_i.$$
>
> The reciprocal of the average interarrival time, $1/\bar{r}$, is the *arrival rate*; the reciprocal of the average service time, $1/\bar{s}$, is the *service rate*.

[*]The equation $\bar{r} = a_n/n$ follows from Definition 1.2.3 and the assumption that $a_0 = 0$.

Example 1.2.3

For the $n = 10$ jobs in Example 1.2.2, $\bar{r} = a_n/n = 320/10 = 32.0$ and $\bar{s} = 34.7$. If time in this example is measured in seconds, then the average interarrival time is 32.0 seconds per job and the average service time is 34.7 seconds per job. The corresponding arrival rate is $1/\bar{r} = 1/32.0 \cong 0.031$ jobs per second; the service rate is $1/\bar{s} = 1/34.7 \cong 0.029$ jobs per second. In this particular example, the server is not quite able to process jobs at the rate they arrive, on average.

Definition 1.2.5   For the first n jobs, the *average delay* in the queue and the *average wait* in the service node are, respectively,

$$\bar{d} = \frac{1}{n} \sum_{i=1}^{n} d_i \qquad \text{and} \qquad \bar{w} = \frac{1}{n} \sum_{i=1}^{n} w_i.$$

Recall that $w_i = d_i + s_i$ for all i. Therefore, the average time spent in the service node will be the sum of the average times spent in the queue and in service—that is,

$$\bar{w} = \frac{1}{n} \sum_{i=1}^{n} w_i = \frac{1}{n} \sum_{i=1}^{n} (d_i + s_i) = \frac{1}{n} \sum_{i=1}^{n} d_i + \frac{1}{n} \sum_{i=1}^{n} s_i = \bar{d} + \bar{s}.$$

The point here is that it is sufficient to compute any two of the statistics $\bar{w}$, $\bar{d}$, and $\bar{s}$. The third statistic can then be computed from the other two, if needed.

Example 1.2.4

For the data in Example 1.2.2, $\bar{d} = 26.7$ and $\bar{s} = 34.7$. Therefore, the average wait in the node is $\bar{w} = 26.7 + 34.7 = 61.4$.

In subsequent chapters, we will construct increasingly more complex discrete-event simulation models. Because it is never easy to verify and validate a complex model, it is desirable to be able to apply as many *consistency checks* to the output data as possible. For example, although program ssq1 is certainly not a complex discrete-event simulation model, it is desirable in this program to accumulate $\bar{w}$, $\bar{d}$, and $\bar{s}$ *independently*. Then, from the program output, the equation $\bar{w} = \bar{d} + \bar{s}$ can be used as a consistency check.

Time-Averaged Statistics.   The three statistics $\bar{w}$, $\bar{d}$, and $\bar{s}$ are *job-averaged* statistics—the data is averaged over all jobs. Job averages are easy to understand; they are just traditional arithmetic averages. We now turn to another type of statistic that is equally meaningful, *time-averaged*. Time-averaged statistics may be less familiar, however, because they are defined by an area under a curve (i.e., by integration instead of by summation).

Time-averaged statistics for a single-server service node are defined in terms of three additional variables. At any time $t > 0$:

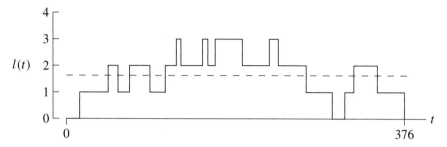

Figure 1.2.6   Number of jobs in the service node.

- $l(t) = 0, 1, 2, \ldots$ is the number of jobs in the service node at time t;
- $q(t) = 0, 1, 2, \ldots$ is the number of jobs in the queue at time t;
- $x(t) = 0, 1$ is the number of jobs in service at time t.

By definition, $l(t) = q(t) + x(t)$ for any $t > 0$.

The three functions $l(\cdot)$, $q(\cdot)$, and $x(\cdot)$ are *piecewise constant*—for example, a display of $l(t)$ versus t will consist of a sequence of constant segments with unit-height step discontinuities as illustrated in Figure 1.2.6. (This figure corresponds to the data in Example 1.2.2. The dashed line represents the time-averaged number in the node—see Example 1.2.6.)

The step discontinuities are positive at the arrival times and negative at the departure times. The corresponding figures for $q(\cdot)$ and $x(\cdot)$ can be deduced from the following fact: $q(t) = 0$ and $x(t) = 0$ if and only if $l(t) = 0$; otherwise, $q(t) = l(t) - 1$ and $x(t) = 1$.

Definition 1.2.6   Over the time interval $(0, \tau)$ the *time-averaged number in the node* is

$$\bar{l} = \frac{1}{\tau} \int_0^\tau l(t)\, dt.$$

Similarly, the *time-averaged number in the queue* and the *time-averaged number in service* are

$$\bar{q} = \frac{1}{\tau} \int_0^\tau q(t)\, dt \qquad \text{and} \qquad \bar{x} = \frac{1}{\tau} \int_0^\tau x(t)\, dt.$$

Because $l(t) = q(t) + x(t)$ for all $t > 0$, it follows that

$$\bar{l} = \bar{q} + \bar{x}.$$

Example 1.2.5

For the data in Example 1.2.2 (with $\tau = c_{10} = 376$), the three time-averaged statistics are $\bar{l} = 1.633$, $\bar{q} = 0.710$, and $\bar{x} = 0.923$. These values can be computed by calculating the areas associated with the integrals given in Definition 1.2.6 or by exploiting a mathematical relationship between the job-averaged statistics $\bar{w}$, $\bar{d}$, and $\bar{s}$ and the time-averaged statistics $\bar{l}$, $\bar{q}$, and $\bar{x}$, as will be illustrated in Example 1.2.6.

The equation $\bar{l} = \bar{q} + \bar{x}$ is the time-averaged analog of the job-averaged equation $\bar{w} = \bar{d} + \bar{s}$. As we will see in later chapters, time-averaged statistics have the following important characterizations.

- If we were to observe (sample) the number in, for example, the service node at many different times chosen *at random* between 0 and τ and then calculate the arithmetic average of all these observations, the result should be close to $\bar{l}$.
- Similarly, the arithmetic average of many random observations of the number in the queue should be close to $\bar{q}$ and the arithmetic average of many random observations of the number in service (0 or 1) should be close to $\bar{x}$.
- $\bar{x}$ must lie in the closed interval $[0, 1]$.

Definition 1.2.7 The time-averaged number in service $\bar{x}$ is also known as the server *utilization*. The reason for this terminology is that $\bar{x}$ represents the proportion of time that the server is busy.

Equivalently, if one particular time is picked *at random* between 0 and τ, then $\bar{x}$ is the probability that the server is busy at that time. If $\bar{x}$ is close to 1.0 then the server is busy most of the time and correspondingly large values of $\bar{l}$ and $\bar{q}$ will be produced. On the other hand, if the utilization is close to 0.0, then the server is idle most of the time and the values of $\bar{l}$ and $\bar{q}$ will be small. The case study, presented later, is an illustration.

Little's Equations. One important issue remains—how are job averages and time averages related? Specifically, how are $\bar{w}$, $\bar{d}$, and $\bar{s}$ related to $\bar{l}$, $\bar{q}$, and $\bar{x}$?

In the particular case of an infinite-capacity FIFO service node that begins and ends in an idle state, the following theorem provides an answer to this question. (See Exercise 1.2.7 for a generalization of this theorem to any queue discipline.)

Theorem 1.2.1 (Little, 1961) If the queue discipline is FIFO, the service-node capacity is infinite, and the server is idle both initially (at $t = 0$) and immediately after the departure of the nth job (at $t = c_n$), then

$$\int_0^{c_n} l(t)\,dt = \sum_{i=1}^n w_i \quad \text{and} \quad \int_0^{c_n} q(t)\,dt = \sum_{i=1}^n d_i \quad \text{and} \quad \int_0^{c_n} x(t)\,dt = \sum_{i=1}^n s_i.$$

Proof. For each job $i = 1, 2, \ldots$, define an *indicator function* $\psi_i(t)$ that is 1 when the ith job is in the service node and is 0 otherwise:

$$\psi_i(t) = \begin{cases} 1 & a_i < t < c_i \\ 0 & \text{otherwise.} \end{cases}$$

Then

$$l(t) = \sum_{i=1}^{n} \psi_i(t) \qquad 0 < t < c_n,$$

and so

$$\int_0^{c_n} l(t)\,dt = \int_0^{c_n} \sum_{i=1}^{n} \psi_i(t)\,dt = \sum_{i=1}^{n} \int_0^{c_n} \psi_i(t)\,dt = \sum_{i=1}^{n}(c_i - a_i) = \sum_{i=1}^{n} w_i.$$

The other two equations can be derived in a similar way. $\qquad\qquad\qquad\qquad$ □

Example 1.2.6

Figure 1.2.7 illustrates Little's first equation for the data in Example 1.2.2. The top step function denotes the cumulative number of arrivals to the service node; the bottom step function denotes the cumulative number of departures from the service node. The vertical distance between the two step-functions at any time t is $l(t)$, which was plotted in Figure 1.2.6. The wait times are indicated as the horizontal distances between the risers. In this figure, it is easy to see that

$$\int_0^{376} l(t)\,dt = \sum_{i=1}^{10} w_i = 614.$$

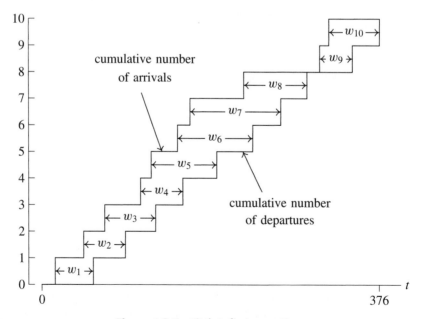

Figure 1.2.7 Little's first equation.

Little's equations provide a valuable link between the job-averaged statistics $\overline{w}$, $\overline{d}$, and $\overline{s}$ and the time-averaged statistics $\overline{l}$, $\overline{q}$, and $\overline{x}$. In Definition 1.2.6, let $\tau = c_n$; then, from Theorem 1.2.1, we have

$$c_n\overline{l} = \int_0^{c_n} l(t)\,dt = \sum_{i=1}^n w_i = n\overline{w},$$

and so $\overline{l} = (n/c_n)\overline{w}$. Similarly, $c_n\overline{q} = n\overline{d}$ and $c_n\overline{x} = n\overline{s}$. Therefore,

$$\overline{l} = \left(\frac{n}{c_n}\right)\overline{w} \qquad\text{and}\qquad \overline{q} = \left(\frac{n}{c_n}\right)\overline{d} \qquad\text{and}\qquad \overline{x} = \left(\frac{n}{c_n}\right)\overline{s},$$

which explains how $\overline{w}$, $\overline{d}$, and $\overline{s}$ are related to $\overline{l}$, $\overline{q}$, and $\overline{x}$. These important equations relate to *steady-state* statistics and *Little's equations*—for more detail, see Chapter 8.

Example 1.2.7
For the data in Example 1.2.2, the last ($n = 10$) job departs at $c_n = 376$. From Example 1.2.4, $\overline{w} = 61.4$, and therefore $\overline{l} = (10/376)\,61.4 \cong 1.633$. Similarly, the time-averaged numbers in the queue and in service are $\overline{q} = (10/376)\,26.7 \cong 0.710$ and $\overline{x} = (10/376)\,34.7 \cong 0.923$, respectively.

1.2.4 Computational Model

As discussed previously, using Algorithm 1.2.1 in conjunction with some statistics gathering logic makes it a straightforward programming exercise to produce a computational model of a single-server FIFO service node with infinite capacity. The ANSI C program `ssq1` is an example. Like all of the software presented in this book, this program is designed with readability and extendibility considerations.

Program `ssq1`. Program `ssq1` reads arrival and service time data from the disk file `ssq1.dat`. This is a text file that consists of arrival times $a_1, a_2, \ldots, a_n$ and service times $s_1, s_2, \ldots, s_n$ for $n = 1000$ jobs, in the format

$$
\begin{array}{cc}
a_1 & s_1 \\
a_2 & s_2 \\
\vdots & \vdots \\
a_n & s_n\,.
\end{array}
$$

In Chapter 3, we will free this trace-driven program from its reliance on external data by using randomly generated arrival and service times instead.

Because the queue discipline is FIFO, program `ssq1` does not need to use a queue data structure. In Chapter 5, we will consider non-FIFO queue disciplines and some corresponding priority-queue data structures that can be used at the computational-model level.

Program `ssq1` computes the average interarrival time $\overline{r}$, the average service time $\overline{s}$, the average delay $\overline{d}$, and the average wait $\overline{w}$. In Exercise 1.2.2, you are asked to modify this program so that it will also compute the time-averaged statistics $\overline{l}$, $\overline{q}$, and $\overline{x}$.

Example 1.2.8

For the datafile `ssq1.dat`, the observed arrival rate $1/\overline{r} \cong 0.10$ is significantly less than the observed service rate $1/\overline{s} \cong 0.14$. If you modify `ssq1` to compute $\overline{l}$, $\overline{q}$, and $\overline{x}$, you will find that $1 - \overline{x} \cong 0.28$, and so the server is idle 28% of the time. Despite this significant idle time, enough jobs are delayed so that the average number in the queue is nearly 2.0.

Traffic Intensity.   The ratio of the arrival rate to the service rate is commonly called the *traffic intensity*. From the equations in Definition 1.2.4 and Theorem 1.2.1, it follows that the observed traffic intensity is the ratio of the observed arrival rate to the observed service rate:

$$\frac{1/\overline{r}}{1/\overline{s}} = \frac{\overline{s}}{\overline{r}} = \frac{\overline{s}}{a_n/n} = \left(\frac{c_n}{a_n}\right)\overline{x}.$$

Therefore, provided that the ratio c_n/a_n is close to 1.0, the traffic intensity and utilization will be nearly equal. In particular, if the traffic intensity is less than 1.0 and n is large, then it is reasonable to expect that the ratio $c_n/a_n = 1 + w_n/a_n$ will be close to 1.0. We will return to this issue in later chapters. For now, we close with an example illustrating how relatively sensitive the service-node statistics $\overline{l}$, $\overline{q}$, $\overline{w}$, and $\overline{d}$ are to changes in the utilization and how nonlinear this dependence can be.

Case Study.   Sven & Larry's Ice Cream Shoppe is a thriving business that can be modeled as a single-server queue. The owners are considering adding additional flavors and cone options, but are concerned about the effect of the resultant increase in service times on queue length. They decide to use a trace-driven simulation to assess the impact of the longer service times associated with the additional flavors and cone options.

The file `ssq1.dat` represents 1000 customer interactions at Sven & Larry's. The file consists of arrival times of groups of customers and the group's corresponding service times. The service times vary significantly with the number of customers in each group.

The dependence of the average queue length $\overline{q}$ on the utilization $\overline{x}$ is illustrated in Figure 1.2.8. This figure was created by systematically increasing or decreasing each service time in the datafile `ssq1.dat` by a common multiplicative factor, thereby causing both $\overline{x}$ and $\overline{q}$ to change correspondingly. The $(\overline{x}, \overline{q}) \cong (0.72, 1.88)$ point circled corresponds to the data in `ssq1.dat`; the point immediately to its right, for example, corresponds to the same data with each service time multiplied by 1.05. The next point to the right corresponds to each service time multiplied by 1.10. From this figure, we see that even a modest increase in service times will produce a significant increase in the average queue length. A nonlinear relationship between $\overline{x}$ and $\overline{q}$ is particularly pronounced for utilizations near $\overline{x} = 1$. A 15% increase in the service times from their current values results in a 109% increase in the time-averaged number of customers in queue; a 30% increase in the service times from their current values results in a 518% increase in the time-averaged number of customers in queue.

Sven & Larry need to assess the impact of the increased service time associated with new flavors and cones on their operation. If the service times increase by only a modest amount, say 5% or 10% above the current times, the average queue length will grow modestly. The popularity of the new flavors and cone options might, however, also increase the arrival rate—potentially exacerbating the problem with long lines. If queues grow to the point where the owners believe

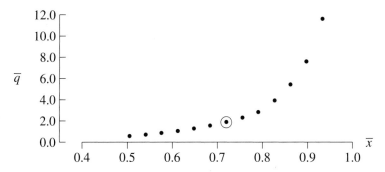

Figure 1.2.8   Average queue length vs. utilization.

that customers are taking their ice cream business elsewhere, they should consider hiring a second server. A separate analysis would be necessary to estimate the probability that an arriving group of customers will balk (never enter the queue) or renege (depart from the queue after entering) as a function of the queue's length.

Graphics Considerations.   Figure 1.2.8 presents "raw" simulation output data—that is, each • represents a computed $(\overline{x}, \overline{q})$ point. Because there is nothing inherently discrete about either $\overline{x}$ or $\overline{q}$, many additional points could have been computed and displayed to produce an (essentially) continuous $\overline{q}$ versus $\overline{x}$ curve. In this case, however, additional computations seem redundant; few would question the validity of the smooth curve produced by connecting the •'s with lines, as illustrated in Figure 1.2.9. The nonlinear dependence of $\overline{q}$ on $\overline{x}$ is evident, particularly as $\overline{x}$ approaches 1.0 and the corresponding increase in $\overline{q}$ becomes dramatic.

Perhaps because we were taught to do this as children, there is a natural tendency to *always* "connect the dots" (interpolate) when presenting a discrete set of experimental data. (The three most common interpolating functions are linear, quadratic, and spline functions.) Before taking such artistic liberties, however, consider the following guidelines.

* If the data have essentially no uncertainty and if the resulting interpolating curve is smooth, then there is little danger in connecting the dots, provided that the original dots are left in the figure to remind the reader that some artistic license was used.

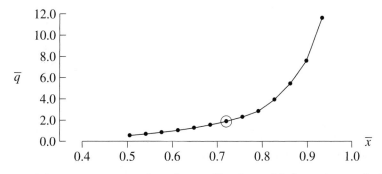

Figure 1.2.9   Average queue length vs. utilization, with linear interpolation.

- If the data have essentially no uncertainty but the resulting interpolating curve is not smooth, then more dots need to be generated to achieve a graphics scale at which smooth interpolation is reasonable.
- If the dots correspond to uncertain (noisy) data, then interpolation is not justified; instead, either approximation should be used in place of interpolation, or the temptation to super-impose a continuous curve should be resisted completely.

These guidelines presume that the data is not inherently discrete. If the data is inherently discrete then it is illogical and potentially confusing to superimpose a continuous (interpolating or approx-imating) curve. Example 1.3.7 in the next section is an illustration of data that is inherently discrete.*

1.2.5 Exercises

1.2.1   How would you use the table in Example 1.2.2 to construct the associated $l(t)$ versus t figure—that is, to construct an algorithm that will compute *in order* the interlaced arrival and departure times that define the points at which $l(t)$ changes? (Avoid storing $a_1, a_2, \ldots, a_n$ and $c_1, c_2, \ldots, c_n$ as two arrays, linked lists, or external disk files and then merging the two into one because of memory and CPU considerations for large n.)

1.2.2   (*a*) Modify program `ssq1` to output the additional statistics $\bar{l}, \bar{q}$, and $\bar{x}$. (*b*) As in the case study, use this program to compute a table of $\bar{l}, \bar{q}$, and $\bar{x}$ for the traffic intensities 0.6, 0.7, 0.8, 0.9, 1.0, 1.1, and 1.2. (*c*) Comment on how $\bar{l}, \bar{q}$, and $\bar{x}$ depend on traffic intensity. (*d*) Relative to the case study, if it is decided that $\bar{q}$ greater than 5.0 is not acceptable, what systematic increase in service times would be acceptable? Report your results to *d.dd* precision.

1.2.3   (*a*) Modify program `ssq1` by adding the capability to compute the maximum delay, the number of jobs in the service node at a specified time (known at compile time), and the proportion of jobs delayed. (*b*) What was the maximum delay experienced? (*c*) How many jobs were in the service node at $t = 400$, and how does the computation of this number relate to the proof of Theorem 1.2.1? (*d*) What proportion of jobs were delayed, and how does this proportion relate to the utilization?

1.2.4   Complete the proof of Theorem 1.2.1.

1.2.5   If the traffic intensity is less than 1.0, use Theorem 1.2.1 to argue why, for large n, you would expect to find that $\bar{l} \cong \lambda \, \bar{w}$, $\bar{q} \cong \lambda \, \bar{d}$, and $\bar{x} \cong \lambda \, \bar{s}$, where the observed arrival rate is $\lambda = 1/\bar{r}$.

1.2.6   The text file `ac.dat` consists of the arrival times $a_1, a_2, \ldots, a_n$ and the departure times $c_1, c_2, \ldots, c_n$ for $n = 500$ jobs, in the format

$$
\begin{array}{cc}
a_1 & c_1 \\
a_2 & c_2 \\
\vdots & \vdots \\
a_n & c_n
\end{array} .
$$

---

*Those interested in an excellent discussion and illustration of graphics considerations are encouraged to read the classic *The Visual Display of Quantitative Information* (Tufte, 2001). The author discusses clarity of presentation through uncluttered graphics that maximize information transmission with minimal ink. The accurate display of simulation output will be stressed throughout this text.

(a) If these times are for an initially idle single-server FIFO service node with infinite capacity, calculate the average service time, the server's utilization, and the traffic intensity. (b) Be explicit: For $i = 1, 2, \ldots, n$, how does s_i relate to a_{i-1}, a_i, c_{i-1}, and c_i?

1.2.7 State and prove a theorem analogous to Theorem 1.2.1, but valid for *any* queue discipline. *Hint*: In place of c_n use $\tau_n = \max\{c_1, c_2, \ldots, c_n\}$. For a conservative server, prove that τ_n is *independent* of the queue discipline.

1.2.8 (a) Much as in Exercise 1.2.2, modify program ssq1 to output the additional statistics $\bar{l}$, $\bar{q}$, and $\bar{x}$. (b) By using the arrival times in the file ssq1.dat and an appropriate *constant* service time in place of the service times in the file ssq1.dat, use the modified program to compute a table of $\bar{l}$, $\bar{q}$, and $\bar{x}$ for the traffic intensities 0.6, 0.7, 0.8, 0.9, 1.0, 1.1, and 1.2. (c) Comment on how $\bar{l}$, $\bar{q}$, and $\bar{x}$ depend on the traffic intensity.

1.2.9 (a) Work Exercises 1.2.2 and 1.2.8. (b) Compare the two tables produced and explain (or conjecture) why the two tables are different. Be specific.

1.3 A SIMPLE INVENTORY SYSTEM

The inputs to program ssq1—namely, the arrival times and the service times—can have any positive real value; they are *continuous* variables. In some models, however, the input variables are inherently *discrete*. That is the case with the (trace-driven) discrete-event simulation model of a simple inventory system constructed in this section. As in the previous section, we begin with a conceptual model, then move to a specification model and, finally, to a computational model.

1.3.1 Conceptual Model

> **Definition 1.3.1**   An *inventory system* consists of a facility that distributes items from its current inventory to its customers in response to a (typically random) customer demand, as illustrated in Figure 1.3.1. Moreover, the demand is integer-valued (discrete); customers do not want a portion of an item.* Because there is a *holding cost* associated with items in inventory, it is undesirable for the inventory level to be too high. On the other hand, if the inventory level is too low, the facility is in danger of incurring a *shortage cost* whenever a demand occurs that cannot be met.

Figure 1.3.1 Simple inventory system diagram.

---

*Some inventory systems distribute "items" that are not inherently discrete, for example, a service station that sells gasoline. With minor modifications, the model developed in this section is applicable to these inventory systems as well.

As a policy, the inventory level is reviewed periodically and new items are then (and only then) ordered from a supplier, if necessary.* When items are ordered, the facility incurs an *ordering cost* that is the sum of a fixed *setup cost* independent of the amount ordered plus an *item cost* proportional to the number of items ordered. This *periodic inventory review policy* is defined by two parameters, conventionally denoted s and S.

- s is the *minimum* inventory level—if, at the time of review, the current inventory level is below the threshold s, then an order will be placed with the supplier to replenish the inventory. If the current inventory level is at or above s, then no order will be placed.
- S is the *maximum* inventory level—when an order is placed, the amount ordered is the number of items required to bring the inventory up to the level S.
- The (s, S) parameters are constant in time, with $0 \leq s < S$.

A discrete-event simulation model can be used to compute the cost of operating the facility. In some cases, the values of s and S are fixed; if so, the cost of operating the facility is also fixed. In other cases, if at least one of the (s, S) values (usually s) is not fixed, the cost of operating the facility can be modified and it is natural to search for values of (s, S) at which the cost of operating the facility is minimized.

To complete the conceptual model of this simple (one type of item) inventory system, we make four additional assumptions: (*a*) *Back-ordering* (backlogging) is possible—the inventory level can become negative in order to model customer demands not immediately satisfied. (*b*) There is no *delivery lag*—an order placed with the supplier will be delivered immediately. Usually, this is an unrealistic assumption; it will be removed in Chapter 3. (*c*) The *initial* inventory level is S. (*d*) The *terminal* inventory level is S.

Example 1.3.5, presented later in this section, describes an automobile dealership as an example of an inventory system with back-ordering and no delivery lag. In this example the periodic inventory review occurs each week. The value of S is fixed, the value of s is not.

1.3.2 Specification Model

The following variables provide the basis for a specification model of a simple inventory system. Time begins at $t = 0$ and is measured in a coordinate system in which the inventory review times are $t = 0, 1, 2, 3, \ldots$, with the convention that the ith time interval begins at time $t = i - 1$ and ends at $t = i$.

- The inventory level at the *beginning* of the ith time interval is an integer, l_{i-1}.
- The amount ordered (if any) at time $t = i - 1$ is an integer, $o_{i-1} \geq 0$.
- The demand quantity *during* the ith time interval is an integer, $d_i \geq 0$.

---

*An alternate to the periodic inventory review policy is a *transaction reporting inventory policy*. With this policy, inventory review occurs after *each* demand instance. Because inventory review occurs more frequently, significantly more labor may be required to implement a transaction reporting inventory policy. (The scanners at a grocery store, however, require no extra labor.) The transaction reporting policy has the desirable property that, for the same value of s, it is less likely for the inventory system to experience a shortage.

Because we have assumed that back-ordering is possible, if the demand during the ith time interval is greater than the inventory level at the beginning of the interval (plus the amount ordered, if any), then the inventory level at the end of the interval will be negative.

The inventory level is reviewed at $t = i - 1$. If l_{i-1} is greater than or equal to s, then no items are ordered, and so $o_{i-1} = 0$. If, instead, l_{i-1} is less than s, then $o_{i-1} = S - l_{i-1}$ items are ordered to replenish inventory. In this case, because we have assumed there is no delivery lag, ordered items are delivered immediately (at $t = i - 1$), thereby restoring inventory to the level S. In either case, the inventory level at the end of the ith time interval is diminished by d_i. Therefore, as summarized by Algorithm 1.3.1, with $l_0 = S$, the inventory orders $o_0, o_1, o_2, \ldots$ and the corresponding inventory levels $l_1, l_2, \ldots$ are defined by

$$o_{i-1} = \begin{cases} 0 & l_{i-1} \geq s \\ S - l_{i-1} & l_{i-1} < s \end{cases} \qquad \text{and} \qquad l_i = l_{i-1} + o_{i-1} - d_i.$$

Note that $l_0 = S > s$ and so o_0 must be zero; accordingly, only $o_1, o_2, \ldots$ are of interest.

Algorithm 1.3.1 If the demands $d_1, d_2, \ldots$ are known, then this algorithm computes the discrete-time evolution of the inventory level for a simple (s, S) inventory system with back-ordering and no delivery lag.

```
l₀ = S;                         // the initial inventory level is S
i = 0;
while ( more demand to process ) {
    i++;
    if (l_{i-1} < s)
        o_{i-1} = S - l_{i-1};
    else
        o_{i-1} = 0;
    d_i = GetDemand();
    l_i = l_{i-1} + o_{i-1} - d_i;
}
n = i;
o_n = S - l_n;
l_n = S;                        // the terminal inventory level is S
return l₁, l₂, ..., l_n and o₁, o₂, ..., o_n;
```

Example 1.3.1

Let $(s, S) = (20, 60)$ and apply Algorithm 1.3.1 to process $n = 12$ time intervals of operation with the following input demand schedule:

i	1	2	3	4	5	6	7	8	9	10	11	12
input d_i	30	15	25	15	45	30	25	15	20	35	20	30

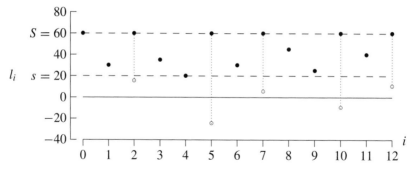

Figure 1.3.2   Inventory levels.

As illustrated in Figure 1.3.2, the time evolution of the inventory level typically features several intervals of decline, followed by an increase when an order is placed (indicated by the vertical dotted line) and, because there is no delivery lag, is immediately delivered.

At the end of the last interval (at $t = n = 12$), an order for $o_n = 50$ inventory units is placed. The immediate delivery of this order restores the inventory level at the end of the simulation to the initial inventory level $S = 60$, as shown in Figure 1.3.2.

1.3.3 Output Statistics

As with the development of program `ssq1` in the previous section, we must address the issue of what statistics should be computed to measure the performance of a simple inventory system. As always, our objective is to analyze these statistics and, by so doing, better understand how the system operates.

Definition 1.3.2   The *average demand* and *average order* are, respectively,

$$\overline{d} = \frac{1}{n} \sum_{i=1}^{n} d_i \qquad \text{and} \qquad \overline{o} = \frac{1}{n} \sum_{i=1}^{n} o_i.$$

Example 1.3.2
For the data in Example 1.3.1, $\overline{d} = \overline{o} = 305/12 \cong 25.42$ items per time interval. As explained next, these two averages *must* be equal.

The terminal condition in Algorithm 1.3.1 is that, at the end of the nth time interval, an order is placed to return the inventory to its initial level. Because of this terminal condition, independent of the value of s and S, the average demand $\overline{d}$ and the average order $\overline{o}$ must be equal—that is, over the course of the simulated period of operation, all demand is satisfied (although not immediately, if back-ordering occurs). Therefore, if the inventory level is the same at the beginning and end of the simulation then the average "flow" of items into the facility from the supplier, $\overline{o}$, must have been equal to the average "flow" of items out of the facility to the customers, $\overline{d}$. With respect to the flow of items into and out of the facility, the inventory system is said to be *flow balanced*, as illustrated in Figure 1.3.3.

Figure 1.3.3   Flow balance.

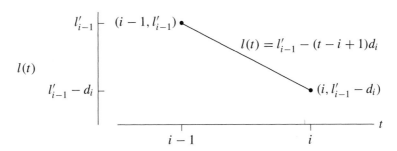

Figure 1.3.4   Piecewise-linear inventory levels.

Average Inventory Level.   The holding cost and shortage cost are proportional to time-averaged inventory levels. To compute these averages, it is necessary to know the inventory level for *all* t, not just at the inventory review times. Therefore, we *assume* that the *demand rate* is constant between review times, so that the continuous-time evolution of the inventory level is *piecewise linear*, as illustrated in Figure 1.3.4.

Definition 1.3.3   If the demand rate is constant between review times, then, at any time t in the ith time interval, the inventory level is $l(t) = l'_{i-1} - (t - i + 1)d_i$, as illustrated in Figure 1.3.5.

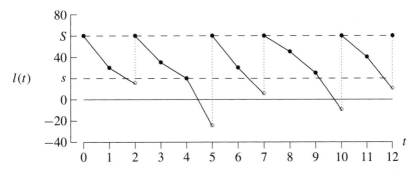

Figure 1.3.5   Linear inventory level in time interval i.

In this figure and in related figures and equations elsewhere in this section, $l'_{i-1} = l_{i-1} + o_{i-1}$ represents the inventory level *after* inventory review. Accordingly, $l'_{i-1} \geq s$ for all i. (For the figure in Example 1.3.1, the ∘'s and •'s represent l_{i-1} and l'_{i-1}, respectively.)

The equation for $l(t)$ is the basis for calculating the time-averaged inventory level for the ith time interval.* There are two cases to consider. If $l(t)$ remains nonnegative over the ith time interval, then there is only a time-averaged *holding level* integral to evaluate:

$$\bar{l}_i^+ = \int_{i-1}^{i} l(t)\, dt.$$

If, instead, $l(t)$ becomes negative at some time τ interior to the ith interval, then, in addition to a time-averaged holding-level integral, there is also a time-averaged *shortage-level* integral to evaluate. In this case, the two integrals are

$$\bar{l}_i^+ = \int_{i-1}^{\tau} l(t)\, dt \qquad \text{and} \qquad \bar{l}_i^- = -\int_{\tau}^{i} l(t)\, dt.$$

No Back-Ordering. The inventory level $l(t)$ remains nonnegative throughout the ith time interval if and only if the inventory level at the end of this interval is nonnegative, as in Figure 1.3.7. Therefore, there is no shortage during the ith time interval if and only if $d_i \le l'_{i-1}$. In this case, the time-averaged holding-level integral for the ith time interval can be evaluated as the area of a trapezoid:

$$\bar{l}_i^+ = \int_{i-1}^{i} l(t)\, dt = \frac{l'_{i-1} + (l'_{i-1} - d_i)}{2} = l'_{i-1} - \frac{1}{2} d_i \qquad \text{and} \qquad \bar{l}_i^- = 0.$$

With Back-Ordering. The inventory level becomes negative at some point τ in the ith time interval if and only if $d_i > l'_{i-1}$, as illustrated in Figure 1.3.8.

---

*Because the inventory level at any time is an *integer*, the figure in Definition 1.3.3 is technically incorrect. Instead, rounding to an integer value should be used to produce the inventory-level time history illustrated in Figure 1.3.6. ($\lfloor z \rfloor$ is the floor function; $\lfloor z + 0.5 \rfloor$ is z rounded to the nearest integer.)

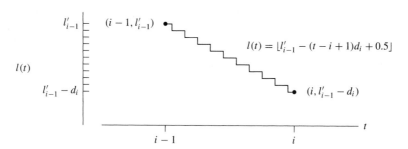

Figure 1.3.6 Piecewise-constant inventory level in time interval i.

It can be shown, however, that rounding has *no* effect on the value of $\bar{l}_i^+$ and $\bar{l}_i^-$.

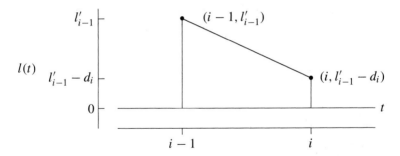

Figure 1.3.7   Inventory level in time interval *i*, with no back-ordering.

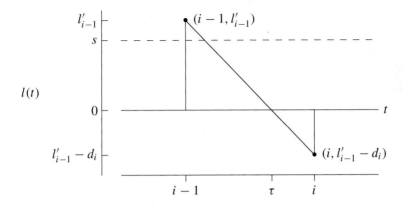

Figure 1.3.8   Inventory level in time interval *i*, with back-ordering.

By using similar triangles, it can be shown that $\tau = i - 1 + (l'_{i-1}/d_i)$. In this case, the time-averaged holding-level integral and shortage-level integral for the ith time interval can each be evaluated as the areas of triangles:

$$\bar{l}_i^+ = \int_{i-1}^{\tau} l(t)\,dt = \cdots = \frac{(l'_{i-1})^2}{2d_i} \qquad \text{and} \qquad \bar{l}_i^- = -\int_{\tau}^{i} l(t)\,dt = \cdots = \frac{(d_i - l'_{i-1})^2}{2d_i}.$$

The time-averaged holding level and shortage level for each time interval can be summed over all intervals, with the resulting sums being divided by the number of intervals. Consistent with Definition 1.3.4, the result represents the average number of items "held" and "short" respectively, with the average taken over all time intervals.

Definition 1.3.4   The *time-averaged holding level* and the *time-averaged shortage level* are, respectively,

$$\bar{l}^+ = \frac{1}{n}\sum_{i=1}^{n}\bar{l}_i^+ \qquad \text{and} \qquad \bar{l}^- = \frac{1}{n}\sum_{i=1}^{n}\bar{l}_i^-.$$

It is potentially confusing to define the time-averaged shortage level as a *positive* number, as we have done in Definition 1.3.3. In particular, it would be a mistake to compute the time-averaged inventory level as the sum of $\bar{l}^{+}$ and $\bar{l}^{-}$. Instead, the *time-averaged inventory level* is the difference

$$\bar{l} = \frac{1}{n} \int_0^n l(t)\, dt = \bar{l}^{+} - \bar{l}^{-}.$$

The proof of this result is left as an exercise.

Example 1.3.3
For the data in Example 1.3.1, $\bar{l}^{+} = 31.74$ and $\bar{l}^{-} = 0.70$. Therefore, over the 12 time intervals, the average number of items held was 31.74, the average number of items short was 0.70, and the average inventory level was 31.04.

1.3.4 Computational Model

Algorithm 1.3.1 is the basis for program `sis1`—a trace-driven computational model of a simple inventory system.

Program `sis1`.   Program `sis1` computes five statistics: $\bar{d}, \bar{o}, \bar{l}^{+}, \bar{l}^{-}$, and the order frequency $\bar{u}$, which is

$$\bar{u} = \frac{\text{number of orders}}{n}.$$

Because the simulated system is flow balanced, $\bar{o} = \bar{d}$. It would be sufficient for program `sis1` to compute just one of these two statistics; the independent computation of both $\bar{o}$ and $\bar{d}$ is desirable, however, because it provides an important consistency check for a (flow-balanced) simple inventory system.

Example 1.3.4
Program `sis1` reads input data corresponding to $n = 100$ time intervals from the file `sis1.dat`. With the inventory-policy parameter values $(s, S) = (20, 80)$ the results (with *dd.dd* precision) are

$$\bar{o} = \bar{d} = 29.29 \qquad \bar{u} = 0.39 \qquad \bar{l}^{+} = 42.40 \qquad \bar{l}^{-} = 0.25.$$

As with program `ssq1`, in Chapter 3 we will free program `sis1` from its reliance on external data by using randomly generated demand data instead.

1.3.5 Operating Cost

> **Definition 1.3.5**   In conjunction with the four statistics $\bar{o}$, $\bar{u}$, $\bar{l}^+$, and $\bar{l}^-$, a facility's cost of operation is determined by four constants:
>
> - c_{item}—the (unit) cost of a new item;
> - c_{setup}—the setup cost associated with placing an order;
> - c_{hold}—the cost to hold one item for one time interval;
> - c_{short}—the cost of being short one item for one time interval.

Case Study.   Consider a hypothetical automobile dealership that uses a weekly periodic inventory review policy. The facility is the dealer's showroom, service area, and surrounding storage lot, and the items that flow into and out of the facility are new cars. The supplier is the manufacturer of the cars, and the customers are people convinced by clever advertising that their lives will be improved significantly if they purchase a new car from this dealer.

Example 1.3.5

Suppose space in the facility is limited to a maximum of, say, $S = 80$ cars. (This is a small dealership.) Every Monday morning, the dealer's inventory of cars is reviewed; if the inventory level at that time is below a threshold—say, $s = 20$—then enough new cars are ordered from the supplier to restock the inventory to level S.*

- The (unit) cost to the dealer for each new car ordered is $c_{item} = \$8000$.
- The setup cost associated with deciding what cars to order (color, model, options, etc.) and arranging for additional bank financing (this is not a rich automobile dealer) is $c_{setup} = \$1000$, independent of the number ordered.
- The holding cost (interest charges primarily) to the dealer, per week, to have a car sit unsold in the facility is $c_{hold} = \$25$.
- The shortage cost to the dealer, per week, of failing to have a car in inventory is hard to evaluate because, in our model, we have assumed that *all* demand will ultimately be satisfied. Therefore, any customer who wants to buy a new car, even if none is available, will agree to wait until next Monday when new cars arrive. Thus the shortage cost to the dealer is primarily in goodwill. Our dealer realizes, however, that in this situation customers can buy from another dealer; so, when a shortage occurs, the deal is sweetened by agreeing to pay "shorted" customers $100 cash *per day* when they come back on Monday to pick up their new car. This means that the cost of being short one car for one week is $c_{short} = \$700$.

---

*There will be some, perhaps quite significant, delivery lag, but that is ignored for now in our model. In effect, we are assuming that this dealer is located adjacent to the supplier and that the supplier responds immediately to each order.

Definition 1.3.6   A simple inventory system's average operating costs *per time interval* are defined as follows:

- item cost: $c_{item} \cdot \bar{o}$;
- setup cost: $c_{setup} \cdot \bar{u}$;
- holding cost: $c_{hold} \cdot \bar{l}^{+}$;
- shortage cost: $c_{short} \cdot \bar{l}^{-}$.

The average total cost of operation per time interval is the sum of these four costs. This sum multiplied by the number of time intervals is the total cost of operation.

Example 1.3.6

From the statistics in Example 1.3.4 and the constants in Example 1.3.5, for our auto dealership, the average costs are as follows:

- the item cost is $\$8000 \cdot 29.29 = \$234,320$;
- the setup cost is $\$1000 \cdot 0.39 = \$390$;
- the holding cost is $\$25 \cdot 42.40 = \$1060$;
- the shortage cost is $\$700 \cdot 0.25 = \$175$.

Each of these costs is a per-week average.

Cost Minimization.   Although the inventory-system statistic of primary interest is the average total cost per time interval, it is important to know the four components of this total cost. By varying the value of s (and possibly S), it seems reasonable to expect that an optimal (minimal-average-cost) periodic-inventory-review policy can be designed for which these components are properly balanced.

In a search for optimal (s, S) values, because $\bar{o} = \bar{d}$ and $\bar{d}$ depends only on the demand sequence, it is important to note that the item cost is *independent* of (s, S). Therefore, the only cost that can be controlled by adjusting the inventory-policy parameters is the sum of the average setup, holding, and shortage costs. In Example 1.3.7, this sum is called the average *dependent cost*. For reference, in Example 1.3.6, the average dependent cost is $\$390 + \$1060 + \$175 = \$1625$ per week.

If S and the demand sequence are fixed, and if s is systematically increased, say from 0 to some large value less than S, then we expect to see the following.

- Generally, the average setup cost and holding cost will increase with increasing s.
- Generally, the average shortage cost will decrease with increasing s.
- Generally, the average total cost will have a 'U' shape indicating the presence of one (or more) optimal value(s) of s.

Example 1.3.7 is an illustration.

Example 1.3.7

With S fixed at 80, a modified version of program sis1 was used to study how the total cost relates to the value of s. The cost constants in Example 1.3.5 were used to compute the average setup, holding, shortage, and dependent costs for a range of s values from 1 to 40. As illustrated in Figure 1.3.9, the minimum average dependent cost is approximately $1550 at $s = 22$.

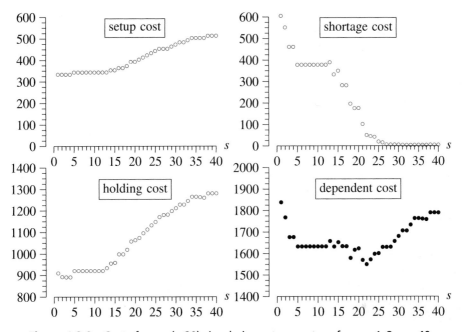

Figure 1.3.9   Costs for an $(s, 80)$ simple inventory system for $s = 1, 2, \ldots, 40$.

As in the case study concerning the ice cream shop, the "raw" simulation output data is presented. In this case, however, because the parameter s is inherently integer-valued (and so there is no "missing" output data at, for example, $s = 22.5$) no interpolating or approximating curve is superimposed. [For a more general treatment of issues surrounding simulation optimization, see Andradóttir (1998).]

1.3.6 Exercises

1.3.1   Verify that the results in Example 1.3.1 and the averages in Examples 1.3.2 and 1.3.3 are correct.

1.3.2   (*a*) Using the cost constants in Example 1.3.5, modify program sis1 to compute all four components of the total average cost per week. (*b*) These four costs could differ somewhat from the numbers in Example 1.3.6—why? (*c*) By constructing graphs like those in Example 1.3.7, explain the trade-offs involved in concluding that $s = 22$ is the optimum value (when $S = 80$). (d) Comment on how well defined this optimum is.

1.3.3 Suppose that the inventory level $l(t)$ has a constant rate of change over the time interval $a \le t \le b$ and that both $l(a)$ and $l(b)$ are integers. (a) Prove that

$$\int_a^b l(t)\, dt = \int_a^b \lfloor l(t) + 0.5 \rfloor\, dt = \frac{1}{2}(b - a)(l(a) + l(b)).$$

(b) What is the value of this integral if $l(t)$ is truncated rather than rounded (i.e., if the 0.5 is omitted in the second integral)?

1.3.4 (a) Construct a table or figure similar to Figure 1.3.7, but for $S = 100$ and $S = 60$. (b) How does the minimum cost value of s seem to depend on S? (See Exercise 1.3.2.)

1.3.5 Prove that, if there is no delivery lag, and if $d_i \le s$ for $i = 1, 2, \ldots, n$, then $\bar{l}^- = 0$.

1.3.6 (a) Provided that there is no delivery lag, prove that, if $S - s < d_i \le S$ for $i = 1, 2, \ldots, n$, then $\bar{l}^+ = S - \bar{d}/2$. (b) What is the value of $\bar{l}^-$ and $\bar{u}$ in this case?

1.3.7 Use Definitions 1.3.3 and 1.3.4 to prove that the average inventory level equation

$$\bar{l} = \frac{1}{n} \int_0^n l(t)\, dt = \bar{l}^+ - \bar{l}^-$$

is correct. *Hint*: use the $(\cdot)^+$ and $(\cdot)^-$ functions defined for any integer (or real number) x as

$$x^+ = \frac{|x| + x}{2} \qquad \text{and} \qquad x^- = \frac{|x| - x}{2}$$

and recognize that $x = x^+ - x^-$.

1.3.8 (a) Modify program sis1 so that the demands are first read into a circular array, then read out of that array, as needed, during program execution. (b) By experimenting with different starting locations for reading the demands from the circular array, explore how sensitive the program's statistical output is to the order in which the demands occur.

1.3.9 (a) Consider a variant of Exercise 1.3.8, where you use a conventional (noncircular) array and randomly *shuffle* the demands within this array before the demands are then read out of the array, as needed, during program execution. (b) Repeat for at least 10 different random shuffles and explore how sensitive the program's statistical output is to the order in which the demands occur.

2
Random-Number Generation

The material in this chapter presumes some knowledge of integer arithmetic and elementary number theory; for background material, see Appendix B.

Discrete-event and Monte Carlo simulation cannot be done correctly without access to a good random-number generator. Ideally, computer users should be able to assume the existence of a good system-supplied generator that can be used as a "black box" like other standard, reliable mathematical functions (sqrt, sin, exp, etc.). Unfortunately, history suggests this is frequently a false assumption. For that reason, the first two sections of this chapter provide a comprehensive discussion of an easily understood random-number-generation algorithm that can be used with confidence.

Section 2.1 is introductory, beginning with a conceptual model of the two-parameter algorithm as equivalent to drawing, at random, from an urn whose contents are determined by the choice of parameters. Given an appropriate choice of parameter values, Section 2.2 is primarily concerned with the important issue of correct algorithm implementation—that is, a software implementation of the random-number-generation algorithm is developed that is correct and efficient, and is portable to essentially any contemporary computing system. A more comprehensive reference on random-number generation than the presentation here is L'Ecuyer (2006). This software implementation is then used in Section 2.3 as the basis for introducing Monte Carlo simulation as it relates to the estimation of probabilities.

Section 2.4 provides Monte Carlo examples that are more involved and interesting than those in the third section. Unlike the material in the first two sections, which is specific to a particular random-number-generation algorithm, Section 2.5 provides a more general discussion of random-number-generation algorithms.

2.1 LEHMER RANDOM-NUMBER GENERATORS: INTRODUCTION

Programs `ssq1` and `sis1` both require input data from an external source. Because of this, the usefulness of these programs is limited by the amount of available input data. What if more input data is needed? Or, what if the model is changed; can the input data be modified accordingly? Or, what can be done if only a small amount of input data is available or, perhaps, none at all?

In each case, the answer is to use a *random-number generator*. By convention, each call to the random-number generator will produce a real-valued result between 0.0 and 1.0 that becomes the inherent source of randomness for a discrete-event simulation model. As is illustrated later in this chapter and is consistent with user-defined stochastic models of arrival times, service times, demand amounts, and so on, the random-number generator's output can be converted to a *random variate* via an appropriate mathematical transformation. The random variates are used to approximate some probabilistic element (e.g., service times) of a real-world system for a discrete-event simulation model.

2.1.1 Random-Number Generation

Historically, three types of random-number generators have been advocated for computational applications: (*a*) 1950's-style table-look-up generators—for example, the RAND Corporation table of a million random digits; (*b*) hardware generators—for example, thermal "white noise" devices; and (*c*) algorithmic (software) generators. Of these three types, only algorithmic generators have achieved widespread acceptance. The reason for this is that only algorithmic generators have the potential to satisfy *all* of the following generally well-accepted random-number-generation criteria:

- *randomness*—ability to produce output that passes all reasonable statistical tests of randomness;
- *controllability*—ability to reproduce its output, if desired;
- *portability*—ability to produce the same output on a wide variety of computer systems;
- *efficiency*—speed, with minimal computer-resource requirements;
- *documentation*—theoretical analysis and extensive testing.

In addition, as will be discussed and illustrated in Chapter 3, it should be easy to partition the generator's output into multiple "streams."

Definition 2.1.1   An *ideal* random-number generator is a function—say, Random—with the property that *each* assignment

u = Random();

will produce a real-valued (floating-point) number u between 0.0 and 1.0 in such a way that any value in the interval $0.0 < u < 1.0$ is *equally likely* to occur. A *good* random-number generator produces results that are statistically indistinguishable, for all practical purposes, from those produced by an ideal generator.

What we will do in this chapter is construct, from first principles, a good random-number generator that will satisfy all the criteria listed previously. We begin with the following conceptual model.

- Choose a *large* positive integer m. This defines the set $\mathcal{X}_m = \{1, 2, \ldots, m-1\}$.
- Fill a (conceptual) urn with the elements of $\mathcal{X}_m$.
- Each time a random number u is needed, draw an integer x "at random" from the urn and let $u = x/m$.

Each draw *simulates* a realization (observation, sample) of an independent, identically distributed (*iid*) sequence of so-called *Uniform*(0, 1) random variables. Because the possible values of u are $1/m, 2/m, \ldots, 1 - 1/m$, it is important for m to be so large that the possible values of u will be densely distributed between 0.0 and 1.0. Note that the values $u = 0.0$ and $u = 1.0$ are impossible. As will be discussed in later chapters, excluding these two extreme values is important for avoiding problems associated with certain random-variate-generation algorithms.

Ideally, we would like to draw from the urn independently and *with* replacement. If so, then each of the $m - 1$ possible values of u would be equally likely to be selected on each draw. For practical reasons, however, we will use a random-number-generation algorithm that simulates drawing from the urn *without* replacement. Fortunately, if m is large and the number of draws is small relative to m, then the distinction between drawing with and without replacement is largely irrelevant. To turn the conceptual urn model into a specification model, we will use a time-tested algorithm suggested by Lehmer (1951).

2.1.2 Lehmer's Algorithm

Definition 2.1.2   *Lehmer's algorithm* for random-number generation is defined in terms of two fixed parameters

- *modulus m*, a fixed large *prime* integer
- *multiplier a*, a fixed integer in $\mathcal{X}_m$

and the subsequent generation of the integer sequence $x_0, x_1, x_2, \ldots$ via the iterative equation

- $x_{i+1} = g(x_i)$ $i = 0, 1, 2, \ldots$

where the function $g(\cdot)$ is defined for all $x \in \mathcal{X}_m = \{1, 2, \ldots, m-1\}$ as

- $g(x) = ax \bmod m$

and the *initial seed* x_0 is chosen from the set $\mathcal{X}_m$. The *modulus function* mod gives the remainder when the first argument (ax in this case) is divided by the second argument (the modulus m in this case). It is defined more carefully in Appendix B. A random-number generator based on Lehmer's algorithm is called a *Lehmer* generator. (This random-number generator is also referred to as a special case of a *linear congruential generator*.)

The mod (remainder) operator causes the value of $g(x)$ to be always an integer between 0 and $m - 1$; however, it is important to exclude 0 as a possible value. Note that $g(0) = 0$, so, if 0 ever occurs in the sequence $x_0, x_1, x_2, \ldots$, then *all* the terms in the sequence will be 0 from that point on. Fortunately, if (a, m) is chosen consistent with Definition 2.1.2, it can be shown that $g(x) \neq 0$ for all $x \in \mathcal{X}_m$, and so $g: \mathcal{X}_m \to \mathcal{X}_m$. This guaranteed avoidance of 0 is part of the reason for choosing m to be prime. Because $g: \mathcal{X}_m \to \mathcal{X}_m$, it follows (by induction) that if $x_0 \in \mathcal{X}_m$ then $x_i \in \mathcal{X}_m$ for all $i = 0, 1, 2, \ldots$.

Lehmer's algorithm represents a good example of the elegance of simplicity. The genius of Lehmer's algorithm is that *if* the multiplier a and prime modulus m are properly chosen, the resulting sequence $x_0, x_1, x_2, \ldots, x_{m-2}$ will be statistically indistinguishable from a sequence drawn at random, albeit without replacement, from $\mathcal{X}_m$. Indeed, it is *only* in the sense of *simulating* this random draw that the algorithm is random; there is actually *nothing* random about Lehmer's algorithm, except possibly the choice of the initial seed. For this reason Lehmer generators are sometimes called *pseudo*-random.[*]

An intuitive explanation of why Lehmer's algorithm simulates randomness is based on the observation that, if two large integers, a and x, are multiplied and the product divided by another large integer, m, then the remainder, $g(x) = ax \bmod m$, is "likely" to take on any value between 0 and $m - 1$, as illustrated in Figure 2.1.1.

This is somewhat like going to the grocery store with no change, only dollars, and buying many identical items. After you pay for the items, the change you will receive is likely to be any value between 0¢ and 99¢. [That is, a is the price of the item in cents, x is the number of items purchased, and $m = 100$. The analogy is not exact, however, because the change is $m - g(x)$, not $g(x)$].

Modulus and Multiplier Considerations.   To construct a Lehmer generator, standard practice is to choose the large prime modulus m first, then choose the multiplier a. The choice of m is dictated, in part, by system considerations. For example, on a computer system that supports 32-bit, two's-complement integer arithmetic, $m = 2^{31} - 1$ is a natural choice, because it is the largest possible positive integer and it happens to be prime. (For 16-bit integer arithmetic we are not so fortunate, however, because $2^{15} - 1$ is not prime. See Exercise 2.1.6. Similarly, for 64-bit integer arithmetic $2^{63} - 1$ is not prime. See Exercise 2.1.10.) Given m, the subsequent choice of a must be made with great care. The following example is an illustration.

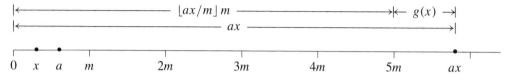

Figure 2.1.1   Lehmer generator geometry.

---

[*]Lehmer generators are also known as (take a deep breath) prime-modulus multiplicative linear congruential pseudo-random-number generators, abbreviated PMMLCG.

Example 2.1.1

As a tiny example, consider the prime modulus $m = 13.$*

- If the multiplier is $a = 6$ and the initial seed is $x_0 = 1$ then the resulting sequence of x's is

$$1, 6, 10, 8, 9, 2, 12, 7, 3, 5, 4, 11, 1, \ldots$$

where, as the ellipses indicate, the sequence is actually *periodic* because it begins to cycle (with a *full* period of length $m - 1 = 12$) when the initial seed reappears. The point is that any 12 consecutive terms in this sequence appear to have been drawn at random, without replacement, from the set $\mathcal{X}_{13} = \{1, 2, \ldots, 12\}$.

- If the multiplier is $a = 7$ and the initial seed is $x_0 = 1$ then the resulting full-period sequence of x's is

$$1, 7, 10, 5, 9, 11, 12, 6, 3, 8, 4, 2, 1, \ldots.$$

Randomness is, like beauty, only in the eye of the beholder. Because of the $12, 6, 3$ and $8, 4, 2, 1$ patterns, however, most people would consider this second sequence to be "less random" than the first.

- If the multiplier is $a = 5$ then either

$$1, 5, 12, 8, 1, \ldots \qquad \text{or} \qquad 2, 10, 11, 3, 2, \ldots \qquad \text{or} \qquad 4, 7, 9, 6, 4, \ldots$$

will be produced, depending on the initial seed $x_0 = 1, 2,$ or 4. This type of less-than-full-period behavior is clearly undesirable because, in terms of our conceptual model of random-number generation, this behavior corresponds to first partitioning the set $\mathcal{X}_m$ into several disjoint subsets (urns), then selecting one subset and thereafter drawing exclusively from it.

Example 2.1.1 illustrates two of the three central issues that must be resolved when choosing (a, m).

- Does the function $g(\cdot)$ generate a full-period sequence?
- If a full-period sequence is generated, how random does the sequence appear to be?

The third central issue is implementation.

- Can the $ax \bmod m$ operation be evaluated efficiently and correctly for all $x \in \mathcal{X}_m$?

For Example 2.1.1, the issue of implementation is trivial. However, for realistically large values of a and m, a portable, efficient implementation in a high-level language is a substantive issue, because of potential integer overflow associated with the ax product. We will return to this implementation issue in the next section. In the remainder of this section, we will concentrate on the full-period issue.

---

*For a hypothetical computer system with 5-bit two's-complement integer arithmetic, the largest positive integer would be $2^4 - 1 = 15$. In this case $m = 13$ would be the largest possible prime and, in that sense, a natural choice for the modulus.

Full-Period Considerations. From the definition of the mod function (see Appendix B), if $x_{i+1} = g(x_i)$, then there exists a nonnegative integer $c_i = \lfloor ax_i/m \rfloor$ such that

$$x_{i+1} = g(x_i) = ax_i \bmod m = ax_i - mc_i \qquad i = 0, 1, 2, \ldots.$$

Therefore (by induction),

$$x_1 = ax_0 - mc_0$$

$$x_2 = ax_1 - mc_1 = a^2 x_0 - m(ac_0 + c_1)$$

$$x_3 = ax_2 - mc_2 = a^3 x_0 - m(a^2 c_0 + ac_1 + c_2)$$

$$\vdots$$

$$x_i = ax_{i-1} - mc_{i-1} = a^i x_0 - m(a^{i-1} c_0 + a^{i-2} c_1 + \cdots + c_{i-1}).$$

Because $x_i \in \mathcal{X}_m$, we have $x_i = x_i \bmod m$. Moreover, $(a^i x_0 - mc) \bmod m = a^i x_0 \bmod m$, and this fact is independent of the value of the integer $c = a^{i-1} c_0 + a^{i-2} c_1 + \cdots + c_{i-1}$. Therefore, we have proven the following theorem.

Theorem 2.1.1   If the sequence $x_0, x_1, x_2, \ldots$ is produced by a Lehmer generator with multiplier a and modulus m, then

$$x_i = a^i x_0 \bmod m \qquad i = 0, 1, 2, \ldots.$$

Although it would be an eminently bad idea to compute x_i by first computing a^i, particularly if i is large, Theorem 2.1.1 has significant theoretical importance that is due, in part, to the fact that x_i can be written equivalently as

$$x_i = a^i x_0 \bmod m = [(a^i \bmod m)(x_0 \bmod m)] \bmod m = [(a^i \bmod m) x_0] \bmod m$$

for $i = 0, 1, 2, \ldots$ via Theorem B.2 in Appendix B. In particular, this is true for $i = m - 1$. Therefore, because m is prime and $a \bmod m \neq 0$, from Fermat's little theorem (see Appendix B) it follows that $a^{m-1} \bmod m = 1$, and so $x_{m-1} = x_0$. This observation is the key to proving the following theorem. The details of the proof are left as an exercise.

Theorem 2.1.2   If $x_0 \in \mathcal{X}_m$ and the sequence $x_0, x_1, x_2, \ldots$ is produced by a Lehmer generator with multiplier a and (prime) modulus m, then there is a positive integer p with $p \leq m - 1$ such that $x_0, x_1, x_2, \ldots, x_{p-1}$ are all different and

$$x_{i+p} = x_i \qquad i = 0, 1, 2, \ldots.$$

That is, the sequence $x_0, x_1, x_2, \ldots$ is *periodic* with *fundamental period* p. In addition, $(m - 1) \bmod p = 0$.

The significance of Theorem 2.1.2 is profound. If we pick *any* initial seed $x_0 \in \mathcal{X}_m$ and start to generate the sequence $x_0, x_1, x_2, \ldots$, then we are guaranteed that the initial seed will reappear. In addition, the first instance of reappearance is guaranteed to occur at some index p that is either $m - 1$ or an integer divisor of $m - 1$, and, from that index on, the sequence will cycle through the same p distinct values, as illustrated:

$$\underbrace{x_0, x_1, \ldots, x_{p-1}}_{\text{period}}, \underbrace{x_0, x_1, \ldots, x_{p-1}}_{\text{period}}, x_0, \ldots.$$

Consistent with Definition 2.1.3, we are interested in choosing multipliers for which the fundamental period is $p = m - 1$.

Full-Period Multipliers

> **Definition 2.1.3**   The sequence $x_0, x_1, x_2, \ldots$ produced by a Lehmer generator with modulus m and multiplier a has a *full period* if and only if the fundamental period p is $m - 1$. If the sequence has a full period, then a is said to be a *full-period multiplier* relative to m.*

If the sequence $x_0, x_1, x_2, \ldots$ has a full period, then any $m - 1$ consecutive terms in the sequence have been drawn without replacement from the set $\mathcal{X}_m$. In effect, a full-period Lehmer generator creates a virtual *circular list* with $m - 1$ distinct elements. The initial seed provides a starting list element; subsequent calls to the generator traverse the list.

Example 2.1.2
From Example 2.1.1, if $m = 13$ then $a = 6$ and $a = 7$ are full-period multipliers ($p = 12$) and $a = 5$ is not ($p = 4$). The virtual circular lists (with traversal in a clockwise direction) corresponding to these two full-period multipliers are shown in Figure 2.1.2.

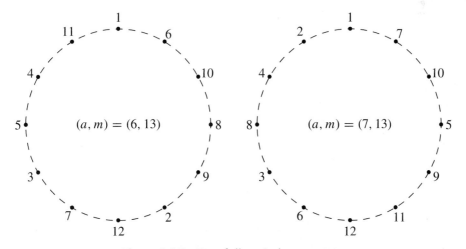

Figure 2.1.2 Two full-period generators.

---

*A full-period multiplier a relative to m is also said to be a *primitive root* of m.

Algorithm 2.1.1   This algorithm, based upon Theorem 2.1.2, can be used to determine whether a is a full-period multiplier relative to the prime modulus m.

```
p = 1;
x = a;
while (x != 1) {
    p++;
    x = (a * x) % m;                      // beware of a * x overflow
}
if (p == m - 1)
    // a is a full-period multiplier
else
    // a is not a full-period multiplier
```

If m is not prime, Algorithm 2.1.1 might not halt. It provides a slow-but-sure $O(m)^*$ test for a full-period multiplier. The algorithm employs $x_0 = 1$ as an arbitrary initial seed (note that $x_1 = a$) and uses the recursive generation of new values of x until the initial seed reappears. Before starting to search for full-period multipliers, however, it would be good to know that they exist and with what frequency—that is, given a prime modulus m, how many corresponding full-period multipliers are there? The following theorem provides the answer to this question.

Theorem 2.1.3   If m is prime and $p_1, p_2, \ldots, p_r$ are the (unique) *prime factors* of $m - 1$, then the number of full-period multipliers in $\mathcal{X}_m$ is

$$\frac{(p_1 - 1)(p_2 - 1)\ldots(p_r - 1)}{p_1 p_2 \ldots p_r}(m - 1).$$

Example 2.1.3
If $m = 13$, then $m - 1 = 2^2 \cdot 3$. From the equation in Theorem 2.1.3, this prime modulus has $\frac{(2-1)(3-1)}{2 \cdot 3}(13 - 1) = 4$ full-period multipliers: $a = 2, 6, 7$, and 11.

Example 2.1.4
The Lehmer generator used in this book has the (Mersenne, i.e., of the form $2^k - 1$, where k is a positive integer) prime modulus $m = 2^{31} - 1 = 2\,147\,483\,647$. Because the prime decomposition of $m - 1$ is

$$m - 1 = 2^{31} - 2 = 2 \cdot 3^2 \cdot 7 \cdot 11 \cdot 31 \cdot 151 \cdot 331,$$

from the equation in Theorem 2.1.3, the number of full-period multipliers is

$$\left(\frac{1 \cdot 2 \cdot 6 \cdot 10 \cdot 30 \cdot 150 \cdot 330}{2 \cdot 3 \cdot 7 \cdot 11 \cdot 31 \cdot 151 \cdot 331}\right)(2 \cdot 3^2 \cdot 7 \cdot 11 \cdot 31 \cdot 151 \cdot 331) = 534\,600\,000.$$

---

*A function f is called "order m" [written $O(m)$] if there exist real positive constants c_1 and c_2 independent of m such that $c_1 m \leq f(m) \leq c_2 m$.

Therefore, for this prime modulus, approximately 25% of the multipliers between 1 and $m - 1$ are full-period multipliers.

By using Algorithm 2.1.1 (and *a lot* of computer time), it can be shown that, if $m = 2^{31} - 1$ then $a = 2, 3, 4, 5, 6$ are not full-period multipliers, but that $a = 7$ is. Remarkably, once *one* full-period multiplier has been found, in this case $a = 7$, then *all* the others can be found by using the following $O(m)$ algorithm. This algorithm presumes the availability of a function gcd that returns the greatest common divisor of two positive integers (as described in Appendix B).

Algorithm 2.1.2   Given the prime modulus m and any full-period multiplier a, the following algorithm generates all the full-period multipliers relative to m.*

```
i = 1;
x = a;
while (x != 1) {
    if (gcd(i, m - 1) == 1)
                    // x is a full-period multiplier equal to a^i mod m
    i++;
    x = (a * x) % m;                        // beware of a * x overflow
}
```

Algorithm 2.1.2 is based on Theorem 2.1.4, which establishes a one-to-one correspondence between integers $i \in \mathcal{X}_m$ that are relatively prime to $m - 1$ and full-period multipliers $a^i \bmod m \in \mathcal{X}_m$. The equation in Theorem 2.1.3 counts both the number of full-period multipliers *and* the number of integers in $\mathcal{X}_m$ that are relatively prime to $m - 1$. The proof of Theorem 2.1.4 is left as an (advanced) exercise.

Theorem 2.1.4   If a is any full-period multiplier relative to the prime modulus m, then each of the integers

$$a^i \bmod m \in \mathcal{X}_m \qquad\qquad i = 1, 2, 3, \dots, m - 1$$

is also a full-period multiplier relative to m if and only if the integer i has no prime factors in common with the prime factors of $m - 1$ (i.e., i and $m - 1$ are relatively prime; see Appendix B).

Example 2.1.5

If $m = 13$, then, from Example 2.1.3, there are 4 integers between 1 and 12 that are relatively prime to 12. They are $i = 1, 5, 7, 11$. From Example 2.1.1, $a = 6$ is a full-period multiplier relative to 13; from Theorem 2.1.4, the 4 full-period multipliers relative to 13 are, therefore,

$$6^1 \bmod 13 = 6, \qquad 6^5 \bmod 13 = 2, \qquad 6^7 \bmod 13 = 7, \qquad 6^{11} \bmod 13 = 11.$$

---

*See Exercises 2.1.4 and 2.1.5.

Equivalently, if we had known that $a = 2$ is a full-period multiplier relative to 13, we could have used Algorithm 2.1.2 with $a = 2$ to calculate the full-period multipliers to be

$$2^1 \bmod 13 = 2, \qquad 2^5 \bmod 13 = 6, \qquad 2^7 \bmod 13 = 11, \qquad 2^{11} \bmod 13 = 7.$$

Example 2.1.6

If $m = 2^{31} - 1$, then from Example 2.1.4, there are $534\,600\,000$ integers between 1 and $m - 1$ that are relatively prime to $m - 1$. The first few of these are $i = 1, 5, 13, 17, 19$. As was discussed previously, $a = 7$ is a full-period multiplier relative to this modulus; hence, from Algorithm 2.1.2,

$$7^1 \bmod 2147483647 = 7$$

$$7^5 \bmod 2147483647 = 16807$$

$$7^{13} \bmod 2147483647 = 252246292$$

$$7^{17} \bmod 2147483647 = 52958638$$

$$7^{19} \bmod 2147483647 = 447489615$$

are full-period multipliers relative to $2^{31} - 1 = 2\,147\,483\,647$. Note that the full-period multiplier 16807 is a *logical* choice; 7 is the smallest full-period multiplier and 5 is the smallest integer (other than 1) that is relatively prime to $m - 1$. However, that the multiplier 16807 is a logical choice does not mean that the resulting sequence it generates is adequately "random." We will have more to say about this in the next section.

Standard Algorithmic Generators.   Many of the standard algorithmic generators in use today are of one of the following three *linear congruential* types (Law and Kelton, 2000, pages 406–412):

- *mixed*: $g(x) = (ax + c) \bmod m$ with $m = 2^b$ ($b = 31$ typically), $a \bmod 4 = 1$ and $c \bmod 2 = 1$. All integer values of $x \in \{0\} \cup \mathcal{X}_m$ are possible, and the period is m.
- *multiplicative* with $m = 2^b$: $g(x) = ax \bmod m$ with $m = 2^b$ ($b = 31$ typically) and $a \bmod 8 = 3$ or $a \bmod 8 = 5$. To achieve the maximum period, which is only $m/4$, x must be restricted to the odd integers in $\mathcal{X}_m$.
- *multiplicative* with m prime: $g(x) = ax \bmod m$ with m prime and a a full-period multiplier. All integer values of $x \in \mathcal{X}_m = \{1, 2, \ldots, m - 1\}$ are possible and the period is $m - 1$.

Of these three, specialists generally consider the third generator, which has been presented in this chapter, to be best. Random-number generation remains an area of active research, however, and reasonable people will probably always disagree about the *best* algorithm for random-number generation. Section 2.5 contains a more general discussion of random-number generators.

For now, we will avoid the temptation to be drawn further into the "what is the best possible random-number generator" debate. Instead, as discussed in the next section, we will use a Lehmer generator with modulus $m = 2^{31} - 1$ and a corresponding full-period multiplier $a = 48271$ carefully selected to provide generally acceptable randomness *and* facilitate efficient software implementation. With this generator in hand, we can then turn our attention to the main topic of this book—the modeling, simulation, and analysis of discrete-event stochastic systems.

2.1.3 Exercises

2.1.1   For the tiny Lehmer generator defined by $g(x) = ax \bmod 127$, find all the full-period multipliers. (*a*) How many are there? (*b*) What is the smallest multiplier?

2.1.2   Prove Theorem 2.1.2.

2.1.3   Prove that, if (a, m) are chosen consistent with Definition 2.1.2, then $g: \mathcal{X}_m \to \mathcal{X}_m$ is a bijection (one-to-one and onto).

2.1.4   (*a*) Prove that, if (a, m) are chosen consistent with Definition 2.1.2, then $g(x) \neq 0$ for all $x \in \mathcal{X}_m$, and so $g: \mathcal{X}_m \to \mathcal{X}_m$. (*b*) In addition, relative to this definition, prove that there is nothing to be gained by considering integer multipliers outside of $\mathcal{X}_m$. Consider the two cases $a \geq m$ and $a \leq 0$.

2.1.5   (*a*) Except for the special case $m = 2$, prove that $a = 1$ cannot be a full-period multiplier. (*b*) What about $a = m - 1$?

2.1.6   In ANSI C, an `int` is guaranteed to hold all integer values between $-(2^{15} - 1)$ and $2^{15} - 1$ inclusive. (*a*) What is the largest prime modulus in this range? (*b*) How many corresponding full-period multipliers are there and what is the smallest one?

2.1.7   Prove Theorem 2.1.4.

2.1.8   (*a*) Evaluate $7^i \bmod 13$ and $11^i \bmod 13$ for $i = 1, 5, 7, 11$. (*b*) How does this relate to Example 2.1.5?

2.1.9   (*a*) Verify that the list of five full-period multipliers in Example 2.1.6 is correct. (*b*) What are the next five elements in this list?

2.1.10   (*a*) What is the largest prime modulus less than or equal to $2^{63} - 1$? (*b*) How many corresponding full-period multipliers are there? (*c*) How does this relate to the use of Lehmer random-number generators on computer systems that support 64-bit integer arithmetic?

2.1.11   For the first few prime moduli, this table lists the number of full-period multipliers and the smallest full-period multiplier. Add the next ten rows to this table.

prime modulus m	number of full-period multipliers	smallest full-period multiplier a
2	1	1
3	1	2
5	2	2
7	2	3
11	4	2
13	4	2

2.2 LEHMER RANDOM-NUMBER GENERATORS: IMPLEMENTATION

Recall that one good reason to choose $m = 2^{31} - 1$ as the modulus for a Lehmer random-number generator is that virtually all contemporary computer systems support 32-bit two's-complement integer arithmetic and that, on such systems, $2^{31} - 1$ is the largest possible prime. Consistent with that observation, a portable and efficient algorithmic implementation of an $m = 2^{31} - 1$ Lehmer generator that is valid (provided that all integers between $-m$ and m can be represented exactly) is developed in this section.

We can use ANSI C for the implementation, because 32-bit signed integer arithmetic is supported in a natural way—that is, the ANSI C type `long` is required to be valid for all integers between LONG_MIN and LONG_MAX inclusive and, although the values are implementation dependent, LONG_MAX and LONG_MIN are required to be at least $2^{31} - 1$ and at most $-(2^{31} - 1)$ respectively.[*]

2.2.1 Implementation

If there is no guarantee that integers larger than m can be represented exactly, then the implementation issue of *potential integer overflow* must be addressed. Note that, for any full-period Lehmer generator, the product ax could be no larger than $a(m - 1)$. Therefore, unless values of t as large as $a(m - 1)$ can be represented exactly, it is *not* possible to evaluate $g(x) = ax \bmod m$ in the "obvious" way by first computing the intermediate product $t = ax$ and then computing $t \bmod m = t - \lfloor t/m \rfloor m$.

> **Example 2.2.1**
>
> If $(a, m) = (48271, 2^{31} - 1)$ then $a(m - 1) \cong 1.47 \times 2^{46}$. Therefore, it would not be possible to implement this (a, m) Lehmer generator in the obvious way without access to a register that is at least 47 bits wide to store the intermediate product t. This is true even though $t \bmod m$ is no more than 31 bits wide. (If $a = 16807$ then a 46-bit register would be required to hold the intermediate product.)

Type Considerations.   If we wish to implement a Lehmer generator with $m = 2^{31} - 1$ in ANSI C and do it in the obvious way, then the type declaration of t will dictate the number of bits available to store the intermediate $t = ax$ product. Correspondingly, if t is (naturally) an *integer* type, then the integer division t/m can be used to evaluate $\lfloor t/m \rfloor$; or if t is a *floating-point* type, then a floating-point division t/m followed by a floating-point-to-integer cast can be used to evaluate $\lfloor t/m \rfloor$.

> **Example 2.2.2**
>
> If the variable t is declared to be a `long` and $m = 2^{31} - 1$, then the obvious implementation will be correct only if LONG_MAX is $a(m - 1)$ or larger. Most contemporary computer systems do not support integers this large; thus, for a Lehmer generator with $m = 2^{31} - 1$, the obvious implementation is *not* a viable algorithm option.

> **Example 2.2.3**
>
> If the variable t is declared to be the ANSI C floating-point type `double`, then the obvious implementation might be correct, provided that the multiplier a is not too large. That is, `double` is generally consistent with the IEEE 754 64-bit floating-point

---

[*]The macros LONG_MIN and LONG_MAX are defined in the ANSI C library `<limits.h>`. The type `long` is a shorthand representation for the type `long int`. LONG_LONG is supported in C99, the new (maybe latest) ANSI/ISO Standard for C for 64-bit integers. This option is appropriate only if the hardware supports 64-bit arithmetic (e.g., Apple, AMD).

standard, which specifies a 53-bit mantissa (including the sign bit); if $m = 2^{31} - 1$, a mantissa this large allows for t to be much larger than m. So it might be possible to implement a $m = 2^{31} - 1$ Lehmer generator with a sufficiently small multiplier in the obvious way by doing the *integer* calculations in *floating-point* arithmetic. However, when portability is required and an efficient integer-based implementation is possible, only the unwise would use a floating-point implementation instead.

Algorithm Development.   Consistent with the previous examples, it is desirable to have an integer-based implementation of Lehmer's algorithm that will port to any system that supports the ANSI C type `long`. This can be done if no integer calculation produces an intermediate or final result larger than $m = 2^{31} - 1$ in magnitude. With this constraint in mind, we must be prepared to do some algorithm development.

If it were possible to factor the modulus as $m = aq$ for some integer q then $g(x)$ could be written as $g(x) = ax \bmod m = a(x \bmod q)$, enabling us to do the mod *before* the multiply and thereby avoid the potential overflow problem. In this case, the largest possible value of $ax \bmod m$ would be $a(q - 1) = m - a$, which is less than m. Of course, if m is prime, no such factorization is possible. It is always possible, however, to "approximately factor" m as $m = aq + r$ with $q = \lfloor m/a \rfloor$ and $r = m \bmod a$. As demonstrated in this section, if the *remainder r* is small relative to the *quotient q*, specifically if $r < q$, then this (q, r) decomposition of m provides the basis for an algorithm to evaluate $g(x)$ in such a way that integer overflow is avoided.

Example 2.2.4
If $(a, m) = (48271, 2^{31} - 1)$, then the quotient is $q = \lfloor m/a \rfloor = 44488$ and the remainder is $r = m \bmod a = 3399$. Similarly, if $a = 16807$, then $q = 127773$ and $r = 2836$. In both cases, $r < q$.

For all $x \in \mathcal{X}_m = \{1, 2, \ldots, m - 1\}$, define the two functions

$$\gamma(x) = a(x \bmod q) - r\lfloor x/q \rfloor \qquad \text{and} \qquad \delta(x) = \lfloor x/q \rfloor - \lfloor ax/m \rfloor.$$

Then, for any $x \in \mathcal{X}_m$,

$$\begin{aligned}
g(x) &= ax \bmod m \\
&= ax - m\lfloor ax/m \rfloor \\
&= ax - m\lfloor x/q \rfloor + m\lfloor x/q \rfloor - m\lfloor ax/m \rfloor \\
&= ax - (aq + r)\lfloor x/q \rfloor + m\delta(x) \\
&= a(x - q\lfloor x/q \rfloor) - r\lfloor x/q \rfloor + m\delta(x) \\
&= a(x \bmod q) - r\lfloor x/q \rfloor + m\delta(x) \\
&= \gamma(x) + m\delta(x).
\end{aligned}$$

An efficient, portable implementation of a Lehmer random-number generator is based upon this alternate representation of $g(x)$ and on the following theorem.

Theorem 2.2.1   If $m = aq + r$ is prime, $r < q$, and $x \in \mathcal{X}_m$, then $\delta(x) = \lfloor x/q \rfloor - \lfloor ax/m \rfloor$ is either 0 or 1. Moreover, with $\gamma(x) = a(x \bmod q) - r\lfloor x/q \rfloor$,

- $\delta(x) = 0$ if and only if $\gamma(x) \in \mathcal{X}_m$;
- $\delta(x) = 1$ if and only if $-\gamma(x) \in \mathcal{X}_m$.

Proof. First observe that, if u and v are real numbers with $0 < u - v < 1$, then the integer difference $\lfloor u \rfloor - \lfloor v \rfloor$ is either 0 or 1. Therefore, because $\delta(x) = \lfloor x/q \rfloor - \lfloor ax/m \rfloor$, the first part of the theorem is true if we can show that

$$0 < \frac{x}{q} - \frac{ax}{m} < 1.$$

Rewriting the center of the inequality as

$$\frac{x}{q} - \frac{ax}{m} = x\left(\frac{1}{q} - \frac{a}{m}\right) = x\left(\frac{m - aq}{mq}\right) = \frac{xr}{mq},$$

and because $r < q$,

$$0 < \frac{xr}{mq} < \frac{x}{m} \le \frac{m-1}{m} < 1,$$

which establishes that, if $x \in \mathcal{X}_m$, then $\delta(x)$ is either 0 or 1. To prove the second part of this theorem, recall that, if $x \in \mathcal{X}_m$, then $g(x) = \gamma(x) + m\delta(x)$ is in $\mathcal{X}_m$. Therefore, if $\delta(x) = 0$, then $\gamma(x) = g(x)$ is in $\mathcal{X}_m$; conversely, if $\gamma(x) \in \mathcal{X}_m$, then $\delta(x)$ cannot be 1, for, if it were, then $g(x) = \gamma(x) + m$ would be $m + 1$ or larger, which contradicts $g(x) \in \mathcal{X}_m$. Similarly, if $\delta(x) = 1$, then $-\gamma(x) = m - g(x)$ is in $\mathcal{X}_m$; conversely, if $-\gamma(x) \in \mathcal{X}_m$, then $\delta(x)$ cannot be 0, for, if it were, then $g(x) = \gamma(x)$ would not be in $\mathcal{X}_m$. □

Algorithm.   The key to avoiding overflow in the evaluation of $g(x)$ is that the calculation with the potential to cause overflow, the ax product, be "trapped" in $\delta(x)$. This is important because, from Theorem 2.2.1, the value of $\delta(x)$ can be deduced from the value of $\gamma(x)$, and it can be shown that $\gamma(x)$ can be computed without overflow—see Exercise 2.2.5. This comment and Theorem 2.2.1 can be summarized with the following algorithm.

Algorithm 2.2.1   If $m = aq + r$ is prime, $r < q$, and $x \in \mathcal{X}_m$, then $g(x) = ax \bmod m$ can be evaluated as follows without producing any intermediate or final values larger than $m - 1$ in magnitude.

```
t = a * (x % q) - r * (x / q);                          // t = γ(x)
if (t > 0)
    return (t);                                         // δ(x) = 0
else
    return (t + m);                                     // δ(x) = 1
```

Modulus Compatibility

> **Definition 2.2.1**   The multiplier a is *modulus-compatible* with the prime modulus m if and only if the remainder $r = m \bmod a$ is less than the quotient $q = \lfloor m/a \rfloor$.

If a is modulus-compatible with m, then Algorithm 2.2.1 can be used as the basis for an implementation of a Lehmer random-number generator. In particular, if the multiplier is modulus-compatible with $m = 2^{31} - 1$, then Algorithm 2.2.1 can be used to implement the corresponding Lehmer random-number generator in such a way that it will port to any system that supports 32-bit integer arithmetic. From Example 2.2.4, for example, the full-period multiplier $a = 48271$ is modulus-compatible with $m = 2^{31} - 1$.

In general, there are no modulus-compatible multipliers beyond $(m - 1)/2$. Moreover, as the following example illustrates, modulus-compatible multipliers are much more densely distributed on the low end of the $1 \le a \le (m - 1)/2$ scale.

Example 2.2.5

The (tiny) modulus $m = 401$ is prime. Figure 2.2.1 illustrates, on the first line, the 38 associated modulus-compatible multipliers. On the second line are the 160 full-period multipliers and the third line illustrates the ten multipliers (3, 6, 12, 13, 15, 17, 19, 21, 23, and 66) that are *both* modulus-compatible and full-period.

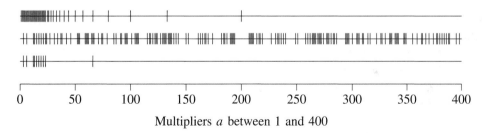

Figure 2.2.1   Modulus-compatible full-period multipliers for $m = 401$.

If you have a very high-resolution graphics device, a very long piece of paper, a magnifying glass, and a few hours of CPU time, you are encouraged to make the corresponding figure for $m = 2^{31} - 1$. (For this modulus, there are 92 679 modulus-compatible multipliers. Of these, 23 093 are also full-period—see Example 2.2.6.)

Modulus compatibility is closely related to the smallness of the multiplier. In particular, a is defined to be "small" if and only if $a^2 < m$. If a is small, then a is modulus-compatible with m—see Exercise 2.2.4. From this result, it follows, for example, that all the multipliers between 1 and 46340 inclusive are modulus-compatible with $m = 2^{31} - 1$. Thus smallness is sufficient to guarantee modulus compatibility. Smallness is not a necessary condition, however. In particular, $a = 48271$ is modulus-compatible with $m = 2^{31} - 1$, but it is not small. Similarly, of the 38 modulus-compatible multipliers for $m = 401$ in Example 2.2.5, only about half (those between 1 and 20) satisfy the $a^2 < m$ small-multiplier test.

As illustrated in Example 2.2.5, for a given prime modulus there are relatively few associated full-period, modulus-compatible multipliers. A mechanical way to find one of these is to start with the small (and thus modulus-compatible) multipliers $a = 2$, $a = 3$, and so on, and continue until the first full-period multiplier is found—see Algorithm 2.1.1. Given that we have found one full-period, modulus-compatible multiplier, the following $O(m)$ algorithm can be used to generate *all* the others. This algorithm, an extension of Algorithm 2.1.2, presumes the availability of a function gcd that returns the greatest common divisor of two positive integers (as in Appendix B).

Algorithm 2.2.2 Given the prime modulus m and any associated full-period, modulus-compatible multiplier a, the following algorithm generates all the full-period, modulus-compatible multipliers relative to m.

```
i = 1;
x = a;
while (x != 1) {
    if ((m % x < m / x) and (gcd(i, m - 1) == 1))
        // x is a full-period modulus-compatible multiplier
    i++;
    x = g(x);        // use Algorithm 2.2.1 to evaluate g(x) = ax mod m
}
```

Example 2.2.6

Using Algorithm 2.2.2 with $m = 2^{31} - 1 = 2147483647$, $a = 7$, and a lot of CPU cycles we find that there are a total of 23 093 associated full-period, modulus-compatible multipliers, the first few of which are

$$7^1 \bmod 2147483647 = 7$$

$$7^5 \bmod 2147483647 = 16807$$

$$7^{113039} \bmod 2147483647 = 41214$$

$$7^{188509} \bmod 2147483647 = 25697$$

$$7^{536035} \bmod 2147483647 = 63295.$$

Of these, the multiplier $a = 16807$ deserves special mention. It was first suggested by Lewis, Goodman, and Miller (1969), largely because it was easy to prove (as we have done) that it is a full-period multiplier. Since then it has become something of a "minimal" standard (Park and Miller, 1988).

In retrospect, $a = 16807$ was such an obvious choice that it seems unlikely to be the best possible full-period multiplier relative to $m = 2^{31} - 1$ and, indeed, subsequent testing for randomness has verified that, at least in theory, other full-period multipliers generate (slightly) more random sequences. Although several decades of generally favorable user experience with $a = 16807$ is not easily ignored, we will use $a = 48271$ instead.

Randomness. For a given (prime) modulus m, from among all the full-period, modulus-compatible multipliers relative to m, we would like to choose the one that generates the "most random" sequence. As suggested in Example 2.1.1, however, there is no simple and universal definition of randomness, and so it should come as no surprise to find that there is less than complete agreement on what this metric should be.

Example 2.2.7

To the extent that there is agreement on a randomness metric, it is based on a fundamental geometric characteristic of all Lehmer generators, first established by Marsaglia (1968), that, unfortunately, "random numbers fall mainly in the planes." That is, in 2-space for example, the points

$$(x_0, x_1), (x_1, x_2), (x_2, x_3), \ldots$$

all fall on a finite, and possibly small, number of parallel lines to form a *lattice* structure as illustrated in Figure 2.2.2 for $(a, m) = (23, 401)$ and $(66, 401)$. If a simulation model of a nuclear power plant, for example, required two consecutive small random numbers (e.g., both smaller than 0.08) for a particular rare event (e.g., meltdown), this would *never* occur for the $(66, 401)$ generator, which has a void in the southwest corner of the graph. [Using the axiomatic approach to probability, the exact probability is $(0.08)(0.08) = 0.0064$.]

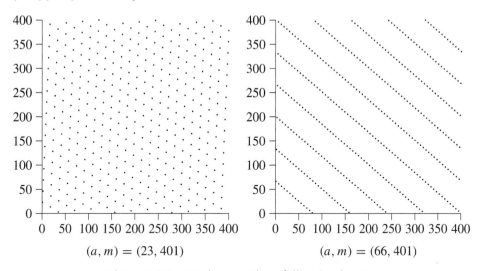

$(a, m) = (23, 401)$ $(a, m) = (66, 401)$

Figure 2.2.2   Random numbers falling in planes.

In general, for any integer $k \geq 2$, the points

$$(x_0, x_1, \ldots, x_{k-1}), (x_1, x_2, \ldots, x_k), (x_2, x_3, \ldots, x_{k+1}), \ldots$$

form a lattice structure. The best Lehmer-generator randomness metrics are based on analyzing (numerically, not visually) the *uniformity* of this lattice structure in k-space for small values of k. Knuth (1998) gives one popular randomness metric of this type known as the *spectral test*.

According to these metrics, for $m = 401$, the multiplier $a = 23$ would be a much better full-period, modulus-compatible multiplier than $a = 66$.

ANSI C Implementation. The kind of theoretical testing for randomness illustrated in Example 2.2.7 has been done for all of the 23 093 full-period, modulus-compatible multipliers relative to $m = 2^{31} - 1$. Of these, the winner is $a = 48271$ with $q = \lfloor m/a \rfloor = 44488$ and $r = m \bmod a = 3399$.*

Example 2.2.8

A Lehmer random-number generator with $(a, m) = (48\,271, 2^{31} - 1)$ can be implemented correctly, efficiently, and portably in ANSI C as follows.

```
double Random(void)
{
    const    long A     = 48271;                    // multiplier
    const    long M     = 2147483647;               // modulus
    const    long Q     = M / A;                    // quotient
    const    long R     = M % A;                    // remainder
      static long state = 1;
             long t     = A * (state % Q) - R * (state / Q);
      if (t > 0)
         state = t;
      else
         state = t + M;
      return ((double) state / M);
}
```

With minor implementation-dependent modifications, the random-number generator in Example 2.2.8 is the basis for all the simulated stochastic results presented in this book. There are three important points relative to this particular implementation.

- The static variable `state` is used to hold the current *state* of the random-number generator, initialized to 1. There is nothing magic about 1 as an initial state (seed); any value between 1 and 2147483646 could have been used.
- Because `state` is a static variable, the state of the generator will be retained between successive calls to `Random`, as must be the case for the generator to operate properly. Moreover, because the scope of `state` is local to the function, the state of the generator is protected—it cannot be changed in any way other than by a call to `Random`.
- If the implementation is correct, tests for randomness are redundant; if the implementation is incorrect, it should be discarded. A standard way of testing for a correct implementation is based on the fact that, if the initial value of `state` is 1, then, after 10 000 calls to `Random` the value of `state` should be 399 268 537.

---

*In addition to "in theory" testing, the randomness associated with this (a, m) pair has also been subjected to a significant amount of "in practice" empirical testing—see Section 10.1.

Example 2.2.9

As a potential alternative to the generator in Example 2.2.8, the random-number generator in the ANSI C library <stdlib.h> is the function rand. The intent of this function is to simulate drawing at random from the set $\{0, 1, 2, \ldots, m-1\}$ with m required to be at least 2^{15}. That is, rand returns an int between 0 and RAND_MAX inclusive, where the macro constant RAND_MAX (defined in the same library) is required to be at least $2^{15} - 1 = 32767$. To convert the integer value returned by rand to a floating-point number between 0.0 and 1.0 (consistent with Definition 2.1.1), it is conventional to use an assignment like

u = (double) rand() / RAND_MAX;

Note, however, that the ANSI C standard does *not* specify the details of the algorithm on which this generator is based. Indeed, the standard does not even require the output to be random! For scientific applications, it is generally a good idea to avoid using rand, as indicated in Section 13.15 of Summit (1995).

Random-Number Generation Library.   The random-number generation library used in this course is based upon the implementation considerations developed in this section. This library is defined by the header file "rng.h" and is recommended as a replacement for the standard ANSI C library functions rand and srand, particularly in simulation applications where the statistical goodness of the random-number generator is important. The library provides the following capabilities.

- double Random(void) —This is the Lehmer random-number generator in Example 2.2.8. We recommended it as a replacement for the standard ANSI C library function rand.
- void PutSeed(long seed) —This function can be used to initialize or reset the current state of the random-number generator. We recommended it as a replacement for the standard ANSI C library function srand. If seed is positive, then that value becomes the current state of the generator. If seed is 0, then the user is prompted to set the state of the generator interactively via keyboard input. If seed is negative, then the state of the generator is set by the system clock.*
- void GetSeed(long *seed) —This function can be used to get the current state of the random-number generator.*
- void TestRandom(void) —This function can be used to test for a correct implementation of the library.

Although we recommend the use of the multiplier 48271, and that is the value used in the library rng, as discussed in Example 2.2.6 the 16807 multiplier is something of a minimal standard. Accordingly, the library is designed so that it is easy to use 16807 as an alternative to 48271.

---

*See Section 2.3 for more discussion about the use of PutSeed and GetSeed.

Example 2.2.10

As an example of the use of `Random` and `PutSeed`, this algorithm was used to create the two scatterplots illustrated in Figure 2.2.3 for two different values of the initial seed.

```
seed = 123456789;                          // or 987654321
PutSeed(seed);
x₀ = Random();
for (i = 0; i < 400; i++) {
    xᵢ₊₁ = Random();
    Plot(xᵢ, xᵢ₊₁);                        // a generic graphics function
}
```

Unlike the lattice structure so obvious in Example 2.2.7, the (x_i, x_{i+1}) pairs in this case appear to be random with no lattice structure evident, as is desirable. (For a more direct comparison, all the integer-valued coordinates in Example 2.2.7 would need to be normalized to 1.0 via division by $m = 401$. That would be a purely cosmetic change, however.)

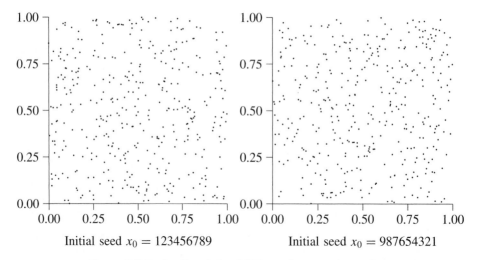

Initial seed $x_0 = 123456789$ Initial seed $x_0 = 987654321$

Figure 2.2.3   Scatterplots of 400 random number pairs.

Consistent with the discussion in Example 2.2.7, it should be mentioned that, in a sense, the appearance of randomness in Example 2.2.10 is an illusion. If *all* of the possible pairs of (x_i, x_{i+1}) points were generated (there are $m - 1 = 2^{31} - 2$ of these), and if it were somehow possible to plot all of these pairs of points at a fine enough scale to avoid blackening the page, then, much as in the figures in Example 2.2.7, a microscale lattice structure would be evident.* This observation is one of the reasons why we distinguish between *ideal* and *good* random-number generators (per Definition 2.1.1).

---

*Contemplate what size graphics device and associated dpi (dots per inch) resolution would be required to actually do this.

Example 2.2.11

Plotting consecutive, overlapping random-number pairs (x_i, x_{i+1}) from a full-period Lehmer generator with $m = 2^{31} - 1$ would blacken the unit square, obscuring the lattice structure. The fact that any tiny square contained in the unit square will exhibit approximately the same appearance is exploited in the algorithm below, where *all* of the random numbers are generated by Random(), but only those that fall in the square with opposite corners $(0, 0)$ and $(0.001, 0.001)$ are plotted. One would expect that approximately $(0.001)(0.001)(2^{31} - 2) \cong 2147$ of the points would fall in the tiny square.

```
seed = 123456789;
PutSeed(seed);
x₀ = Random();
for (i = 0; i < 2147483646; i++) {
    xᵢ₊₁ = Random();
    if ((xᵢ < 0.001) and (xᵢ₊₁ < 0.001)) Plot(xᵢ, xᵢ₊₁);
}
```

The results of the implementation of the algorithm are displayed in Figure 2.2.4 for the multipliers $a = 16807$ (on the left) and $a = 48271$ (on the right). The random numbers produced by $a = 16807$ fall in just 17 nearly vertical parallel lines, whereas the random numbers produced by $a = 48271$ fall in 47 parallel lines. These scatterplots provide further evidence for our choice of $a = 48271$ over $a = 16807$ in the random-number generator provided in the library rng.

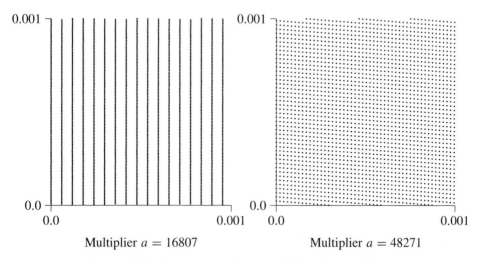

$$\text{Multiplier } a = 16807 \qquad \text{Multiplier } a = 48271$$

Figure 2.2.4   Scatterplots for $a = 16807$ and $a = 48271$.

We have shown that, as we zoom in on the unit square, the (unwanted) lattice structure of the (x_i, x_{i+1}) pairs produced by our *good* random-number generator becomes more apparent. It is appropriate at this juncture to contemplate what would happen with an *ideal* random-number generator. An ideal generator exhibits no lattice behavior as you zoom in on increasingly tiny squares. Like a fractal or a scale-free network, all scatterplots will look like those in Figure 2.2.3, assuming more points are generated as you zoom in.

Other Multipliers.   Recall that, for the modulus $m = 2^{31} - 1 = 2\,147\,483\,647$, there are $534\,600\,000$ multipliers a that result in a full-period random-number generator. Of these, $23\,093$ are modulus-compatible. Although the modulus-compatible, full-period multiplier $a = 48271$ is used in library `rng`, are there other multipliers that are recommended? One rather tedious way of approaching this question is to execute the statistical tests for random-number generators outlined in Section 10.1 on all of the generators. A second, more efficient approach is to consult the experts who perform research in this area. Fishman (2001, pages 428–445) notes that $a = 16807$, $a = 630\,360\,016$, and $a = 742\,938\,285$ are employed by the discrete-event simulation languages SIMAN, SIMSCRIPT II.5, and GPSS/H, respectively, ordered here by improving lattice structure. In addition to $a = 742\,938\,285$, he also suggests four other multipliers with similar performance in terms of parallel hyperplanes in 2, 3, 4, 5, and 6 dimensions: $a = 950\,706\,376$, $a = 1\,226\,874\,159$, $a = 62\,089\,911$, and $a = 1\,343\,714\,438$. Working out whether these multipliers are modulus-compatible is left as an exercise.

Good Generator—Nonrepresentative Subsequences.   In any simulation run, only a subset of the random numbers produced by a generator is typically used. Suppose, for example, that only 20 random numbers were needed for a simulation. If you had the *extreme* misfortune of choosing the seed $x_0 = 109\,869\,724$ with the generator in the library `rng`, the resulting 20 random numbers (to only 0.*dd* precision) are

$$
\begin{array}{cccccccccc}
0.64 & 0.72 & 0.77 & 0.93 & 0.82 & 0.88 & 0.67 & 0.76 & 0.84 & 0.84 \\
0.74 & 0.76 & 0.80 & 0.75 & 0.63 & 0.94 & 0.86 & 0.63 & 0.78 & 0.67.
\end{array}
$$

Is there something wrong with this generator to have produced 20 consecutive random numbers exceeding 0.62? Certainly not. It will occur only once in a blue moon, but this particular seed resulted in this rather unique sequence of random numbers.[*] This sequence of 20 consecutive random numbers might initially seem analogous to an *outlier* from statistics. Statisticians sometimes discard outliers as being unrepresentative, but it is *never* appropriate to do so in simulation. The random-number generator will have both a few sequences of unusually high random numbers and a few sequences of unusually low random numbers, as it should. Analogously, if we were to perform the experiment of tossing a fair coin 20 times for a large number of replications, there should be a few rare cases when all 20 tosses come up heads (or tails).

This string of 20 large random numbers highlights the importance of *replicating* a simulation many times so as to average out these unusual cases. Replication will be considered in detail in Chapter 8.

---

[*]The probability of 20 consecutive random numbers exceeding 0.62, according to the axiomatic approach to probability, is $(0.38)^{20} \cong 4 \cdot 10^{-9}$.

The Curse of Fast CPUs.   When discrete-event simulation was in its infancy, the time required to cycle through a full-period Lehmer generator with modulus $m = 2^{31} - 1 = 2\,147\,483\,647$ was measured in days. This quickly shrank to hours, and now it takes only a few minutes on a desktop machine. Before long, the time to complete a cycle of such a generator will be measured in seconds. The problem of *cycling*, that is, reusing the same random numbers in a single simulation, must be avoided because of the resultant dependency in the simulation output.

Extending the period length of Lehmer generators has been an active research topic within the simulation community for many years. One simple solution is to run all simulations on 64-bit machines. Since the largest prime number less than $2^{63} - 1$ is $2^{63} - 25 = 9\,223\,372\,036\,854\,775\,783$, the period is lengthened from about $2.14 \cdot 10^9$ to $9.22 \cdot 10^{18}$. In this case, cycle times are measured in years. Obviously, the increasing availability of machines supporting 64-bit integer arithmetic is beneficial to the simulation community. Since it is too early to know if 64-bit machines will become the norm, other portable techniques on 32-bit machines for achieving longer periods have been developed.

One technique for achieving a longer period is to use a *multiple recursive generator*, where our usual $x_{i+1} = ax_i \bmod m$ is replaced by

$$x_{i+1} = (a_1 x_i + a_2 x_{i-1} + \cdots + a_q x_{i-q}) \bmod m.$$

These generators can produce periods as long as $m^q - 1$ if the parameters are chosen properly. A second technique for extending the period is to use a *composite generator*, where several Lehmer generators can be combined in a manner to extend the period and improve statistical behavior. A third technique is to use a *Tausworthe*, or *shift-register* generator, where the modulo function is applied to bits, rather than large integers. The interested reader should consult Chapter 6 of Bratley, Fox, and Schrage (1987), Chapter 5 of Lewis and Orav (1989), Chapter 7 of Law and Kelton (2000), Chapter 9 of Fishman (2001), Gentle (2003), and L'Ecuyer, Simard, Chen, and Kelton (2002) for more details.

Any of these generators that have been well-tested by multiple researchers yield two benefits: longer periods and better statistical behavior (e.g., fewer hyperplanes). The cost is always the same: increased CPU time.

2.2.2   Exercises

2.2.1   Prove that, if u and v are real numbers with $0 < u - v < 1$, then the integer difference $\lfloor u \rfloor - \lfloor v \rfloor$ is either 0 or 1.

2.2.2   If $g(\cdot)$ is a Lehmer generator (full-period or not), then there must exist an integer $x \in \mathcal{X}_m$ such that $g(x) = 1$. (*a*) Why? (*b*) Use the $m = aq + r$ decomposition of m to derive an $O(a)$ algorithm that will solve for this x. (*c*) If $a = 48271$ and $m = 2^{31} - 1$, then what is x? (*d*) Same question if $a = 16807$.

2.2.3   Derive a $O\big(\log(a)\big)$ algorithm to solve Exercise 2.2.2. *Hint*: a and m are relatively prime.

2.2.4   Prove that, if a, m are positive integers and if a is "small" in the sense that $a^2 < m$, then $r < q$, where $r = m \bmod a$ and $q = \lfloor m/a \rfloor$.

2.2.5   Write $\gamma(x) = \alpha(x) - \beta(x)$, where $\alpha(x) = a(x \bmod q)$ and $\beta(x) = r\lfloor x/q \rfloor$, with $m = aq + r$ and $r = m \bmod a$. Prove that, if $r < q$, then for all $x \in \mathcal{X}_m$, both $\alpha(x)$ and $\beta(x)$ are in $\{0, 1, 2, \ldots, m - 1\}$.

2.2.6   Is Algorithm 2.2.1 valid if m is not prime? If not, how should it be modified?

2.2.7   (*a*) Implement a correct version of `Random` that uses floating-point arithmetic and do a timing study. (*b*) Comment.

2.2.8   Prove that, if a, x, q are positive integers, then $ax \bmod aq = a(x \bmod q)$.

2.2.9   You have been hired as a consultant by *XYZ Inc.* to assess the market potential of a relatively inexpensive *hardware* random-number generator they might develop for simulation-computing applications. List all the technical reasons you can think of to convince them this is a bad idea.

2.2.10   There are exactly 400 points in each of the figures in Example 2.2.7. (*a*) Why? (*b*) How many points would there be if a were not a full-period multiplier?

2.2.11   Let m be the largest prime modulus less than or equal to $2^{15} - 1$. (See Exercise 2.1.6.) (*a*) Compute all the corresponding modulus-compatible full-period multipliers. (*b*) Comment on how this result relates to random-number generation on systems that support 16-bit integer arithmetic only.

2.2.12   (*a*) Prove that, if m is prime with $m \bmod 4 = 1$, then a is a full-period multiplier if and only if $m - a$ is also a full-period multiplier. (*b*) What if $m \bmod 4 = 3$?

2.2.13   If $m = 2^{31} - 1$, compute the $x \in \mathcal{X}_m$ for which $7^x \bmod m = 48\,271$.

2.2.14   The lines on the scatterplot in Figure 2.2.4 associated with the multiplier $a = 16807$ appear to be vertical. Argue that the lines must *not* be vertical, using the fact that $(a, m) = (16807, 2^{31} - 1)$ is a full-period generator.

2.2.15   Figure out whether the multipliers associated with $m = 2^{31} - 1$ given by Fishman (2001) are modulus-compatible: $a = 630\,360\,016$, $a = 742\,938\,285$, $a = 950\,706\,376$, $a = 1\,226\,874\,159$, $a = 62\,089\,911$, and $a = 1\,343\,714\,438$.

2.3 MONTE CARLO SIMULATION

In this section, we will consider Monte Carlo simulation. (See the taxonomy in Figure 1.1.1.) Specifically, the discussion will focus on the estimation of one or more probabilities by using the functions in the library `rng` to implement an experimental technique whose validity is based on what is known as the *frequency* theory of probability.

2.3.1 Probability

There are two approaches to probability, both of which will be illustrated in this section. One approach is experimental (empirical), the other is theoretical (axiomatic).

Empirical Probability

Definition 2.3.1   If a random experiment is repeated n times and if n_a is the number of times event $\mathcal{A}$ occurs ($n_a \leq n$), then the *relative frequency* of occurrence of event $\mathcal{A}$ is n_a/n. The *frequency theory of probability* asserts that this relative frequency converges to the probability of $\mathcal{A}$ as $n \to \infty$

$$\Pr(\mathcal{A}) = \lim_{n \to \infty} \frac{n_a}{n}.$$

Based as it is on the acquisition of experimental data, Definition 2.3.1 is also called the *empirical* or *experimental* definition of probability. By its reliance on arbitrarily many replications of the random experiment, the equation in Definition 2.3.1 is related to the strong and weak *laws of large numbers*. Although the frequency theory of probability has limited theoretical importance, it is the cornerstone of experimental statistics and simulation. Monte Carlo simulation uses the frequency theory of probability in a natural way. The idea is simple yet profound. A computational model is built that uses a random-number generator to simulate the random experiment. The simulated experiment is then repeated many times (n) and the number of times event $\mathcal{A}$ occurs (n_a) is recorded. If n is large, the ratio n_a/n is a good *point estimate* of the probability $\Pr(\mathcal{A})$.

Axiomatic Probability. In contrast to the frequency theory, the formal study of probability is usually based on the *axiomatic* theory of probability. This is the familiar set-theoretic approach to probability, in which the primary emphasis is on the mathematical construction of a sample space with an associated probability function $\Pr(\mathcal{A})$ defined for all events $\mathcal{A}$ in the sample space. The axiomatic theory of probability and the frequency theory of probability are best viewed as complementary. Doing one brings insight to the other. The best solution to any probability problem is a mathematical solution established via the axiomatic method and verified experimentally via an independent Monte Carlo simulation. Many probability problems, however, are just too hard to solve mathematically. For those problems, Monte Carlo simulation may be the only viable approach. The axiomatic and frequency theories of probability are also complementary in the use of the latter to explain the former—that is, the significance of a mathematically derived probability $\Pr(\mathcal{A})$ is commonly explained by interpreting the probability as the relative frequency with which the event $\mathcal{A}$ would occur if the random experiment were to be repeated many times.

Example 2.3.1
Roll two dice and observe the up faces. This classic random experiment is simple enough that it can be analyzed by using a special case of the axiomatic approach: Construct a finite sample space with all points equally likely, then to evaluate $\Pr(\mathcal{A})$, count the points in $\mathcal{A}$ and divide by the cardinality of the sample space. That is, think of the dice as distinguishable (say, one is green and the other is red) and construct the sample space as a set of $6^2 = 36$ ordered pairs of the form (a, b), where both a and b can take on any integer value between 1 and 6:

$$
\begin{array}{cccccc}
(1,1) & (1,2) & (1,3) & (1,4) & (1,5) & (1,6) \\
(2,1) & (2,2) & (2,3) & (2,4) & (2,5) & (2,6) \\
(3,1) & (3,2) & (3,3) & (3,4) & (3,5) & (3,6) \\
(4,1) & (4,2) & (4,3) & (4,4) & (4,5) & (4,6) \\
(5,1) & (5,2) & (5,3) & (5,4) & (5,5) & (5,6) \\
(6,1) & (6,2) & (6,3) & (6,4) & (6,5) & (6,6).
\end{array}
$$

If the dice are fair, the 36 points in the sample space are equally likely, each with probability $1/36$. If the two up faces are summed, an integer-valued random variable, say X, is defined, having as possible values 2 through 12, inclusive. The probability associated with each possible value of the sum can be found by counting points in the sample space. For example, there are six points in the sample space corresponding to the sum 7 (this is $\mathcal{A}$). Thus, if two dice are rolled, the probability that the up faces will sum to 7 is $\Pr(X = 7) = 6/36 = 1/6$. In this way, the following table of probabilities can be constructed:

sum, x	2	3	4	5	6	7	8	9	10	11	12
$\Pr(X = x)$	1/36	2/36	3/36	4/36	5/36	6/36	5/36	4/36	3/36	2/36	1/36

As a consistency check, note that the probabilities are nonnegative and sum to 1.

Interpretation.   Because the (axiomatic) probability of rolling a 7 is $1/6$, in 6000 replications of the two-dice experiment, we expect to see a total of 7 occur approximately 1000 times. This is the relative-frequency interpretation of the mathematically derived (axiomatic) probability $1/6$; in the "long run," 7 will occur $1/6$ of the time. In the "short run," the relative frequency of 7's can be quite different from $1/6$, a fact that is well known to any gambler.

Alternatively, if it were not so easy to figure out the probability of rolling a 7 via the axiomatic approach—for example, if the dice were not fair—then $\Pr(X = 7)$ could be estimated by replicating the experiment many times and calculating the relative frequency of occurrence of 7's. Doing this with a random-number generator is an excellent simple example of what Monte Carlo simulation is all about. (See Exercise 2.3.4.)

Probability estimation via Monte Carlo simulation will be illustrated in this section with two examples of classic probability problems. In both cases, the axiomatic solution and Monte Carlo estimates are compared. Before we do so, however, a few important definitions are needed.

2.3.2 Random Variates

> **Definition 2.3.2**   A random *variate* is an algorithmically generated realization of a random *variable*.*
>
> If Random is a good random-number generator, a *Uniform*$(0, 1)$ random variate is what is generated by the assignment u = Random().

Uniform Random Variates.   Given the ability to generate a *Uniform*$(0, 1)$ random variate, what if we need a random variate, say x, that is *Uniform*(a, b)? That is, suppose a and b are real-valued parameters with $a < b$ and that we want to generate values of x in such a way that all real numbers between a and b are equally likely. How should this be done?

The answer is to first use u = Random() to generate a *Uniform*$(0, 1)$ random variate u, and to then transform from u to x via the equation $x = a + (b - a)u$, as shown in Figure 2.3.1. Values of u between 0.0 and 1.0 are mapped linearly to values of x between a and b, as illustrated by this set of equivalences:

$$0 < u < 1 \iff 0 < (b - a)u < b - a$$
$$\iff a < a + (b - a)u < a + (b - a)$$
$$\iff a < x < b.$$

---

*See Chapters 6 and 7 for a discussion of discrete and continuous random variables.

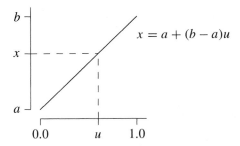

Figure 2.3.1   Geometry associated with *Uniform(a, b)* variate generation.

Definition 2.3.3   This ANSI C function generates a *Uniform*(a, b) random variate.

```
double Uniform(double a, double b)                      // use a < b
{
    return (a + (b - a) * Random());
}
```

Example 2.3.2
A point x is selected at random in the interval (a, b) to form two subintervals of length $x - a$ and $b - x$. What is the probability that the larger subinterval is more than twice the length of the smaller subinterval? For particular values of a, b, the relative-frequency (Monte Carlo) approach to probability is easily applied in this case by generating x as a *Uniform*(a, b) random variate. The details are left as Exercise 2.3.10.

Equilikely Random Variates.   *Uniform*$(0, 1)$ random variates can also be used to generate integer-valued random variates. In particular, let a and b be integer-valued parameters with $a < b$, and suppose we want to generate values of an integer-valued random variate x in such a way that all integers between a and b inclusive are equally likely. In this case, x is an *Equilikely*(a, b) random variate. How can this random variate be generated?

If u = Random() then x can be generated via the transformation $x = a + \lfloor (b - a + 1)u \rfloor$. That is,

$$0 < u < 1 \iff 0 < (b - a + 1)u < b - a + 1$$

$$\iff 0 \le \lfloor (b - a + 1)u \rfloor \le b - a$$

$$\iff a \le a + \lfloor (b - a + 1)u \rfloor \le b$$

$$\iff a \le x \le b$$

and so values of u between 0.0 and 1.0 are mapped linearly in a discrete manner to integer values of x between a and b inclusive. The geometric representation of this transformation is identical to the corresponding *Uniform* (a, b) geometry shown in Figure 2.3.1, except that the diagonal line will stairstep, in keeping with the discrete character of the transformation.

Definition 2.3.4   This ANSI C function generates an *Equilikely* (a, b) random variate.

```
long Equilikely(long a, long b)                          // use a < b
{
   return (a + (long) ((b - a + 1) * Random()));
}
```

Example 2.3.3
To generate a random variate x that simulates rolling two fair dice and summing the resulting up faces, use

```
x = Equilikely(1, 6) + Equilikely(1, 6);
```

Note that this is *not* equivalent to

```
x = Equilikely(2, 12);
```

Example 2.3.4
To select an element x at random from the array $a[0], a[1], \ldots, a[n-1]$, use

```
i = Equilikely(0, n - 1);   // pick an array index at random
x = a[i];
```

See Section 6.5 for more discussion of this notion.

We are now ready to consider two complete classic Monte Carlo simulation examples. The first makes use of *Equilikely* (a, b) random variates, the other makes use of *Uniform* (a, b) random variates. Many other types of random variates are considered in later chapters, primarily in discrete-event-simulation applications.

2.3.3 Galileo's Dice

Example 2.3.5
Three fair dice are rolled and the random variable X is the sum of the three up faces. Which sum is more likely, a 9 or a 10? This example is alleged to have been solved by Galileo when asked by a gambler to explain why 10's seemed to appear more often than 9's as the sum of three dice. The axiomatic theory of probability can be used to solve this problem in several ways; the most direct, albeit tedious, approach is based on an extension of Example 2.3.1. That is, the sample space for the three-dice random experiment can be constructed as a set of $6^3 = 216$ equally likely possible outcomes. By listing all 216 of these and counting the ones that yield a sum of 9 or 10,

respectively (axiomatic does not necessarily mean elegant), it can be verified that the probabilities are

$$\Pr(X = 9) = \frac{25}{216} \cong 0.116 \qquad \text{and} \qquad \Pr(X = 10) = \frac{27}{216} = 0.125.$$

So, a 10 is slightly more likely than a 9; in 2160 rolls of three dice, we expect to see a 10 approximately 270 times and a 9 approximately 20 times less.

Program galileo.   As an alternative, Monte Carlo approach to Example 2.3.5, we can use the sum of three *Equilikely*(1, 6) random variates to simulate the rolling of three dice and use the computed relative frequencies of 9's and 10's to estimate these two probabilities. As it turns out, it is no more difficult to compute the relative frequencies of all the other possible sums as well and, in this way, estimate the probability of each possible sum between 3 and 18. The program galileo does exactly that.

The simplicity of program galileo is compelling—that is the appeal of simulation. The drawback of Monte Carlo simulation, however, is that the $n \to \infty$ limit operation in Definition 2.3.1 can only be approximated; as a result, program galileo can only produce probability *estimates*. Moreover, as the following example illustrates, relative-frequency probability estimates can converge slowly and erratically. Indeed, the $n \to \infty$ limit in Definition 2.3.1 is *not* the traditional type of limit studied in calculus. That is, there is no *guarantee* that larger values of n will always produce more accurate probability estimates. There is, however, the *expectation* that this will be so. The mathematics involved in making this statement more precise is beyond the scope of this book.

Given that relative-frequency probability estimates based on a finite number of replications (finite n) have an inherent uncertainty, an important issue remains—how accurate are these estimates? For a probability-estimation application with modest accuracy requirements, generally 1000 to 10 000 replications is sufficient. For example, in Chapter 8, we will see that, to achieve an estimate of the three-dice probability $\Pr(X = 10) = 0.125$ that is accurate to ± 0.01 (with 95% confidence), approximately 4400 replications are required. For applications with more stringent accuracy requirements, even more replications would be required to reduce the uncertainty to an acceptable level.

For now, we will largely avoid the issue of accuracy versus the number of replications, except to observe that personal computers and workstations are inexpensive, time on them is free for many people, and it is reasonable to expect that, as n becomes larger, the accuracy of the probability estimates will tend to improve.

Example 2.3.6
Figure 2.3.2 illustrates the relatively slow and somewhat erratic convergence of relative-frequency probability estimates for the three-dice random experiment. Three sequences of $\Pr(X = 10)$ estimates are shown, corresponding to the three rng initial seeds indicated. A point is plotted for $n = 20, 40, \ldots, 1000$. The horizontal line represents the axiomatic probability 0.125. A figure for $\Pr(X = 9)$ estimates would be similar.

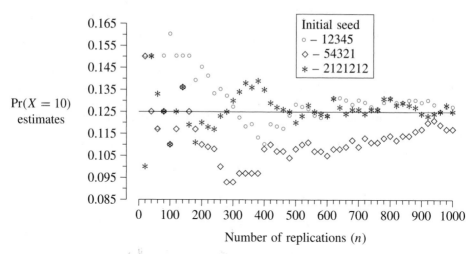

Figure 2.3.2   Program `galileo` output.

Only probability estimates up to the first 1000 replications are shown. Do not interpret this to mean that there is something magic about 1000. There is an inherent uncertainty in any Monte Carlo probability estimate. The magnitude of the uncertainty is dictated, in part, by the number of replications; increasing the number of replications reduces the uncertainty.

Multiple Initial Seeds.   Because program `galileo` uses `PutSeed(0)` to initialize the state of the random-number generator, each time the program is run the user is prompted to supply an initial seed. As Example 2.3.6 illustrates, different initial seeds will cause different sequences of three-dice results to be generated and, in that way, different probability estimates will be produced. It is *always* good practice to run a Monte Carlo simulation program multiple times, using a different initial seed for each run, and pay close attention to how the results vary from run to run. If the variability is too large, the number of replications (n) per run should be increased (perhaps a lot—see Chapter 8).

2.3.4 Geometric Applications

The Monte Carlo-simulation solution to the three-dice problem was based on the use of *Equilikely* (a, b) random variates. The use of this discrete random variate is common in Monte Carlo simulation because of the large variety of stochastic models based on a "select *at random* from the finite set $\{a, \ldots, b\}$" characterization. For comparison, we now consider continuous Monte Carlo simulation examples with geometric applications based on the use of *Uniform* (a, b) random variates to simulate the selection of a point "*at random* in the interval (a, b)."

Example 2.3.7

Many Monte Carlo-simulation applications that use *Uniform*(a, b) random variates involve the experimental solution of geometry-based probability problems. To solve these problems, it is typical to use a *Uniform*(a, b) random variate to position points in two or more dimensions at random, subject to geometric constraints.

For example, to generate a point (x, y) at random inside a rectangle with opposite corners at (α_1, β_1) and (α_2, β_2) as illustrated on the left-hand side of Figure 2.3.3, use

```
x = Uniform(α₁, α₂);
y = Uniform(β₁, β₂);
```

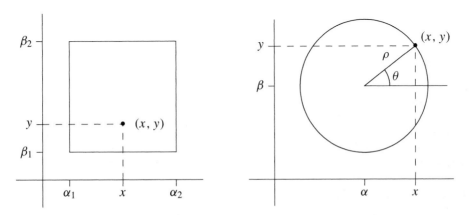

Figure 2.3.3   Generating random points inside a rectangle and on the circumference of a circle.

The bivariate random variate (x, y) so generated is *Uniform* within the rectangle—all locations are equally likely. Similarly, to generate a point (x, y) at random on the *circumference* of a circle with radius ρ and center (α, β), as illustrated on the right-hand side of Figure 2.3.3, use

```
θ = Uniform(-π, π);
x = α + ρ * cos(θ);
y = β + ρ * sin(θ);
```

The resulting bivariate random variate (x, y) is *Uniform* on the circumference of the circle. In both cases the phrase "at random" is a synonym for *Uniform*, properly interpreted relative to the geometry of the figure.

Example 2.3.8

It might seem natural to generalize the second part of Example 2.3.7 to generate a point (x, y) at random *interior* to the circle of radius ρ centered at (α, β) as follows:

```
θ = Uniform(-π, π);
r = Uniform(0, ρ);
x = α + r * cos(θ);
y = β + r * sin(θ);
```

This algorithm is *not* correct, however, because the distribution of points so generated will not be uniform interior to the circle. Instead, points will be more densely distributed close to the center than close to the circumference, as is illustrated in Example 2.3.9.

Acceptance–Rejection.   A correct approach to generating a point at random interior to a circle is to generate a point at random within a circumscribing square, then either *accept* or *reject* the point on the basis of whether it is within the circle. That is, points are generated uniformly in a circumscribing square until a point falls within the circle, as summarized by the following algorithm.

Algorithm 2.3.1   This algorithm uses the *acceptance–rejection* technique to generate a point (x, y) at random (uniformly) interior to a circle of radius ρ centered at (α, β).

```
do {
    x = Uniform(-ρ, ρ);
    y = Uniform(-ρ, ρ);
} while (x * x + y * y >= ρ * ρ);          // rejection
x = α + x;
y = β + y;
return (x, y);
```

The probability of acceptance in the `do-while` loop in Algorithm 2.3.1 is the ratio of the area of the circle to the area of the circumscribed square, which is

$$p = \frac{\pi \rho^2}{4\rho^2} = \frac{\pi}{4} \cong 0.785.$$

Therefore, the algorithm will loop once with probability p, or twice with probability $p(1 - p)$, or three times with probability $p(1 - p)^2$, etc., so that the expected (average) number of passes through the loop per point generated is

$$p + 2p(1 - p) + 3p(1 - p)^2 + 4p(1 - p)^3 + \cdots.$$

It can be shown that this infinite geometric series converges to $1/p = 4/\pi \cong 1.273$ (see Section 6.1) and so the expected number of passes through the `do-while` loop in Algorithm 2.3.1 is reasonably small.

Although classic and efficient, Algorithm 2.3.1 is somewhat esthetically unsatisfactory because it "wastes" at least two calls to Random with probability $1 - p = 1 - \pi/4 \cong 0.215$. That is, it seems reasonable to expect that there is a *synchronized* algorithm that generates the (x, y) pairs using *exactly* two calls to Random. Such an algorithm exists; see Exercise 2.3.9.

Example 2.3.9

The scatterplot on the left-hand side of Figure 2.3.4 was generated using the (incorrect) algorithm in Example 2.3.8. The increased density of points near the origin is evident. In contrast, the figure on the right-hand side of Figure 2.3.4 was generated using the (correct) acceptance–rejection technique in Algorithm 2.3.1. (For both figures, the number of (x, y) points is 400, generated with an rng initial seed of 12345.)

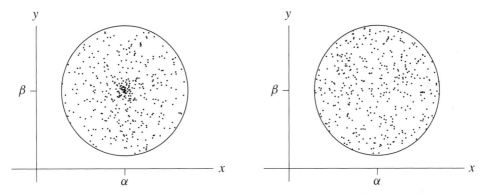

Figure 2.3.4   Incorrect and correct generation of points interior to a circle.

2.3.5 Buffon's Needle

The *Uniform(a, b)* random variate Monte Carlo simulation example in this section is the classic *Buffon's needle problem*, which is one of the oldest known problems in geometric probability. Buffon was a French naturalist of the eighteenth century who first posed and solved the problem that bears his name. The solution of this problem suggests a Monte Carlo technique for estimating π.

Example 2.3.10

Suppose that an infinite family of infinitely long vertical lines are spaced one unit apart in the (x, y) plane. If a needle of length $r > 0$ (and negligible width) is dropped at random onto the plane, what is the probability that it will land crossing at least one line?

One realization of this experiment is shown in Figure 2.3.5. In the particular realization depicted, the needle does not cross one or more of the vertical lines. Obviously, the probability of one or more crossings depends on three quantities: the length of the needle r, the horizontal position of one end of the needle (we arbitrarily choose the left-hand endpoint), and the angle orientation of the needle from horizontal.

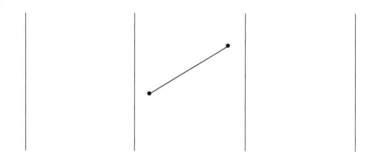

Figure 2.3.5   Buffon needle realization.

Without loss of generality, we can assume that the needle falls with its *left-hand* endpoint between the two parallel lines defined by $x = 0$ and $x = 1$. Let u be the x-coordinate of the left-hand endpoint of the needle and let θ be the angle of the needle relative to the x-axis. The right-hand endpoint of the needle is to the right of $x = u$, as illustrated in Figure 2.3.6.

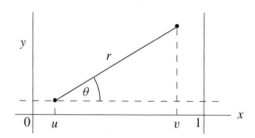

Figure 2.3.6   Buffon needle geometry.

The x-coordinate of the *right-hand* endpoint of the needle is $v = u + r \cos\theta$ and the needle crosses at least one line if and only if $v > 1$. If the phrase "dropped at random" is interpreted (modeled) to mean that u and θ are independent $Uniform(0, 1)$ and $Uniform(-\pi/2, \pi/2)$ random variables respectively, then the dropping of the needle can be simulated by generating two $Uniform(a, b)$ random variates, as in program buffon.

Program buffon.   Program buffon is a Monte Carlo simulation that can be used to estimate the probability that the needle will cross at least one line. Note, in particular, the use of

```
PutSeed(-1);                         // any negative integer will do
GetSeed(&seed);                      // trap the value of the initial seed
   ⋮
printf("with an initial seed of %ld", seed);
```

as the mechanism for initializing the random-number generator. By using this approach, each time the program is run, a different initial seed will be supplied by the system clock and printed along with other program output.

Interestingly, an inspection of program `buffon` illustrates how to solve the problem by using the axiomatic approach. That is, this program generates points (θ, u) at random (uniformly) throughout the rectangle illustrated in Figure 2.3.7.

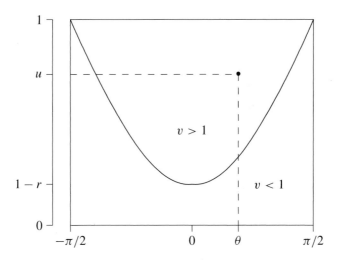

Figure 2.3.7   Sample space of all (θ, u) values.

The probability that the needle will cross at least one line is equivalent to the probability that a (θ, u) point (indicated by the '•') will fall in the shaded region with the curved boundary defined by the equation $u = 1 - r \cos(\theta)$. Because all points within the rectangle are equally likely, the probability of interest is the area of the shaded region divided by the area of the rectangle. If $0 < r \le 1$, as illustrated, then the area of the shaded region is

$$\pi - \int_{-\pi/2}^{\pi/2} \left(1 - r \cos\theta\right) d\theta = r \int_{-\pi/2}^{\pi/2} \cos\theta \, d\theta = \cdots = 2r.$$

Therefore, because the area of the rectangle is π, the probability that the needle will cross at least one line is $2r/\pi$. The case $r > 1$ is left as an exercise.

Disclaimer.   The simplicity of programs `galileo` and `buffon` is compelling. This simplicity is possible because the associated static models are so simple. That simplicity makes it possible to effectively short-circuit the first three model-development steps in Algorithm 1.1.1. Do not be misled by this into thinking that a similar short-circuiting is desirable for more complicated static systems or any *dynamic* stochastic system.

2.3.6 Exercises

2.3.1   (*a*) Derive the theoretical value of the probability of Buffon's needle crossing at least one line when $r > 1$. (*b*) Use Monte Carlo simulation to verify the correctness of your result when $r = 2$.

2.3.2   A rod of length 1 is broken at random into three pieces. (*a*) Use Monte Carlo simulation to estimate the probability that the resulting three pieces form a triangle. (*b*) Explain your interpretation of "at random" in this case. (*c*) What is the axiomatic probability?

2.3.3   A fair coin is tossed once. If it comes up heads, a fair die is rolled, and you are paid the number showing in dollars. If it comes up tails, two fair dice are rolled, and you are paid the sum of the two numbers showing in dollars. Let X be the amount won. Enumerate all the possible values of X and use Monte Carlo simulation to estimate the probability of each.

2.3.4   Suppose that each die in a pair of dice is loaded (un-fair) in such a way that the 6-face is four times as likely as the opposite 1-face and each of the other four faces are twice as likely as the 1-face. (*a*) Use Monte Carlo simulation to estimate the probability that, if the dice are rolled, the sum of the two up-faces will be 7. (*b*) What is the axiomatic probability?

2.3.5   (*a*) If two points are selected at random on the circumference of a circle of radius ρ, use Monte Carlo simulation to estimate the probability that the distance between the points is greater than ρ. (*b*) How does this probability depend on ρ?

2.3.6   You and I and eight other people are divided, at random, into two groups, each of size five. Use Monte Carlo simulation to estimate the probability that we will both end up in the same group.

2.3.7   Three dice are rolled and the largest of the three up faces is recorded. Let X be this value. The possible values of X are $1, 2, \ldots, 6$. Use Monte Carlo simulation to estimate the probability of each possible value.

2.3.8   Three identical boxes each contain two compartments. The first box has a \$10 bill in each compartment. The second box has a \$5 bill in each compartment. The third box has a \$10 bill in one compartment and a \$5 bill in the other compartment. A box is selected at random and one of compartments of that box is selected at random and is opened. If the compartment contains a \$10 bill, use Monte Carlo simulation to estimate the probability that the other compartment also contains a \$10 bill.

2.3.9   (*a*) Correct the algorithm in Example 2.3.8. (*b*) How would you test this new algorithm for correctness? *Hint*: The assignment r = Uniform(0, ρ) must be altered to make large values of r more likely than small values.

2.3.10   (*a*) Use Monte Carlo simulation to estimate the probability in Example 2.3.2. (*b*) Verify that the axiomatic approach yields the same probability. (*c*) How does this probability depend on a and b?

2.3.11   (*a*) Modify program galileo to estimate the probability of rolling a sum of 8 when five fair dice are tossed. (*b*) Compare the estimated probability with the value obtained by the axiomatic method.

2.3.12   Consider the intuitive (and incorrect) algorithm for generating a point uniformly in the interior of a circle given in Example 2.3.8. Using this algorithm, find the probability of being within a of the center of the circle, where $0 < a < \rho$. If the point generated was truly a random point in the interior of the circle, what should this probability equal?

2.3.13   Use program buffon to estimate π.

2.3.14   Use Monte Carlo simulation to estimate the average distance between two points generated randomly in the interior of a unit square.

2.3.15   Use Monte Carlo simulation to estimate the average shortest circuit (a path that begins and ends at one arbitrary point) among four points generated randomly in the interior of a unit square.

2.3.16   Modify program `buffon` to estimate π 30 times, using $10\,000$ needle tosses per replication. Summarize the results of the 30 replications in a table or an appropriate figure.

2.3.17   Consider the following *two-dimensional* Buffon needle problem. Suppose that an infinite family of infinitely long vertical and horizontal lines are spaced one unit apart in the (x, y) plane, in a similar fashion to square tiles on the floor of a large room. A needle of length $r < 1$ (and negligible width) is dropped at random onto the plane. If an observer considered only the vertical crossings, the method for estimating π would be identical to the development in Section 2.3.5. Likewise, if an observer considered only the horizontal crossings, the method for estimating π would also be identical to the development in Section 2.3.5. What if the observer considered both vertical and horizontal crossings? Develop a conceptual model for estimating π when both types of crossings are considered simultaneously. Is this an appropriate technique for halving the number of needle tosses necessary to achieve the same precision?

2.4 MONTE CARLO SIMULATION EXAMPLES

This section presents four applications of Monte Carlo simulation that are designed to augment the elementary examples considered in Section 2.3. The applications have been chosen to highlight the diversity of problems that can be addressed by Monte Carlo simulation and have been arranged in increasing order of the complexity of their implementation. The problems are (1) estimating the probability that the determinant of a 3×3 matrix of random numbers having a particular sign pattern is positive (program `det`), (2) estimating the probability of winning in Craps (program `craps`), (3) estimating the probability that a hatcheck girl will return all of the hats to the wrong owners when she returns n hats at random (program `hat`) and (4) estimating the mean time to complete a stochastic activity network (program `san`). Although the axiomatic approach to probability can be used to solve some of these problems, a minor twist in the assumptions associated with the problems often sinks an elegant axiomatic solution. A minor twist in the assumptions typically does not cause serious difficulties with the Monte Carlo simulation approach.

2.4.1 Random Matrices

Although the elementary definitions associated with matrices may be familiar to readers who have taken a course in linear algebra, we begin by reviewing some elementary definitions associated with matrices.

Matrices and Determinants

> **Definition 2.4.1**   A *matrix* is a collection, including possible repetitions, of real or complex numbers arranged in a rectangular array.

Uppercase letters are typically used to denote matrices, with subscripted lowercase letters used for their entries. The first subscript denotes the row where the entry resides and the second subscript denotes the column. Thus a generic $m \times n$ array A containing m rows and n columns is written as:

$$A = \begin{bmatrix} a_{11} & a_{12} & \cdots & a_{1n} \\ a_{21} & a_{22} & \cdots & a_{2n} \\ \vdots & \vdots & \ddots & \vdots \\ a_{m1} & a_{m2} & \cdots & a_{mn} \end{bmatrix}.$$

There are many operations (e.g., addition, multiplication, inversion), quantities (e.g., eigenvalues, trace, rank), and properties (e.g., positive definiteness) associated with matrices. We consider just one particular quantity associated with a matrix: its determinant, a single number associated with a square $(n \times n)$ matrix A, typically denoted* as $|A|$ or det A.

Definition 2.4.2   The *determinant* of a 2×2 matrix

$$A = \begin{bmatrix} a_{11} & a_{12} \\ a_{21} & a_{22} \end{bmatrix}$$

is

$$|A| = \begin{vmatrix} a_{11} & a_{12} \\ a_{21} & a_{22} \end{vmatrix} = a_{11}a_{22} - a_{21}a_{12}.$$

The determinant of a 3×3 matrix can be defined as a function of the determinants of 2×2 submatrices in the following manner.

Definition 2.4.3   The *determinant* of a 3×3 matrix A is

$$|A| = \begin{vmatrix} a_{11} & a_{12} & a_{13} \\ a_{21} & a_{22} & a_{23} \\ a_{31} & a_{32} & a_{33} \end{vmatrix} = a_{11} \begin{vmatrix} a_{22} & a_{23} \\ a_{32} & a_{33} \end{vmatrix} - a_{12} \begin{vmatrix} a_{21} & a_{23} \\ a_{31} & a_{33} \end{vmatrix} + a_{13} \begin{vmatrix} a_{21} & a_{22} \\ a_{31} & a_{32} \end{vmatrix}.$$

Random Matrices.   Matrix theory traditionally emphasizes matrices that consist of real or complex constants. But what if the entries of a matrix are random variables? Such matrices are referred to as "stochastic" or "random" matrices.

Although a myriad of questions can be asked concerning random matrices, our emphasis here will be limited to the following question: If the entries of a 3×3 matrix are independent random numbers with positive diagonal entries and negative off-diagonal entries, what is the probability

---

*The bars around A associated with the notation $|A|$ have nothing to do with absolute value.

that the matrix has a positive determinant? That is, find the probability that

$$\begin{vmatrix} +u_{11} & -u_{12} & -u_{13} \\ -u_{21} & +u_{22} & -u_{23} \\ -u_{31} & -u_{32} & +u_{33} \end{vmatrix} > 0,$$

where the u_{ij}'s are independent random numbers. This question is rather vexing using the axiomatic approach to probability due to the appearance of some of the random numbers multiple times on the right-hand side of Definition 2.4.3.*

Having completed the conceptual formulation of this problem, we can go directly to the implementation step, since the algorithm is straightforward. The ANSI C program det generates random matrices in a loop, counts the number of these matrices that have a positive determinant, and prints the ratio of this count to the number of replications.

In order to estimate the probability with some precision, it is reasonable to make one long run. We should ensure, however, that we do not recycle random numbers. Since nine random numbers are used to generate the 3×3 random matrix in each replication, we should use fewer than $(2^{31} - 1)/9 \cong 239\,000\,000$ replications. For an initial random-number generator seed of 987654321 and 200 000 000 replications, the program returns an estimated probability of a positive determinant as 0.05017347.

The point estimate for the probability has been reported to seven significant digits in the preceding paragraph. How many of the leading digits are significant, and how many of the trailing digits are sampling variability (noise) that should not be reported? One way to address this question is to rerun the simulation with the same number of replications and seed, but a different multiplier a. (Simply using a new seed with the user random number generator will not be effective, because the random numbers will be recycled by the large number of replications.) Using $a = 16807$ and $a = 41214$, the estimates of the probability of a positive determinant are 0.050168935 and 0.05021236, respectively. Making a minimum of three runs in Monte Carlo and discrete-event simulation is good practice because it allows you to see the spread of the point estimates. Using common sense or some statistical techniques that will be developed in subsequent chapters, it is reasonable to report the point estimate to just three significant digits: 0.0502. Obtaining a fourth digit of accuracy pushes the limits of independence and uniformity of the random-number generator.

As illustrated in the exercises, this Monte Carlo simulation solution can be extended to cover other random variables as entries in the matrix or larger matrices.

2.4.2 Craps

The gambling game known as "craps" provides a slightly more complicated Monte Carlo simulation application. The game involves tossing a pair of fair dice one or more times and observing the total number of spots showing on the up faces. If a 7 or 11 is tossed on the first roll, the

---

*This question is of interest to matrix theorists as it is the first example of a probability that cannot be calculated easily. A positive determinant in this case is equivalent to the matrix being of a special type called an *M-matrix* (Horn and Johnson, 1990). A positive determinant in a matrix with this particular sign pattern (positive diagonal entries and negative off-diagonal entries) corresponds to having all three 2×2 principal minors (determinants of submatrices determined by deleting the same numbered row and column) being positive.

player wins immediately. If a 2, 3, or 12 is tossed on the first roll, the player loses immediately. If any other number is tossed on the first roll, this number is called the "point." The dice are rolled repeatedly until the point is tossed (and the player wins) or a 7 is tossed (and the player loses).

The problem is conceptually straightforward. The algorithm development is slightly more complicated than in the previous example involving the 3×3 random matrix. An *Equilikely*$(1, 6)$ random variate is used to model the roll of a single fair die. The algorithm that follows uses N for the number of replications of the game of craps. The variable *wins* counts the number of wins and the `do-while` loop is used to simulate the player attempting to make the point when more than one roll is necessary to complete the game.

Algorithm 2.4.1 This algorithm estimates the probability of winning at craps using N replications.

```
wins = 0;
for (i = 1; i <= N; i++) {
   roll = Equilikely(1, 6) + Equilikely(1, 6);
   if (roll == 7 or roll == 11)
      wins++;
   else if (roll != 2 and roll != 3 and roll != 12) {
      point = roll;
      do {
         roll = Equilikely(1, 6) + Equilikely(1, 6);
         if (roll == point) wins++;
      } while (roll != point and roll != 7)
   }
}
return (wins / N);
```

The algorithm has been implemented in program `craps`. The function `Roll` returns the outcome when a pair of dice is tossed by summing the results of two calls to *Equilikely*$(1, 6)$. The `switch` statement with several `case` prefixes (for 2 .. 12) has been used to identify the result of the game based on the outcome of the first roll. The program was executed for 10 000 replications, with the seeds 987654321, 123456789, and 555555555, yielding as estimates of winning a game 0.497, 0.485, and 0.502, respectively. Since 0.5 is an important probability to both the house and the player, clearly 10 000 replications is inadequate. Unlike the previous example, the solution using the axiomatic approach is known in this case—it is $244/495 \cong 0.4929$, which seems to be consistent with the Monte Carlo simulation results. The fact that the probability of winning is slightly less than 0.5 lures gamblers to the game, yet assures that the house will extract money from gamblers in the long run.

2.4.3 Hatcheck Girl Problem

A hatcheck girl at a fancy restaurant collects n hats and returns them at random. What is the probability that everyone receives the wrong hat?

Monte Carlo simulation can be used to solve this classic probability problem, given, for example, in Ross (2006). We begin by numbering the hats $1, 2, \ldots, n$, which is suitable for algorithm development and computer implementation. The approach will be to generate one vector containing the hats in the order that they are checked and another for the order that they are returned. Since the order that they are checked is arbitrary, we can assume that the order that the hats are checked is $1, 2, \ldots, n$. Next, it is necessary to determine the meaning of returning the hats "at random." One reasonable interpretation is that, as each customer finishes dining and requests a hat from the hatcheck girl, she chooses one of the remaining hats in an equally likely manner. If this is the case, then all of the $n!$ permutations are equally likely. When $n = 3$, for example, the $3! = 6$ permutations:

$$1, 2, 3 \quad 1, 3, 2 \quad 2, 1, 3 \quad 2, 3, 1 \quad 3, 1, 2 \quad 3, 2, 1$$

are equally likely. Only the permutations $2, 3, 1$ and $3, 1, 2$ correspond to everyone leaving the restaurant with the wrong hat (since the hats are checked in as $1, 2, 3$), so the probability that all three hats went to the wrong owner is $2/6 = 1/3$.

Thus the algorithmic portion of solving this problem via Monte Carlo simulation reduces to devising an algorithm for generating a random permutation that represents the order that the hats are returned.

One way to generate an element in a random permutation is to pick one of the n hats randomly and return it to the customer if it has not already been returned. Generating a random permutation by this "obvious" algorithm is inefficient because: (1) it spends a substantial amount of time on checking to see whether the hat has already been selected, and (2) it can require several randomly generated hats in order to produce one that has not yet been returned. Both of these detractors are particularly pronounced for the last few customers that retrieve their hats. Fortunately, there is a very clever algorithm for generating a random permutation that avoids these difficulties. Although it will be presented in detail in Section 6.5, the reader is encouraged to think carefully about the $O(n)$ algorithm below, which is used to generate a random permutation of the array a with elements a[0], a[1], ..., a[n - 1] from the array a with elements input *in any order.*

```
for (i = 0; i < n - 1; i++) {
   j = Equilikely(i, n - 1);
   hold = a[j];
   a[j] = a[i];                          // swap a[i] and a[j]
   a[i] = hold;
}
```

The Monte Carlo simulation to estimate the probability that all n customers will receive the wrong hat is implemented in program hat. The function Shuffle is used to generate a random permutation, using the algorithm given above, and the function Check is used to determine whether any customers left the restaurant with their own hat. When run for 10 000 replications with $n = 10$ hats and initial seeds 987654321, 123456789, and 555555555, the estimates of the probability that all customers leave with the wrong hats are 0.369, 0.369, and 0.368, respectively.

A question of interest here is the behavior of this probability as n increases. Intuition might suggest that the probability goes to one in the limit as $n \to \infty$, but this is not the case. One way to approach this problem is to simulate for increasing values of n and fashion a guess based on the results of multiple long simulations. One problem with this approach is that you can never be sure that n is large enough. For this problem, the axiomatic approach is a better and more elegant way to determine the behavior of this probability as $n \to \infty$.

The hatcheck girl problem can be solved by using the axiomatic approach to probability. The probability of no matches is:

$$1 - \left(1 - \frac{1}{2!} + \frac{1}{3!} - \cdots + (-1)^{n+1}\frac{1}{n!}\right)$$

(Ross, 2006, pages 44–46), which is $16\,481/44\,800 \cong 0.3679$ when $n = 10$. This result helps to validate the implementation of the program hat for the case of $n = 10$ hats.*

Since the Taylor expansion of e^{-x} around $x = 0$ is

$$e^{-x} = 1 - x + \frac{x^2}{2!} - \frac{x^3}{3!} + \cdots,$$

the limiting probability (as the number of hats $n \to \infty$) of no matches is $1/e \cong 0.36787944$. This axiomatic approach is clearly superior to making long Monte Carlo simulation runs for various increasing values of n in order to estimate this limiting probability.

2.4.4 Stochastic Activity Networks

The fourth and final example of a Monte Carlo simulation study is more comprehensive than the first three. We consider the estimation of the mean time to complete a *stochastic activity network*, where arcs in the network represent activities that must be completed according to prescribed precedences.

Project Management and Networks. *Project management* is a field of study devoted to the analysis and management of projects. Project management analyzes a project that typically occurs once (such as building a stadium, landing a man on the moon, building a dam, laying out a job shop) whereas *production management* involves overseeing the manufacture of many items (such as automobiles on an assembly line). Production management is more forgiving than project management, in that there are multiple opportunities to correct faulty design decisions midstream. Production management is repetitive in nature, whereas project management considers only a single undertaking.

Projects consist of activities. One aspect of project management is the sequencing of these activities. When constructing a stadium, for example, the foundation must be poured before the seats can be installed. *Precedence relationships* can be established between the various activities.

---

*The fact that all three simulation experiments yielded probability estimates slightly *above* the theoretical value, however, would provide some impetus to make a few more runs. In an analogous fashion, flipping three heads in a row is not sufficient evidence to conclude that a coin is biased; a prudent person would certainly make a few more flips just to be on the safe side.

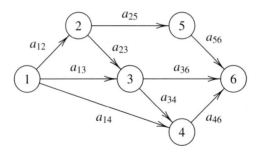

Figure 2.4.1   Six-node, nine-activity network.

Certain activities cannot be started until others have been completed. Project managers often use *networks* to illustrate the precedence relationships graphically.

Generally speaking, a network is a collection of nodes and arcs. In our setting, the arcs represent activities and the nodes are used to delay the beginning of outgoing activities until all of the incoming activities have been completed. The network shown in Figure 2.4.1, adopted from Pritsker (1995, pages 216–221), has six nodes and nine activities. The arc connecting nodes i and j is denoted by a_{ij}. Activity start times are constrained: No activity emanating from a given node can start until all activities that enter that node have been completed. In the network in Figure 2.4.1, activity a_{46}, for example, may not start until activities a_{14} *and* a_{34} have both been completed.

Activity Durations and Critical Paths.   The nodes and arcs in an activity network can be used to mark certain points in the time evolution of the project:

- the nodes can mark the maximum completion times (e.g., the end time of the latest activity completed for entering activities and the beginning time of all exiting activities);
- the arcs in an activity network can represent activity durations.

Figure 2.4.2 shows an example of a deterministic activity network, where the integers shown near each arc are the time to complete each activity. If these activity durations are fixed constants known with certainty, how long will it take to complete this network?

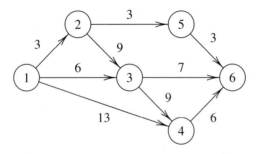

Figure 2.4.2   Deterministic activity network.

To answer this question, we must introduce the concept of *paths*. A path is an ordered set (sequence) of arcs leading in succession from node 1 to node n. Label the paths $\pi_1, \pi_2, \ldots, \pi_r$. Thus, arc $a_{ij} \in \pi_k$ means that arc a_{ij} is along the path π_k. The *length* of path π_k, denoted L_k, is the sum of all of the activity durations corresponding to the arcs $a_{ij} \in \pi_k$.

Returning to the network in Figure 2.4.2, there are $r = 6$ paths in the network:

k	Node sequence	π_k	L_k
1	$1 \rightarrow 3 \rightarrow 6$	$\{a_{13}, a_{36}\}$	13
2	$1 \rightarrow 2 \rightarrow 3 \rightarrow 6$	$\{a_{12}, a_{23}, a_{36}\}$	19
3	$1 \rightarrow 2 \rightarrow 5 \rightarrow 6$	$\{a_{12}, a_{25}, a_{56}\}$	9
4	$1 \rightarrow 4 \rightarrow 6$	$\{a_{14}, a_{46}\}$	19
5	$1 \rightarrow 3 \rightarrow 4 \rightarrow 6$	$\{a_{13}, a_{34}, a_{46}\}$	21
6	$1 \rightarrow 2 \rightarrow 3 \rightarrow 4 \rightarrow 6$	$\{a_{12}, a_{23}, a_{34}, a_{46}\}$	27

The ordering of the six paths is arbitrary. Assume that time is measured in days. The longest of these paths, namely π_6, which has a total duration of $L_6 = 27$ days, is known as the *critical path*. An algorithm known as the *critical path method* (CPM) can be used to identify, and compute the length of, the critical path for a large network. In general, we denote the critical path as π_c which is the path with the longest length, $L_c = \max\{L_1, L_2, \ldots, L_r\}$. The length of the critical path determines the time to complete the entire network. Any path i whose length is shorter than the critical path (that is, $L_i < L_c$) *can* be delayed without lengthening the duration of the project. Activity a_{56}, for example, from the network in Figure 2.4.2 is on only one path, π_3. Its duration could be changed from 3 days to as long as $27 - 9 + 3 = 21$ days without extending the duration of the project.

The analysis of a project with deterministic activity durations, such as the one in Figure 2.4.2, is straightforward. In most practical applications, however, activity durations are not deterministic. We consider the analysis of a *stochastic activity network*, where the activity durations are positive random variables. In practice, contracts often have penalties for projects that exceed a deadline and bonuses for projects that are completed early. Hence, the accurate analysis of the time to complete a project can be quite important.* We begin with the conceptual development of the problem, where the notation necessary to develop an algorithm is defined. We will alternate between general notation and our sample network.

Conceptual Development. Activity networks have a single source node labeled 1 and a single terminal node labeled n, where n is the number of nodes in the network. Node 1 has one or more arcs leaving it, and node n only has arcs entering it. All other nodes must have at least one arc entering and at least one arc leaving. There are m arcs (activities) in the network.

Each arc a_{ij} has a positive random activity duration Y_{ij}. Node j has a random time value T_j which denotes the time of completion of all activities entering node j, $j = 1, 2, \ldots, m$. The random time T_n is therefore the completion time of the entire network.

---

*The analysis of stochastic activity networks is often performed by a method known as the *project evaluation and review technique* (PERT), which yields the *approximate* distribution of the time to complete the network based on a result from probability theory known as the *central limit theorem*. Our development here estimates the *exact* distribution of the time to complete the network via simulation.

In the deterministic network in Figure 2.4.2, there are always one or more critical paths. In a stochastic activity network, a path is critical with a certain probability. For each realization of a stochastic network, a path is the critical path with some probability $p(\pi_k) = \Pr(\pi_k \equiv \pi_c)$, $k = 1, 2, \ldots, r$.

We need to transform the network given in Figure 2.4.1 to a mathematical entity appropriate for algorithm development. A matrix is well suited for this task, because two subscripts can be used to define arc a_{ij} and ease the computer implementation. We use a well-known tool known as a node–arc incidence matrix. A node–arc incidence matrix is an $n \times m$ matrix N, where each row represents a node and each column represents an arc. Let

$$N[i, j] = \begin{cases} 1 & \text{arc } j \text{ leaves node } i \\ -1 & \text{arc } j \text{ enters node } i \\ 0 & \text{otherwise.} \end{cases}$$

We now apply the general notation developed so far to specific questions associated with a specific network. For the network in Figure 2.4.3, use Monte Carlo simulation to estimate the following:

- the mean time to complete the network;
- the probability that each path is the critical path.

We assume that each activity duration is a uniform random variate having the lower limit zero and the upper limit given in Figure 2.4.2 [e.g., Y_{12} has a *Uniform*(0, 3) distribution].

It might seem initially that this problem is best addressed by discrete-event simulation, rather than Monte Carlo simulation, complete with an event calendar to keep track of the timing of the activities. The following mathematical relationship, however, can be used to analyze any network of this type by looking at just one activity at a time and exploiting the use of recursion in a programming language. If T_1 is defined to be 0.0, then

$$T_j = \max_{i \in \mathcal{B}(j)} \{T_i + Y_{ij}\},$$

for $j = 2, 3, \ldots, n$, where the set $\mathcal{B}(j)$ is the set of all nodes immediately *before* node j. In the six-node network with $j = 6$, for example,

$$T_6 = \max\{T_3 + Y_{36}, T_4 + Y_{46}, T_5 + Y_{56}\}.$$

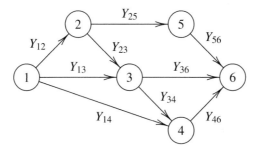

Figure 2.4.3   Stochastic activity network.

This observation allows us to write a function that returns a T_j value by recursively calling for the completion times of all nodes before node j.

The 6×9 matrix

$$N = \begin{bmatrix} 1 & 1 & 1 & 0 & 0 & 0 & 0 & 0 & 0 \\ -1 & 0 & 0 & 1 & 1 & 0 & 0 & 0 & 0 \\ 0 & -1 & 0 & -1 & 0 & 1 & 1 & 0 & 0 \\ 0 & 0 & -1 & 0 & 0 & -1 & 0 & 1 & 0 \\ 0 & 0 & 0 & 0 & -1 & 0 & 0 & 0 & 1 \\ 0 & 0 & 0 & 0 & 0 & 0 & -1 & -1 & -1 \end{bmatrix}$$

is a node–arc incidence matrix of the network with $n = 6$ nodes and $m = 9$ arcs in Figure 2.4.3. The number of 1's in each row indicates the number of arcs exiting a particular node and the number of -1's indicates the number of arcs entering a particular node. There is exactly one 1 and one -1 in each column and their positions indicate the beginning and ending node of one particular activity.

Finally, we discuss the appropriate point estimators for the two measures of performance prior to presenting the algorithm. The point estimator for the mean time to complete the network, for example, is the sample mean of the T_n's generated using the algorithm to follow (which is generated by the recursive expression for T_j given earlier). The point estimator for the probability $p(\pi_k)$ that path π_k is a critical path is the fraction of the networks generated that have path π_k as the critical path, $k = 1, 2, \ldots, r$.

Algorithm.   The recursive algorithm named T (for *Time*) that follows generates a single time to completion T_j for some node j, given that the network is represented by the node–arc incidence matrix N. The realization of the stochastic activity durations Y_{ij} associated with each arc a_{ij} are generated *prior* to the call to T. All that is needed to complete the algorithm is to embed this code in a loop for a particular number of replications and then compute point estimates for the measures of performance listed in the previous paragraph.

The two (global) parameters for this algorithm are the $n \times m$ node–arc incidence matrix N and one realization of the activity durations Y_{ij} generated as random variates. The procedure T has a single argument j, the node whose completion time is of interest (typically n, to estimate the time to complete the entire network). When T is passed the argument 1, it will return the completion time 0.0, which is assumed to be the beginning time of the network.

Algorithm 2.4.2   This algorithm returns a random time to complete all activities prior to node j for a single stochastic activity network with node–arc incidence matrix N.

```
k = 1;                        // initialize index for columns of N
l = 0;                // initialize index for predecessors to node j
tmax = 0.0;        // initialize longest time of all paths to node j
while (l < |B(j)|) {     // loop through predecessor nodes to node j
   if (N[j][k] == -1) {
                          // if column k of N has arc entering node j
```

```
        i = 1;                    // begin search for predecessor node
        while (N[i][k] != 1) {    // while i not a predecessor index
            i++;                              // increment i
        }
        t = T_i + Y_{ij}    // recursive call: t is completion time of a_{ij}
        if (t >= t_max) t_max = t;      // choose largest completion time
        l++;                          // increment predecessor index
    }
    k++;                                      // increment column index
}
return (t_max);                            // return completion time T_j
```

In most cases, this algorithm is called with argument n, so that a realization of the time to complete the entire network T_n is generated. In order to estimate $p(\pi_k)$, for $k = 1, 2, \ldots, r$, each path's length must be calculated for each realization; then the fraction of times that the length is the longest is calculated.

Implementation. The Monte Carlo simulation program san implements the recursive algorithm for generating nrep replications of the stochastic activity network completion times and estimates the mean time to complete the network and the probability that each of the six paths is the critical path. The axiomatic approach to probability does not provide an analytic solution for comparison.

When T is called with the argument $j = 6$, one realization of the nine uniform Y_{ij}'s is generated, then the algorithm given above is called to find T_6. The recursive calls associated with the predecessors (for node 6, the predecessor nodes are 3, 4, and 5) and subsequent recursive calls will be made to their predecessors. When the algorithm is called to generate the T_6's, the order of the recursive calls associated with the node–arc incidence matrix given above is T_6, T_3, T_1, T_2, T_1, and so on.

The simulation was run for 10 000 network realizations with the initial seeds 987654321, 123456789, and 555555555. The mean times to complete the network were 14.64, 14.59, and 14.57, respectively. The point estimates for $p(\pi_k)$ for each of the three runs are given in the table that follows, where $\hat{p}_1(\pi_k)$, $\hat{p}_2(\pi_k)$, and $\hat{p}_3(\pi_k)$ denote the estimates of $p(\pi_k)$ for the three initial seeds and $\hat{p}_a(\pi_k)$ denotes the average of the three estimates.

k	π_k	$\hat{p}_1(\pi_k)$	$\hat{p}_2(\pi_k)$	$\hat{p}_3(\pi_k)$	$\hat{p}_a(\pi_k)$
1	$\{a_{13}, a_{36}\}$	0.0168	0.0181	0.0193	0.0181
2	$\{a_{12}, a_{23}, a_{36}\}$	0.0962	0.0970	0.0904	0.0945
3	$\{a_{12}, a_{25}, a_{56}\}$	0.0013	0.0020	0.0013	0.0015
4	$\{a_{14}, a_{46}\}$	0.1952	0.1974	0.1907	0.1944
5	$\{a_{13}, a_{34}, a_{46}\}$	0.1161	0.1223	0.1182	0.1189
6	$\{a_{12}, a_{23}, a_{34}, a_{46}\}$	0.5744	0.5632	0.5801	0.5726

The path through the nodes $1 \to 2 \to 3 \to 4 \to 6$ is the most likely of the paths to be the critical path. It is the critical path for 57.26% of the 30 000 realizations of the network generated.

2.4.5 Exercises

2.4.1   Modify program det so that all $2^{31} - 1$ possible matrices associated with the random-number generator with $(a, m) = (48271, 2^{31} - 1)$ are generated.

2.4.2   Modify program det so as to test the other random variates as entries: (a) *Uniform*(0, 2), (b) *Uniform*(0.1, 1), (c) *Exponential*(1). (See Section 3.1 for the algorithm to generate an *Exponential*(1) random variate.)

2.4.3   Modify program det so that it estimates the probability that an $n \times n$ matrix of random numbers with positive diagonal and negative off-diagonal signs has a positive determinant for $n = 2, 3, 4, 5$. Use 1 000 000 replications and the initial seed 987654321.

2.4.4   Show that the exact probability of winning at craps is 244/495.

2.4.5   Dice can be *loaded* with weights, so that one particular outcome is more likely (e.g., rolling a six)—and as a result the outcome on the opposite side of the die (e.g., rolling a one) becomes less likely. Assuming that one and six are on opposite faces, two and five are on opposite faces, and three and four are on opposite faces, is there a way to load a die so that the probability of winning at craps exceeds 0.5?

2.4.6   You make one $1000 bet on a game of craps. I make 1000 $1 bets. Find the probability that each of us comes out ahead. (The result should show you why a casino would prefer many small bets to few large bets.)

2.4.7   Modify program craps to estimate the mean number of rolls required per game.

2.4.8   Modify program hat to estimate, for $i = 1, 2, \ldots, 20$, (a) the average number of hats returned correctly, (b) the average number of hats returned prior to the first hat returned correctly, conditioned on the correct return of one or more hats.

2.4.9   Show that the exact probability that the hatcheck girl will return all of the n hats to the improper person is

$$1 - \left(1 - \frac{1}{2!} + \frac{1}{3!} - \cdots + (-1)^{n+1} \frac{1}{n!} \right).$$

2.4.10   Modify program san to estimate the mean time to complete the network of *Uniform*(0, 1) activity durations given below.

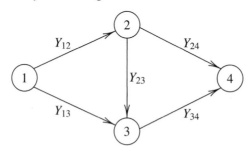

2.4.11   Modify program san to store intermediate-node completion times, so that no redundant recursive calls are made.

2.4.12   Write a program bubble that estimates the mean number of comparisons and the mean number of swaps required to bubble-sort a list of n numbers in random order. Test your program for $n = 2, 3, \ldots, 20$.

2.4.13   A *traveling salesman* has his home in a certain city represented by a random position on the unit square. (Any other shaped country could be modeled by use of the acceptance–rejection technique illustrated in Algorithm 2.3.1.) The salesman wishes to travel to each of the $n - 1$ cities, also uniformly distributed on the unit square, where he does business, then, after visiting all of them, returns home. Develop an algorithm and write a Monte Carlo simulation program `tsp` that estimates the average distance for the salesman to take the shortest route that visits each city exactly once and returns home. Code `tsp` to measure distance between cities: (*a*) as the crow flies (Euclidean), and (*b*) only on North–South and East–West roads (rectilinear, as in a large city). This *traveling salesman problem* (Hillier and Lieberman, 2005) plays a central role in combinatorics and in a branch of operations research known as "discrete optimization."

2.4.14   *Bootstrapping* is a statistical technique that is used to analyze a data set without making any assumptions about the population distribution (for example, that the population is bell-shaped). Efron and Tibshirani (1993, page 11) provide the rat-survival data given below. Seven rats are given a treatment; their survival times, given in days, are 16, 23, 38, 94, 99, 141, and 197. Nine other rats constitute a control group; their survival times are 10, 27, 30, 40, 46, 51, 52, 104, and 146 days. The median of the treatment group is 94 days; the median of the control group is 46 days. Do the medians differ by enough to conclude that there is a statistically significant difference between them? The *bootstrapping* approach to estimating the variability of the median for the treatment group is to generate B bootstrap samples, each of which consists of seven samples drawn with replacement from 16, 23, 38, 94, 99, 141, and 197. Draw $B = 1000$ such samples, using an *Equilikely*$(1, 7)$ random variate to determine the index of the value sampled, and perform a similar sampling of $B = 1000$ values from the control group. Perform whatever analysis you deem appropriate to evaluate whether there is a statistically significant difference between the median survival times.

2.4.15   During World War II, Axis Forces unwisely numbered their tanks sequentially from 1. As the Allied Forces captured tanks, they recorded the serial numbers in an effort to determine enemy strength by estimating the total number of tanks. Assume that four tanks, numbered 952, 1923, 1927, and 2000, have been captured. If an *Equilikely*$(1, N)$ random variate is used to model the number of tanks, where the total number of tanks N is unknown, here are four reasonable and intuitive ways to estimate the number of tanks:

- **Maximum likelihood.** Choose the estimate that maximizes the likelihood of observing 2000 as the largest tank number. Obviously any estimate less than 2000 is impossible (since 2000 was observed), so using $\hat{N}_1 = 2000$ as an estimate would yield the probability of observing 2000 as the largest tank number as $1/2000$. Any larger value, such as 2007, will have a lower probability (that is, $1/2007$).

- **Mean matching.** The theoretical mean of the *Equilikely*$(1, N)$ distribution is $(N + 1)/2$, so the theoretical mean can be equated to the sample mean:

$$\frac{N + 1}{2} = \frac{952 + 1923 + 1927 + 2000}{4}.$$

Solving for N yields $\hat{N}_2 = 3400$.

- **Gaps.** Four tanks were captured, so the average gap between the data values is

$$\frac{2000 - 4}{4} = \frac{1996}{4} = 499$$

tanks. Thus, if consistent gaps were maintained, an estimate for the highest numbered tank is $\hat{N}_3 = 2000 + 499 + 1 = 2500$.

- **Minimum-Variance Unbiased Estimator.** For sampling with replacement (not a brilliant military strategy) from an *Equilikely*$(1, N)$ distribution, Hogg, McKean, and Craig (2005, page 388) suggest the estimator

$$\hat{N}_4 = \frac{Y^{n+1} - (Y-1)^{n+1}}{Y^n - (Y-1)^n},$$

where n is the sample size and Y is the largest data value. For our data, $\hat{N}_4 = 2499.375$.

Implement a Monte Carlo simulation program that evaluates these four very different estimates for the number of tanks, using criteria of your choice.

2.5 FINITE-STATE SEQUENCES

At a high level of abstraction, a common deterministic system model consists of a *state space* that defines the set of possible system states and a corresponding *state-transition function* that determines how the system will evolve from state to state. Because the system is deterministic, there is no uncertainty about state transitions: When the state of the system changes, the new state is uniquely defined by the prior state of the system.

As illustrated in Section 2.1, this simple deterministic model is directly applicable to the computational process associated with random-number generation. In this kind of application, the size (cardinality) of the state space is typically large but *finite*. Because the state space is finite, as the system transitions from state to state it is inevitable that eventually some previously occurring state will occur again. At that point, a *cycle* has occurred and, because each state determines its successor uniquely, the cycle will then repeat endlessly. The primary focus of this section is the development and implementation of three algorithms to find when this cycle begins and, when it does, the length of the cycle so produced.

2.5.1 Background

Definition 2.5.1   A deterministic finite-state system is defined in terms of

- a *finite* state space (set) $\mathcal{X}$ with $|\mathcal{X}|$ elements (typically $|\mathcal{X}|$ is very large),
- a state-transition function $g : \mathcal{X} \to \mathcal{X}$, and
- finite-state sequences $x_0, x_1, x_2, \ldots$ generated by selecting x_0 (perhaps at random) from $\mathcal{X}$ and then defining $x_{i+1} = g(x_i)$ for $i = 0, 1, 2, \ldots$.

Example 2.5.1

As a simple example, suppose that $\mathcal{X}$ is the set of all integers between 0 and 9999 ($|\mathcal{X}| = 10\,000$). Thus, the states are the four-digit nonnegative integers, with the understanding that, if a state x is between 0 and 999, it will be made into a four-digit integer by affixing 0's on the left as necessary. Define the state-transition function $g(\cdot)$, using the now infamous *midsquares* method first suggested by von Neumann and Metropolis in the 1940's as a possible way to generate random numbers. It defines the transition from state to state as follows: Take the current state, a four-digit number, and square it. Write the result as an eight-digit number by affixing 0's on the left if necessary. The next state is then the four-digit number defined by extracting the *middle* four digits. For example, if the current state is $x_i = 1726$, then $x_i^2 = 02\underline{9790}76$ and so $x_{i+1} = 9790$. You should convince yourself that the four-digit midsquares method is equivalent to applying the following state-transition function $g(x) = \lfloor x^2/100 \rfloor \bmod 10000$. To illustrate, the finite-state sequences corresponding to the initial states $x_0 = 1726, 6283, 5600$ are

$$1726, 9790, 8441, 2504, 2700, 2900, 4100, 8100, 6100, 2100, 4100, \ldots$$

$$6283, 4760, 6576, 2437, 9389, 1533, 3500, 2500, 2500, \ldots$$

$$5600, 3600, 9600, 1600, 5600, \ldots$$

Definition 2.5.2   The sequence $x_0, x_1, x_2, \ldots$ is *ultimately periodic* if and only if there is a *period* $p \geq 1$ and *starting index* $s \geq 0$ such that $x_{i+p} = x_i$ for all $i \geq s$.

As is consistent with the following theorem, all three finite-state sequences in Example 2.5.1 are ultimately periodic. Indeed, because of Theorem 2.5.1, dealing with any finite-state sequence means that the existence of (s, p) is guaranteed.

Theorem 2.5.1   For any finite-state space $\mathcal{X}$, state-transition function $g : \mathcal{X} \to \mathcal{X}$, and initial state $x_0 \in \mathcal{X}$, the corresponding infinite sequence of states $x_0, x_1, x_2, \ldots$ defined by $x_{i+1} = g(x_i)$ for $i = 0, 1, 2, \ldots$ is ultimately periodic. Moreover, there are a unique *fundamental* (smallest) period $p \geq 1$ and a unique associated starting index $s \geq 0$ such that

- $s + p \leq |\mathcal{X}|$;
- if $s > 0$, the first s (*transient*) states $x_0, x_1, \ldots, x_{s-1}$ are all different and never reappear in the infinite sequence;
- the next p (*periodic*) states $x_s, x_{s+1}, \ldots, x_{s+p-1}$ are all different, but each reappears in the infinite sequence according to the periodic pattern $x_{i+p} = x_i$ for $i \geq s$.

Proof.   Because $\mathcal{X}$ is finite and $g : \mathcal{X} \to \mathcal{X}$, from the pigeonhole principle at least two of the first $|\mathcal{X}| + 1$ states in the sequence $x_0, x_1, x_2, \ldots$ must be the same. That is, there are indices s and t with $0 \leq s < t \leq |\mathcal{X}|$ such that $x_s = x_t$. Because any set of nonnegative integers has a least element, there is no loss of generality if it is assumed that s and t are the *smallest* such integers.

This defines s. Moreover, if $p = t - s$, then $x_{p+s} = x_t = x_s$; so p is the fundamental (smallest) period. The remaining details, including the inductive proof that $x_{p+i} = x_i$ for all $i \geq s$, are left as an exercise. □

Example 2.5.2

In general, the (s, p) pair depends on x_0. For Example 2.5.1:

- if the initial state is $x_0 = 1726$, so that

$$1726, \ 9790, \ 8441, \ 2504, \ 2700, \ 2900, \ 4100, \ 8100, \ 6100, \ 2100, \ 4100, \ldots,$$

  then $(s, p) = (6, 4)$;
- if the initial state is $x_0 = 6283$, so that

$$6283, \ 4760, \ 6576, \ 2437, \ 9389, \ 1533, \ 3500, \ 2500, \ 2500, \ \ldots,$$

  then $(s, p) = (7, 1)$;
- if the initial state is $x_0 = 5600$, so that

$$5600, \ 3600, \ 9600, \ 1600, \ 5600, \ \ldots,$$

  then $(s, p) = (0, 4)$.

Three algorithms for computing (s, p) are presented in this section. The first two of these algorithms are "obvious," the third (and best) is not.

2.5.2 Algorithms

The first two algorithms are based on the observation that (s, p) can be identified as follows.

- Pick x_0 and then generate $x_1 = g(x_0)$. If $x_0 = x_1$ then $(s, p) = (0, 1)$.
- If x_0 and x_1 are different, then generate $x_2 = g(x_1)$. If $x_0 = x_2$ then $(s, p) = (0, 2)$, else if $x_1 = x_2$ then $(s, p) = (1, 1)$.
- If x_0, x_1, and x_2 are all different, then generate $x_3 = g(x_2)$. If $x_0 = x_3$ then $(s, p) = (0, 3)$, else if $x_1 = x_3$ then $(s, p) = (1, 2)$, else if $x_2 = x_3$ then $(s, p) = (2, 1)$.
- If x_0, x_1, x_2, x_3 are all different, then generate $x_4 = g(x_3)$ and compare, etc.

Eventually, this algorithm will terminate with an index $1 \leq t \leq |\mathcal{X}|$ and corresponding state x_t such that the t states $x_0, x_1, \ldots, x_{t-1}$ are all different but exactly one of these states, say x_s, is equal to x_t:

$$\underbrace{x_0, x_1, x_2, x_3, \ldots, x_s, \ldots, x_{t-1}}, x_t, \ldots$$
$$\text{all are different, but } x_s = x_t$$

At this point s and $p = t - s$ are both identified. Indeed, this algorithm is a constructive proof of Theorem 2.5.1.

Memory-Intensive Algorithm.   To implement the algorithm outlined above, note that, if $|\mathcal{X}|$ is small enough, then all the intermediate states $x_0, x_1, x_2, \ldots, x_t$ can be stored in a linked list as they are generated. This concept is summarized by the following algorithm, which has $(s + p)$ storage requirements and makes $(s + p)$ calls to $g(\cdot)$.

Algorithm 2.5.1   Given the state-transition function $g(\cdot)$ and initial state x_0, this algorithm identifies the fundamental pair (s, p).

```
x₀ =initial state;                              // initialize the list
t = 0;
s = 0;
while (s == t) {                                // while no match found
    xₜ₊₁ = g(xₜ);                               // add a state to the list
    t++;
    s = 0;
    while (xₛ != xₜ)                            // traverse the list
        s++;
}
p = t - s;                  // period is the distance between matches
return s, p;
```

There is an important time–space trade-off in Algorithm 2.5.1. When the algorithm terminates, $(s + p)$ could be as large as $|\mathcal{X}|$. In many important applications, $|\mathcal{X}|$ is large, or unknown, relative to the amount of available storage for the intermediate states. In this case, storing all the intermediate states might be impossible; if so, as an alternative, it will be necessary to trade time for space by generating these states *repeatedly* as necessary, once for each comparison.

Computationally Intensive Algorithm.   Algorithm 2.5.1 can be modified to eliminate the need to store all the intermediate states by observing that only *three* states need to be stored—the initial state x_0 and the current values of x_s and x_t. In the following algorithm, these three states are denoted $x.o$, $x.s$, and $x.t$, respectively. This algorithm has low storage requirements and is easy to program; however, the number of calls to $g(\cdot)$ is $s + (p + s)(p + s + 1)/2$.

Algorithm 2.5.2   Given the state-transition function $g(\cdot)$ and initial state $x.o$, this algorithm identifies the corresponding fundamental pair (s, p).

```
x.t = x.o;
t = 0;
s = 0;
while (s == t) {
    x.t = g(x.t);
```

```
    t++;
    x.s = x.o;
    s = 0;
    while (x.s != x.t) {
        x.s = g(x.s);
        s++;
    }
}
p = t - s;
return s, p;
```

Example 2.5.3
Algorithm 2.5.1 or 2.5.2 applied to the six-digit midsquares state-transition function

$$g(x) = \lfloor x^2/1000 \rfloor \bmod 1000000$$

yields the following results (albeit slowly):

initial state x_0	141138	119448	586593	735812	613282
fundamental pair (s, p)	(296, 29)	(428, 210)	(48, 13)	(225, 1)	(469, 20)

The time complexity of Algorithm 2.5.2 will be dominated by calls to the state-transition function $g(\cdot)$. For many practical applications, for example analyzing the (fundamental) period of a (good) random-number generator, $s + (s + p)(s + p + 1)/2$ is so large that the time complexity (speed) of Algorithm 2.5.2 is unacceptable. For that reason, we are motivated to look for a better algorithm for finding (s, p)—an algorithm with the best features of Algorithms 2.5.1 and 2.5.2.

A More Efficient Algorithm.   The development of this better algorithm is based on the following theorem, which provides two important characterizations of periods, fundamental or not. One characterization is that the starting index s has the property that $x_{2q} = x_q$ if and only if q is a period at least as large as s; the other is that the fundamental period p has the property that the positive integer q is a period if and only if q is an integer multiple of p. The proof of this theorem is left as an exercise.

Theorem 2.5.2   If the finite-state sequence $x_0, x_1, x_2, \ldots$ has the fundamental pair (s, p), and if $q > 0$, then:

- q is a period and $q \geq s$ if and only if $x_{2q} = x_q$;
- q is a period if and only if $q \bmod p = 0$.

Given the fundamental pair (s, p), the index

$$q = \begin{cases} p & s = 0 \\ \lceil s/p \rceil p & s > 0 \end{cases}$$

plays an important role in the development of Algorithm 2.5.3: q is either p or an integer multiple of p, so it follows from Theorem 2.5.2 that q is a period of the infinite sequence. In addition, if $s > 0$, then $q = \lceil s/p \rceil p$, and so

$$s = (s/p)p \le \lceil s/p \rceil p = q \qquad \text{and} \qquad q = \lceil s/p \rceil p < (1 + s/p)p = p + s.$$

Similarly, if $s = 0$, then $q = p$. Therefore, the period q satisfies the inequality $s \le q \le p + s$. Because $q \ge s$, it follows from Theorem 2.5.2 that $x_{2q} = x_q$.

Now, suppose that $x_{2i} = x_i$ for *any* integer $i > 0$. It follows from Theorem 2.5.2 that i is a period with $i \ge s$ and that there exists an integer $\alpha \ge 1$ such that $i = \alpha p \ge s$. If $s > 0$, then, from the condition $\alpha p \ge s$, it follows that $\alpha \ge \lceil s/p \rceil$. If we multiply both sides of this inequality by p, it follows that $i = \alpha p \ge \lceil s/p \rceil p = q$. Similarly, if $s = 0$, then $q = p$, and, because $i = \alpha p \ge p$, it follows that $i \ge q$. Therefore, independent of the value of s, if $x_i = x_{2i}$ for some $i > 0$, then $i \ge q \ge s$.

From the previous discussion, we see that, although q might not be the smallest (fundamental) period, it is at least the smallest period that is not smaller than s. This is summarized by the following theorem.

Theorem 2.5.3   Given the finite-state sequence $x_0, x_1, x_2, \dots$ with fundamental pair (s, p), the smallest index $i > 0$ such that $x_{2i} = x_i$ is $i = q$, where

$$q = \begin{cases} p & s = 0 \\ \lceil s/p \rceil p & s > 0. \end{cases}$$

Moreover, $i = q$ is a period that satisfies the inequality $s \le q \le s + p$.

Theorem 2.5.3 is the basis for a three-step algorithm, with at most $(5s + 4p)$ calls to $g(\cdot)$ and minimal storage requirements, that will find (s, p). This algorithm is based on the "race-track analogy" in step (1).

(1) *To determine q and x_q*—think of the two sequences

$$\begin{array}{cccccc} x_0, & x_1, & x_2, & x_3, & \dots, & x_i, & \dots \\ x_0, & x_2, & x_4, & x_6, & \dots, & x_{2i}, & \dots \end{array}$$

as points (states) that start together and then move around a race track with the second sequence moving twice as fast as the first. From Theorem 2.5.3, the second sequence will catch the first sequence for the first time at $i = q$. Since $q \le s + p$, the number of steps in this race is no more than $s + p$. Thus, there is an algorithm for finding q and x_q with no more than $(3s + 3p)$ calls to $g(\cdot)$. Moreover, the only states that are needed in this race are the current values of x_i and x_{2i}, so the storage requirements for this step are minimal.

(2) *To determine s and x_s* —because q is a period, $x_s = x_{s+q}$. Therefore, the starting index s and corresponding state x_s can be calculated from a knowledge of x_q by beginning with the states x_0, x_q and generating the successor states

$$x_0, \quad x_1, \quad x_2, \quad x_3, \quad \ldots, \quad x_i, \quad \ldots$$
$$x_q, \quad x_{1+q}, \quad x_{2+q}, \quad x_{3+q}, \quad \ldots, \quad x_{i+q}, \quad \ldots$$

The starting index s is the first value of $i \geq 0$ for which $x_i = x_{i+q}$. The number of calls to $g(\cdot)$ for this step is $2s$, and the storage requirements are minimal.

(3) *To determine p* —once s, q, and x_s have been found, the fundamental period p can be calculated in the following way:

(a) If $2s \leq q$ (this is the usual case), then $p = q$. That is, because $q \leq s + p$, it follows that $s + q \leq 2s + p$. Therefore, if $2s \leq q$, then $s + q \leq q + p$, which is equivalent to $s \leq p$. If $s \leq p$, however, then $q = p$.

(b) If $2s > q$, then, because $x_s = x_{s+p}$, a linear search can be used to find p. That is, p can be found by starting with x_s and generating the successor states x_{s+i}. The fundamental period p is then the first value of $i > 0$ for which $x_s = x_{s+i}$. The number of calls to $g(\cdot)$ for this step is p, with minimal storage requirements.

Algorithm 2.5.3 is an implementation of this three-step approach.

Algorithm 2.5.3 Given the state-transition function $g(\cdot)$ and initial state $x.o$, this algorithm identifies the corresponding fundamental pair (s, p).

```
q   = 1;
x.q = g(x.o);
z   = g(x.q);
while (x.q != z) {                    // step 1:  determine q and x.q
    q++;
    x.q = g(x.q);
    z   = g(g(z));
}
s   = 0;
x.s = x.o;
z   = x.q;
while (x.s != z) {                    // step 2:  determine s and x.s
    s++;
    x.s = g(x.s);
    z   = g(z);
}
if (2 * s <= q)                       // step 3:  determine p
    p = q;
else {
```

```
    p = 1;
    z = g(x.s);
    while (x.s != z) {
      p++;
      z = g(z);
    }
  }
  return s, p;
```

Example 2.5.4
As a continuation of Example 2.5.3, Algorithm 2.5.3, when applied to the six-digit midsquares state-transition function

$$g(x) = \lfloor x^2/1000 \rfloor \bmod 1000000,$$

yields the following results:

initial state	141138	119448	586593	735812	613282
speed up	34	72	7	21	48

The "speed up" row indicates the efficiency of Algorithm 2.5.3 relative to Algorithm 2.5.2; for example, for $x_0 = 141138$, Algorithm 2.5.3 is approximately 34 times faster than Algorithm 2.5.2.

Full-Period Sequences.   As suggested previously, a primary application for Algorithm 2.5.3 is to analyze the period of a random-number generator. Algorithm 2.5.3 is attractive in this application because it is efficient and (s, p) can be found without any knowledge of the "details" of $g(\cdot)$. All that is needed to apply Algorithm 2.5.3 is a function that implements $g(\cdot)$ and, perhaps, lots of CPU cycles.

As with the discussion in Sections 2.1 and 2.2, when analyzing the period of a random-number generator, we are primarily concerned with *full-period* sequences. In the generalized context of this section, that idea is expressed in terms of (s, p) by the following definition and associated theorem.

Definition 2.5.3   The function $g : \mathcal{X} \to \mathcal{X}$ is a *full-period* state-transition function, and the corresponding sequence $x_0, x_1, x_2, \ldots$ has a *full period*, if and only if $p = |\mathcal{X}|$.

Note that $p = |\mathcal{X}|$ is possible only if $s = 0$. That is, $s = 0$ is a necessary (but not sufficient) condition for $g(\cdot)$ to be a full-period state-transition function. Of course, if (somehow) you *know* that $s = 0$, then there is no need to use Algorithm 2.5.3 to test for a full period. Instead, it is sufficient to use the following variant of Algorithm 2.1.1:

```
p = 1;
x = g(x_0);
```

```
while (x != x₀) {
    p++;
    x = g(x);
}
return p;
```

If $p = |\mathcal{X}|$ then $g(\cdot)$ is a full-period state-transition function.

Example 2.5.5

If $g(\cdot)$ is a Lehmer generator (full period or not), then, from Theorem 2.1.2, for any initial state we know that $s = 0$. Therefore (with $x_0 = 1$), the use of Algorithm 2.1.1 is justified as a test for full period.

The following theorem is an immediate consequence of Definition 2.5.3 and Theorem 2.5.1. The proof is left as an exercise.

Theorem 2.5.4   The infinite finite-state sequence $x_0, x_1, x_2, \ldots$ has a full period if and only if each possible state appears exactly once in *any* $|\mathcal{X}|$ consecutive terms of the sequence.

As an exercise, you are encouraged to construct your own many-digit state-transition function, using a variety of obscure bit-level and byte-level operations to "randomize" the output. You will be impressed with how difficult it is to construct a full-period state-transition function. In that regard, see Algorithm K in Knuth (1998).

2.5.3 State-Space Partition

Provided that (a, m) are properly chosen, Lehmer generators (1) have full period, (2) have been extensively studied, (3) provide multiple streams, and (4) produce output that is generally considered to be acceptably random for discrete-event simulation purposes. Given all of that, one could ask why anyone would be interested in other types of generators, particularly if they are not full-period generators. The discussion in this section is, however, not focused on random-number generators exclusively, and certainly not on Lehmer generators. Instead, the focus is on general state-transition functions and corresponding finite-state sequences that might not (and, generally, will not) be full period. Correspondingly, in the remainder of this section, we will study a state-space partition that is naturally associated with state-transition functions that are *not* full period. The study of this state-space partition is facilitated by introducing the idea of "jump" functions, defined as follows.

Jump Functions.   If $g\colon \mathcal{X} \to \mathcal{X}$, then, for all $x \in \mathcal{X}$, inductively define the *jump function* $g^j(x)$ as follows:

$$g^j(x) = \begin{cases} x & j = 0 \\ g\big(g^{j-1}(x)\big) & j = 1, 2, 3, \ldots. \end{cases}$$

Because $g: \mathcal{X} \to \mathcal{X}$, it follows that $g^j : \mathcal{X} \to \mathcal{X}$ for any $j \geq 0$. That is, the jump function $g^j(\cdot)$ is also a state-transition function. Moreover, if $g(\cdot)$ generates the sequence $x_0, x_1, x_2, x_3, \ldots$, then $g^j(\cdot)$ generates the sequence $x_0, x_j, x_{2j}, x_{3j}, \ldots$. In terms of the race-track analogy, if $j \geq 2$, then the state-transition function $g^j(\cdot)$ runs the race j times faster than $g(\cdot)$, by jumping ahead j states at a time. This important idea is used in Chapter 3 to construct a Lehmer random-number generator that can create and maintain multiple *streams* of random numbers.

Periodic and Transient States.   The following definition is the key to studying the state-space partition that is naturally associated with a state-transition function that is not full period.

Definition 2.5.4   Given a finite-state space $\mathcal{X}$ and a state-transition function $g : \mathcal{X} \to \mathcal{X}$, then $x \in \mathcal{X}$ is a *periodic* (or recurrent) state if and only if there exists a positive integer j with $0 < j \leq |\mathcal{X}|$ such that $g^j(x) = x$. If $x \in \mathcal{X}$ is not a periodic state, then it is said to be a *transient* state.

Theorem 2.5.1 is the basis for this definition. If $x = x_0$ is the initial state in a finite-state sequence, then either $s = 0$, in which case x is a periodic state because $g^p(x) = x$, or $s > 0$, in which case x is a transient state because $g^j(x) \neq x$ for all $j = 1, 2, \ldots$.

Definition 2.5.5   If $x \in \mathcal{X}$ is a periodic state, then the *recurrent class* corresponding to x is the set

$$\mathcal{R} = \{g^i(x) : i = 0, 1, 2, \ldots, j - 1\},$$

where j is the smallest positive integer such that $g^j(x) = x$. Note that $|\mathcal{R}| = j$.

From Theorem 2.5.1, there must be at least one periodic state. Therefore, there must be at least one recurrent class. Moreover, if $x \in \mathcal{X}$ is a periodic state with corresponding recurrent class $\mathcal{R}$ and if $x' \in \mathcal{X}$ is a periodic state with corresponding recurrent class $\mathcal{R}'$, then either $\mathcal{R} = \mathcal{R}'$ or $\mathcal{R} \cap \mathcal{R}' = \emptyset$. That is, the set of periodic states is *partitioned* into recurrent classes. Equivalently, as Examples 2.5.6 and 2.5.7 illustrate, if there are multiple recurrent classes, then the classes are disjoint—each periodic state belongs to exactly one recurrent class.

For reference, note that, if $g(\cdot)$ is a full-period state-transition function, then all states are periodic, and there is only one recurrent class, which is $\mathcal{X}$. If $g(\cdot)$ is not a full-period state-transition function then, generally, there will be transient states and more than one recurrent class, but not necessarily.

Example 2.5.6

The one-digit midsquares state-transition function

$$g(x) = \lfloor x^2 \rfloor \bmod 10 = x^2 \bmod 10$$

has four periodic states (0, 1, 5, 6) and four transient states (2, 3, 4, 7, 8, 9). The periodic states are partitioned into four recurrent classes: {0}, {1}, {5}, and {6}, as illustrated in Figure 2.5.1.

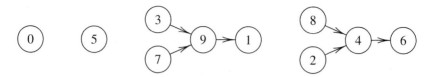

Figure 2.5.1   Periodic and transient states.

Example 2.5.7

For the four-digit midsquares state-transition function, it can be shown that there are 17 periodic states and 9983 transient states. The periodic states are partitioned into eight recurrent classes: $\{2100, 4100, 8100, 6100\}$, $\{0\}$, $\{100\}$, $\{2500\}$, $\{7600\}$, $\{1600, 5600, 3600, 9600\}$, $\{540, 2916, 5030, 3009\}$, and $\{3792\}$.

Although the recurrent classes contain periodic states only, from Theorem 2.5.1, any transient state $x \in \mathcal{X}$ is "connected to" a recurrent class in the sense that there is a positive integer j such that $g^j(x)$ is a periodic state. This observation is summarized by the following theorem.

Theorem 2.5.5   If $x \in \mathcal{X}$ is a transient state, then there exists a unique recurrent class $\mathcal{R}$ such that $g^j(x) \in \mathcal{R}$ for some j with $0 < j < |\mathcal{X}|$. In this case, x is said to be *connected to* $\mathcal{R}$.

It follows from Theorem 2.5.5 that each recurrent class has associated with it a (perhaps empty) set of transient states that are connected to it. In this way, each recurrent class can be naturally enlarged to an associated "cluster."

Definition 2.5.6   For each recurrent class $\mathcal{R}$, the associated recurrent class *cluster* is

$$\mathcal{C} = \mathcal{R} \cup \{\text{all the transient states that are connected to } \mathcal{R}\}.$$

There is one cluster for each recurrent class, the clusters are disjoint, and each state (transient or periodic) is in exactly one cluster. Equivalently, the clusters partition the state space. This is an important conceptual result, particularly as it relates to understanding the behavior of bad random-number generators.

Relative to the conceptual model that random-number generation corresponds to drawing at random from an urn, the clusters correspond to urns, the initial state of the generator determines the urn (cluster) from which *all* subsequent draws will be made, and the recurrent class in the cluster defines the periodic cycle into which the output of the generator must ultimately collapse. If the initial state of the generator is selected at random, the relative cardinalities of the clusters correspond to the probabilities that the clusters (urns) will be selected.

Example 2.5.8

As a continuation of Example 2.5.7, for the four-digit midsquares state-transition function, there are eight clusters. The recurrent class associated with each cluster is illustrated in the table below, along with the size of each cluster. The table is sorted by decreasing cluster size.

cluster size	recurrent class
6291	{2100, 4100, 8100, 6100}
1968	{0}
1360	{1600, 5600, 3600, 9600}
130	{2500}
104	{100}
86	{540, 2916, 5030, 3009}
60	{7600}
1	{3792}

As a source of random numbers, the four-digit midsquares method is a nonstarter.

It is an interesting exercise to construct a time-efficient algorithm that will generate the state-space decomposition into recurrent class clusters. In that regard, see Exercise 2.5.4.

2.5.4 Exercises

2.5.1   The BASICA random-number generator for the original "true blue" IBM PC was defined by the state-transition function

$$g(x) = (a(x \bmod 2^{16}) + c) \bmod 2^{24}$$

with $a = 214013$ and $c = 13\,523\,655$. (a) If $g: \mathcal{X} \to \mathcal{X}$ with $\mathcal{X} = \{0, 1, \ldots, m-1\}$, what is m? (b) Find (s, p) for the initial seeds 0, 1, 12, 123, 1234, and the last four digits of your telephone number. (c) For each initial seed, how does $|\mathcal{X}| = m$ relate to p? (d) Comment.

2.5.2   Determine (s, p) for both of the following state-transition functions

```
long g(long x)                                    // start with x = 777
{
x = (123 * x + 1) % 123456;
return x;
}
long g(long x)                  // start with x = 37911 and x = 1
{
x = (9806 * x + 1) % 131071;
return x;
}
```

2.5.3   Prove or disprove the following two conjectures. Given that $|\mathcal{X}|$ is finite, $g: \mathcal{X} \to \mathcal{X}$, $x_0 \in \mathcal{X}$, and (s, p) are the values returned by Algorithm 2.5.3: (a) if $s = 0$ and $p = |\mathcal{X}|$, then the function g is a bijection (one-to-one and onto); (b) if the function g is a bijection, then $s = 0$ and $p = |\mathcal{X}|$.

2.5.4   Given a finite-state space $\mathcal{X}$ and a state-transition function $g: \mathcal{X} \to \mathcal{X}$, construct an efficient algorithm that partitions the state space into clusters. In particular, the algorithm should output the size (cardinality) of each cluster, the smallest periodic state in each cluster, and the size of each associated recurrent class. (*a*) What is the output of this algorithm when applied to the six-digit midsquares state-transition function? (*b*) Same question for the state-transition function in Exercise 2.5.1. (*c*) Comment on both of these state-transition functions as a possible source of random numbers. (*d*) What is the time complexity (relative to $g(\cdot)$ evaluations) and space complexity of your algorithm?

2.5.5   Consider a large round 12-hour clock with an hour hand, a minute hand, and 3600 marks (seconds) around the circumference of the clock. Both hands move in discrete steps from one mark to the next. The marks define the set $\mathcal{X} = \{0, 1, \ldots, 3599\}$. Assume that both hands are "geared" so that a manual movement of the hour hand by one mark forces the minute hand to move by 12 marks. Generate a sequence of integers $x_0, x_1, x_2, \ldots$ in $\mathcal{X}$ as follows. Start with the hands in an arbitrary location (time) and let $x_0 \in \mathcal{X}$ be the location (mark) of the minute hand. To generate x_1, move the *hour* hand clockwise to the *minute* hand mark (x_0), allowing the minutes hand to spin accordingly to a new mark; this new minute hand mark is $x_1 \in \mathcal{X}$. To generate x_2, move the hour hand to x_1 and let $x_2 \in \mathcal{X}$ be the new location of the minute hand, etc. What do you think of this as a source of random numbers? ("Not much" is not a sufficient answer.)

2.5.6   Same as Exercise 2.5.5, except for a clock with 360 000 marks around the circumference.

3

Discrete-Event Simulation

The *trace-driven* single-server service-node model and simple inventory system formulated in Chapter 1 were limited by their dependence on an external data source for generating their stochastic elements. This chapter links the models presented in Chapter 1 with the random-number generation algorithms in Chapter 2 to create discrete-event simulation models that are free of any reliance on external data sources.

We return to the single-server service-node model and the simple inventory system model from Sections 1.2 and 1.3. The single-server service node required arrival times and the associated service times. The simple inventory system required demands.

This chapter uses the Lehmer random-number generators developed in Chapter 2 to free discrete-event simulation models from their dependence on the trace-driven approach. Section 3.1 introduces exponential and geometric variates and uses them in the single-server service-node and (s, S) simple inventory system models. Section 3.2 tackles the problem of *multiple-stream generators* that are used to effectively provide a separate random-number generator for each stochastic element in a discrete-event simulation model. Section 3.3 extends the two elementary models in Section 3.1 to three slightly more complicated models: (*i*) a single-server service node *with immediate feedback*, (*ii*) a simple inventory system *with delivery lag*, and (*iii*) a single-server machine shop.

3.1 DISCRETE-EVENT SIMULATION

As discussed in Chapter 1, `ssq1` and `sis1` are examples of trace-driven discrete-event simulation programs. By definition, a trace-driven simulation relies on input data from an external source to supply recorded realizations of naturally occurring stochastic processes. Total reliance on such external data limits the applicability of a discrete-event simulation program, naturally inhibiting the user's ability to do "what if" studies. Given this limitation, a general discrete-event simulation

99

objective is to develop methods for using a random-number generator to convert a trace-driven discrete-event simulation program to a discrete-event simulation program that is not dependent on external data. This chapter provides several examples.

3.1.1 Single-Server Service Node

Relative to the single-server service-node model from Chapter 1, two stochastic assumptions are needed to free program `ssq1` from its reliance on external data. One assumption relates to the arrival times, the other assumption relates to the service times. We consider the service times first, using a *Uniform*(a, b) random variate model.

> **Example 3.1.1**
> Suppose that time is measured in minutes in a single-server service-node model and that all that is known about the service time is that it is random with possible values between 1.0 and 2.0. We know the range of possible values, but we are otherwise in such a state of ignorance about the stochastic behavior of this server that we are unwilling to say some service times (between 1.0 and 2.0) are more likely than others. In this case we have modeled service time as a *Uniform*(1.0, 2.0) random variable. Accordingly, a random-variate service time, say s, can be generated via the assignment
>
> s = Uniform(1.0, 2.0);

Exponential Random Variates. A *Uniform*(a, b) random variable has the property that all values between a and b are equally likely. In most applications, this is an unrealistic assumption; instead, some values will be more likely than others. Specifically, there are many discrete-event simulation applications that require a continuous random variate, say x, that can take on any positive value, but in such a way that small values of x are more likely than large values.

To generate such a random variate we need a *nonlinear* transformation that maps values of the random number u between 0.0 and 1.0 to values of x between 0.0 and ∞ and does so by "stretching" large values of u much more so than small values. Although a variety of such nonlinear transformation are possible, for example $x = u/(1 - u)$, perhaps the most common is $x = -\mu \ln(1 - u)$, where $\mu > 0$ is a parameter that controls the rate of stretching and $\ln(\cdot)$ is the natural logarithm (base e).* As will be explained in Chapter 7, this transformation generates what is known as an *Exponential*(μ) random variate.

It is often appropriate to generate interarrival times as *Exponential*(μ) random variates, as we will do later in Example 3.1.2. In some important theoretical situations, service times are also modeled this way (see Section 8.5). The geometry associated with the *Exponential*(μ) random-variate transformation $x = -\mu \ln(1 - u)$ is illustrated in Figure 3.1.1.

---

*The function $\log(x)$ in the ANSI C library `<math.h>` represents the mathematical natural log function $\ln(x)$. In the equation $x = -\mu \ln(1 - u)$ do not confuse the positive, real-valued parameter μ with the *Uniform*(0, 1) random variate u.

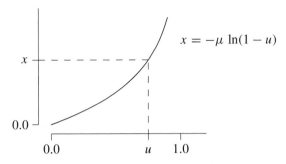

Figure 3.1.1   Exponential-variate-generation geometry.

By inspection, we see that the transformation is monotone increasing and, for any value of the parameter $\mu > 0$, the interval $0 < u < 1$ is mapped one-to-one and onto the interval $0 < x < \infty$. That is,

$$0 < u < 1 \iff 0 < (1 - u) < 1$$

$$\iff -\infty < \ln(1 - u) < 0$$

$$\iff 0 < -\mu \ln(1 - u) < \infty$$

$$\iff 0 < x < \infty.$$

Definition 3.1.1   This ANSI C function generates an *Exponential*(μ) random variate.[*]

```
double Exponential(double μ)                                // use μ > 0.0
{
    return (-μ * log(1.0 - Random()));
}
```

The statistical significance of the parameter μ is that, if repeated calls to the function Exponential(μ) are used to generate a random variate sample $x_1, x_2, \ldots, x_n$, then, in the limit as $n \to \infty$, the sample mean (average) of this sample will converge to μ. In the same sense, repeated calls to the function Uniform(a, b) will produce a sample whose mean converges to $(a + b)/2$, and repeated calls to the function Equilikely(a, b) will also produce a sample whose mean converges to $(a + b)/2$.

Example 3.1.2
In a single-server service-node simulation, to generate a sequence of random-variate *arrival* times $a_1, a_2, \ldots, a_n$ with an average interarrival time that will converge to μ

---

[*]The ANSI C standard says that log(0.0) may produce "a range error." This possibility is naturally avoided in the function Exponential because the largest possible value of Random is less than 1.0. See Exercise 3.1.3.

as $n \to \infty$, it is common to generate *Exponential*(μ) *interarrival* times (see Definition 1.2.3) and then (with $a_0 = 0$) create the arrival times by the assignment

$$a_i \;=\; a_{i-1} \;+\; \texttt{Exponential}(\mu)\,; \qquad\qquad i = 1, 2, \ldots, n\,.$$

As will be discussed in Chapter 7, this use of an *Exponential*(μ) random variate corresponds naturally to the idea of jobs arriving *at random* with an arrival rate that will converge to $1/\mu$ as $n \to \infty$. Similarly, as in Example 3.1.1, to generate a random-variate sequence of service times $s_1, s_2, \ldots, s_n$ equally likely to lie anywhere between a and b (with $0 \le a < b$) and with an average service time that converges to $(a + b)/2$, the assignment

$$s_i \;=\; \texttt{Uniform}(a,\ b)\,; \qquad\qquad i = 1, 2, \ldots, n$$

can be used. The average service time $(a + b)/2$ corresponds to the service rate $2/(a + b)$.

Program ssq2. Program `ssq2` is based on the two stochastic modeling assumptions in Example 3.1.2.* Program `ssq2` is an extension of program `ssq1`, in that the arrival times and service times are generated randomly (rather than by relying on a trace-driven input) and that a complete set of first-order statistics $\bar{r}, \bar{w}, \bar{d}, \bar{s}, \bar{l}, \bar{q}, \bar{x}$ is generated. Note that each time the function `GetArrival()` is called, the static variable `arrival`, which represents an *arrival* time, is incremented by a call to the function `Exponential(2.0)`, which generates an *interarrival* time.

Because program `ssq2` generates stochastic data as needed, there is essentially no restriction on the number of jobs that can be processed. Therefore, the program can be used to study the *steady-state* behavior of a single-server service node; by experimenting with an increasing number of jobs processed, one can investigate whether the service-node statistics will converge to constant values, *independent* of the choice of the `rng` initial seed and the initial state of the service node. Steady-state behavior—whether it can be achieved and, if so, how many jobs it will take to do so—is an important issue that will be explored briefly in this chapter and in more detail in Chapter 8.

Program `ssq2` can also be used to study the *transient* behavior of a single-server service node. The idea in this case is to fix the number of jobs processed at some finite value and run (replicate) the program repeatedly with the initial state of the service node fixed, changing *only* the `rng` initial seed from run to run. In this case, replication will produce a natural variation in the service-node statistics that is consistent with the fact that, for a fixed number of jobs, the service-node statistics are *not* independent of the initial seed or of the initial state of the service node. Transient behavior and its relation to steady-state behavior will be considered in Chapter 8.

---

*In the jargon of queuing theory (see Section 8.5), program `ssq2` simulates what is known as an $M/G/1$ queue. (See either Kleinrock, 1975, 1976, or Gross and Harris, 1985.)

Steady-State Statistics

Example 3.1.3

If the *Exponential*(μ) interarrival time parameter is set to $\mu = 2.0$, so that the steady-state arrival rate is $1/\mu = 0.5$; and if the *Uniform*(a, b) service time parameters are set to $a = 1.0$ and $b = 2.0$, respectively, so that the steady-state service rate is $2/(a + b) \cong 0.67$; then, to $d.dd$ precision, the theoretical steady-state statistics that the analytic-model (exact, not estimates by simulation) are (per Gross and Harris, 1985)

$\bar{r}$	$\bar{w}$	$\bar{d}$	$\bar{s}$	$\bar{l}$	$\bar{q}$	$\bar{x}$
2.00	3.83	2.33	1.50	1.92	1.17	0.75

Therefore, although the server is busy only 75% of the time ($\bar{x} = 0.75$), "on average" there are approximately two jobs ($\bar{l} = 23/12 \cong 1.92$) in the service node, and a job can expect to spend more time ($\bar{d} = 23/6 - 3/2 \cong 2.33$ time units) in the queue than in service ($\bar{s} = 1.50$ time units). As illustrated in Figure 3.1.2 for the average wait, the number of jobs that must be pushed through the service node to achieve these steady-state statistics is large. To produce this figure, program ssq2 was modified to print the accumulated average wait every 20 jobs. The results are presented for three choices of the rng initial seed.* The solid, horizontal line at height $23/6 \cong 3.83$ represents the steady-state value of $\bar{w}$.

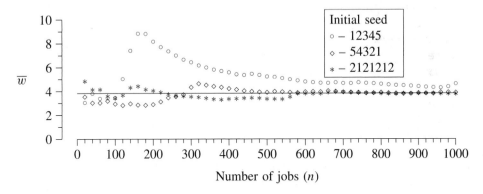

Figure 3.1.2 Average wait times.

In Example 3.1.3, convergence of $\bar{w}$ to the steady-state value 3.83 is slow and erratic and is very much dependent on the random-variate sequence of stochastic arrival and service times, as manifested by the choice of initial seed. Note, for example, the dramatic rise in average wait beginning at about job 100 associated with the rng initial seed 12345. You are encouraged to

---

*There is nothing special about these three initial seeds. Do not fall into the common trap of thinking that some rng initial seeds are necessarily better than others.

add diagnostic printing and additional statistics-gathering to program `ssq2` to better understand what combination of chance events occurred to produce this rise.

The *Uniform*(a, b) service-time model in Example 3.1.3 might be unrealistic. Service times seldom "cut off" beyond a minimum and maximum value. More detail will be given in subsequent chapters on how to select realistic distributions for input models.

Example 3.1.3 shows that the stochastic character of the arrival times and service times, as manifested by the choice of `rng` initial seed, has a significant effect on the transition-to-steady-state behavior of a single-server service node. This example also illustrates the use of the library `rng` to conduct *controlled* "what if" experiments. Studies like this are the stuff of discrete-event simulation. Additional examples of this kind of experimentation and of the selection of the values of μ, a, and b are presented in later chapters.

Geometric Random Variates.   As discussed in Chapter 2, an *Equilikely*(a, b) random variate is the discrete analog of a continuous *Uniform*(a, b) random variate. Consistent with this characterization, one way to generate an *Equilikely*(a, b) random variate is to generate a *Uniform*$(a, b+1)$ random variate instead (note that the upper limit is $b+1$, not b) and then convert (cast) the resulting floating-point result to an integer. In other words, if a and b are integers with $a < b$ and if x is a *Uniform*$(a, b+1)$ random variate, then $\lfloor x \rfloor$ is an *Equilikely*(a, b) random variate.

Given the analogy between *Uniform*(a, b) and *Equilikely*(a, b) random variates, it is reasonable to expect that there is a discrete analog to a continuous *Exponential*(μ) random variate—and there is. Specifically, if x is an *Exponential*(μ) random variate, let y be the *discrete* random variate defined by $y = \lfloor x \rfloor$. For a better understanding of this discrete random variate, let $p = \Pr(y \neq 0)$ denote the probability that y is not equal to zero. Since the random variate x is generated as $x = -\mu \ln(1-u)$ with u a *Uniform*$(0, 1)$ random variate, $y = \lfloor x \rfloor$ will be nonzero if and only if $x \geq 1$. Equivalently,

$$x \geq 1 \iff -\mu \ln(1-u) \geq 1$$
$$\iff \ln(1-u) \leq -1/\mu$$
$$\iff 1-u \leq \exp(-1/\mu),^{*}$$

and so $y \neq 0$ if and only if $1 - u \leq \exp(-1/\mu)$. Like u, $1 - u$ is also a *Uniform*$(0, 1)$ random variate. Moreover, for any $0 < p < 1$, the condition $1 - u \leq p$ is true with probability p (as noted in Section 7.1). Therefore, $p = \Pr(y \neq 0) = \exp(-1/\mu)$.

If x is an *Exponential*(μ) random variate, and if $y = \lfloor x \rfloor$ with $p = \exp(-1/\mu)$, then it is conventional to call y a *Geometric*(p) random variate. (See Chapter 6.) Moreover, it is conventional to use p rather than $\mu = -1/\ln(p)$ to define y directly, by the equation

$$y = \lfloor \ln(1-u)/\ln(p) \rfloor.$$

---

*The function $\exp(x)$ in the ANSI C library `<math.h>` represents the mathematical exponential function $\exp(x) = e^{x}$.

Definition 3.1.2   This ANSI C function generates a *Geometric*(p) random variate*

```
long Geometric(double p)                                    // use 0.0 < p < 1.0
{
   return ((long) (log(1.0 - Random()) / log(p)));
}
```

In addition to its significance as $\Pr(y \neq 0)$, the parameter p is also related to the mean of a *Geometric*(p) sample. Specifically, if repeated calls to the function `Geometric(p)` are used to generate a random-variate sample $y_1, y_2, \ldots, y_n$, then, in the limit as $n \to \infty$, the mean of this sample will converge to $p/(1 - p)$. Note that, if p is close to 0.0, then the mean will be close to 0.0. At the other extreme, if p is close to 1.0, then the mean will be large.

In the following example, a *Geometric*(p) random variate is used as part of a *composite* service-time model. In this example, the parameter p has been adjusted to make the average service time match that of the *Uniform*(1.0, 2.0) server in program `ssq2`.

Example 3.1.4

Usually, one has sufficient information to argue that a *Uniform*(a, b) random-variate service-time model is not appropriate; instead, a more sophisticated model is justified. Consider a hypothetical server that, as in program `ssq2`, processes a stream of jobs arriving, at random, with a steady-state arrival rate of 0.5 jobs per minute. The service requirement associated with each arriving job has two stochastic components:

- the *number* of service tasks is one plus a *Geometric*(0.9) random variate;
- the *time* (in minutes) per task is, independently for each task, a *Uniform*(0.1, 0.2) random variate.

In this case, program `ssq2` would need to be modified by including the function `Geometric` from Definition 3.1.2 and changing the function `GetService` to the following.

```
double GetService(void)
{
   long    k;
   double  sum   = 0.0;
   long    tasks = 1 + Geometric(0.9);

   for (k = 0; k < tasks; k++)
      sum += Uniform(0.1, 0.2);
   return (sum);
}
```

With this modification, the population steady-state statistics from the analytic model, to *d.dd* precision, are

---

*Note that `log(0.0)` is avoided in this function, because the largest possible value returned by `Random` is less than 1.0.

$\bar{r}$	$\bar{w}$	$\bar{d}$	$\bar{s}$	$\bar{l}$	$\bar{q}$	$\bar{x}$
2.00	5.77	4.27	1.50	2.89	2.14	0.75

Program `ssq2` will produce results that converge to these values for a very long run. When comparing them with the steady-state results in Example 3.1.3, note that, although the arrival rate $1/\bar{r} = 0.50$, the service rate $1/\bar{s} = 0.67$, and the utilization $\bar{x} = 0.75$ are the same, the other four statistics are significantly larger than in Example 3.1.3. This difference illustrates the sensitivity of the performance measures to the service-time distribution. This highlights the importance of using an accurate service-time model. (See Exercise 3.1.5.)

3.1.2 Simple Inventory System

Example 3.1.5

In a simple-inventory-system simulation, to generate a random variate sequence of demands d_1, d_2, d_3, ... equally likely to have any integer value between a and b inclusive and with an average of $(a + b)/2$, use

$$d_i = \text{Equilikely(a, b);} \qquad i = 1, 2, 3, \ldots.$$

Recognize, however, that the previous discussion about *Uniform*(a, b) service times as an unrealistic model also applies here in the discrete case. The modeling assumption that all the demands between a and b are equally likely is probably unrealistic; some demands should be more likely than others. As an alternative model, we could consider generating the demands as *Geometric*(p) random variates. In this particular case, however, a *Geometric*(p) demand model is probably not very realistic either. In Chapter 6, we will consider alternative models that might be more appropriate.

Program `sis2`.   Program `sis2` is based on the *Equilikely*(a, b) stochastic modeling assumption in Example 3.1.5. This program is an extension of program `sis1`, in that the demands are generated randomly (rather than by relying on a trace-driven input). Consequently, in a manner analogous to that illustrated in Example 3.1.3, program `sis2` can be used to study the transition-to-steady-state behavior of a simple inventory system.

Example 3.1.6

If the *Equilikely*(a, b) demand parameters are set to $(a, b) = (10, 50)$, so that the average demand per time interval is $(a + b)/2 = 30$, then, with $(s, S) = (20, 80)$, the (approximate) steady-state statistics that program `sis2` will produce are

$\bar{d}$	$\bar{o}$	$\bar{u}$	$\bar{l}^{+}$	$\bar{l}^{-}$
30.00	30.00	0.39	42.86	0.26

As illustrated in Figure 3.1.3 (using the same three `rng` initial seeds used in Example 3.1.3) for the average inventory level $\bar{l} = \bar{l}^{+} - \bar{l}^{-}$, at least several hundred time intervals must be simulated to approximate these steady-state statistics. To produce Figure 3.1.3, program `sis2` was modified to print the accumulated value of $\bar{l}$ every 5 time intervals.

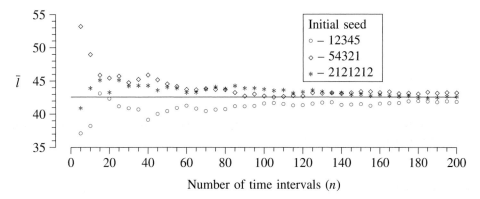

Figure 3.1.3   Average inventory level.

Optimal Steady-State System Performance

Example 3.1.7

As an extension of Example 1.3.7, a modified version of program `sis2` was used to simulate $n = 100$ weeks (about two years) of automobile-dealership operation, with $S = 80$ and values of s varied from 0 to 60. The cost parameters from Example 1.3.5 were used to determine the variable part of the dealership's average weekly cost of operation—that is, as discussed in Section 1.3, the (large) part of the average weekly cost that is proportional to the average weekly order $\bar{o}$ [and therefore is independent of (s, S)] was ignored, and only the dependent cost was computed. The results are presented with ∘'s representing the $n = 100$ averages. As illustrated in Figure 3.1.4, there is a relatively well-defined minimum with an optimal value of s somewhere between 20 and 30.

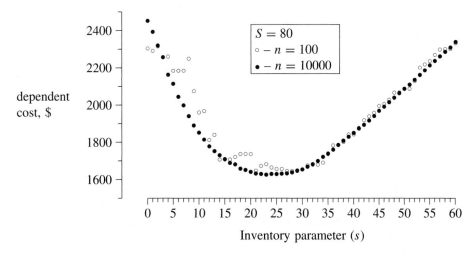

Figure 3.1.4   Dependent cost for (s, S) inventory system.

The same initial seed (12345) was used for each value of s. Fixing the initial seed guarantees that *exactly* the same sequence of demands will be processed; therefore, any changes in the resulting system statistics are due to changes in s only. As in Example 3.1.6, the demands were generated as *Equilikely*$(10, 50)$ random variates. To compute steady-state statistics, all the modeling assumptions upon which this simulation is based would have to remain valid for many years. For that reason, steady-state statistics are not very meaningful in this example. Indeed, it is questionable to assume that the demand distribution, cost parameters, and inventory policy will remain constant for even two years. Steady-state performance statistics are fashionable, however, and represent an interesting limiting case. For that reason, averages based on $n = 10\,000$ weeks of operation (approximately 192 years) are presented as •'s for comparison. These estimated steady-state averages have the attractive feature that they vary smoothly with s and hence the optimum steady-state value of s is relatively well defined. In a world that (on average) never changes, if $S = 80$, the auto dealer should use the inventory threshold $s = 23$. This example illustrates the more general problem, called *stochastic optimization*, of optimizing a function that is not known with certainty. (See Fu, 2006, for more details.)

3.1.3 Statistical Considerations

The statistical analysis of simulation-generated data is discussed in Chapters 4 and 8. In anticipation of that discussion, we note that Example 3.1.7 illustrates two important ideas: *variance reduction* and *robust estimation*. We consider variance reduction first, particularly as it relates to simulation output.

Variance Reduction. Because the statistical output produced by any discrete-event simulation program depends on the sequence of random variates generated as the program executes, the output will always have some inherent uncertainty. For example, in Examples 3.1.3 and 3.1.6, the statistical output depends on the value of the rng initial seed, particularly if the number of jobs or the number of simulated time intervals is small. Therefore, if several initial seeds are used (a policy we certainly advocate), there will be a natural *variance* in any computed statistic.

The statistical tools used to quantify this variance/uncertainty will be developed in Chapter 8. For now, the key point to be made is that using *common random numbers*, as in Example 3.1.7 when $n = 100$, is an intuitive approach to reducing the variance in computed system statistics. This example of variance reduction (using common random numbers) is consistent with the time-honored approach to experimental scientific "what if" studies where, if possible, all variables except one are fixed. Fixing the initial seed at 12345 in Example 3.1.7 isolates the variability of the performance measure (dependent cost). This technique is generally known in statistics as *blocking*. Other variance reduction strategies are considered by Bucklew (2004), Juneja and Shahabuddin (2006), Lemieux (2006), and Szechtman (2006).

Robust Estimation. The optimal (minimal cost) estimated steady-state threshold value $s = 23$ in Example 3.1.7 is a robust estimate because other values of s close to 23 yield essentially the same cost. Therefore, the impact of using one of these alternative values of s in place of $s = 23$ would be slight. In a more general sense, it is desirable for an optimal value to be robust to *all* the assumptions upon which the discrete-event simulation model is based. Relative to Example 3.1.7, this means that we would be interested in learning what happens to the optimal value of s when, for example, S is varied about 80 or when the average demand (per week) is varied about 30, or when there are delivery lags. The estimate $s = 23$ is robust only if it remains close to the minimal-cost operating-policy level when all these assumptions are altered. In general, robust estimators are insensitive to model assumptions.

3.1.4 Exercises

3.1.1   (*a*) Modify program `ssq2` to use *Exponential* (1.5) service times. (*b*) Process a relatively large number of jobs, say 100 000, and report what changes this produces relative to the statistics in Example 3.1.3. (*c*) Explain (or conjecture) why some statistics change and others do not.

3.1.2   (*a*) Relative to the steady-state statistics in Example 3.1.3 and the statistical equations in Section 1.2, list all of the consistency checks that should be applicable. (*b*) Verify that all of these consistency checks are valid.

3.1.3   (*a*) Given that the Lehmer random-number generator used in the library `rng` has the modulus $2^{31} - 1$, what are the largest and smallest possible numerical values (as a function of μ) that the function `Exponential`(μ) can return? (*b*) Comment on this relative to the theoretical expectation that an *Exponential* (μ) random variate can have an arbitrarily large value and a value arbitrarily close to zero.

3.1.4   (*a*) Conduct a transition-to-steady-state study like that in Example 3.1.3 except for a service-time model that is *Uniform* (1.3, 2.3). Be specific about the number of jobs that seem to be required to produce steady-state statistics. (*b*) Comment.

3.1.5   (*a*) Verify that the mean service time in Example 3.1.4 is 1.5. (*b*) Verify that the steady-state statistics in Example 3.1.4 seem to be correct. (*c*) Note that the arrival rate, service rate, and utilization are the same as those in Example 3.1.3, yet all the other statistics are larger than those in Example 3.1.3. Explain (or conjecture) why this is so. Be specific.

3.1.6   (*a*) Modify program `sis2` to compute data like that in Example 3.1.7. Use the functions `PutSeed` and `GetSeed` from the library `rng` in such a way that *one* initial seed is supplied by the system clock, is printed as part of the program's output, and then is automatically used to generate the same demand sequence for all values of s. (*b*) For $s = 15, 16, \ldots, 35$, create a figure (or table) similar to the one in Example 3.1.7. (*c*) Comment.

3.1.7   (*a*) Relative to Example 3.1.5, if instead the random-variate sequence of demands is generated as

$$d_i \;=\; \texttt{Equilikely(5, 25)} \;+\; \texttt{Equilikely(5, 25)} \qquad\qquad i = 1, 2, 3, \ldots,$$

then demonstrate that, when compared with those in Example 3.1.6, some of the steady-state statistics will be the same and others will not. (*b*) Explain why this is so.

3.1.8   Modify program `sis2` to simulate the operation of a simple inventory system *with a delivery lag*. (*a*) Specifically, assume that, if an order is placed at time $t = i - 1$, then the order will arrive at the later time $t = i - 1 + \delta_i$, where the delivery lag δ_i is a *Uniform* (0, 1) random variate, independent of the size of the order. (*b*) What are the equations for $\bar{l}_i^+$ and $\bar{l}_i^-$? (*c*) Using the same parameter values as in Example 3.1.7, calculate that value of *s* for which the average dependent cost is least. Compare this result with that obtained in Example 3.1.7. (*d*) It is important to have the *same* sequence of demands for all values of *s*, with and without a lag. Why? How did you accomplish this? (*e*) Discuss what you did to convince yourself that the modification is correct. (*f*) For both $n = 100$ and $n = 10\,000$, produce a table of dependent costs corresponding to $s = 10, 20, 30, 40, 50, 60$.

3.2 MULTIPLE-STREAM LEHMER RANDOM-NUMBER GENERATORS

A typical discrete-event simulation model will have many stochastic components. When this model is implemented at the computational level, the statistical analysis of system performance is often facilitated by having a unique source of randomness for each stochastic component. Although it might seem that the best way to meet this need for multiple sources of randomness is to create multiple random-number generators, there is a simpler and better approach—use *one* random-number generator to generate multiple "streams" of random numbers, using multiple initial seeds as entry points (one for each stochastic system component). To be consistent with this approach, in this section we extend the Lehmer random-number generation algorithm from Chapter 2 by adding the ability to partition the generator's output sequence into multiple subsequences (streams).

3.2.1 Streams

The library `rng` provides a way to partition the random-number generator's output into multiple streams by establishing multiple states for the generator, one for each stream. As illustrated by the following example, the function `PutSeed` can be used to set the state of the generator with the current state of the stream before generating a random variate appropriate to the corresponding stochastic component, and the function `GetSeed` can be used to retrieve the revised state of the stream after the random variate has been generated.

Example 3.2.1
The program `ssq2` has two stochastic components, the arrival process and the service process, represented by the functions `GetArrival` and `GetService` respectively. To create a different stream of random numbers for each component, it is sufficient to allocate a different Lehmer-generator state variable to each function. This is illustrated by modifying `GetService` from its original form in `ssq2`, which is

```
double GetService(void)                        // original form
{
    return (Uniform(1.0, 2.0));
}
```

to the multiple-stream form indicated, which uses the static variable x to represent the current state of the service-process stream, initialized to 123456789.

```
double GetService(void)                    // multi-stream form
{
    double s;
    static long x = 123456789;
                            // use your favorite initial seed
    PutSeed(x);             // set the state of the generator
    s = Uniform(1.0, 2.0);
    GetSeed(&x);                  // save the new generator state
    return (s);
}
```

Example 3.2.2
As in the previous example, the function GetArrival should be modified similarly, with a corresponding static variable to represent the current state of the arrival process stream, but initialized to a *different* value—that is, the original form of GetArrival in program ssq2,

```
double GetArrival(void)                        // original form
{
    static double arrival = START;
    arrival += Exponential(2.0);
    return (arrival);
}
```

should be modified to something like

```
double GetArrival(void)                     // multi-stream form
{
    static double arrival = START;
    static long x = 987654321;
                            // use an appropriate initial seed
    PutSeed(x);             // set the state of the generator
    arrival += Exponential(2.0);
    GetSeed(&x);                  // save the new generator state
    return (arrival);
}
```

As in Example 3.2.1, in the multiple-stream form, the static variable x represents the current state of the arrival process stream, initialized in this case to 987654321. Note that there is nothing magic about this initial state (relative to 123456789), and, indeed, it could even fail to be a particularly good choice—more about that point later in this section.

If GetService and GetArrival are modified as in Examples 3.2.1 and 3.2.2, then the arrival times will be drawn from one stream of random numbers and the service times will be drawn from another stream. Provided the two streams don't overlap, in this way the arrival process and service process will be uncoupled.* As the following example illustrates, the cost of this uncoupling in terms of execution time is modest.

Example 3.2.3
The parameter LAST in program ssq2 was changed to process 1 000 000 jobs, and the execution time to process this many jobs was recorded. (A large number of jobs was used to get an accurate time comparison.) Program ssq2 was then modified as in Examples 3.2.1 and 3.2.2 and used to process 1 000 000 jobs, with the execution time recorded. The increase in the (already small) execution time was about 20%.

Jump Multipliers. As illustrated in the previous examples, the library rng can be used to support the allocation of a unique stream of random numbers to each stochastic component in a discrete-event simulation program. There is, however, a potential problem with this approach—the assignment of initial seeds. Each stream requires a unique initial state that should be chosen to produce *disjoint* streams, but, if multiple initial states are picked at whim, there is no convenient way to guarantee that the streams are disjoint; some of the initial states could be just a few calls to Random away from one another. With this limitation of the library rng in mind, we now turn to the issue of constructing a random-number generation library called rngs that is a multiple-stream version of the library rng. We begin by recalling two key points from Section 2.1.

- A Lehmer random-number generator is defined by the function

$$g(x) = ax \bmod m,$$

  where the modulus m is a large prime integer, the full-period multiplier a is modulus-compatible with m, and $x \in \mathcal{X}_m = \{1, 2, \ldots, m - 1\}$.
- If $x_0, x_1, x_2, \ldots$ is an infinite periodic sequence in $\mathcal{X}_m$ generated by $g(x) = ax \bmod m$, then each x_i is related to x_0 by the equation

$$x_i = a^i x_0 \bmod m \qquad i = 1, 2, \ldots.$$

The following theorem is the key to creating the library rngs. The proof is left as an exercise.

---

*Also, because the scope of the two-stream state variables (both are called x) is local to their corresponding functions, the use of PutSeed in program ssq2 to initialize the generator can, and should, be eliminated from main.

Theorem 3.2.1   Given a Lehmer random-number generator defined by $g(x) = ax$ mod m and any integer j with $j = 1, 2, \ldots, m - 1$, the associated *jump function* is

$$g^j(x) = (a^j \bmod m) \, x \bmod m$$

and has the *jump multiplier* a^j mod m. For any $x_0 \in \mathcal{X}_m$, if the function $g(\cdot)$ generates the sequence $x_0, x_1, x_2, \ldots$, then the jump function $g^j(\cdot)$ generates the sequence $x_0, x_j, x_{2j}, \ldots$.

Example 3.2.4
If $m = 31$, $a = 3$, and $j = 6$, then the jump multiplier is

$$a^j \bmod m = 3^6 \bmod 31 = 16.$$

Starting with $x_0 = 1$, the function $g(x) = 3x$ mod 31 generates the sequence

$$\underline{1}, 3, 9, 27, 19, 26, \underline{16}, 17, 20, 29, 25, 13, \underline{8}, 24, 10, 30, 28, 22, \underline{4}, 12, 5, 15, 14, 11, \underline{2}, 6, \ldots,$$

while the jump function $g^6(x) = 16x$ mod 31 generates the sequence of underlined terms,

$$1, 16, 8, 4, 2, \ldots.$$

Note that the first sequence is $x_0, x_1, x_2, \ldots$ and the second sequence is $x_0, x_6, x_{12}, \ldots$.

The previous example illustrates that, once the jump multiplier a^j mod m has been computed—and this is a one-time cost—then the jump function $g^j(\cdot)$ provides a mechanism to jump from x_0 to x_j to x_{2j}, etc. If j is chosen properly, then the jump function can be used in conjunction with a user-supplied initial seed to "plant" additional initial seeds, each separated one from the next by j calls to Random. In this way, disjoint streams can be created automatically, with the initial state of each stream dictated by the choice of just *one* initial state.

Example 3.2.5
There are approximately 2^{31} possible values in the full period of our standard $(a, m) = (48271, 2^{31} - 1)$ Lehmer random-number generator. Therefore, if we wish to maintain $256 = 2^8$ streams of random numbers (the choice of 256 is largely arbitrary), it is natural to partition the periodic sequence of possible values into 256 disjoint subsequences, each of equal length. This is accomplished by finding the largest value of j less than $2^{31}/2^8 = 2^{23} = 8\,388\,608$ such that the associated jump multiplier 48271^j mod m is modulus-compatible with m. Because this jump multiplier is modulus-compatible, the jump function

$$g^j(x) = (48271^j \bmod m) \, x \bmod m$$

can be implemented by Algorithm 2.2.1. This jump function can then be used in conjunction with one user-supplied initial seed to efficiently plant the other 255 additional initial seeds, each separated one from the next by $j \cong 2^{23}$ steps.* Planting the additional seeds this way minimizes the possibility of stream overlap.

Maximal Modulus-Compatible Jump Multipliers

> **Definition 3.2.1**   Given a Lehmer random-number generator with (prime) modulus m, full-period modulus-compatible multiplier a, and a requirement for s disjoint streams as widely separated as possible, the *maximal* jump multiplier is $a^j \bmod m$, where j is the largest integer less than $\lfloor m/s \rfloor$ such that $a^j \bmod m$ is modulus-compatible with m.

Example 3.2.6

Consistent with Definition 3.2.1 and with $(a, m) = (48271, 2^{31} - 1)$, a table of maximal modulus-compatible jump multipliers can be constructed for 1024, 512, 256, and 128 streams, as illustrated.

# of streams s	$\lfloor m/s \rfloor$	jump size j	jump multiplier $a^j \bmod m$
1024	2097151	2082675	97070
512	4194303	4170283	44857
256	8388607	8367782	22925
128	16777215	16775552	40509

Computation of the corresponding table for $a = 16807$ (the minimal standard multiplier) is left as an exercise.

Library `rngs`.   The library `rngs` is an upward-compatible multiple-stream replacement for the library `rng`. The library `rngs` can be used as an alternative to `rng` in any of the programs presented earlier by replacing

```
#include "rng.h"
```

with

```
#include "rngs.h"
```

As configured, `rngs` provides for 256 streams, indexed from 0 to 255, with 0 as the default stream. Although the library is designed so that all streams will be initialized to default values if necessary, the recommended way to initialize all streams is by using the function `PlantSeeds`. Only one stream is *active* at any time; the other 255 are *passive*. The function `SelectStream` is used to define the active stream. If the default stream is used exclusively, so that 0 is *always* the active stream, then the library `rngs` is functionally equivalent to the library `rng` in the sense

---

*Because j is less than $2^{31}/2^8$, the last planted initial seed will be more than j steps from the first.

that `rngs` will produce *exactly* the same `Random` output as `rng` (for the same initial seed, of course).

The library `rngs` provides six functions, the first four of which correspond to analogous functions in the library `rng`.

- `double Random(void)` —This is the Lehmer random-number generator used throughout this book.
- `void PutSeed(long x)` —This function can be used to set the state of the active stream.
- `void GetSeed(long *x)` —This function can be used to get the state of the active stream.
- `void TestRandom(void)` —This function can be used to test for a correct implementation of the library.
- `void SelectStream(int s)` —This function can be used to define the active stream (i.e., the stream from which the next random number will come). The active stream will remain as the source of future random numbers until another active stream is selected by calling `SelectStream` with a different stream index `s`.
- `void PlantSeeds(long x)` —This function can be used to set the state of all the streams by "planting" a sequence of states (seeds), one per stream, with all states dictated by the state of the default stream. The following convention is used to set the state of the default stream:

  if `x` is positive then `x` is the state;
  if `x` is negative then the state is obtained from the system clock;
  if `x` is 0 then the state is to be supplied interactively.

3.2.2 Examples

The following examples illustrate how to use the library `rngs` to allocate a separate stream of random numbers to each stochastic component of a discrete-event simulation model. We will see additional illustrations of how to use `rngs` in this and later chapters. From this point on, `rngs` will be the basic random-number-generation library used for *all* the discrete-event simulation programs in this book.

Example 3.2.7

As a superior alternative to the multiple-stream generator approach in Examples 3.2.1 and 3.2.2, the functions `GetArrival` and `GetService` in program `ssq2` can be modified to use the library `rngs`:

```
double GetArrival(void)
{
   static double arrival = START;
   SelectStream(0);                          // this line is new
   arrival += Exponential(2.0);
   return (arrival);
}
```

```
double GetService(void)
{
   SelectStream(2);                        // this line is new
   return (Uniform(1.0, 2.0));
}
```

The other modifications are to include `"rngs.h"` in place of `"rng.h"` and to use the function `PlantSeeds(123456789)` in place of `PutSeed(123456789)` to initialize the streams.*

   If program `ssq2` is modified consistent with Example 3.2.7, then the arrival process will be *uncoupled* from the service process. That is important because we might want to study what happens to system performance if, for example, the `return` in the function `GetService` is replaced with

```
return (Uniform(0.0, 1.5) + Uniform(0.0, 1.5));
```

Although two calls to `Random` are now required to generate each service time, this new service process "sees" *exactly* the same job arrival sequence as did the old service process. This kind of uncoupling provides a desirable variance-reduction technique when discrete-event simulation is used to compare the performance of different systems.

A Single-Server Service Node With Multiple Job Types.   A meaningful extension to the single-server service-node model is *multiple job types*, each with its own arrival and service process. This model extension is easily accommodated at the conceptual level; each arriving job carries a job type that determines the kind of service provided when the job enters service. Similarly, provided that the queue discipline is FIFO, the model extension is straightforward at the specification level. Therefore, using program `ssq2` as a starting point, we can focus on the model extension at the implementation level. Moreover, we recognize that, to facilitate the use of common random numbers, the library `rngs` can be used with a different stream allocated to each of the stochastic arrival and service processes in the model. The following example is an illustration.

Example 3.2.8
Suppose that there are two job types arriving independently, one with *Exponential*(4.0) interarrival times and *Uniform*(1.0, 3.0) service times and the other with *Exponential*(6.0) interarrival times and *Uniform*(0.0, 4.0) service times. In this case, the arrival-process generator in program `ssq2` can be modified as follows:

```
double GetArrival(int *j)                 // j denotes job type
{
   const  double mean[2]    = {4.0, 6.0};
   static double arrival[2] = {START, START};
```

---

*Note that there is nothing magic about the use of `rngs` stream 0 for the arrival process and stream 2 for the service process—any two different streams can be used. In particular, if even more separation between streams is required, then, for example, streams 0 and 128 could be used.

```
static int    init     = 1;
        double temp;
if (init) {                // initialize the arrival array
  SelectStream(0);
  arrival[0] += Exponential(mean[0]);
  SelectStream(1);
  arrival[1] += Exponential(mean[1]);
  init         = 0;
}
if (arrival[0] <= arrival[1])
  *j = 0;                       // next arrival is job type 0
else
  *j = 1;                       // next arrival is job type 1
temp = arrival[*j];                    // next arrival time
SelectStream(*j);          // use stream j for job type j
arrival[*j] += Exponential(mean[*j]);
return (temp);
}
```

Note that GetArrival returns the next arrival time *and* the job type as an index, with value 0 or 1 as appropriate.

Example 3.2.9

As a continuation of Example 3.2.8, the corresponding service-process generator in program ssq2 can be modified as follows:

```
double GetService(int j)
{
   const double min[2] = {1.0, 0.0};
   const double max[2] = {3.0, 4.0};
   SelectStream(j + 2);   // use stream j + 2 for job type j
   return (Uniform(min[j], max[j]));
}
```

Relative to Example 3.2.9, note that the job-type index j is used in GetService to insure that the service time corresponds to the appropriate job type. Also, rngs streams 2 and 3 are allocated to job types 0 and 1, respectively. In this way, all four simulated stochastic processes are uncoupled. Thus, the random-variate model corresponding to any one of these four processes could be changed without altering the generated sequence of random variates corresponding to the other three processes.

Consistency Checks.   Beyond the modifications in Examples 3.2.8 and 3.2.9, some job-type-specific statistics-gathering needs to be added in main to complete the modification of program `ssq2` to accommodate multiple job types. To *d.dd* precision, the (theoretical) steady-state statistics are

$\bar{r}$	$\bar{w}$	$\bar{d}$	$\bar{s}$	$\bar{l}$	$\bar{q}$	$\bar{x}$
2.40	7.92	5.92	2.00	3.30	2.47	0.83

The details associated with modifying program `ssq2` to produce corresponding estimates are left in Exercise 3.2.4. How do we know these values are correct?

In addition to $\bar{w} = \bar{d} + \bar{s}$ and $\bar{l} = \bar{q} + \bar{x}$, the three following intuitive consistency checks give us increased confidence in these (estimated) steady-state results:

*   Both job types have the average service time 2.00, so that $\bar{s}$ should be 2.00. The corresponding service rate is 0.5.
*   The arrival rates of job types 0 and 1 are 1/4 and 1/6, respectively. Intuitively, the net arrival rate should then be $1/4 + 1/6 = 5/12$, which corresponds to $\bar{r} = 12/5 = 2.40$.
*   The steady-state utilization should be the ratio of the arrival rate to the service rate, which is $(5/12)/(1/2) = 5/6 \cong 0.83$.

3.2.3 Exercises

3.2.1   (*a*) Construct the $a = 16807$ version of the table in Example 3.2.6. (*b*) What is the $O(\cdot)$ time complexity of the algorithm you used?

3.2.2   (*a*) Prove that, if m is prime, $a = 1, 2, \ldots, m - 1$, and $a^* = a^{m-2}$ mod m, then

$$a^* a \text{ mod } m = 1.$$

Now define

$$g(x) = ax \text{ mod } m \qquad \text{and} \qquad g^*(x) = a^* x \text{ mod } m$$

for all $x \in \mathcal{X} = \{1, 2, \ldots, m - 1\}$. (*b*) Prove that the functions $g(\cdot)$ and $g^*(\cdot)$ generate the same sequence of states, except in *opposite* orders. (*c*) Comment on the implication of this relative to full-period multipliers. (*d*) If $m = 2^{31} - 1$ and $a = 48271$, what is a^*?

3.2.3   Modify program `ssq2` as suggested in Example 3.2.7 to create two programs that differ only in the function `GetService`. For one of these programs, use the function as implemented in Example 3.2.7; for the other program, use the code given below.

```
double GetService(void)
{
    SelectStream(2);                          // this line is new
    return (Uniform(0.0, 1.5) + Uniform(0.0, 1.5));
}
```

(*a*) For both programs, verify that *exactly* the same average interarrival time is produced. (Print the average with *d.dddddd* precision.) Note that the average service time is approximately the same in both cases, as is the utilization, yet the service-node statistics $\bar{w}, \bar{d}, \bar{l}$, and $\bar{q}$ are different. (*b*) Why?

3.2.4   Modify program `ssq2` as suggested in Examples 3.2.8 and 3.2.9. (*a*) What proportion of processed jobs are of type 0? (*b*) What are $\bar{w}$, $\bar{d}$, $\bar{s}$, $\bar{l}$, $\bar{q}$, and $\bar{x}$ for each job type? (*c*) What did you do to convince yourself that your results are valid? (*d*) Why are $\bar{w}$, $\bar{d}$, and $\bar{s}$ the same for both job types, while $\bar{l}$, $\bar{q}$, and $\bar{x}$ are different?

3.2.5   Prove Theorem 3.2.1.

3.2.6   Same as Exercise 3.2.3, but using the `GetService` function in Example 3.1.4 instead of the `GetService` function in Exercise 3.2.3.

3.2.7   Suppose there are three job types arriving independently to a single-server service node. The interarrival times and service times have the following characterization:

job type	interarrival times	service times
0	*Exponential*(4.0)	*Uniform*(0.0, 2.0)
1	*Exponential*(6.0)	*Uniform*(1.0, 2.0)
2	*Exponential*(8.0)	*Uniform*(1.0, 5.0)

(*a*) What is the proportion of processed jobs for each type? (*b*) What are $\bar{w}$, $\bar{d}$, $\bar{s}$, $\bar{l}$, $\bar{q}$, and $\bar{x}$ for each job type? (*c*) What did you do to convince yourself that your results are valid? (Simulate at least 100 000 processed jobs.)

3.3 DISCRETE-EVENT SIMULATION EXAMPLES

In this section, we will consider three discrete-event system models, each of which is an extension of a model considered previously. The three models are (*i*) a single-server service node *with immediate feedback*, (*ii*) a simple inventory system *with delivery lag*, and (*iii*) a single-server machine shop.

3.3.1 Single-Server Service Node with Immediate Feedback

We begin by considering an important extension of the single-server service-node model first introduced in Section 1.2. Consistent with the following definition (based on Definition 1.2.1), the extension is *immediate feedback* — the possibility that the service a job just received was incomplete or otherwise unsatisfactory and, if so, that the job feeds back immediately to request service once again.

> **Definition 3.3.1**   A single-server service node *with immediate feedback* consists of a server plus its queue with a feedback mechanism, as illustrated in Figure 3.3.1.

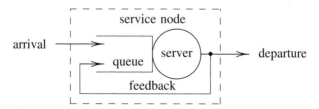

Figure 3.3.1   Single-server service node with immediate feedback.

Jobs arrive at the service node, generally at random, seeking service. When service is provided, the time involved is generally also random. At the completion of service, jobs either depart the service node (forever) or immediately feed back and once again seek service. The service node operates as follows: as each job arrives, if the server is busy then the job enters the queue, otherwise the job immediately enters service; as each job completes service, either a departure or a feedback occurs, generally at random. When a *departure* occurs, if the queue is empty, then the server becomes idle; otherwise, another job is selected from the queue to enter service immediately. When a *feedback* occurs, if the queue is empty, then the job immediately reenters service; otherwise, the job enters the queue, and then one of the jobs in the queue is selected to immediately enter service. At any instant in time, the state of the server will either be busy or idle and the state of the queue will be either empty or not empty. If the server is idle then the queue must be empty; if the queue is not empty then the server must be busy.

Note the distinction between the two events "completion of service" and "departure." If there is no feedback these two events are equivalent; if feedback is possible, then it is important to make a distinction. When the distinction is important, the completion-of-service event is more fundamental, because, at the completion of service, either a departure event or a feedback event then occurs. This kind of "which event comes first" causal reasoning is important at the conceptual model-building level.

Model Considerations.   When feedback occurs, we assume that the job joins the queue (if any) in a way that is consistent with the queue discipline. For example, if the queue discipline is FIFO, then a fed-back job would receive no priority; it would join the queue at the end, in effect becoming indistinguishable from an arriving job. Of course, other feedback queue disciplines are possible, the most common of which involves assigning a priority to jobs that are fed back. If feedback is possible, the default assumption in this book is that a fed-back job will join the queue according to the queue discipline and that a new service time will be required, independent of any prior service provided. Similarly, the default assumption is that the decision to depart or feed back is random with *feedback probability* β, as illustrated in Figure 3.3.2.

In addition to β, the other two parameters that characterize the stochastic behavior of a single-server service node with immediate feedback are the *arrival rate* λ and the *service rate* ν. As in Definition 1.2.5, $1/\lambda$ is the average interarrival time, and $1/\nu$ is the average service time.*

As each job completes service, it departs the service node with probability $1 - \beta$ or feeds back with probability β. According to this model, feedback is independent of past history, and

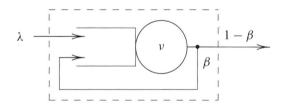

Figure 3.3.2   Single-server service node with feedback probability β.

---

*The use of the symbol ν to denote the service rate is nonstandard. Instead, the usual convention is to use the symbol μ. See Section 8.5 for more discussion of arrival rates, service rates, and our justification for the use of ν in place of μ.

so a job could feed back more than once. Indeed, in theory, a job can feed back arbitrarily many times—see Exercise 3.3.1. Typically, β is close to 0.0, indicating that feedback is a rare event. This is not a universal assumption, however, and so a well-written discrete-event simulation program should accommodate any probability of feedback in the range $0.0 \leq \beta < 1.0$. At the computational model-building level, feedback can be modeled with a Boolean-valued function, as illustrated.

```
int GetFeedback(double beta)              // use 0.0 <= beta < 1.0
{
    SelectStream(2);                      // use rngs stream 2 for feedback
    if (Random() < beta)
      return (1);                                    // feedback occurs
    else
      return (0);                                    // no feedback
}
```

Statistical Considerations.   If properly interpreted, the mathematical variables and associated definitions in Section 1.2 remain valid if immediate feedback is possible. The interpretation required is that the index $i = 1, 2, 3, \ldots$ counts jobs that enter the service node; once indexed in this way, a fed-back job is not counted again. Because of this indexing, all the job-averaged statistics defined in Section 1.2 remain valid *provided* delay times, wait times, and service times are incremented each time a job is fed back. For example, the average wait is the sum of the waits experienced by all the jobs that enter the service node, divided by the number of such jobs; each time a job is fed back it contributes an additional wait to the sum of waits, but it does *not* cause the number of jobs to be increased. Similarly, the time-averaged statistics defined in Section 1.2 also remain valid if feedback is possible.

The key feature of immediate feedback is that jobs from outside the system are merged with jobs from the feedback process. In this way, the (steady-state) request-for-service rate is larger than λ by the positive additive factor $\beta \bar{x} \nu$. As is illustrated later in Example 3.3.2, if there is no corresponding increase in service rate, this increase in the request-for-service rate will cause job-averaged and time-averaged statistics to increase from their nonfeedback values. This increase is intuitive—if you are entering a grocery store and the check-out queues are already long, you certainly do not want to see customers reentering these queues because they just realized they were short-changed at check-out or forgot to buy a gallon of milk.

Note that indexing by arriving jobs will cause the average service time $\bar{s}$ to increase as the feedback probability increases. In this case, do not confuse $\bar{s}$ with the reciprocal of the service rate; $1/\nu$ is the (theoretical) average service time *per service request*, irrespective of whether that request is by an arriving job or by a fed-back job.

Algorithm and Data-Structure Considerations

Example 3.3.1
Consider the following arrival times, service times, and completion times for the first nine jobs entering a single-server FIFO service node with immediate feedback. (For simplicity, all times are integers.)

job index	1	2	3	4	5	·	6	·	7	8	·	9	···
arrival/feedback	1	3	4	7	10	**13**	14	**15**	19	24	**26**	30	···
service	9	3	2	4	7	5	6	3	4	6	3	7	···
completion	10	**13**	**15**	19	**26**	31	37	**40**	44	50	53	**60**	···

The boldface times correspond to jobs that were fed back. For example, the second job completed service at time 13 and immediately fed back. At the computational level, note that some algorithm and data structure are necessary to insert fed-back jobs into the arrival stream. [An inspection of the yet-to-be-inserted feedback times 15 and 26 reveals that the job fed back at time 15 must be inserted in the arrival stream after job 6 (which arrived at time 14) and before job 7 (which arrived at time 19).]

The reader is encouraged to extend the specification model of a single-server service node in Section 1.2 to account for immediate feedback; then extend this model at the computational level by starting with program ssq2 and using a different rngs stream for each stochastic process. Example 3.3.1 provides insight into an algorithmic extension and associated data structure that can be used to accomplish this.

Example 3.3.2
Program ssq2 was modified to account for immediate feedback. Consistent with the stochastic modeling assumptions in Example 3.1.3, the arrival process has *Exponential*(2.0) random-variate interarrival times corresponding to the fixed arrival rate $\lambda = 0.50$; the service process has *Uniform*(1.0, 2.0) random-variate service times corresponding to the fixed service rate $\nu \cong 0.67$; and the feedback probability is $0.0 \le \beta < 1.0$. To illustrate the effect of feedback, the modified program was used to simulate the operation of a single-server service node with nine different values of levels of feedback, varied from $\beta = 0.0$ (no feedback) to $\beta = 0.20$. In each case, 100 000 arrivals were simulated. Utilization $\bar{x}$ as a function of β is illustrated on the left-hand side of Figure 3.3.3; the average number in the queue $\bar{q}$ as a function of β is illustrated on the right-hand side.

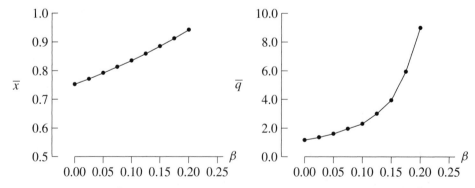

Figure 3.3.3   Utilization and average number in queue as a function of β.

As the probability of feedback increases, the utilization increases from the steady-state value of $\bar{x} = 0.75$ (when there is no feedback) toward the maximum possible value, $\bar{x} = 1.0$. If this $\bar{x}$ versus β figure is extrapolated, it appears that saturation ($\bar{x} = 1.0$) is achieved as $\beta \to 0.25$.

Flow Balance and Saturation.   The observation that saturation occurs as β approaches 0.25 is an important consistency check that is based on steady-state *flow-balance* considerations: Jobs flow *into* the service node at the average rate λ. To remain flow balanced, jobs must flow *out* of the service node at the same average rate. Because the average rate at which jobs flow out of the service node is $\bar{x}(1 - \beta)\nu$, flow balance is achieved when $\lambda = \bar{x}(1 - \beta)\nu$. Saturation is achieved when $\bar{x} = 1$; this happens as $\beta \to 1 - \lambda/\nu = 0.25$. In Example 3.3.2, we see that, in a way consistent with saturation, the average number in the queue increases dramatically as β increases, becoming effectively infinite as $\beta \to 0.25$.

3.3.2 Simple Inventory System with Delivery Lag

The second discrete-event system model we will consider in this section represents an important extension of the periodic-review simple-inventory-system model first introduced in Section 1.3. The extension is *delivery lag* (or *lead time*): An inventory replacement order placed with the supplier will not be delivered immediately; instead, there will be a lag between the time an order is placed and the time the order is delivered. Unless it is stated otherwise, this lag is assumed to be random and independent of the amount ordered.

If there are no delivery lags, then a typical inventory time history looks like the one in Section 1.3, reproduced in Figure 3.3.4 for convenience, with jump discontinuities possible only at the (integer-valued) times of inventory review.

If delivery lags are possible, a typical inventory time history would have jump discontinuities at arbitrary times, as illustrated in Figure 3.3.5. (The special-case order at the terminal time $t = 12$ is assumed to be delivered with zero lag.)

Unless we state otherwise, we assume that any order placed at the beginning of a time interval (at times $t = 2, 5, 7$, and 10, in this case) will be delivered before the end of the time interval.

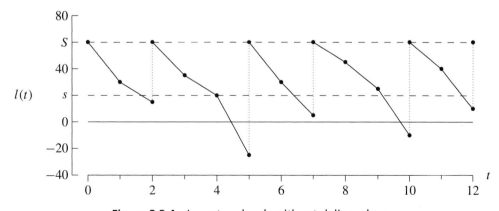

Figure 3.3.4   Inventory levels without delivery lags.

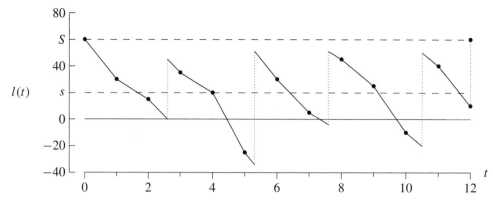

Figure 3.3.5 Inventory levels with delivery lags.

With this assumption, there is *no* change in the simple inventory system model at the specification level. (See Algorithm 1.3.1.) There is, however, a significant change in how the system statistics are computed. For those time intervals in which a delivery lag occurs, the time-averaged holding and shortage integrals in Section 1.3 must be modified.

Statistical Considerations.   As in Section 1.3, l_{i-1} denotes the inventory level at the beginning of the ith time interval (at $t = i - 1$), and d_i denotes the amount of demand during this interval. Consistent with the model in Section 1.3, the demand rate is assumed to be constant between review times. Given d_i, l_{i-1}, and this assumption, there are two cases to consider.

If $l_{i-1} \geq s$, then, because no order is placed at $t = i - 1$, the inventory decreases at a constant rate throughout the interval, with no "jump" in level. The inventory level at the end of the interval is $l_{i-1} - d_i$. In this case, the equations for $\bar{l}_i^{\,+}$ and $\bar{l}_i^{\,-}$ in Section 1.3 remain valid (with l_{i-1} in place of l'_{i-1}).

If $l_{i-1} < s$, then an order is placed at $t = i - 1$ that later causes a jump in the inventory level (when the order is delivered). In this case, the equations for $\bar{l}_i^{\,+}$ and $\bar{l}_i^{\,-}$ in Section 1.3 must be modified: If $l_{i-1} < s$, then an order for $S - l_{i-1}$ items is placed at $t = i - 1$ and a *delivery lag* $0 < \delta_i < 1$ occurs, during which time the inventory level drops at a constant rate to $l_{i-1} - \delta_i d_i$. When the order is delivered at $t = i - 1 + \delta_i$, the inventory level jumps to $S - \delta_i d_i$. During the remainder of the interval, the inventory level drops at the same constant rate to its final level $S - d_i$ at $t = i$. All of this is summarized with the observation that, in this case, the inventory-level time history during the ith time interval is defined by one vertical and two parallel lines, as illustrated in Figure 3.3.6.*

Depending on the location of the four line-segment endpoints indicated by •'s, with each location measured relative to the line $l(t) = 0$, either triangular or trapezoidal figures will be generated. To compute the time-averaged holding level $\bar{l}_i^{\,+}$ and the time-averaged shortage level $\bar{l}_i^{\,-}$ (see Definition 1.3.3), it is necessary to calculate the area of each figure. The details are left as an exercise.

---

*Note that $\delta_i d_i$ must be integer-valued.

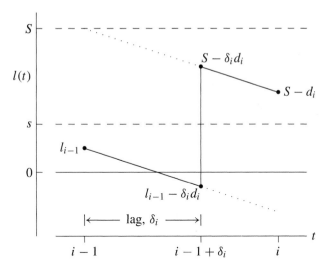

Figure 3.3.6 Inventory level during time interval i when an order is placed.

Consistency Checks. When system models are extended, it is fundamentally important to verify that the extended model is consistent with the parent model (the model before extension). This is usually accomplished by setting system parameters to special values. For example, if the feedback probability is set to zero, an extended computational model that simulates a single-server service node *with* feedback reduces to a parent computational model of a single-server service node *without* feedback. At the computational level, the usual way to make this kind of consistency check is to compare output system statistics and verify that, with the extension removed, the output statistics produced by the extended model agree with the output statistics produced by the parent model. Use of the library rngs facilitates this kind of comparison. In addition to these "extension removal" consistency checks, it is also good practice to check for intuitive "small-perturbation" consistency. For example, if the feedback probability is small, but nonzero, the average number in the queue should be slightly larger than its feedback-free value. The following example applies this idea to a simple-inventory-system model with delivery lag.

Example 3.3.3
For a simple inventory system with delivery lag, we adopt the convention that δ_i is defined for all $i = 1, 2, \ldots, n$ with $\delta_i = 0.0$ if and only if there is no order placed at the beginning of the ith time interval (that is, if $l_{i-1} \geq s$). If an order is placed then $0.0 < \delta_i < 1.0$. With this convention, the stochastic time evolution of a simple inventory system with delivery lag is driven by the two n-point stochastic sequences $d_1, d_2, \ldots, d_n$ and $\delta_1, \delta_2, \ldots, \delta_n$. The simple inventory system is *lag-free* if and only if $\delta_i = 0.0$ for all $i = 1, 2, \ldots, n$; if $\delta_i > 0.0$ for at least one i, then the system is not lag-free. Relative to the five system statistics in Section 1.3, if the inventory parameters (S, s) are fixed, then, even if the delivery lags are small, the following points are valid.

- The average order $\bar{o}$, average demand $\bar{d}$, and relative frequency of setups $\bar{u}$ are exactly the same, regardless of whether the system is lag-free.
- If the system is not lag-free, then the time-averaged holding level $\bar{l}^{+}$ will decrease as compared with the lag-free value.
- If the system is not lag-free, then the time-averaged shortage level $\bar{l}^{-}$ will either remain unchanged or it will increase as compared with the lag-free value.

At the computational level, these three points provide valuable consistency checks for a simple-inventory-system discrete-event simulation program.

Delivery Lag.   If the statistics-gathering logic in program `sis2` is modified to be consistent with the previous discussion, then the resulting program will provide a computational model of a simple inventory system with delivery lag. To complete this modification, a stochastic model of delivery lag is needed. In the absence of information to the contrary, we assume that each delivery lag is an independent *Uniform*(0, 1) random variate.

Example 3.3.4

Program `sis2` was modified to account for *Uniform*(0, 1) random variate delivery lags, independent of the size of the order. As an extension of the automobile dealership example (Example 3.1.7), this modified program was used to study the effect of delivery lag. With $S = 80$, the average weekly cost was computed for a range of inventory threshold values s between 20 and 60. To avoid clutter, only steady-state cost estimates (based on $n = 10\,000$ time intervals) are illustrated. For comparison, the corresponding lag-free cost values from Example 3.1.7 are also illustrated in Figure 3.3.7.

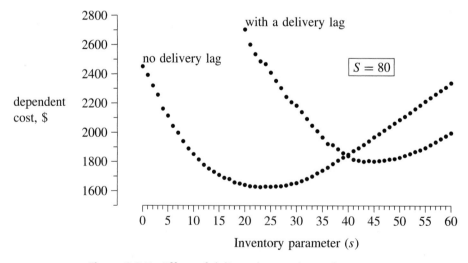

Figure 3.3.7   Effect of delivery lag on dependent cost.

Figure 3.3.7 shows that the effect of delivery lag is profound; the U-shaped cost-versus-s curve is shifted up and to the right. Because of this shift, the optimum (minimum-cost) value of s is increased by approximately 20 automobiles, and the corresponding minimum weekly cost is increased by almost \$200.*

The shift in the U-shaped curve in Example 3.3.4 is consistent with the second and third points in Example 3.3.3: Delivery lags cause $\bar{l}_i^+$ to decrease and $\bar{l}_i^-$ to increase (or remain the same). Because the holding-cost coefficient is $C_{\text{hold}} = \$25$ and the shortage-cost coefficient is $C_{\text{short}} = \$700$ (see Example 1.3.5), delivery lags will cause holding costs to decrease a little for all values of s and will cause shortage costs to increase a lot, but only for small values of s. The shift in the U-shaped curve is the result.

Examples 3.3.2 and 3.3.4 present results corresponding to significant extensions of the two canonical system models used throughout this book. The reader is strongly encouraged to work through the details of these extensions at the computational level and reproduce the results in Examples 3.3.2 and 3.3.4.

3.3.3 Single-Server Machine Shop

The third discrete-event simulation model considered in this section is a single-server machine shop. This is a simplified version of the multiple-server machine-shop model used in Section 1.1 to illustrate model-building at the conceptual, specification, and computational levels. (See Example 1.1.1.)

A single-server machine-shop model is essentially identical to the single-server service-node model first introduced in Section 1.2, except for one important difference. The service-node model is *open*, in the sense that an effectively infinite number of jobs is available to arrive from "outside" the system and, after service is complete, return to the "outside." In contrast, the machine-shop model is *closed*: There is only a finite number of machines (jobs) that are part of the system—as illustrated in Figure 3.3.8, there is no "outside."

In more detail, there is a finite population of statistically identical machines, all of which are initially in an *operational* state (so the server is initially idle and the queue is empty). Over

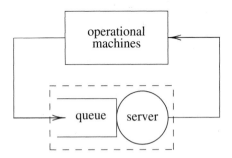

Figure 3.3.8   Single-server machine-shop system diagram.

---

*Because of the dramatic shift in the optimum value of s from the lag-free value, $s \cong 23$, to the with-lag value, $s \cong 43$, we see that the optimal value of s is *not* robust with respect to model assumptions about the delivery lag.

time, these machines fail, independently, at which time they enter a *broken* state and request repair service at the single-server service node.* Once repaired, a machine immediately re-enters an operational state and remains in this state until it fails again. Machines are repaired in the order in which they fail, without interruption. Correspondingly, the queue discipline is FIFO, nonpreemptive, and conservative. There is no feedback: The repair is done properly the first time.

To make the single-server machine-shop model specific, we assume that the service (repair) time is a *Uniform*(1.0, 2.0) random variate, that there are M machines, and that the time a machine spends in the operational state is an *Exponential*(100.0) random variate. All times are in hours. On the basis of 100 000 simulated machine failures, we want to estimate the steady-state time-averaged number of operational machines and the server utilization as functions of M.

Program ssms.   Program ssms simulates the single-server machine shop described in this section. This program is similar to program ssq2, but with two important differences.

- The library rngs is used to provide an independent source of random numbers both to the simulated machine-failure process and to the machine-repair process.
- The failure process is defined by the array failure, which represents the time of the next failure for each of the M machines.

The time-of-next-failure list (array) is not maintained in sorted order; it must be searched completely each time a machine failure is simulated. The efficiency of this $O(M)$ search could be a problem for large M. Exercise 3.3.7 investigates computational-efficiency improvements associated with an alternative algorithm and its associated data structure.

Example 3.3.5
Because the time-averaged number in the service node $\bar{l}$ represents the time-averaged number of broken machines, $M - \bar{l}$ represents the time-averaged number of operational machines. Program ssms was used to estimate $M - \bar{l}$ for values of M between 20 and 100, as illustrated in Figure 3.3.9.

As expected, for small values of M, the time-averaged number of operational machines is essentially M. This is consistent with low server utilization and a correspondingly small value of $\bar{l}$. The angled dashed line indicates the ideal situation where all machines are continuously operational (i.e., $\bar{l} = 0$). Also, as expected, for large values of M the time-averaged number of operational machines is essentially constant, independent of M. This is consistent with a saturated server (utilization of 1.0) and a correspondingly large value of $\bar{l}$. The horizontal dashed line indicates that this saturated-server constant value is approximately 67.

---

*Conceptually, the machines move along the network arcs indicated, from the operational pool into and out of service and then back to the operational pool. In practice, the machines are usually stationary and the server moves to the machines. The time for the server to move from one machine to another is either negligible or is modeled as a part of the service time.

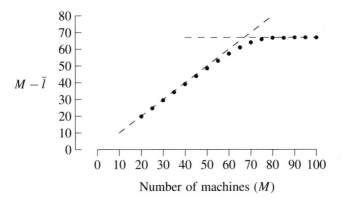

Figure 3.3.9   Time-averaged number of operational machines as a function of the number of machines.

Distribution Parameters.   The parameters used in the distributions in the models presented in this section ($\lambda = 0.5$ for the arrival rate and $\beta = 0.20$ for the feedback probability in the single-server service node with feedback) have been chosen rather arbitrarily, in order to allow us to focus on the simulation modeling and its associated algorithms. Chapter 9 on "Input Modeling" focuses on techniques for estimating realistic parameters from data.

3.3.4 Exercises

3.3.1   Let β be the probability of feedback and let the integer-valued random variable X be the number of times a job feeds back. (*a*) For $x = 0, 1, 2, \ldots$ what is $\Pr(X = x)$? (*b*) How does this relate to the discussion of acceptance/rejection in Section 2.3?

3.3.2   (*a*) Relative to Example 3.3.2, based on 100 000 arrivals, generate a table of $\bar{x}$ and $\bar{q}$ values for β from 0.00 to 0.24 in steps of 0.02. (*b*) What data structure did you use and why? (*c*) Discuss how external arrivals are merged with fed-back jobs.

3.3.3   For the model of a single-server service node with feedback presented in this section, there is nothing to prevent a fed-back job from colliding with an arriving job. Is this a model deficiency that needs to be fixed, and, if so, how would you fix it?

3.3.4   Modify program ssq2 to account for a *finite* service-node capacity. (*a*) For the capacities 1, 2, 3, 4, 5, and 6, construct a table of the estimated steady-state probability of rejection. (*b*) Also, construct a similar table if the service-time distribution is changed to be *Uniform*(1.0, 3.0). (*c*) Comment on how the probability of rejection depends on the service process. (*d*) How did you convince yourself these tables are correct?

3.3.5   Verify that the results in Example 3.3.4 are correct. Provide a table of values corresponding to the figure in this example.

3.3.6   (*a*) Relative to Example 3.3.5, construct a figure or table illustrating how $\bar{x}$ (utilization) depends on M. (*b*) If you extrapolate linearly from small values of M, at what value of M will saturation ($\bar{x} = 1$) occur? (*c*) Can you provide an empirical argument or equation to justify this value?

3.3.7   In program ssms, the time-of-next-failure list (array) is not maintained in sorted order; the list must be searched completely each time another machine failure is simulated. As an alternative, implement an algorithm and associated sorted data structure, to investigate whether a significant improvement in computational efficiency can be obtained. (You might need to simulate a huge number of machine failures to get an accurate estimate of computational-efficiency improvement.)

3.3.8   (*a*) Relative to Example 3.3.2, compare a FIFO queue discipline with a priority queue discipline where fed-back jobs go to the head of the queue (i.e., reenter service immediately). (*b*) Is the following conjecture true or false: Although statistics for the fed-back jobs change, system statistics do not change?

3.3.9   (*a*) Repeat Exercise 3.3.7 using $M = 120$ machines, with the time a machine spends in the operational state increased to an *Exponential* (200.0) random variate. (*b*) Use $M = 180$ machines with the time spent in the operational state increased accordingly. (*c*) What does the $O(\cdot)$ computational complexity of your algorithm seem to be?

4

Statistics

The first three sections in this chapter consider the computation and interpretation of univariate-data sample statistics. If the size of the sample is small, then essentially all that can be done is compute the sample mean and standard deviation. Welford's algorithm for computing these quantities is presented in Section 4.1. In addition, Chebyshev's inequality is derived and is used to illustrate how the sample mean and standard deviation are related to the distribution of the data in the sample.

If the size of the sample is not small, then a sample-data histogram can be computed and then used to analyze the distribution of the data in the sample. Two algorithms are presented in Section 4.2 for computing a discrete-data histogram. A variety of examples are presented as well. Similarly, a continuous-data histogram algorithm and a variety of examples are presented in Section 4.3. In both of these sections concerning histograms, the empirical cumulative distribution function is presented as an alternative way to graphically present a data set. The fourth section deals with the computation and interpretation of paired-data sample statistics, in both paired-correlation and autocorrelation applications. Correlation is apparent in many discrete-event simulations. The wait times of adjacent jobs in a single-server node, for example, tend to be above their means and below their means together—that is, they are "positively" correlated.

4.1 SAMPLE STATISTICS

Discrete-event simulations generate experimental data—frequently, *a lot* of experimental data. To facilitate the analysis of all this data, it is conventional to compress the data into a handful of meaningful statistics. We have already seen examples of this in Sections 1.2 and 3.1, where job averages and time averages were used to characterize the performance of a single-server service node. Similarly, in Sections 1.3 and 3.1, averages were used to characterize the performance of a simple inventory system. This kind of "within-the-run" data generation and statistical computation (see programs `ssq2` and `sis2`, for example) is common in discrete-event simulation. Although less common, there is also a "between-the-runs" kind of statistical analysis commonly used in discrete-event simulation: A discrete-event simulation program can be used to simulate the same stochastic system repeatedly—all you need to do is change the initial seed for the random-number generator from run to run. This process is known as *replication*.

In either case, each time a discrete-event simulation program is used to generate data, it is important to appreciate that this data is only a *sample* from that much larger *population* that would be produced, in the first case, if the program were run for a much longer time or, in the second case, if more replications were used. Analyzing a sample and then inferring something about the population from which the sample was drawn is the essence of statistics. We begin by defining the sample mean and sample standard deviation.

4.1.1 Sample Mean and Sample Standard Deviation

Definition 4.1.1   Given the sample $x_1, x_2, \ldots, x_n$ (either continuous or discrete data), the *sample mean* is

$$\bar{x} = \frac{1}{n} \sum_{i=1}^{n} x_i.$$

The *sample variance* is the average of the squared differences about the sample mean,

$$s^2 = \frac{1}{n} \sum_{i=1}^{n} (x_i - \bar{x})^2.$$

The *sample standard deviation* is the positive square root of the sample variance, $s = \sqrt{s^2}$.

The sample mean is a measure of *central tendency* of the data values. The sample variance and sample standard deviation are measures of *dispersion*—the spread of the data about the sample mean. The sample standard deviation has the same "units" as the data and the sample mean. For example, if the data has the units *sec*, then so also do the sample mean and standard deviation. The sample variance would have the units *sec*^{2}. Although the sample variance is more amenable to mathematical manipulation (because it is free of the square root), the sample standard deviation is typically the preferred measure of dispersion, since it has the same units as the data. (Because $\bar{x}$ and s have the same units, the ratio $s/\bar{x}$, known as the *coefficient of variation*, has no units. This statistic is commonly used only if the data is inherently nonnegative, although it is inferior to s as a measure of dispersion because a common shift in the data results in a change in $s/\bar{x}$.)

For statistical reasons that will be explained further in Chapter 8, a common alternative definition of the sample variance (and thus the sample standard deviation) is

$$\frac{1}{n-1} \sum_{i=1}^{n} (x_i - \bar{x})^2 \qquad \text{rather than} \qquad \frac{1}{n} \sum_{i=1}^{n} (x_i - \bar{x})^2.$$

Provided that *one* of these definitions is used consistently for *all* computations, the choice of which equation to use for s^2 is largely a matter of taste, one based on the statistical considerations listed below. There are three reasons that the $1/(n-1)$ form of the sample variance appears almost universally in introductory statistics books:

- The sample variance is undefined when $n = 1$ when using the $1/(n-1)$ form. This is intuitive in the sense that a single observed data value indicates nothing about the spread of the distribution from which the data value is drawn.
- The $1/(n-1)$ form of the sample variance is an *unbiased estimate* of the population variance when the data values are drawn independently from a population with a finite mean and variance. The designation of an estimator as "unbiased" in statistical terminology implies that the sample average of many such sample variances converges to the population variance.
- The statistic

$$\sum_{i=1}^{n}(x_i - \overline{x})^2$$

  has $n - 1$ "degrees of freedom." It is common practice in statistics (particularly in a sub-field known as *analysis of variance*) to divide a statistic by its degrees of freedom.

Despite these compelling reasons, why do we still use the $1/n$ form of the sample variance as given in Definition 4.1.1 consistently throughout the book? Here are five reasons:

- The sample size n is typically *large* in discrete-event simulation, making the difference between the results obtained by using the two definitions small.
- The unbiased property associated with the $1/(n-1)$ form of the sample variance applies only when observations are independent. Observations within a simulation run are typically not independent in discrete-event simulation.
- In the unlikely case that only $n = 1$ observation is collected, the algorithms presented in this text using the $1/n$ form are able to compute a numeric value (zero) for the sample variance. The reader should understand that a value of zero for the sample variance in the case $n = 1$ does *not* imply that there is no variability in the population.
- The $1/n$ form of the sample variance enjoys status as a "plug-in" (details given in Chapter 6) estimate of the population variance. It also is the "maximum likelihood estimate" (details given in Chapter 9) when sampling from bell-shaped populations.
- Authors of many higher-level books on mathematical statistics prefer the $1/n$ form of the sample variance.

The relationship between the sample mean and the sample standard deviation is summarized by Theorem 4.1.1. The proof is left as an exercise.

Theorem 4.1.1 In a *root-mean-square* (rms) sense, the sample mean is that unique value that best fits the sample $x_1, x_2, \ldots, x_n$, and the sample standard deviation is the corresponding smallest possible rms value. In other words, if the rms value associated with any value of x is

$$d(x) = \sqrt{\frac{1}{n}\sum_{i=1}^{n}(x_i - x)^2}$$

then the smallest possible rms value of $d(x)$ is $d(\overline{x}) = s$, and this value is achieved if and only if $x = \overline{x}$.

Example 4.1.1

Theorem 4.1.1 is illustrated in Figure 4.1.1 for a random-variate sample of size $n = 50$ that was generated by using a modified version of program buffon. The data corresponds to 50 observations of the x-coordinate of the right-hand endpoint of a unit-length needle dropped at random. (See Example 2.3.10.) The 50 points in the sample are indicated with hash marks.

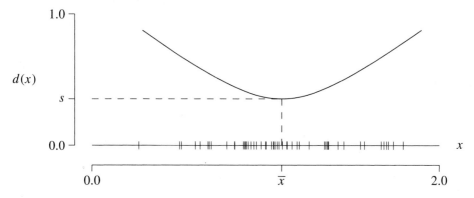

Figure 4.1.1   Sample mean minimizes rms.

For this sample, $\overline{x} \cong 1.095$ and $s \cong 0.354$. Per Theorem 4.1.1, the smallest value of $d(x)$ is $d(\overline{x}) = s$, as illustrated.

The portion of the figure in Figure 4.1.1 that looks like

is known as a univariate *scatter diagram*—a convenient way to visualize a sample, provided the sample size is not too large. Later in this chapter we will discuss *histograms*—a superior way to visualize a univariate sample if the sample size is sufficiently large.

Chebyshev's Inequality.   To better understand how the sample mean and standard deviation are related, appreciate that the number of points in a sample that lie within k standard deviations of the mean can be bounded by Chebyshev's inequality. The derivation of this fundamental result is based on a simple observation—the points that make the largest contribution to the sample standard deviation are those that are most distant from the sample mean.

Given the sample $\mathcal{S} = \{x_1, x_2, \ldots, x_n\}$ with mean $\overline{x}$ and standard deviation s, and given a parameter $k > 1$, define the set*

$$\mathcal{S}_k = \{x_i \mid \overline{x} - ks < x_i < \overline{x} + ks\}$$

---

*Some values in the sample may occur more than once. Therefore, $\mathcal{S}$ is actually a *multiset*. For the purposes of counting, however, the elements of $\mathcal{S}$ are treated as distinguishable, so that the cardinality (size) of $\mathcal{S}$ is $|\mathcal{S}| = n$.

consisting of all those $x_i \in S$ for which $|x_i - \overline{x}| < ks$. For $k = 2$ and the sample in Example 4.1.1, the set S_2 is defined by the hash marks that lie within the interval of width $2ks = 4s$ centered on $\overline{x}$, as shown by the portion of Figure 4.1.1 given below.

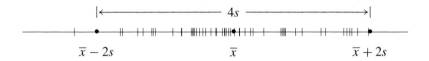

$$\overline{x} - 2s \qquad\qquad\qquad \overline{x} \qquad\qquad\qquad \overline{x} + 2s$$

Let $p_k = |S_k|/n$ be the *proportion* of x_i that lie within $\pm ks$ of $\overline{x}$. This proportion is the probability that an x_i selected at random from the sample will be in S_k. From the definition of s^2, we have that

$$ns^2 = \sum_{i=1}^{n}(x_i - \overline{x})^2 = \sum_{x_i \in S_k}(x_i - \overline{x})^2 + \sum_{x_i \in \overline{S}_k}(x_i - \overline{x})^2,$$

where the set $\overline{S}_k$ is the complement of S_k. If the contribution to s^2 from all the points close to the sample mean (the points in S_k) is ignored, then, because the contribution to s^2 from each point in $\overline{S}_k$ (the points far from the sample mean) is at least $(ks)^2$, the previous equation becomes the following inequality

$$ns^2 \geq \sum_{x_i \in \overline{S}_k}(x_i - \overline{x})^2 \geq \sum_{x_i \in \overline{S}_k}(ks)^2 = |\overline{S}_k|(ks)^2 = n(1 - p_k)k^2 s^2.$$

If ns^2 is eliminated, the result is *Chebyshev's inequality*:

$$p_k \geq 1 - \frac{1}{k^2} \qquad\qquad (k > 1).$$

(This inequality says *nothing* for $k \leq 1$.) From this inequality, it follows that, in particular, $p_2 \geq 0.75$. That is, *for any sample*, at least 75% of the points in the sample must lie within $\pm 2s$ of the sample mean.

For the data illustrated by the scatter diagram, the true percentage of points within $\pm 2s$ of the sample mean is 98%—that is, for the sample in Example 4.1.1 (as for most samples), the 75% in the $k = 2$ form of Chebyshev's inequality is very conservative. Indeed, it is common to find approximately 95% of the points in a sample within $\pm 2s$ of the sample mean. (See Exercise 4.1.7.)

The primary issue here is not the accuracy (or lack thereof) of Chebyshev's inequality. Instead, Chebyshev's inequality and practical experience with actual data suggest that the $\overline{x} \pm 2s$ interval defines the "effective width" of a sample.* As a rule of thumb, most but not all of the points in

---

*There is nothing magical about the value $k = 2$. Other values like $k = 2.5$ and $k = 3$ are sometimes offered as alternatives to define the $\overline{x} \pm ks$ interval. To the extent that there is a standard, however, $k = 2$ is it. This issue will be revisited in Chapter 8.

a sample will fall within this interval. With this in mind, when analyzing experimental data it is common to look for *outliers*—values so far from the sample mean, for example $\pm 3s$ or more, that they must be viewed with suspicion.

Linear Data Transformations. It is sometimes the case that the output data generated by a discrete-event simulation, for example the wait times in a single-server service node, are statistically analyzed in one system of units (say seconds), but later there is a need to convert to a different system of units (say minutes). Usually this conversion of units is a *linear* data transformation and, if so, the change in system statistics can be computed directly, without any need to reprocess the converted data.

For example, if $x_i' = ax_i + b$ for $i = 1, 2, \ldots, n$, then how do the sample mean $\bar{x}$ and standard deviation s of the x-data relate to the sample mean $\bar{x}'$ and standard deviation s' of the x'-data? The answer is that

$$\bar{x}' = \frac{1}{n} \sum_{i=1}^{n} x_i' = \frac{1}{n} \sum_{i=1}^{n} (ax_i + b) = \frac{a}{n} \left(\sum_{i=1}^{n} x_i \right) + b = a\bar{x} + b$$

and

$$(s')^2 = \frac{1}{n} \sum_{i=1}^{n} (x_i' - \bar{x}')^2 = \frac{1}{n} \sum_{i=1}^{n} (ax_i + b - a\bar{x} - b)^2 = \frac{a^2}{n} \sum_{i=1}^{n} (x_i - \bar{x})^2 = a^2 s^2.$$

Therefore

$$\bar{x}' = a\bar{x} + b \qquad \text{and} \qquad s' = |a|s.$$

Example 4.1.2

Suppose the sample $x_1, x_2, \ldots, x_n$ is measured in seconds. To convert to minutes, the transformation is $x_i' = x_i/60$, so that if, say, $\bar{x} = 45$ (seconds) and $s = 15$ (seconds), then

$$\bar{x}' = \frac{45}{60} = 0.75 \text{ (minutes)} \qquad \text{and} \qquad s' = \frac{15}{60} = 0.25 \text{ (minutes)}.$$

Example 4.1.3

Some people, particularly those used to analyzing so-called "normally distributed" data, like to *standardize* the data by subtracting the sample mean and dividing the result by the sample standard deviation. That is, given a sample $x_1, x_2, \ldots, x_n$ with sample mean $\bar{x}$ and standard deviation s, the corresponding standardized sample is computed as

$$x_i' = \frac{x_i - \bar{x}}{s} \qquad i = 1, 2, \ldots, n,$$

for $n > 1$. It follows that the sample mean of the standardized sample is $\bar{x}' = 0$ and its sample standard deviation is $s' = 1$. Standardization is commonly used to avoid potential numerical problems associated with samples containing very large or very small values.

Nonlinear Data Transformations.   Although nonlinear data transformations are difficult to analyze in general, there are times when data is used to create a Boolean (two-state) outcome. In such cases, the *value* of x_i is not as important as the *effect*—does x_i cause something to occur, or not? In particular, if $\mathcal{A}$ is a fixed set, and if, for $i = 1, 2, \ldots, n$, we define the nonlinear data transformation

$$x_i' = \begin{cases} 1 & x_i \in \mathcal{A} \\ 0 & \text{otherwise,} \end{cases}$$

then what are $\bar{x}'$ and s'? To answer this question, let the proportion of x_i that fall in $\mathcal{A}$ be

$$p = \frac{\text{the number of } x_i \text{ in } \mathcal{A}}{n};$$

then

$$\bar{x}' = \frac{1}{n} \sum_{i=1}^{n} x_i' = p$$

and

$$(s')^2 = \frac{1}{n} \sum_{i=1}^{n} (x_i' - p)^2 = \cdots = (1 - p)(-p)^2 + p(1 - p)^2 = p(1 - p).$$

Therefore,

$$\bar{x}' = p \qquad \text{and} \qquad s' = \sqrt{p(1 - p)}.$$

Example 4.1.4
For a single-server service node, let $x_i = d_i$ be the delay experienced by the ith job, and let $\mathcal{A}$ be the set of positive real numbers. Then x_i' is 1 if and only if $d_i > 0$, and p is the proportion of jobs (out of n) that experience a nonzero delay. If, as in Exercise 1.2.3, $p = 0.723$, then $\bar{x}' = 0.723$, and $s' = \sqrt{(0.723)(0.277)} = 0.448$.

Example 4.1.5
As illustrated in Section 2.3, a Monte Carlo simulation used to estimate a probability ultimately involves the generation of a sequence of 0's and 1's, with the probability estimate p being the ratio of the number of 1's to the number of trials. Thus p is the sample mean of this sequence of 0's and 1's and $\sqrt{p(1 - p)}$ is the sample standard deviation.

4.1.2 Computational Considerations

One significant computational drawback to calculating the sample standard deviation via the equation

$$s = \sqrt{\frac{1}{n} \sum_{i=1}^{n} (x_i - \bar{x})^2}$$

is that it requires a *two-pass* algorithm: The sample must be scanned once to calculate the sample mean by accumulating the $\sum x_i$ partial sums and then scanned a second time to calculate the sample standard deviation by accumulating the $\sum (x_i - \overline{x})^2$ partial sums. This two-pass approach is usually undesirable in discrete-event simulation because it creates a need to temporarily store or re-create the *entire* sample, which is undesirable when n is large.

Conventional One-Pass Algorithm.   There is an alternate, mathematically equivalent, equation for s^2 (and thus s) that can be implemented as a *one-pass* algorithm. The derivation of this equation is as follows:

$$s^2 = \frac{1}{n} \sum_{i=1}^{n} (x_i - \overline{x})^2$$

$$= \frac{1}{n} \sum_{i=1}^{n} (x_i^2 - 2\overline{x}x_i + \overline{x}^2)$$

$$= \left(\frac{1}{n} \sum_{i=1}^{n} x_i^2 \right) - \left(\frac{2}{n}\overline{x} \sum_{i=1}^{n} x_i \right) + \left(\frac{1}{n} \sum_{i=1}^{n} \overline{x}^2 \right)$$

$$= \left(\frac{1}{n} \sum_{i=1}^{n} x_i^2 \right) - 2\overline{x}^2 + \overline{x}^2$$

$$= \left(\frac{1}{n} \sum_{i=1}^{n} x_i^2 \right) - \overline{x}^2.$$

If this alternate equation is used, the sample mean and standard deviation can be calculated in one pass through the sample by accumulating the $\sum x_i$ and $\sum x_i^2$ partial sums, thereby eliminating the need to store the sample. As each new observation arises in a Monte Carlo or discrete-event simulation, only three memory locations are needed in order to store $\sum x_i$, $\sum x_i^2$, and the number of observations to date. There is, however, a potential problem with this one-pass approach—floating-point round-off error. To avoid overflow, the sums are typically accumulated with floating-point arithmetic, even if the data is integer-valued. Floating-point arithmetic is problematic in this case, because the sample variance is ultimately calculated as the difference of two quantities that could be *very* large relative to their difference—that is, if the true value of s is tiny relative to $|\overline{x}|$, then accumulated floating-point round-off error could cause an incorrect value for s.

Fortunately, there is an alternative one-pass algorithm (due to B. P. Welford in 1962) that can be used to calculate the sample mean and standard deviation. This algorithm is superior to the conventional one-pass algorithm in that it is much less prone to significant floating-point round-off error. The original algorithm is given in Welford (1962). Chan, Golub, and LeVeque (1983) surveyed other alternative algorithms. Welford's algorithm is based upon the following definition and associated theorem.

Welford's One-Pass Algorithm

Definition 4.1.2 Given the sample $x_1, x_2, x_3, \ldots$, for $i = 1, 2, 3, \ldots$, define the running sample mean and the running sample sum of squared deviations as

$$\overline{x}_i = \frac{1}{i}(x_1 + x_2 + \cdots + x_i)$$

$$v_i = (x_1 - \overline{x}_i)^2 + (x_2 - \overline{x}_i)^2 + \cdots + (x_i - \overline{x}_i)^2,$$

where $\overline{x}_i$ and v_i / i are the sample mean and variance, respectively, of $x_1, x_2, \ldots, x_i$.

Theorem 4.1.2 For $i = 1, 2, 3, \ldots$, the variables $\overline{x}_i$ and v_i in Definition 4.1.2 can be computed recursively by

$$\overline{x}_i = \overline{x}_{i-1} + \frac{1}{i}(x_i - \overline{x}_{i-1})$$

$$v_i = v_{i-1} + \left(\frac{i-1}{i}\right)(x_i - \overline{x}_{i-1})^2$$

with the initial conditions $\overline{x}_0 = 0$ and $v_0 = 0$.

Proof. Both equations in this theorem are valid by inspection if $i = 1$. If $i > 1$, then we can write

$$i\overline{x}_i = x_1 + x_2 + \cdots + x_i$$
$$= (x_1 + x_2 + \cdots + x_{i-1}) + x_i$$
$$= (i - 1)\overline{x}_{i-1} + x_i$$
$$= i\overline{x}_{i-1} + (x_i - \overline{x}_{i-1}),$$

from which the first equation follows via division by i. This establishes the first equation in the theorem. As in the prior discussion of the conventional one-pass algorithm, an alternative equation for v_i is

$$v_i = x_1^2 + x_2^2 + \cdots + x_i^2 - i\overline{x}_i^2 \qquad i = 1, 2, 3, \ldots.$$

From this alternative equation and the first equation in the theorem, if $i > 1$, we can write

$$v_i = (x_1^2 + x_2^2 + \cdots + x_{i-1}^2) + x_i^2 - i\overline{x}_i^2$$
$$= v_{i-1} + (i - 1)\overline{x}_{i-1}^2 + x_i^2 - i\overline{x}_i^2$$
$$= v_{i-1} - i(\overline{x}_i^2 - \overline{x}_{i-1}^2) + (x_i^2 - \overline{x}_{i-1}^2)$$

$$= v_{i-1} - i(\overline{x}_i - \overline{x}_{i-1})(\overline{x}_i + \overline{x}_{i-1}) + (x_i - \overline{x}_{i-1})(x_i + \overline{x}_{i-1})$$

$$= v_{i-1} - (x_i - \overline{x}_{i-1})(\overline{x}_i + \overline{x}_{i-1}) + (x_i - \overline{x}_{i-1})(x_i + \overline{x}_{i-1})$$

$$= v_{i-1} + (x_i - \overline{x}_{i-1})(x_i - \overline{x}_i)$$

$$= v_{i-1} + (x_i - \overline{x}_{i-1})\left(x_i - \overline{x}_{i-1} - \frac{1}{i}(x_i - \overline{x}_{i-1})\right)$$

$$= v_{i-1} + (x_i - \overline{x}_{i-1})^2 - \frac{1}{i}(x_i - \overline{x}_{i-1})^2,$$

which establishes the second equation in the theorem. $\square$

Algorithm 4.1.1 Given the sample $x_1, x_2, x_3, \ldots$, Welford's algorithm for calculating the sample mean $\overline{x}$ and standard deviation s is

```
n = 0;
x = 0.0;
v = 0.0;
while ( more data ) {
    x = GetData();
    n++;
    d = x - x̄;                          // temporary variable
    v = v + d * d * (n - 1) / n;
    x̄ = x̄ + d / n;
}
s = sqrt(v / n);
return n, x̄, s;
```

Program uvs. Algorithm 4.1.1 is a one-pass $O(n)$ algorithm that does not require prior knowledge of the sample size n. The algorithm is not as prone to accumulated floating-point round-off error as is the conventional one-pass algorithm, yet it is essentially as efficient. Program uvs at the end of this section is based on a robust version of Algorithm 4.1.1 designed to avoid failing if applied, by mistake, to an empty sample. This program computes the sample mean and standard deviation as well as the sample minimum and maximum.

Input to program uvs can be integer-valued or real-valued. The input is assumed to be a *text* data file, in a one-value-per-line format with no blank lines in the file. (The file uvs.dat is an example.) To use the program, compile uvs.c to produce the executable file uvs; then, at a command-line prompt, uvs can be used in three ways:

- To have uvs read a disk data file, say uvs.dat, at a command-line prompt, use '<' redirection:

  uvs < uvs.dat

- To have uvs filter the numerical output of a program, say test, at a command-line prompt, use a '|' pipe:

test | uvs

- To use uvs with keyboard input, at a command-line prompt, enter

uvs

  then enter the data, one value per line, being sure to signify an end-of-file as the last line of input [^d (Ctrl-d) in Unix or ^z (Ctrl-z) in Microsoft Windows].

4.1.3 Examples

The two examples given here illustrate the effect of independent and dependent sampling on the sample statistics $\bar{x}$ and s.

Example 4.1.6
If a *Uniform*(a, b) random-variate sample $x_1, x_2, \ldots, x_n$ is generated, then, in the limit as $n \to \infty$, the sample mean and sample standard deviation will converge to the limits indicated:

$$\bar{x} \to \frac{a + b}{2} \qquad \text{and} \qquad s \to \frac{b - a}{\sqrt{12}}.$$

(See Chapter 7 for more details concerning *Uniform*(a, b) random variates.)

Welford's algorithm was used in conjunction with a random-variate sample generated by calls to Random (drawn from stream 0 with an rngs initial seed of 12345). Since the random variates so generated are independent *Uniform*$(0, 1)$, then, as the sample size increases, the sample mean and standard deviation should converge to

$$\frac{0 + 1}{2} = 0.5 \qquad \text{and} \qquad \frac{1 - 0}{\sqrt{12}} \cong 0.2887.$$

In the upper display of Figure 4.1.2, the sample values are indicated with $\circ$'s. The $\bullet$'s show the running values of $\bar{x}$ and indicate the convergence of the corresponding sample mean $\bar{x}$ to 0.5 (the solid, horizontal line). As expected, the $\circ$ and the $\bullet$ overlap when $n = 1$. The plot of $\bar{x}$ may remind students of grade point average (GPA) calculation, which can be influenced significantly during the freshman and sophomore years, but remains fairly constant during the junior and senior years. Similarly, in the lower display in Figure 4.1.2, the $\bullet$'s show the running values of s and indicate the convergence of the sample standard deviation $s = \sqrt{v_n/n}$ to 0.2887 (the solid, horizontal line). As discussed earlier in this section, the sample standard deviation is 0 when $n = 1$. Consistent with the discussion in Section 2.3, the convergence of $\bar{x}$ and s to their theoretical values is not necessarily monotone with increasing n. The final values plotted on each display corresponding to $n = 100$ are $\bar{x} = 0.515$ and $s = 0.292$.

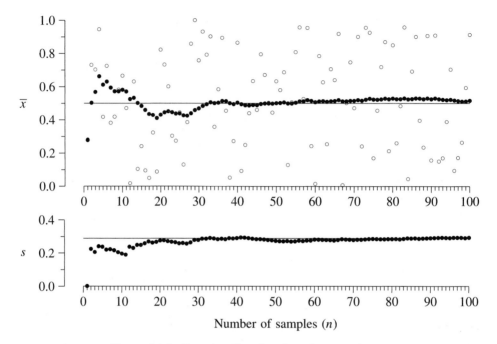

Figure 4.1.2   Running $\bar{x}$ and s of random numbers.

Analogous to Example 4.1.6, if an *Equilikely* (a, b) random-variate sample is generated, then, in the limit as $n \to \infty$, the mean and standard deviation of the sample will converge to the limits indicated:

$$\bar{x} \to \frac{a+b}{2} \qquad \text{and} \qquad s \to \sqrt{\frac{(b-a+1)^2 - 1}{12}}.$$

Similarly, if an *Exponential* (μ) random-variate sample is generated, then

$$\bar{x} \to \mu \qquad \text{and} \qquad s \to \mu$$

as $n \to \infty$; and, for a *Geometric* (p) random-variate sample,

$$\bar{x} \to \frac{p}{1-p} \qquad \text{and} \qquad s \to \frac{\sqrt{p}}{1-p}.$$

(Chapters 6 and 7 contain derivations of many of these results.)

Serial Correlation.   The random variates generated in Example 4.1.6 were *independent* sample values. Informally, independence means that each x_i in the sample $x_1, x_2, \ldots, x_n$ has a value that does not depend in any way on any other point in the sample. When an independent sample is displayed, as in Example 4.1.6, there will not be any deterministic "pattern" or "trend" to the data.

Because repeated calls to Random produce a simulated independent sample, there is no deterministic pattern or trend to the sample in Example 4.1.6. If, however, we were to display the time-sequenced output from a discrete-event simulation, then that sample typically will not be independent. For example, if we display a sample of random waits experienced by consecutive jobs passing through a single-server service node, then a pattern will be evident, because the sample is *not* independent, particularly if the utilization is large: If the job i experiences a large wait, then job $(i + 1)$ is likely to also experience a large wait; an analogous statement is true if the job i experiences a small wait. In statistical terms, the wait times of consecutive jobs have positive *serial correlation* (see Section 4.4 for more details), and, because of this, the wait times in the sample will be dependent (i.e., not independent).

Hence, independence is typically an appropriate assumption for Monte Carlo simulation, as illustrated in Example 4.1.6. It is typically *not* an appropriate assumption for discrete-event simulation, however, as illustrated in the following example.

Example 4.1.7
Program ssq2, with *Exponential* (2) interarrival times and *Uniform* (1, 2) service times, was modified to output the waits $w_1, w_2, \ldots, w_n$ experienced by the first $n = 100$ jobs. To simulate starting the service node in an (approximate) steady state, the initial value of the program variable departure was set to 3.0. The rng initial seed was 12345. For this single-server service node, it can be shown (see Section 8.5) that, as $n \to \infty$, the sample mean and standard deviation of the wait times should converge to 3.83 and 2.83 respectively, as indicated by the solid horizontal lines in Figure 4.1.3.

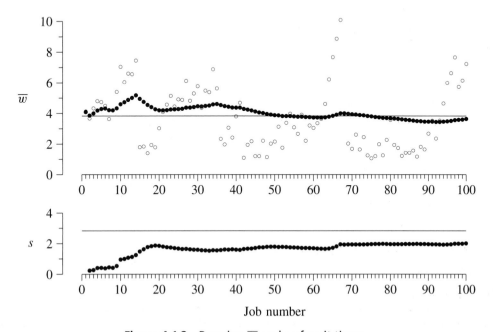

Figure 4.1.3 Running $\overline{w}$ and s of wait times.

As in Example 4.1.6, the sample values are indicated in the upper figure with ∘'s, and the •'s indicate the (unbiased) convergence of the corresponding sample mean to its expected steady-state value.* Unlike the upper figure in Example 4.1.6, this sample reveals a lack-of-independence pattern caused by high positive serial correlation in the sample wait times $w_1, w_2, \ldots, w_n$. A sequence of longer-than-average waits is followed by a sequence of shorter-than-average waits. This pattern is quite different from that in Figure 4.1.2, where each observation is above or below the mean $(1/2)$, as in a sequence of tosses of a fair coin. The running average wait does, however, converge to the theoretical value. Moreover, as illustrated in the lower figure, the high positive serial correlation produces a pronounced *bias* in the sample standard deviation: It consistently underestimates the expected value. In this case, a profound modification to the sample-variance equation (and thus the sample-standard-deviation equation) is required to remove the bias; just replacing $1/n$ with $1/(n-1)$ will not suffice. (See Appendix F.)

4.1.4 Time-Averaged Sample Statistics

As discussed in previous chapters, time-averaged statistics play an important role in discrete-event simulation. The following definition is the time-averaged analog of Definition 4.1.1.

Definition 4.1.3 Given a function $x(t)$ defined for all $0 < t < \tau$ as a realization (sample path) of a stochastic process, the *sample-path mean* is

$$\overline{x} = \frac{1}{\tau} \int_0^\tau x(t)\, dt.$$

The associated *sample-path variance* is

$$s^2 = \frac{1}{\tau} \int_0^\tau (x(t) - \overline{x})^2\, dt$$

and the *sample-path standard deviation* is $s = \sqrt{s^2}$.

The equation for s^2 in Definition 4.1.3 is the two-pass variance equation; the corresponding one-pass equation is

$$s^2 = \left(\frac{1}{\tau} \int_0^\tau x^2(t)\, dt \right) - \overline{x}^2.$$

Any time-averaged statistic changes values only at the distinct event times in a discrete-event simulation, so only three memory locations are needed in order to store the running statistics $\int x(t)\, dt$, $\int x^2(t)\, dt$, and the length of the observation period to date.

---

*If the initial state of the service node had been idle, rather than adjusted to simulate starting the service node in an (approximate) steady state, then there would have been some *initial state bias* in the sample means.

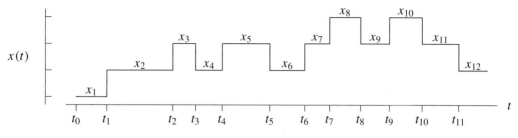

Figure 4.1.4 Generic sample path.

For a discrete-event simulation, a sample path is naturally *piecewise constant*. That is, as illustrated in Figure 4.1.4, there are *event times* $0 = t_0 < t_1 < t_2 < \cdots < t_n = \tau$ and associated state values $x_1, x_2, \ldots, x_n$ on the time intervals $(t_0, t_1], (t_1, t_2], \ldots, (t_{n-1}, t_n]$. The choice of making the left endpoint of these intervals open and the right endpoint of these intervals closed is arbitrary.

If a sample path is piecewise constant, then the sample-path *integral* equations for $\bar{x}$ and s^2 in Definition 4.1.3 can be reduced to *summation* equations, as summarized by Theorem 4.1.3 that follows. The *defining formulas* given in Definition 4.1.3 are useful conceptually, but Theorem 4.1.3 gives the *computational formulas* that are used to compute $\bar{x}$ and s^2 in practice.

Theorem 4.1.3 Given a sample path represented by the piecewise-constant function

$$
x(t) = \begin{cases} x_1 & t_0 < t \le t_1 \\ x_2 & t_1 < t \le t_2 \\ \vdots & \vdots \\ x_n & t_{n-1} < t \le t_n \end{cases}
$$

and the inter-event times $\delta_i = t_i - t_{i-1}$ for $i = 1, 2, \ldots, n$, then, with $t_0 = 0$ and $\tau = t_n$, the sample-path mean is

$$
\bar{x} = \frac{1}{\tau} \int_0^\tau x(t)\, dt = \frac{1}{t_n} \sum_{i=1}^n x_i\, \delta_i
$$

and the sample-path variance is

$$
s^2 = \frac{1}{\tau} \int_0^\tau \left(x(t) - \bar{x} \right)^2 dt = \frac{1}{t_n} \sum_{i=1}^n (x_i - \bar{x})^2 \delta_i = \left(\frac{1}{t_n} \sum_{i=1}^n x_i^2 \delta_i \right) - \bar{x}^2.
$$

Welford's Sample-Path Algorithm. The two-pass and one-pass algorithms are susceptible to floating-point round-off error, as described earlier. Welford's algorithm can be extended to the time-averaged case by defining running sample-path means and running sample-path sums of squared deviations, analogously to the earlier case.

Definition 4.1.4   With reference to Theorem 4.1.3, for $i = 1, 2, \ldots, n$, the sample-path variant of Welford's algorithm is based on the definitions

$$\bar{x}_i = \frac{1}{t_i}(x_1\delta_1 + x_2\delta_2 + \cdots + x_i\delta_i)$$

$$v_i = (x_1 - \bar{x}_i)^2\delta_1 + (x_2 - \bar{x}_i)^2\delta_2 + \cdots + (x_i - \bar{x}_i)^2\delta_i,$$

where $\bar{x}_i$ and v_i/t_i are the sample-path mean and variance, respectively, of the sample path $x(t)$ for $t_0 \leq t \leq t_i$.

Theorem 4.1.4   For $i = 1, 2, \ldots, n$, the variables $\bar{x}_i$ and v_i in Definition 4.1.4 can be computed recursively by

$$\bar{x}_i = \bar{x}_{i-1} + \frac{\delta_i}{t_i}(x_i - \bar{x}_{i-1})$$

$$v_i = v_{i-1} + \frac{\delta_i t_{i-1}}{t_i}(x_i - \bar{x}_{i-1})^2,$$

with the initial conditions $\bar{x}_0 = 0$ and $v_0 = 0$.

4.1.5 Exercises

4.1.1   Prove Theorem 4.1.1 by first proving that $d^2(x) = (x - \bar{x})^2 + s^2$.

4.1.2   Prove Theorem 4.1.1 for the $1/(n-1)$ form of the sample variance.

4.1.3   Relative to Theorem 4.1.2, prove that the sequence $v_1, v_2, \ldots, v_n$ satisfies the inequality $0 = v_1 \leq v_2 \leq \cdots \leq v_{n-1} \leq v_n = ns^2$.

4.1.4   What common sample statistic best fits the sample $x_1, x_2, \ldots, x_n$ in the sense of minimizing

$$d(x) = \frac{1}{n}\sum_{i=1}^{n}|x_i - x|?$$

4.1.5   The statistic q^3 is the so-called sample *skewness*, defined by

$$q^3 = \frac{1}{n}\sum_{i=1}^{n}\left(\frac{x_i - \bar{x}}{s}\right)^3.$$

(*a*) What is the one-pass version of the equation for q^3? (*b*) Extend the conventional (non-Welford) one-pass algorithm to also compute the sample statistic $q = (q^3)^{1/3}$. (*c*) What is the value of q for the data in the file uvs.dat?

4.1.6   Look up the article by Chan, Golub, and LeVeque (1983) in the bibliography. Give a survey of existing algorithms for computing the sample variance, listing the pros and cons of each method.

4.1.7   (a) Generate an *Exponential* (9) random-variate sample of size $n = 100$, and compute the proportion of points in the sample that fall within the intervals $\bar{x} \pm 2s$ and $\bar{x} \pm 3s$. Do this for ten different rngs streams. (b) In each case, compare the results with Chebyshev's inequality. (c) Comment.

4.1.8   Generate a plot similar to that in Figure 4.1.2, but with calls to *Exponential* (17) rather than to Random to generate the variates. State the values to which the sample mean and sample standard deviation will converge.

4.1.9   Prove Theorems 4.1.3 and 4.1.4.

4.1.10 (a) Given $x_1, x_2, \ldots, x_n$ with $a \le x_i \le b$ for $i = 1, 2, \ldots, n$, what are the largest and smallest possible values of $\bar{x}$? (b) Same question for s.

4.1.11 Calculate $\bar{x}$ and s by hand, using the two-pass algorithm, the one-pass algorithm, and Welford's algorithm in the following two cases. (a) Data based on $n = 3$ observations: $x_1 = 1$, $x_2 = 6$, and $x_3 = 2$. (b) The sample path $x(t) = 3$ for $0 < t \le 2$, and $x(t) = 8$ for $2 < t \le 5$, over the time interval $0 < t \le 5$.

4.1.12 The extent to which accumulated round-off error is a potential problem in any calculation involving floating-point arithmetic is determined, in part, by what is known as the computational system's "floating-point precision." One way to find out this precision is by executing the following code fragment.

```
typedef float FP;                           // float or double
const FP   ONE   = 1.0;
const FP   TWO   = 2.0;
       FP   tiny  = ONE;
       int count = 0;
while (ONE + tiny != ONE) {
   tiny  /= TWO;
   count++;
}
```

(a) Experiment with this code fragment on at least three systems, being sure to use both float and double data types for the generic floating-point type FP. (b) Explain what this code fragment does, why the while loop terminates, and what the significance of the variables tiny and count is. Be explicit.

4.2 DISCRETE-DATA HISTOGRAMS

Given a univariate sample $\mathcal{S} = \{x_1, x_2, \ldots, x_n\}$, if the sample size is sufficiently large (say, $n > 50$ or so), then, generally, it is meaningful to do more than just calculate the sample mean and standard deviation. In particular, the sample can be processed to form a *histogram* and thereby provide insight into the distribution of the data. There are two cases to consider: discrete data and continuous data. In this section we deal with the easier case, discrete data; continuous data is considered in the next section.

4.2.1 Discrete-Data Histograms

Definition 4.2.1   Given the discrete-data sample (multiset) $S = \{x_1, x_2, \ldots, x_n\}$, let $\mathcal{X}$ be the set of possible distinct values in S. For each $x \in \mathcal{X}$, the *relative frequency* (*proportion*) is

$$\hat{f}(x) = \frac{\text{the number of } x_i \in S \text{ for which } x_i = x}{n}.$$

A *discrete-data histogram* is a graphical display of $\hat{f}(x)$ versus x.*

The point here is that, if the sample size $n = |S|$ is large and the data is discrete, then it is reasonable to expect that the number of distinct values in the sample will be much smaller than the sample size—values will appear multiple times. A discrete-data histogram is nothing more than a graphical display of the relative frequency with which each distinct value in the sample appears.

Example 4.2.1
As an example of a discrete-data histogram, a modified version of program `galileo` was used to replicate $n = 1000$ rolls of three fair dice (with an `rng` initial seed of 12345) and thereby create a discrete-data sample $S = \{x_1, x_2, \ldots, x_n\}$ with each x_i an integer between 3 and 18 inclusive. (Each x_i is the sum of the three up faces.) Therefore, $\mathcal{X} = \{3, 4, \ldots, 18\}$. The resulting relative frequencies (probability estimates) are plotted as the discrete-data histogram on the left side of Figure 4.1.2. Illustrated on the right is the same histogram with the theoretical probabilities associated with the sum-of-three-dice experiment superimposed as •'s. Each theoretical probability represents the limit to which the corresponding relative frequency will converge as $n \to \infty$.

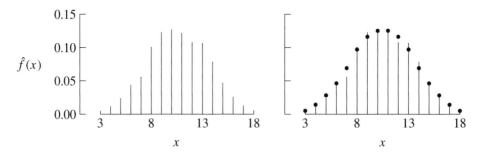

Figure 4.2.1   Histograms for program `galileo`.

No matter how many replications are used in Example 4.2.1, we know a priori that $\mathcal{X} = \{3, 4, \ldots, 18\}$. Because of that, it is natural to use an array to accumulate the relative frequencies,

---

*The reason for the $\hat{f}(x)$ "hat" notation will be explained in Chapter 9.

as in program `galileo`. An array can also be used to accumulate the relative frequencies in the next example. In this case, however, the use of an array is less natural because, although we know that $\mathcal{X} = \{0, 1, 2, \ldots, b\}$ for some value of b, a reasonable value for b, valid for *any* value of n, is not easily established a priori. If an array is not appropriate, a more flexible data structure based on dynamic memory allocation is required to tally relative frequencies. As an illustration, a linked-list implementation of a discrete-data histogram algorithm is presented later in this section.

Example 4.2.2

Suppose that $2n = 2000$ balls are placed *at random* into $n = 1000$ boxes. That is, for each ball, a box is selected at random and the ball is placed in it. The Monte Carlo simulation algorithm given below can be used to generate a random sample $S = \{x_1, x_2, \ldots, x_n\}$, where, for each i, x_i is the number of balls placed in box i.

```
n = 1000;
for (i = 1; i <= n; i++)              // i counts boxes
    x_i = 0;
for (j = 0; j < 2 * n; j++) {         // j counts balls
    i = Equilikely(1, n);             // pick a box at random
    x_i++;                            // then put a ball in it
}
return x_1, x_2, ..., x_n;
```

When 2000 balls are placed into 1000 boxes in this way, then each box will have exactly two balls in it on average; however, some boxes will be empty, some will have just one ball, some will have two balls, some will have three balls, and so on, as illustrated in Figure 4.2.2.

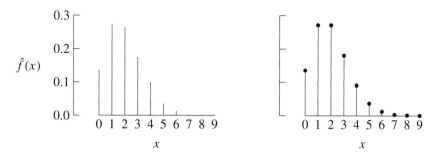

Figure 4.2.2 Histograms of number of balls in boxes.

In this case, for an `rng` initial seed of 12345, $\mathcal{X} = \{0, 1, 2, \ldots, 9\}$. As in Example 4.2.1, the discrete-data histogram is on the left and the figure on the right is the same histogram with theoretical probabilities superimposed as •'s. Each theoretical probability represents the limit to which the corresponding relative frequency will converge as $n \to \infty$. [The relative frequencies $\hat{f}(7) = 0.002$ and $\hat{f}(9) = 0.001$ are too small to be visible in the histogram and, because no boxes ended up with 8 balls, $\hat{f}(8) = 0.000$.]

Note the asymmetric shape of the histogram in Example 4.2.2. This asymmetry is frequently associated with nonnegative data when the mean is small. As an exercise, you should repeat this example in the situation where there are 1000 boxes and 10 000 balls. In that case, the shape of the histogram will change to a distribution symmetric about the expected number of balls in each box, which is 10. The reason for this shift to a symmetric distribution will be discussed in Chapter 6.

Histogram Mean and Standard Deviation

Definition 4.2.2 Given the relative frequencies from Definition 4.2.1, the *discrete-data histogram mean* is

$$\overline{x} = \sum_x x \hat{f}(x)$$

and the associated *discrete-data histogram standard deviation* is

$$s = \sqrt{\sum_x (x - \overline{x})^2 \hat{f}(x)}$$

where the sum is over all $x \in \mathcal{X}$. The discrete-data histogram variance is s^2.

Consistent with their interpretation as probability estimates, the relative frequencies from Definition 4.2.1 are defined so that $\hat{f}(x) \geq 0$ for all $x \in \mathcal{X}$ and

$$\sum_x \hat{f}(x) = 1.$$

Moreover, it follows from the definition of $\mathcal{S}$ and $\mathcal{X}$ that

$$\sum_{i=1}^n x_i = \sum_x x n \hat{f}(x) \qquad \text{and} \qquad \sum_{i=1}^n (x_i - \overline{x})^2 = \sum_x (x - \overline{x})^2 n \hat{f}(x).$$

(In both equations, the two summations compute the same thing, but in a different order.) Therefore, from the equations in Definition 4.1.1 and Definition 4.2.2, it follows that the *sample* mean and standard deviation are mathematically equivalent to the *discrete-data histogram* mean and standard deviation, respectively.

Provided that the relative frequencies have already been computed, it is generally preferable to compute $\overline{x}$ and s via the discrete-data-histogram equations. Moreover, the histogram-based equations have great theoretical significance, in that they provide the motivation for defining the mean and standard deviation of discrete random variables—see Chapter 6.

The equation for s in Definition 4.2.2 is the two-pass version of the standard deviation equation. The mathematically equivalent one-pass version of this equation is

$$s = \sqrt{\left(\sum_x x^2 \hat{f}(x)\right) - \bar{x}^2},$$

where the summation is over all $x \in \mathcal{X}$.

Example 4.2.3
For the data in Example 4.2.1,

$$\bar{x} = \sum_{x=3}^{18} x\, \hat{f}(x) \cong 10.609 \qquad \text{and} \qquad s = \sqrt{\sum_{x=3}^{18}(x - \bar{x})^2 \hat{f}(x)} \cong 2.925.$$

Similarly, for the data in Example 4.2.2,

$$\bar{x} = \sum_{x=0}^{9} x\, \hat{f}(x) = 2.0 \qquad \text{and} \qquad s = \sqrt{\sum_{x=0}^{9}(x - \bar{x})^2 \hat{f}(x)} \cong 1.419.$$

4.2.2 Computational Considerations

As illustrated in Examples 4.2.1 and 4.2.2, in many simulation applications, the discrete data will, in fact, be *integer*-valued. For integer-valued data, the usual way to tally the discrete-data histogram is to use an array. Recognize, however, that the use of an array data structure for the histogram involves the allocation of memory with an associated requirement to know the range of data values—that is, when memory is allocated, it is necessary to assume that $\mathcal{X} = \{a, a + 1, a + 2, \ldots, b\}$ for reasonable integer values of $a < b$. Given a knowledge of a, b, the following one-pass $O(n)$ algorithm is the preferred way to compute a discrete-data histogram for integer-valued data.

Algorithm 4.2.1   Given integers a, b with $a < b$ and integer-valued data $x_1, x_2, \ldots$, the following algorithm computes a discrete-data histogram.

```
long count [b - a + 1];
n = 0;
for (x = a; x <= b; x++)
    count [x - a] = 0;
outliers.lo = 0;
outliers.hi = 0;
while ( more data ) {
```

```
   x = GetData();
   n++;
   if ((a <= x) and (x <= b))
     count[x - a]++;
   else if (a > x)
     outliers.lo++;
   else
     outliers.hi++;
}
return n, count[], outliers;              // f̂(x) is (count[x − a] / n)
```

Outliers. By necessity, Algorithm 4.2.1 allows for the possibility of *outliers*—occasional x_i that fall outside the range $a \leq x_i \leq b$. Generally, with simulation-generated data, there should not be any outliers. An outlier is a data point that some omniscient being considers to be so different from the rest that it should be excluded from any statistical analysis. Although outliers are common with some kinds of experimentally measured data, it is difficult to argue that *any* data generated by a *valid* discrete-event simulation program is an outlier, no matter how unlikely the value may appear to be.

General-Purpose Discrete-Data Histogram Algorithm. Algorithm 4.2.1 is not appropriate as a *general-purpose* discrete-data histogram algorithm, for two reasons.

- If the discrete-data sample is integer-valued and the a, b parameters are not chosen properly, then either outliers will be produced, perhaps without justification, or excessive computational memory (to store the `count` array) could be required needlessly.
- If the data is not integer-valued, then Algorithm 4.2.1 is not applicable.

In either case, we must find an alternative, more general, algorithm that uses dynamic memory allocation and has the ability to handle both real-valued and integer-valued data. The construction of this algorithm is one of several occasions in this book where we will use dynamic memory allocation (in this case, in the form of a linked list) as the basis for an algorithm that is well suited to accommodate the uncertainty and variability naturally associated with discrete-event simulation data.

> **Definition 4.2.3**   A linked-list discrete-data histogram algorithm with general applicability can be constructed by using a *list pointer* `head` and multiple *list nodes* (structures), each consisting of three fields (`value`, `count`, and `next`), as illustrated in Figure 4.2.3.

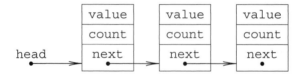

Figure 4.2.3   Linked-list data structure for discrete data.

The association between the first two fields in each list node and the corresponding terms in Definition 4.2.1 is $x \sim$ value and, after all data is processed, $n\hat{f}(x) \sim$ count. The next field is the link (pointer) from one list node to the next.

Algorithm 4.2.2   Given the linked-list data structure in Definition 4.2.3 and the discrete data $x_1, x_2, \ldots, x_n$, a discrete-data histogram is computed by using x_1 to initialize the first list node. Then, for $i = 2, 3, \ldots, n$, as each x_i is read, the list is searched (linearly from the head, by following the links) to see whether a list node with value equal to x_i is already present in the list. If so, the corresponding list node count is increased by 1; otherwise a new list node is added to the end of the list, with value equal to x_i and count equal to 1.

Example 4.2.4
For the discrete data 3.2, 3.7, 3.7, 2.9, 3.7, 3.2, 3.7, 3.2, Algorithm 4.2.2 generates the corresponding linked list illustrated in Figure 4.2.4.

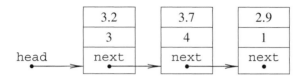

Figure 4.2.4   Sample linked list.

Note that the order of the linked-list nodes is dictated by the order in which the data appears in the sample. As an alternative, it may be better to use the count field and maintain the list in *sorted* order by decreasing relative frequency, or to use the value field and sort by data value—see Exercise 4.2.6.

Program ddh.   The discrete-data histogram program ddh is based on Algorithm 4.2.2 and on the linked-list data structure in Definition 4.2.3. The program has been designed so that the linked list is sorted by value prior to output. This program is valid for both integer-valued and real-valued input, with no artificial outlier check imposed and with no restriction on the sample size. Like program uvs, program ddh supports file redirection, as illustrated by Example 4.2.5 (to follow).

If program ddh is used improperly to tally a large sample that is *not* discrete, the program's execution time can be excessive. The design of program ddh is based on the assumption that, even though the sample size $|\mathcal{S}|$ is essentially arbitrary, the number of distinct values in the sample $|\mathcal{X}|$ is not too large, say a few hundred or less. If the number of distinct values is large, then the $O(|\mathcal{X}|)$ complexity (per sample value) of the function Insert and the one-time $O(|\mathcal{X}|^2)$ complexity of the function Sort will become the source of excessive execution time. Note, however, that a discrete-data sample with more than a few hundred distinct values would be *very* unusual.

Example 4.2.5
Program sis2 computes the statistics $\bar{d}$, $\bar{o}$, $\bar{u}$, $\bar{l}^{+}$, and $\bar{l}^{-}$. If we were interested in a more detailed look at this simple inventory system, we could, for example, construct a

histogram of the inventory level prior to inventory review. This is easily accomplished by eliminating the printing of summary statistics in program `sis2` and modifying the `while` loop in `main` by inserting the one line indicated below.

```
index++;
printf("%ld\n", inventory);              // this line is new
if (inventory < MINIMUM) {
```

If the new program `sis2` is compiled to disk (with $STOP = 10\,000$), the command

```
sis2 | ddh > sis2.out
```

will then produce a discrete-data histogram file `sis2.out` from which an inventory level histogram can be constructed.

Example 4.2.6
As a continuation of Example 4.2.5, the inventory-level histogram is illustrated (x denotes the inventory level prior to review) in Figure 4.2.5.

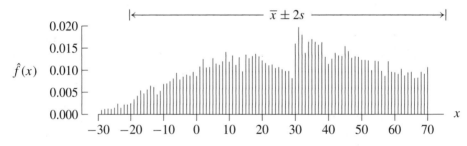

Figure 4.2.5 Inventory level with *Equilikely* demands.

The sample mean and standard deviation for this data are $\bar{x} = 27.63$ and $s = 23.98$. The corresponding $\bar{x} \pm 2s$ interval is indicated. In this case, approximately 98.5% of the data falls in this interval. Therefore, the "at least 75%" estimate guaranteed by Chebyshev's inequality is very conservative: The $\bar{x} \pm 2s$ interval contains almost all the sample data. This is consistent with the discussion in the previous section.

As discussed in Example 3.1.5, the assumption of *Equilikely* (10, 50) random-variate demands in program `sis2` is questionable. Generally, one would expect the extreme values of the demand, in this case 10 and 50, to be much less likely than an intermediate value, such as 30. One way to accomplish this is to use the assignment

```
return (Equilikely(5, 25) + Equilikely(5, 25));
```

in the function `GetDemand`. It can be verified by the axiomatic approach to probability that the theoretical mean and standard deviation of the demand for this assignment are 30 and $\sqrt{220/3} \cong$

8.56, respectively. Moreover, the distribution and associated histogram will have a triangular shape, with a peak at 30. Of course, other stochastic demand models are possible as well. We will investigate this issue in more detail in Chapter 6.

Example 4.2.7
Compare against the histogram in Example 4.2.6; the use of the more realistic discrete triangular random-variate demand model produces a corresponding change in the inventory-level histogram, as illustrated in Figure 4.2.6.

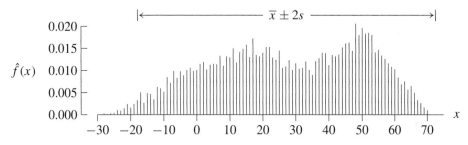

Figure 4.2.6   Inventory level with discrete triangular demands.

This inventory-level histogram is much more tapered at the extreme values, particularly at 70. The sample mean and standard deviation are 27.29 and 22.59, respectively. The corresponding $\bar{x} \pm 2s$ interval is slightly smaller, as illustrated. As in Example 4.2.6, approximately 98.5% of the data lie in this interval.

Accuracy of Point Estimates.   We now consider the accuracy of probability estimates derived by Monte Carlo simulation. This issue was first raised in Chapter 2 and, at that point, largely dismissed. Now we can use program ddh as an experimental tool to study the inherent uncertainty in such probability estimates. The following example uses the Monte Carlo program craps from Section 2.4 to generate 1000 point estimates of the probability of winning in the dice game craps. Because of the inherent uncertainty in any one of these estimates, when many estimates are generated, a natural distribution of values will be produced. This distribution of values can be characterized by a discrete-data histogram; in that way, we can gain significant insight into just how uncertain any one probability estimate is.

Example 4.2.8
A Monte Carlo simulation of the dice game craps was used to generate 1000 estimates of the probability of winning. In the top figure, each estimate is based on just $N = 25$ plays of the game; in the bottom figure, $N = 100$ plays per estimate were used. For either figure, let $p_1, p_2, p_3, \ldots, p_n$ denote the $n = 1000$ estimates, where $p_i = x_i/N$ and x_i is the number of wins in N plays of the game. Since only $N + 1$ values of p_i are possible, it is natural to use a discrete-data histogram in this application, as illustrated in Figure 4.2.7.

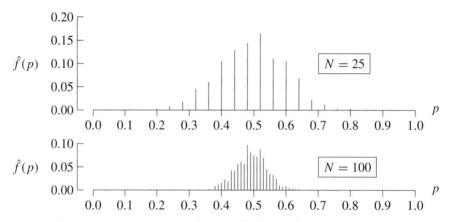

Figure 4.2.7   Histograms of the probability of winning in craps.

In the top histogram, the sample mean is 0.494, the sample standard deviation is 0.102, and there is significant variation about the true probability of winning at craps, which is known to be $244/495 \cong 0.4929$, from the analytic approach to probability. This variation corresponds to an inherently large uncertainty in any one estimate. That is, if just *one* probability estimate were generated based on 25 plays of the game, then our experiment with 1000 replications shows that probability estimates as far away from the true probability as $5/25 = 0.2$ and $20/25 = 0.8$ could be generated. In the bottom histogram, the associated sample mean and standard deviation are 0.492 and 0.048, respectively. Comparison with the top histogram shows that there is still significant variability, but it is reduced. In particular, the standard deviation is reduced by a factor of about two (from 0.102 to 0.048), resulting in a corresponding two-fold reduction in the uncertainty of the second estimate.

The reduction in the sample standard deviation from 0.102 to 0.048 in Example 4.2.8 is good. That a *four*-fold increase in the number of replications (games) is required to produce a *two*-fold reduction in uncertainty, however, is not so good. We will have much more to say about this in Chapter 8.

In Example 4.2.8, the "bell shape" of the discrete-data histogram is important. As will be discussed in Chapter 8, this bell (or Gaussian) shape, which shows up in a wide variety of applications, is accurately described by a well-known mathematical equation, which provides the basis for establishing a probability estimate *interval* within which the true (theoretical) probability can be assumed to lie with high confidence.

4.2.3 Empirical Cumulative Distribution Functions

The histogram is an effective tool for estimating the shape of a distribution. There are occasions, however, when a cumulative version of the histogram is preferred. Two such occasions are when quantiles (e.g., the 90th quantile of a distribution) are of interest and when two or more distributions are to be compared. An *empirical cumulative distribution function*, which simply takes an upward step of $1/n$ at each of the n data values $x_1, x_2, \ldots, x_n$, is easily computed. In the case of discrete data, of course, there will typically be many ties, so the step function will have varying heights to the risers.

Example 4.2.9
The $n = 1000$ estimates of the probability of winning in craps for $N = 100$ plays from Example 4.2.8 range from 0.33 to 0.64. Four *empirical cumulative distribution functions* for this data set that use different formats are plotted in Figure 4.2.8 for $0.3 < p < 0.7$. The top two graphs plot the cumulative probability only at the observed values; the bottom two graphs are step functions plotted for all values of p. The choice among these four styles is largely a matter of taste.

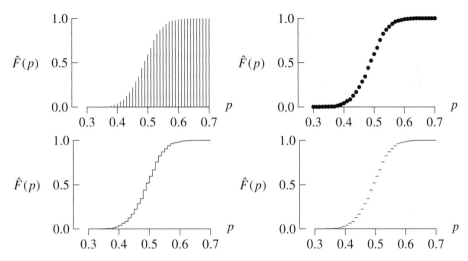

Figure 4.2.8 Empirical cumulative distribution functions.

4.2.4 Exercises

4.2.1   Find the theoretical values of the probabilities, mean, and variance for program `galileo` in Example 4.2.1.

4.2.2   (*a*) Generate the 2000-ball histogram in Example 4.2.2. (*b*) Verify that the resulting relative frequencies $\hat{f}(x)$ approximate the equation

$$\hat{f}(x) \cong \frac{2^x \exp(-2)}{x!} \qquad x = 0, 1, 2, \ldots.$$

(*c*) Generate the corresponding histogram if 10 000 balls are placed, at random, in 1000 boxes. (*d*) Find an equation that seems to fit the resulting relative frequencies well, and illustrate the quality of the fit.

4.2.3   The gap in the histogram in Figure 4.2.2 in Example 4.2.2 corresponding to no boxes holding exactly eight balls brings up an important topic in Monte Carlo and discrete-event simulation: the generation of *rare events*. To use the probabilist's terminology, there is poorer precision for estimating the probability of events out in the fringes (or "tails") of a distribution. (*a*) In theory, could all of the 2000 balls be in just one box? If so, give the probability of this event. If not, explain why not. (*b*) In practice, with our random number generator rng, could all of the 2000 balls be in just one box?

4.2.4   Find the theoretical probabilities required to generate the right-hand plot in Figure 4.2.2.

4.2.5   Program ddh computes the *histogram* mean and standard deviation. If you were to modify this program so that it also computes the *sample* mean and standard deviation by using Welford's algorithm, you would observe that (except perhaps for floating-point round-off errors) these two ways of computing the mean and standard deviation produce identical results. (*a*) If, instead, Algorithm 4.2.1 were used to compute the histogram mean and standard deviation, would they necessarily agree exactly with the sample mean and standard deviation? (*b*) Why or why not?

4.2.6   Although the output of program ddh is sorted by value, this is only a convention. As an alternative, modify program ddh so that the output is sorted by count (in decreasing order).

4.2.7   Use simulation to explore the allegedly "efficiency-increasing" modifications suggested in Example 4.2.4. Be aware that either of these modifications can make the algorithm more complicated for what could be, in return, only a marginal increase in efficiency. Because of its quadratic complexity, beware of simply using a modified version of the function Sort in this application.

4.2.8   How do the inventory-level histograms in Examples 4.2.6 and 4.2.7 relate to the relative frequency of setups $\bar{u}$?

4.2.9   Generate a random-variate demand sample of size $n = 10\,000$ as

d_i = Equilikely(5, 25) + Equilikely(5, 25);

for $i = 1, 2, \ldots, n$. (*a*) Why is the sample mean about 30? (*b*) Why is the sample standard deviation about $\sqrt{220/3}$? (*c*) Why is the shape of the histogram approximately triangular?

4.2.10   A discrete-data histogram of orders is illustrated (corresponding to the demand distribution in Example 4.2.6). From this histogram, by *inspection* estimate the histogram mean and standard deviation.

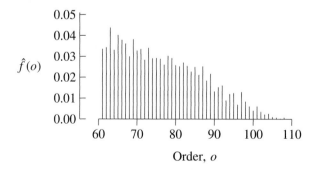

4.2.11   A test is compiled by selecting 12 different questions, at random and without replacement, from a well-publicized list of 120 questions. After studying this list you are able to classify all 120 questions into two classes, I and II. Class I questions are those about which you feel confident; the remaining questions define class II. Assume that your grade probability, conditioned on the class of the problems, is

	A	B	C	D	F
class I	0.6	0.3	0.1	0.0	0.0
class II	0.0	0.1	0.4	0.4	0.1

Each test question is graded on an $A = 4$, $B = 3$, $C = 2$, $D = 1$, $F = 0$ scale and a score of 36 or better is required to pass the test. (*a*) If there are 90 class-I questions in the list, use Monte Carlo simulation and 100 000 replications to generate a discrete-data histogram of scores. (*b*) From this histogram, what is the probability that you will pass the test?

4.2.12   Modify program `ssq2` so that, each time a new job arrives, the program outputs the number of jobs in the service node *prior* to the job's arrival. (*a*) Generate a discrete-data histogram for 10 000 jobs. (*b*) Comment on the shape of the histogram and compare the histogram mean with $\bar{l}$. (*c*) How does $\hat{f}(0)$ relate to $\bar{x}$?

4.3 CONTINUOUS-DATA HISTOGRAMS

As in the previous section, we assume a sample $S = \{x_1, x_2, \ldots, x_n\}$, with n sufficiently large that it is reasonable to do more than just calculate the sample mean and standard deviation. In sharp contrast to the discrete-data situation in the previous section, however, we now consider continuous-data histograms, where the data values $x_1, x_2, \ldots, x_n$ are assumed to be real-valued and generally distinct.

4.3.1 Continuous-Data Histograms

Given a real-valued sample $S = \{x_1, x_2, \ldots, x_n\}$, without loss of generality we can assume the existence of real-valued lower and upper bounds a, b with the property that

$$a \leq x_i < b \qquad i = 1, 2, \ldots, n.$$

This defines an interval of possible values for some random variable X as $\mathcal{X} = [a, b) = \{x \mid a \leq x < b\}$, one that can be partitioned into k equal-width *bins* (k is a positive integer) as

$$[a, b) = \bigcup_{j=0}^{k-1} \mathcal{B}_j = \mathcal{B}_0 \cup \mathcal{B}_1 \cup \cdots \cup \mathcal{B}_{k-1},$$

where the bins are $\mathcal{B}_0 = [a, a + \delta)$, $\mathcal{B}_1 = [a + \delta, a + 2\delta)$, $\ldots$, and the width of each bin is

$$\delta = \frac{b - a}{k},$$

as illustrated on the axis below.

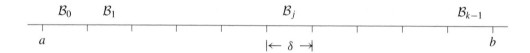

> **Definition 4.3.1**   Given the sample $S = \{x_1, x_2, \ldots, x_n\}$ and the related parameters a, b, and either k or δ, then, for each $x \in [a, b)$, there is a unique bin B_j with $x \in B_j$. The estimated *density* of the random variable X is then
>
> $$\hat{f}(x) = \frac{\text{the number of } x_i \in S \text{ for which } x_i \in B_j}{n\,\delta} \qquad a \le x < b.$$
>
> A *continuous-data histogram* is a "bar" plot of $\hat{f}(x)$ versus x.[*]

As the following example illustrates, $\hat{f}(\cdot)$ is a piecewise-constant function (constant over each bin) with discontinuities at the histogram-bin boundaries. Since simulations tend to produce large data sets, we adopt the graphics convention illustrated in Example 4.3.1 of drawing $\hat{f}(\cdot)$ as a sequence of piecewise-constant horizontal segments connected by vertical lines. This decision is consistent with maximizing the "data-to-ink" ratio (per Tufte, 2001).

Example 4.3.1

As an extension of Example 4.1.1, a modified version of program `buffon` was used to generate a random-variate sample having $n = 1000$ observations of the x-coordinate of the right-hand endpoint of a unit-length needle dropped at random. To form a continuous-data histogram of the sample, the values $a = 0.0$ and $b = 2.0$ are, in this case, obvious choices for lower and upper bounds.[†] The number of histogram bins was selected, somewhat arbitrarily, as $k = 20$, and so $\delta = (b - a)/k = 0.1$. The resulting histogram is illustrated on the left-hand side of Figure 4.3.1.

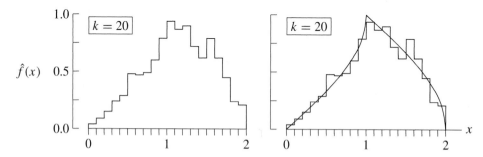

Figure 4.3.1   Histograms for $k = 20$.

---

[*]Compare Definition 4.3.1 with Definition 4.2.1. The bin index is $j = \lfloor (x - a)/\delta \rfloor$. The *density* is the relative frequency of the data in bin B_j *normalized* via a division by δ.

[†]Because the needle has unit length, $0.0 < x_i < 2.0$ for $i = 1, 2, \ldots, n$.

Illustrated on the right-hand side is the same histogram with a continuous curve super-imposed. As will be discussed in Chapter 7, this curve represents the *probability density function* of the right-hand endpoint, which is the limit to which the histogram will converge as the sample size n approaches infinity *and* simultaneously k approaches infinity (or, equivalently, δ approaches zero).

Histogram-Parameter Guidelines.   There is an experimental component to choosing the continuous-data histogram parameters a, b, and either k or δ. These guidelines are certainly not rigid rules.

- For data produced by a (valid) simulation, there should be few, if any, outliers. The bounds a, b should be chosen so that few, if any, data points in the sample are excluded. Of course, as in the discrete-data case, prior knowledge of reasonable values for a, b might not always be easily obtained.
- If k is too large (δ is too small), then the histogram will be too "noisy" and will have potential for exhibiting false features caused by natural sampling variability; if k is too small (δ is too large), then the histogram will be too "smooth" and will have potential for masking significant features—see Example 4.3.2.
- The histogram parameters should always be chosen with the aesthetics of the resulting figure in mind. [For example, for $a = 0$ and $b = 2$, if n is sufficiently large, the choice $k = 20$ ($\delta = 0.1$) would be a better choice than $k = 19$ ($\delta \cong 0.10526$).]
- Typically, $\lfloor \log_2(n) \rfloor \le k \le \lfloor \sqrt{n} \rfloor$, with a bias toward $k \cong \lfloor (5/3)\sqrt[3]{n} \rfloor$ (Wand, 1997).
- Sturges's rule (Law and Kelton, 2000, page 336) suggests $k \cong \lfloor 1 + \log_2 n \rfloor$.

Example 4.3.2
As a continuation of Example 4.3.1, two additional histograms are illustrated in Figure 4.3.2, corresponding to $k = 10$ ($\delta = 0.2$) on the left and to $k = 40$ ($\delta = 0.05$) on the right. The histogram on the right is clearly too noisy, consistent with this choice of k, which violates the second histogram-parameter guideline (i.e., k is too large relative to n). Although this characterization is less clear, the histogram on the left might be too smooth because k is too small. For this sample, the best choice of k seems to be

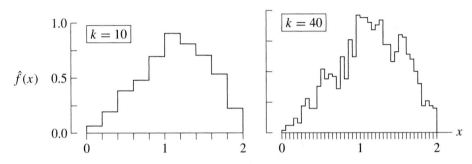

Figure 4.3.2   Histograms for $k = 10$ and $k = 40$.

somewhere between 10 and 20 (using the last of the histogram guidelines, $9 \leq k \leq 31$ and $k \cong \lfloor (5/3)\sqrt[3]{1000} \rfloor = 16$). Since a simulation typically produces a large number of observations, and therefore many histogram bins are required, we avoid the common practice of dropping the vertical lines to the horizontal axis in the histogram graphic, which would visually partition the bins. Including these lines unnecessarily clutters the histogram and obscures the shape of the histogram, particularly for large k.

Histogram Integrals

Definition 4.3.2   As an extension of Definition 4.3.1, for each $j = 0, 1, \ldots, k - 1$, define p_j as the *relative frequency* of points in $S = \{x_1, x_2, \ldots, x_n\}$ that fall into bin B_j. The bins form a partition of $[a, b)$, so each point in S is counted exactly once (assuming no outliers) and so $p_0 + p_1 + \cdots + p_{k-1} = 1$. In addition, define the *bin midpoints*

$$m_j = a + \left(j + \frac{1}{2} \right) \delta \qquad j = 0, 1, \ldots, k - 1$$

as illustrated on the axis below.

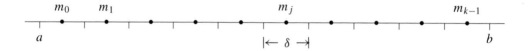

It follows from Definitions 4.3.1 and 4.3.2 that

$$p_j = \delta \hat{f}(m_j) \qquad j = 0, 1, \ldots, k - 1$$

and that $\hat{f}(\cdot)$ is a nonnegative function having unit area:

$$\int_a^b \hat{f}(x)\, dx = \sum_{j=0}^{k-1} \int_{B_j} \hat{f}(x)\, dx = \sum_{j=0}^{k-1} \hat{f}(m_j) \int_{B_j} dx = \sum_{j=0}^{k-1} \left(\frac{p_j}{\delta} \right) \delta = \sum_{j=0}^{k-1} p_j = 1.$$

In addition to proving that $\int_a^b \hat{f}(x)\, dx = 1$, the previous derivation can be extended to the two integrals

$$\int_a^b x \hat{f}(x)\, dx \qquad \text{and} \qquad \int_a^b x^2 \hat{f}(x)\, dx.$$

For the first of these two integrals,

$$\int_a^b x \hat{f}(x)\, dx = \sum_{j=0}^{k-1} \int_{B_j} x \hat{f}(x)\, dx = \sum_{j=0}^{k-1} \hat{f}(m_j) \int_{B_j} x\, dx = \sum_{j=0}^{k-1} \left(\frac{p_j}{\delta} \right) \int_{B_j} x\, dx;$$

and, in analogous fashion, for the second integral,

$$\int_a^b x^2 \hat{f}(x)\, dx = \cdots = \sum_{j=0}^{k-1} \left(\frac{p_j}{\delta}\right) \int_{\mathcal{B}_j} x^2\, dx.$$

In this way, because $\hat{f}(\cdot)$ is piecewise constant, the two integrals over $[a, b)$ are reduced to simple polynomial integration over each histogram bin. In particular, for $j = 0, 1, \ldots, k - 1$,

$$\int_{\mathcal{B}_j} x\, dx = \int_{m_j - \delta/2}^{m_j + \delta/2} x\, dx = \frac{(m_j + \delta/2)^2 - (m_j - \delta/2)^2}{2} = \cdots = m_j \delta,$$

and so the first integral reduces to

$$\int_a^b x \hat{f}(x)\, dx = \sum_{j=0}^{k-1} \left(\frac{p_j}{\delta}\right) \int_{\mathcal{B}_j} x\, dx = \sum_{j=0}^{k-1} m_j\, p_j.$$

Similarly,

$$\int_{\mathcal{B}_j} x^2\, dx = \frac{(m_j + \delta/2)^3 - (m_j - \delta/2)^3}{3} = \cdots = m_j^2 \delta + \frac{\delta^3}{12},$$

and so the second integral reduces to

$$\int_a^b x^2 \hat{f}(x)\, dx = \sum_{j=0}^{k-1} \left(\frac{p_j}{\delta}\right) \int_{\mathcal{B}_j} x^2\, dx = \left(\sum_{j=0}^{k-1} m_j^2\, p_j\right) + \frac{\delta^2}{12}.$$

Therefore, the two integrals $\int_a^b x \hat{f}(x)\, dx$ and $\int_a^b x^2 \hat{f}(x)\, dx$ can be evaluated *exactly* by finite summation. This is significant, because the continuous-data histogram mean and standard deviation are defined in terms of these two integrals.

Histogram Mean and Standard Deviation

Definition 4.3.3   Analogous to the discrete-data equations in Definition 4.2.2 (replacing $\sum$'s with $\int$'s) are the *continuous-data histogram mean* and *standard deviation*, defined as

$$\bar{x} = \int_a^b x \hat{f}(x)\, dx \qquad \text{and} \qquad s = \sqrt{\int_a^b (x - \bar{x})^2 \hat{f}(x)\, dx}.$$

The continuous-data histogram variance is s^2.

The integral equations in Definition 4.3.3 provide the motivation for defining the population mean and population standard deviation of continuous random variables in Chapter 7. From the integral equations derived previously, it follows that a continuous-data histogram mean can be evaluated exactly by finite summation, via the equation

$$\overline{x} = \sum_{j=0}^{k-1} m_j\, p_j.$$

Moreover,

$$s^2 = \int_a^b (x - \overline{x})^2 \hat{f}(x)\, dx = \cdots = \left(\int_a^b x^2 \hat{f}(x)\, dx \right) - \overline{x}^2;$$

and, similarly,

$$\sum_{j=0}^{k-1} (m_j - \overline{x})^2\, p_j = \cdots = \left(\sum_{j=0}^{k-1} m_j^2\, p_j \right) - \overline{x}^2.$$

From these last two equations, it follows that a continuous-data-histogram standard deviation can be evaluated exactly by finite summation via either of the following two equations:*

$$s = \sqrt{\left(\sum_{j=0}^{k-1} (m_j - \overline{x})^2\, p_j \right) + \frac{\delta^2}{12}} \qquad \text{or} \qquad s = \sqrt{\left(\sum_{j=0}^{k-1} m_j^2\, p_j \right) - \overline{x}^2 + \frac{\delta^2}{12}}.$$

In general, the continuous-data *histogram* mean and standard deviation will differ slightly from the *sample* mean and standard deviation, even if there are no outliers. This difference is caused by the *quantization error* associated with the arbitrary binning of continuous data. In any case, this difference should be slight—if the difference is not slight then the histogram parameters a, b, and either k or δ should be adjusted. Although the histogram mean and standard deviation are inferior to the sample mean and standard deviation, there are circumstances where a data analyst is presented with binned data and does not have access to the associated raw data.

Example 4.3.3
For the 1000-point sample in Example 4.3.1, when using $a = 0.0$, $b = 2.0$, and $k = 20$, the difference between the sample and histogram statistics is slight:

	raw data	histogram	histogram with $\delta = 0$
$\overline{x}$	1.135	1.134	1.134
s	0.424	0.426	0.425

---

*There is some disagreement in the literature relative to these two equations. Many authors use these equations, with the $\delta^2/12$ term ignored, to *define* the continuous-data histogram standard deviation.

Moreover, by comparing the last two columns in this table, we see that, in this case, there is essentially no impact from the $\delta^2/12 = (0.1)^2/12 = 1/1200$ term in the computation of the histogram standard deviation.

4.3.2 Computational Considerations

Algorithm 4.3.1   Given the parameters a, b, and k and the real-valued data $x_1, x_2, \ldots$, this algorithm computes a continuous-data histogram.

```
long count[k];
δ = (b - a) / k;
n = 0;
for (j = 0; j < k; j++)
   count[j] = 0;                        // initialize bin counters
outliers.lo = 0;            // initialize outlier counter on (−∞, a)
outliers.hi = 0;            // initialize outlier counter on [b, ∞)
while ( more data ) {
   x = GetData();
   n++;
   if ((a <= x) and (x < b)) {
      j = (long) (x - a) / δ;
      count[j]++;              // increment appropriate bin counter
   }
   else if (a > x)
      outliers.lo++;
   else
      outliers.hi++;
}
return n, count[], outliers;             // p_j is (count[j] / n)
```

The previously derived summation equations for $\bar{x}$ and s can then be used to compute the histogram mean and standard deviation.

If the sample is written to a disk file (with the use of sufficient floating-point precision), then one can experiment with different values for the continuous-data histogram parameters in an *interactive* graphics environment. Program cdh illustrates the construction of a continuous-data histogram for data read from a text file. For an alternative approach, see Exercise 4.3.7.

Example 4.3.4

As in Example 4.1.7, a modified version of program ssq2 was used to generate a sample consisting of the waits $w_1, w_2, \ldots, w_n$ experienced by the first $n = 1000$ jobs. The simulation was initialized to simulate steady state, and the rng initial seed was 12345. Program cdh was used to process this sample with the continuous-data histogram parameters set to $(a, b, k) = (0.0, 30.0, 30)$, as illustrated in Figure 4.3.3.

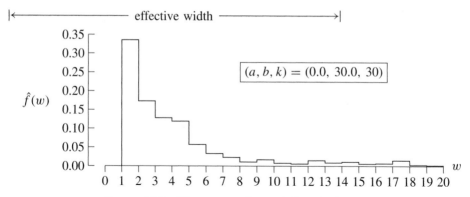

Figure 4.3.3 Histogram of $n = 1000$ wait times.

The histogram mean is 4.57, and the histogram standard deviation is 4.65. The associated two-standard-deviation "effective width" of the sample is 4.57 ± 9.30, as illustrated. As with previous examples, this interval includes most of the points in the sample.[*]

Point Estimation. The issue of sampling variability and how it relates to the uncertainty of any probability estimate derived from a Monte Carlo simulation was considered in the previous section. This issue is sufficiently important to warrant reconsideration here in the context of continuous-data histograms. In particular, recall that, for Example 4.2.8, there was a sample $S = \{p_1, p_2, \ldots, p_n\}$ of $n = 1000$ point estimates of the probability of winning at the game of craps. One figure in Example 4.2.8 corresponds to probability estimates based on 25 plays of the game per estimate; the other figure is based on 100 plays per estimate. For both figures, the samples were displayed as discrete-data histograms, and we were primarily interested in studying how the width of the histograms decreased with an increase in the number of games per estimate.

If we were to study the issue of how the histogram width depends on the number of games in more detail, it would be natural to quadruple the number of games per estimate to 400, 1600, 6400, etc. As the number of games increases, the associated discrete-data histograms will look more nearly continuous as the histogram spikes get closer together. Given that, it is natural to ignore the inherently discrete nature of the probability estimates and, instead, treat the sample as though it were continuous data. That is what is done in the following example.

Example 4.3.5
The sample, $n = 1000$ probability estimates each based on $N = 25$ plays of the game of craps from Example 4.2.8, was processed as a continuous-data histogram with parameters $(a, b, k) = (0.18, 0.82, 16)$. This choice of parameters is matched to the "resolution" of the estimates (which is $\delta = 0.04$), with the center of each histogram bin corresponding to exactly one possible value of an estimate.[†]

---

[*]The interval 4.57 ± 9.30 includes approximately 93% of the sample. There are 21 points in the sample with a value larger than 20.0, the largest of which is 28.2.

[†]Continuous-data histograms are *density* estimates, not *probability* estimates. Thus, values of $\hat{f}(p)$ greater than 1.0 are possible, as in this example, but the histogram value cannot stay at that height too long since $\int_0^1 \hat{f}(p)\, dp = 1$.

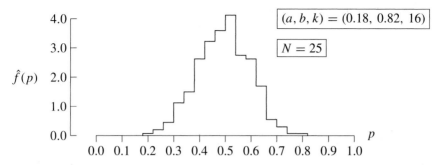

Figure 4.3.4   Histogram of $n = 1000$ estimates of winning at craps from $N = 25$ plays.

In a similar way, the sample having $n = 1000$ probability estimates based on $N = 100$ plays of the game was processed as a continuous-data histogram with parameters $(a, b, k) = (0.325, 0.645, 16)$. This choice of parameters is matched to half the resolution of the estimates ($\delta = 0.02$), with the center of each histogram bin corresponding to the midpoint between exactly two possible values of an estimate, as illustrated in Figure 4.3.5.

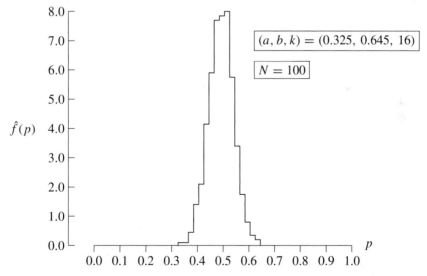

Figure 4.3.5   Histogram of $n = 1000$ estimates of winning at craps from $N = 100$ plays.

As in Example 4.2.8, we see that increasing the number of replications per estimate by a factor of *four* causes the uncertainty in any one probability estimate to decrease by a factor of *two*. The advantage to using a continuous-data histogram representation in this example is that experimentation with more games per estimate can be accommodated naturally. As the number of games per estimate is increased, the histogram will become taller and narrower, always centered near the true value, $244/495 \cong 0.4929$, and always obeying the invariant unit-area requirement, $\int_0^1 \hat{f}(p)\,dp = 1$.

Random Events Yield Exponentially Distributed Inter-Events

Example 4.3.6

As another continuous-data histogram example, pick a positive parameter $t > 0$ and suppose that n calls to the function $\texttt{Uniform}(0, t)$ are used to generate a random-variate sample of n events occurring *at random* in the interval $(0, t)$. If these n event times are then sorted into increasing order, the result is a sequence of event times, $u_1, u_2, \ldots, u_n$, ordered so that $0 < u_1 < u_2 < \cdots < u_n < t$. With $u_0 = 0$, define

$$x_i = u_i - u_{i-1} > 0 \qquad i = 1, 2, \ldots, n,$$

as the inter-event times. Let $\mu = t/n$, and recognize that

$$\bar{x} = \frac{1}{n} \sum_{i=1}^{n} x_i = \frac{u_n - u_0}{n} \cong \frac{t}{n} = \mu;$$

hence, the sample mean of the inter-event times is approximately μ. One might expect that a histogram of the inter-event times will be approximately "bell shaped" and centered at μ. As illustrated in Figure 4.3.6, however, that is *not* the case—the histogram has an *exponential* shape.* In particular, the smallest inter-event times are the most likely.

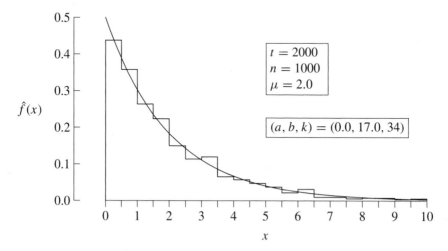

Figure 4.3.6   Histogram of $n = 1000$ inter-event times.

---

*The histogram actually has an even longer tail than is illustrated in Figure 4.3.6. There are eight data points in the sample (out of 1000) having a value larger than 10.

The continuous curve superimposed illustrates that, in this case, $\hat{f}(x) \cong f(x)$, where

$$f(x) = \frac{1}{\mu}\exp(-x/\mu) \qquad x > 0.$$

Indeed, in the limit as $n \to \infty$ and $\delta \to 0$ (with μ held constant), $\hat{f}(x) \to f(x)$ for all $x > 0$.* We will return to this important example in Chapter 7.

4.3.3 Empirical Cumulative Distribution Functions

The fact that parameters, such as the number of bins k, must be chosen by the modeler is a distinct drawback for continuous-data histograms. Two different data analysts could have the extreme misfortune of choosing different binning schemes for the same data set and could produce histograms with somewhat different shapes. This is particularly true if the sampling variability inherent in the data set conspires with the two different binning schemes to accentuate the difference between the two histograms.

An alternative approach to plotting continuous data that avoids arbitrary parameters from the modeler utilizes the *empirical cumulative distribution function*.

Definition 4.3.4   Given the sample $\mathcal{S} = \{x_1, x_2, \ldots, x_n\}$, the estimated *cumulative distribution function* of the random variable X is

$$\hat{F}(x) = \frac{\text{the number of } x_i \in \mathcal{S} \text{ for which } x_i \leq x}{n}.$$

The *empirical cumulative distribution function* is a plot of $\hat{F}(x)$ versus x.

When the $x_1, x_2, \ldots, x_n$ are distinct, the plot of $\hat{F}(x)$ versus x is a step function with an upward step of $1/n$ at each data value; if d values are tied at a particular x-value, the height of the riser on that particular step is d/n.

The empirical cumulative distribution function requires no parameters from the modeler, which means that one data set always produces the same empirical cumulative distribution function.

We now compare the computational complexity and memory requirements of the two graphical procedures. The continuous-data histogram algorithm performs a single pass through the data values, has time complexity $O(n)$, and requires k memory locations. Plotting the empirical cumulative distribution function requires a sort, of time complexity $O(n \log n)$ at best, and all data values must be stored simultaneously, requiring n memory locations.

Example 4.3.7
Consider again the modified version of program $\texttt{buffon}$ that was used to generate a random-variate sample of $n = 50$ observations of the x-coordinate of the right-hand

---

*See, for example, Rigdon and Basu, 2000, pages 50–52, for the details concerning the relationship between the uniform and the exponential distribution in this case.

endpoint of a unit-length needle dropped at random, using as initial seed 123456789. The empirical cumulative distribution function is plotted in Figure 4.3.7 using our choice among the four plotting formats displayed for discrete data in Figure 4.2.8. There is an upward step of $1/50$ at each of the values generated. The minimum and maximum values generated by the program are 0.45688 and 1.79410, respectively, which correspond to the horizontal position of the first and last step of the plot of the empirical cumulative distribution function. In this example, it is possible to compute the theoretical cumulative distribution function using the axiomatic approach to probability. This is the smooth curve superimposed in Figure 4.3.7. The difference between the step function and the smooth curve is due to *random sampling variability*.

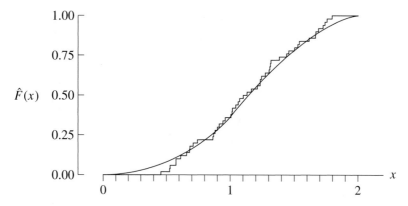

Figure 4.3.7 Theoretical and empirical cumulative distribution functions from 50 replications.

How does one compare the advantages and disadvantages of the continuous-data histogram and empirical cumulative distribution function? The histogram is clearly superior at detecting the *shape* of the distribution of the random quantity of interest. The arbitrary parameters associated with binning are its only downside. Selecting the continuous-data histogram parameters is more of an art than a science, which drives us to an alternative. The empirical cumulative distribution function is nonparametric, and thus less susceptible to the effects of sampling variability; there is no binning. Unfortunately, its shape is less distinct than the continuous-data histogram. It is often used to compare a hypothesized or fitted distribution to a data set via a statistical "goodness-of-fit" test.*

Increased CPU speed makes generating large data sets possible in simulation, putting a strain on the memory and speed associated with plotting an empirical cumulative distribution function. Fortunately, this is typically done only once during a simulation run.

We end this section with an example that combines some of the best features of continuous-data histograms and empirical cumulative distribution functions.

---

*The Kolmogorov–Smirnov, Anderson–Darling, and Cramer–von Mises are three well-known statistical goodness-of-fit tests for continuous data.

Example 4.3.8

If the sample size from the previous example ($n = 50$) were (dramatically) increased to $n = 1\,000\,000\,000$, one would expect the empirical cumulative distribution function to become very smooth and approximate the theoretical curve. We plot an empirical cumulative distribution function in order to take advantage of its nonparametric nature. Unfortunately, plotting the empirical cumulative distribution function requires that we store and sort the one billion right-hand needle endpoints. Since each data value lies on $0 \leq x \leq 2$, we can create a close approximation to the empirical cumulative distribution function by defining, for instance, 200 equal-width cells on $0 \leq x \leq 2$, i.e., $[0, 0.01), [0.01, 0.02), \ldots, [1.99, 2.00).$* Counts associated with these cells are accumulated, and a plot of the cumulative proportions associated with these cells should be virtually identical to a plot from the raw data. The cells take advantage of the fact that the counts can be updated as the data is generated, eliminating the need for storage and sorting. The plot of the cumulative proportions shown in Figure 4.3.8 is much smoother than the plot for $n = 50$ in Figure 4.3.7, because of the huge number of replications. The difference between the plot in Figure 4.3.8 and the true cumulative distribution function would be apparent to someone only with a powerful microscope.

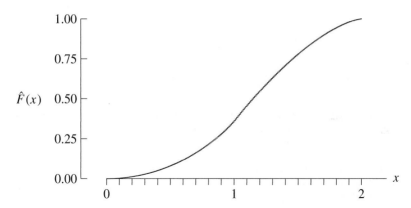

Figure 4.3.8   Approximate empirical cumulative distribution function from one billion replications.

4.3.4 Exercises

4.3.1   (*a*) Use program cdh to construct a continuous-data histogram like the one on the left in Example 4.3.1, but corresponding to a needle of length $r = 2$. (*b*) For this histogram, what is the probability that the needle will cross at least one line? (*c*) What is the corresponding axiomatic probability that a needle of length $r = 2$ will cross at least one line?

4.3.2   Repeat the experiment in Example 4.3.6, but with $t = 5000$ and $n = 2000$. Do not use a bubble sort.

---

*The number of cells chosen in this example (200) is arbitrary. The formulas given in Section 4.3.1 for choosing the number of histogram cells do not apply here. The choice depends on the physical size of the plot, the desired smoothness, and the sample size.

4.3.3 Fill in the $= \cdots =$'s in the derivation of the two equations

$$\int_a^b x \hat{f}(x)\, dx = \cdots = \sum_{j=0}^{k-1} m_j\, p_j \quad \text{and} \quad \int_a^b x^2 \hat{f}(x)\, dx = \cdots = \left(\sum_{j=0}^{k-1} m_j^2\, p_j \right) + \frac{\delta^2}{12}.$$

4.3.4 Generate a random-variate sample $x_1, x_2, \ldots, x_n$ of size $n = 10\,000$ as follows:

```
for (i = 1; i <= n; i++)
    x_i = Random() + Random();
```

(*a*) Use program cdh to construct a 20-bin continuous-data histogram. (*b*) Can you find an equation that seems to fit the histogram density well?

4.3.5 (*a*) As a continuation of Exercise 1.2.6, construct a continuous-data histogram of the service times. (*b*) Compare the *histogram* mean and standard deviation with the corresponding *sample* mean and standard deviation, and justify your choice of the histogram parameters a, b, and either k or δ.

4.3.6 As an extension of Definition 4.3.1, the *cumulative* histogram density is defined as

$$\hat{F}(x) = \int_a^x \hat{f}(t)\, dt \qquad a \le x < b.$$

Derive a finite summation equation for $\hat{F}(x)$.

4.3.7 To have more general applicability, program cdh needs to be restructured to support file redirection, as in programs uvs and ddh. That part is easy. The ultimate objective here, however, should be a large-sample "auto-cdh" program that buffers, say, the first 1000 sample values and then automatically computes good values for the histogram parameters a, b, and either k or δ with a dynamic-data-structure allowance for additional bins to be added at the tails of the histogram and thereby avoid outliers if extreme values occur. (*a*) Construct such a program. (*b*) Discuss the logic you used for computing good values for the histogram parameters.

4.3.8 Show that the theoretical cumulative distribution function superimposed over the empirical cumulative distribution function in Figure 4.3.7 is

$$F(x) = \begin{cases} \dfrac{2 \left(x\, \arcsin(x) + \sqrt{1 - x^2} - 1 \right)}{\pi} & 0 < x < 1 \\[3mm] \dfrac{2(1 - x) \arcsin(x - 1) - 2\sqrt{x\,(2 - x)} + \pi x}{\pi} & 1 < x < 2. \end{cases}$$

4.4 CORRELATION

In a discrete-event simulation model, it is often the case that two random variables are "co-related"—that is, the natural variation in these two variables is somehow coupled. The statistical term for this is *correlation*. For example, in a FIFO single-server service node, there is (positive) correlation between a job's wait in the service node and the wait time of the preceding job. This section examines two types of correlation: *paired* and *serial*.

4.4.1 Paired Correlation

Given a *paired sample* (u_i, v_i) for $i = 1, 2, \ldots, n$, we begin by considering how to compute a statistic that measures the extent to which this sample exhibits correlation.

> **Definition 4.4.1** A display of the paired data (u_i, v_i) for $i = 1, 2, \ldots, n$ as illustrated in Figure 4.4.1 is called a *bivariate scatterplot*.

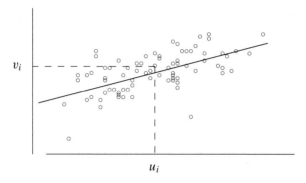

Figure 4.4.1 Bivariate scatterplot.

When a bivariate scatterplot is created, it is sometimes the case that the (u_i, v_i) points lie primarily along and near a line, as is the case in Figure 4.4.1. If this is the case, then it is natural to ask what line "best fits" the scatterplot data. It is in this sense that what we are discussing is called *linear* correlation. Consider the line in the (u, v) plane defined by the equation $au + bv + c = 0$; for each point (u_i, v_i), let d_i be the *orthogonal* distance from this point to the line, as illustrated below.

$$d_i = \frac{|au_i + bv_i + c|}{\sqrt{a^2 + b^2}}$$

From calculus, recall that the equation for d_i is as indicated. Thus we choose the (a, b, c) line parameters that *minimize* the mean-square orthogonal distance

$$D = \frac{1}{n} \sum_{i=1}^{n} d_i^2 = \frac{1}{n\left(a^2 + b^2\right)} \sum_{i=1}^{n} (au_i + bv_i + c)^2.$$

Three-Parameter Minimization. To find the line that best fits the data, or, equivalently, to find the choice of the (a, b, c) line parameters that will minimize D, we begin with the observation that D can be written as*

$$D = \frac{1}{n\left(a^2 + b^2\right)} \sum_{i=1}^{n} \left(a(u_i - \overline{u}) + b(v_i - \overline{v})\right)^2 + \frac{1}{n(a^2 + b^2)} \sum_{i=1}^{n} (a\,\overline{u} + b\,\overline{v} + c)^2,$$

where

$$\overline{u} = \frac{1}{n} \sum_{i=1}^{n} u_i \qquad \text{and} \qquad \overline{v} = \frac{1}{n} \sum_{i=1}^{n} v_i$$

are the sample means of the u and v data respectively. Both terms in the equation for D are nonnegative, since they involve sums of squares. The first term is independent of c; the second term is not. Therefore, to minimize D, the line parameter c should be chosen so as to minimize the second term. By inspection, the minimum value of the second term is zero, achieved by choosing c so that

$$a\,\overline{u} + b\,\overline{v} + c = 0,$$

which eliminates the second term. For any choice of the (a, b) line parameters, if $a\,\overline{u} + b\,\overline{v} + c = 0$, it follows that the line must pass through the point $(u, v) = (\overline{u}, \overline{v})$. In particular, this geometric property is true for the choice of the (a, b) line parameters that will minimize D. The three-parameter minimization problem has been reduced to a two-parameter minimization problem.

Two-Parameter Minimization. Because $c = -a\,\overline{u} - b\,\overline{v}$, the equation $au + bv + c = 0$ can be written equivalently as $a(u - \overline{u}) + b(v - \overline{v}) = 0$. Moreover, the (a, b) line parameters cannot both be zero, so we can simplify the equation for D by assuming that (a, b) are normalized so that $a^2 + b^2 = 1$. The following theorem summarizes this discussion.

Theorem 4.4.1 The line that best fits the data (u_i, v_i) for $i = 1, 2, \ldots, n$ (in a mean-squared orthogonal distance sense) is given by the equation

$$a(u - \overline{u}) + b(v - \overline{v}) = 0,$$

where the (a, b) line parameters are chosen to minimize

$$D = \frac{1}{n} \sum_{i=1}^{n} \left(a(u_i - \overline{u}) + b(v_i - \overline{v})\right)^2$$

subject to the constraint $a^2 + b^2 = 1$.

Covariance and Correlation. We now rewrite the algebraic expression for D in Theorem 4.4.1. We do this by introducing two important (related) statistical measures of how the u's and v's are "co-related." The c symbol defined below differs from the c defined previously.

---

*The details of this derivation are left as an exercise.

> **Definition 4.4.2**   Given the bivariate sample (u_i, v_i) for $i = 1, 2, \ldots, n$, the (linear) *sample covariance* is
>
> $$c = \frac{1}{n} \sum_{i=1}^{n} (u_i - \overline{u})(v_i - \overline{v});$$
>
> and, provided that both s_u and s_v are not zero, the (linear) *sample correlation coefficient* is
>
> $$r = \frac{c}{s_u s_v},$$
>
> where $\overline{u}, \overline{v},$
>
> $$s_u^2 = \frac{1}{n} \sum_{i=1}^{n} (u_i - \overline{u})^2, \qquad \text{and} \qquad s_v^2 = \frac{1}{n} \sum_{i=1}^{n} (v_i - \overline{v})^2$$
>
> are the sample means and sample variances of the u and v data values respectively.[*]

As we will see, the correlation coefficient r measures the "spread" (dispersion) of the u, v data about the line that best fits the data. From Definition 4.4.2, the expression for D can be written in terms of s_u^2, s_v^2, and r as

$$D = \frac{1}{n} \sum_{i=1}^{n} \left(a(u_i - \overline{u}) + b(v_i - \overline{v}) \right)^2$$

$$= \frac{1}{n} \sum_{i=1}^{n} \left(a^2(u_i - \overline{u})^2 + 2ab(u_i - \overline{u})(v_i - \overline{v}) + b^2(v_i - \overline{v})^2 \right)$$

$$= \frac{a^2}{n} \left(\sum_{i=1}^{n} (u_i - \overline{u})^2 \right) + \frac{2ab}{n} \left(\sum_{i=1}^{n} (u_i - \overline{u})(v_i - \overline{v}) \right) + \frac{b^2}{n} \left(\sum_{i=1}^{n} (v_i - \overline{v})^2 \right)$$

$$= a^2 s_u^2 + 2ab r s_u s_v + b^2 s_v^2.$$

Note that $r = 1$ if and only if $D = (as_u + bs_v)^2$ and that $r = -1$ if and only if $D = (as_u - bs_v)^2$. Therefore, if $|r| = 1$, then it is possible to choose (a, b) in such a way that $D = 0$. Indeed, as we will see, the following three conditions are equivalent:

$$|r| = 1 \quad \Longleftrightarrow \quad D = 0 \quad \Longleftrightarrow \quad \textit{all} \text{ the points } (u_i, v_i) \text{ lie on a line.}$$

The covariance equation in Definition 4.4.2 is a two-pass expression. As in Section 4.1, it can be shown that an equivalent one-pass expression for the covariance is

$$c = \frac{1}{n} \left(\sum_{i=1}^{n} u_i v_i \right) - \overline{u}\,\overline{v}.$$

---

[*]The covariance is a generalization of the notion of variance, in the sense that the sample covariance of the (u_i, u_i) data values is s_u^2 and the sample covariance of the (v_i, v_i) data values is s_v^2. Note also that the covariance derives its "dimensions" from u and v. The correlation coefficient is dimensionless.

The derivation is left as an exercise. We will consider the computational significance of this result later in the section.

One-Parameter Minimization.   Let θ be the angle of the line $a(u - \bar{u}) + b(v - \bar{v}) = 0$ as measured counterclockwise relative to the u-axis, as illustrated in Figure 4.4.2. Because $a(u - \bar{u}) + b(v - \bar{v}) = 0$ and $a^2 + b^2 = 1$, the (a, b) line parameters form a unit vector that is *orthogonal* to the line. Therefore the relation between (a, b) and θ, from elementary trigonometry, is

$$a = -\sin\theta \qquad \text{and} \qquad b = \cos\theta.$$

Much as in Theorem 4.4.1, minimizing D is accomplished by rotating the line about the point $(\bar{u}, \bar{v})$ to find the angle θ for which D is smallest. To find this angle, we can write D in terms of θ as

$$
\begin{aligned}
D &= a^2 s_u^2 + 2abc + b^2 s_v^2 \\
&= s_u^2 \sin^2\theta - 2c\sin\theta\cos\theta + s_v^2\cos^2\theta \\
&= s_u^2 \left(1 - \cos^2\theta\right) - c\sin 2\theta + s_v^2\cos^2\theta \\
&\;\;\vdots \\
&= \frac{1}{2}\left(s_u^2 + s_v^2\right) - c\sin 2\theta - \frac{1}{2}\left(s_u^2 - s_v^2\right)\cos 2\theta.
\end{aligned}
$$

The value of θ that minimizes

$$D = \frac{1}{2}\left(s_u^2 + s_v^2\right) - c\sin 2\theta - \frac{1}{2}\left(s_u^2 - s_v^2\right)\cos 2\theta$$

solves the equation

$$\frac{dD}{d\theta} = -2c\cos 2\theta + \left(s_u^2 - s_v^2\right)\sin 2\theta = 0.$$

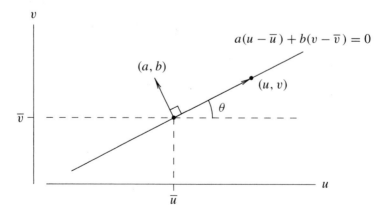

Figure 4.4.2 Defining the angle θ.

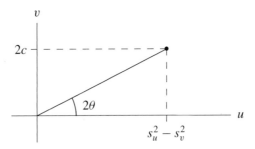

Figure 4.4.3 Finding the value of θ that minimizes D.

This equation can be solved for $\tan 2\theta$ to yield

$$\tan 2\theta = \frac{2c}{s_u^2 - s_v^2} \qquad (s_u \neq s_v),$$

as illustrated in Figure 4.4.3. Therefore, the angle that minimizes D is

$$\theta = \frac{1}{2} \tan^{-1}\left(s_u^2 - s_v^2,\ 2c\right),$$

where $\tan^{-1}(u, v)$ is the usual 4-quadrant inverse tangent function, measured counterclockwise from the positive u-axis.* This discussion is summarized by the following theorem.

Theorem 4.4.2 The line that best fits the data (u_i, v_i) for $i = 1, 2, \ldots, n$, in a mean-squared orthogonal-distance sense, passes through the point $(\overline{u}, \overline{v})$ at the angle

$$\theta = \frac{1}{2} \tan^{-1}\left(s_u^2 - s_v^2,\ 2c\right)$$

measured counterclockwise relative to the positive u-axis. The equation of the line is

$$v = (u - \overline{u}) \tan(\theta) + \overline{v},$$

provided that $\theta \neq \pi/2$. [By convention, $-\pi < \tan^{-1}(u, v) \leq \pi$; hence, $-\pi/2 < \theta \leq \pi/2$.]

Definition 4.4.3 The line that best fits the data (u_i, v_i) for $i = 1, 2, \ldots, n$ is known as the (mean-square orthogonal-distance) *linear-regression* line.[†]

---

*The function `atan2(v, u)` in the ANSI C library `<math.h>` represents the mathematical function $\tan^{-1}(u, v)$. Note the switch of the u and v arguments between the library form and the mathematical form.

[†]The derivation of an alternate mean-square *nonorthogonal*-distance linear-regression line is outlined in the exercises.

As a corollary to Theorem 4.4.2, it can be shown that the smallest possible value of D satisfies

$$2D_{\min} = \left(s_u^2 + s_v^2\right) - \sqrt{\left(s_u^2 - s_v^2\right)^2 + 4r^2 s_u^2 s_v^2},$$

which can be written equivalently as

$$D_{\min} = \frac{2\left(1 - r^2\right) s_u^2 s_v^2}{s_u^2 + s_v^2 + \sqrt{\left(s_u^2 - s_v^2\right)^2 + 4r^2 s_u^2 s_v^2}}.$$

The details of this derivation are left as an exercise. There are three important observations that follow immediately from this equation and from the fact that $D_{\min}$ cannot be negative.

- The correlation coefficient satisfies the inequality $-1 \leq r \leq 1$.
- The closer $|r|$ is to 1, the smaller is the dispersion of the (u, v) data about the regression line and the better is the (linear) fit.
- All the (u_i, v_i) points lie on the regression line, if and only if $D_{\min} = 0$, or, equivalently, if and only if $|r| = 1$.

Example 4.4.1

The scatterplot in Definition 4.4.1, reproduced in Figure 4.4.4, corresponds to 82 student scores on two standardized tests of English verbal skills: the Test of English as a Foreign Language (TOEFL) and Graduate Record Examination (GRE). Although one might hope that the two test scores would be highly correlated, with r close to 1, in this case $r = 0.59$. It is consistent with the considerable scatter that is evident about the linear regression line that the correlation is not particularly high. The consistency between the two tests is certainly less than is desirable.

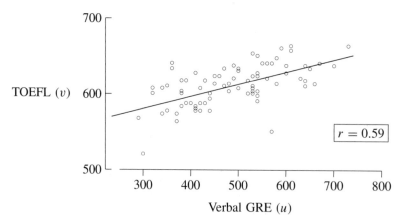

Figure 4.4.4   English-verbal-measures scatterplot.

Significance.   The magnitude of $|r|$ is a measure of the extent to which there is a *linear* relation between the u and v data. Associated terminology is provided by the following definition.

Definition 4.4.4   If $r \neq 0$, then the slope of the regression line is positive ($\theta > 0$) if and only if $r > 0$, and the slope of the regression line is negative ($\theta < 0$) if and only if $r < 0$. If r is close to $+1$ the data is said to be *positively correlated*. If r is close to -1 the data is said to be *negatively correlated*. If r is close to 0 then the data is said to be *uncorrelated*.

Example 4.4.2

A modified version of program ssq2 was used to generate interarrival, service, delay, and wait times for a steady-state sample of 100 jobs passing through an $M/M/1$ service node with arrival rate 1.0 and service rate 1.25. An $M/M/1$ service node has *Exponential* interarrival time, *Exponential* service times, and a single server. By pairing these four times, a total of six bivariate scatterplots could be formed, four of which are illustrated in Figure 4.4.5.

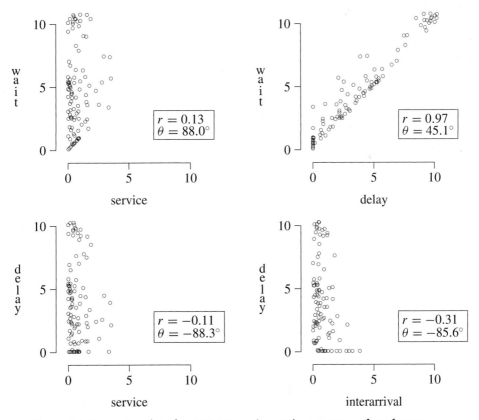

Figure 4.4.5   Scatterplots for $M/M/1$ service node measures of performance.

In this case, we see, as expected, that the strongest positive correlation is between a job's delay and wait. For the other three pairings illustrated, the correlation is weak, if nonzero at all.

The statistical question of how far $|r|$ has to be from 0 to conclude that the bivariate data is actually correlated is a difficult one made more complicated by the fact that the decision depends on the sample size and the joint probability distribution of the two random variables. The smaller the sample size, the larger $|r|$ needs to be before one can safely conclude that the data is correlated.

If the sample size is small, then the value of r is uncertain, in the sense that another sample of the same size could produce a significantly different value of r. For example, for an $M/M/1$ service node, a job's delay and service time are *uncorrelated*. Why? Thus the weak $r = -0.11$ negative correlation indicated in Figure 4.4.5 is not statistically different from zero—another sample of size 100 will produce a *positive* correlation with probability $1/2$.

Computational Considerations.   Let's return to the covariance equation in Definition 4.4.2 and recall from Section 4.1 that there are two ways to calculate a variance (or standard deviation). One method involves *two* passes through the data: the first to evaluate the mean, the second to sum the squares of the deviations about the mean. The other method involves just *one* pass through the data. An analogous result applies to the calculation of the covariance and, therefore, the correlation coefficient. Indeed, we have already observed that there are two mathematically equivalent expressions for the covariance:

$$\underbrace{c = \frac{1}{n} \sum_{i=1}^{n} (u_i - \overline{u})(v_i - \overline{v}),}_{\text{two-pass}} \qquad \text{and} \qquad \underbrace{c = \frac{1}{n} \sum_{i=1}^{n} u_i v_i - \overline{u}\,\overline{v}.}_{\text{one-pass}}$$

For the same reasons discussed in Section 4.1, the one-pass algorithm is virtually always preferred in discrete-event simulation. Moreover, there is an extension to Welford's algorithm (Algorithm 4.1.2) that applies in this case, based on the following theorem.

Theorem 4.4.3   Let $\overline{u}_i$ and $\overline{v}_i$ denote the sample means of $u_1, u_2, \ldots, u_i$ and $v_1, v_2, \ldots, v_i$, respectively, and define

$$w_i = (u_1 - \overline{u}_i)(v_1 - \overline{v}_i) + (u_2 - \overline{u}_i)(v_2 - \overline{v}_i) + \cdots + (u_i - \overline{u}_i)(v_i - \overline{v}_i)$$

for $i = 1, 2, \ldots, n$, where w_i/i is the covariance of the first i data pairs. Then, with the initial condition $w_0 = 0$,

$$w_i = w_{i-1} + \left(\frac{i-1}{i}\right)(u_i - \overline{u}_{i-1})(v_i - \overline{v}_{i-1}) \qquad i = 1, 2, \ldots, n;$$

this provides a one-pass recursive algorithm for computing $c_{uv} = w_n/n$.

Program bvs.   The program bvs is based upon the extended version of Welford's algorithm in Theorem 4.4.3. This program illustrates the calculation of the *bivariate* sample statistics $\bar{u}$, s_u, $\bar{v}$, s_v, and r and of the linear-regression-line angle θ.

4.4.2 Serial Correlation

It is frequently the case that one is interested in the extent to which a set of data is *auto*correlated (e.g., self-correlated). This is particularly true, for example, in a steady-state analysis of the waits experienced by consecutive jobs entering a service node. Intuitively, particularly if the utilization of the service node is high, there will be a high positive correlation between the wait w_i experienced by the ith job and the wait w_{i+1} experienced by the next job. Indeed, there will be a statistically significant positive correlation between w_i and w_{i+j} for some range of small, positive j values.

In general, let $x_1, x_2, \ldots, x_n$ be data which is presumed to represent n *consecutive* observations of some stochastic process whose serial correlation we wish to characterize. In the (u_i, v_i) notation used previously in this section, we pick a (small) fixed positive integer $j \ll n$ and then associate u_i with x_i and v_i with x_{i+j}, as illustrated here.

u			x_1	x_2	x_3	$\cdots$	x_i	$\cdots$	x_{n-j}	x_{n-j+1}	$\cdots$	x_n
v	x_1	$\cdots$	x_j	x_{1+j}	x_{2+j}	x_{3+j}	$\cdots$	x_{i+j}	$\cdots$	x_n		

The integer $j > 0$ is called the *autocorrelation lag* (or *shift*). Although the value $j = 1$ is of primary interest, it is conventional to calculate the serial correlation for a *range* of lag values $j = 1, 2, \ldots, k$ where $k \ll n$.[*]

Because of the lag, we must resolve how to handle the "nonoverlap" in the data at the beginning and end. The standard way to handle this nonoverlap is to do the obvious—ignore the extreme data values. To do so, define the sample autocovariance for lag j, based only on the $n - j$ overlapping values, as

$$c_j = \frac{1}{n-j} \sum_{i=1}^{n-j} (x_i - \overline{x})(x_{i+j} - \overline{x}) \qquad j = 1, 2, \ldots, k,$$

where the sample mean, based on all n values, is

$$\overline{x} = \frac{1}{n} \sum_{i=1}^{n} x_i.$$

The associated autocorrelation is then defined as follows.

---

[*]Serial statistics are commonly known as *auto*-statistics (e.g., *autocovariance* and *autocorrelation*).

Definition 4.4.5   The *sample autocorrelation* for lag j is

$$r_j = \frac{c_j}{c_0} \qquad j = 1, 2, \ldots, k,$$

where the sample variance is

$$c_0 = s^2 = \frac{1}{n} \sum_{i=1}^{n} (x_i - \overline{x})^2.$$

Computational Considerations.   The problem with the "obvious" definition of the sample autocovariance is that an implementation based on this definition would involve a two-pass algorithm. For that reason, it is common to use the following alternative definition of the sample autocovariance. Although this definition is *not* algebraically equivalent to the "obvious" definition, if $j \ll n$ then the numerical difference between these two autocovariance definitions is slight. Because it can be implemented as a one-pass algorithm, Definition 4.4.6 is preferred (in conjunction with Definition 4.4.5).

Definition 4.4.6   The *sample autocovariance* for lag j is

$$c_j = \left(\frac{1}{n-j} \sum_{i=1}^{n-j} x_i x_{i+j} \right) - \overline{x}^2 \qquad j = 1, 2, \ldots, k.$$

Example 4.4.3

A modified version of program `ssq2` was used to generate a sample of waits and services experienced by $10\,000$ consecutive jobs processed through an $M/M/1$ service node, in steady state, with arrival rate 1.0, service rate 1.25, and utilization $1/1.25 = 0.8$. Definitions 4.4.5 and 4.4.6 were used to compute the corresponding sample autocorrelations r_j for $j = 1, 2, \ldots, 50$, in what is commonly known as an *sample autocorrelation function*, or *correlogram*, illustrated in Figure 4.4.6.

As expected, the sample *wait* autocorrelation is positive and high for small values of j, indicating that each job's wait is strongly (auto)correlated with the wait of the next few jobs that follow. Also, as expected, the sample autocorrelation decreases monotonically toward zero as j increases. The rate of decrease might appear to be slower than expected; if the utilization were smaller (larger), the rate of decrease would be higher (lower). It is quite surprising that the wait times of two jobs separated by 49 intervening jobs have a moderately strong positive correlation. Also, as expected, the sample *service* autocorrelation is essentially zero for all values of j, as is consistent with the stochastic independence of the service times.

We select the first three values of the sample autocorrelation function for the wait times, in order to interpret the magnitude of the r_j values. First, $r_0 = 1$ means that there is perfect correlation between each observation and itself, since the lag associated with r_0 is zero. The next sample autocorrelation value, $r_1 = 0.957$, indicates that adjacent (lag-1) jobs, such as job number 15 and job number 16, have a statistically significant

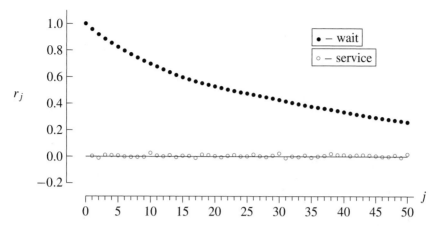

Figure 4.4.6   Sample autocorrelation functions for the $M/M/1$ queue.

strong positive correlation. If the 15th job has a long wait, then the 16th job is almost certain to have a long wait also. Likewise, if the 15th job has a short wait, then the 16th job is almost certain to have a short wait also. Anyone who has waited in a busy queue recognizes this notion intuitively. Finally, consider the estimated lag-two sample autocorrelation $r_2 = 0.918$. That this autocorrelation is not quite as strong as the lag-one sample autocorrelation is due to the increased temporal distance between the wait times. The positive value of r_2 indicates that wait times two jobs apart (e.g., the 29th and the 31st wait times) *tend* to be above the mean wait time together or below the mean wait time together.

Graphical Considerations.   Several formats are common for displaying the sample autocorrelation function. In Figure 4.4.6, we plot the r_j values as points. Another common practice is to draw "spikes" from the horizontal axis to the $r_0, r_1, \ldots, r_k$ values. It is certainly *not* appropriate to connect the points to produce a piecewise-linear function. This would imply that r_j is defined for noninteger values of j —which it is not.

Statistical Considerations.   The previous example indicated that jobs separated by 50 lags have wait times that are positively correlated—but how do we know whether r_{50} differs *significantly* from 0? Leaving out the details, Chatfield (2004, page 56) indicates that an r_j value will fall outside the limits $\pm 2/\sqrt{n}$ with approximate probability 0.95 when the lag j values are uncorrelated. In the previous example with $n = 10\,000$, for instance, these limits are at ± 0.02, indicating the *all* of the wait-time sample-autocorrelation values plotted differ significantly from 0. For the service times, only $r_{10} = 0.022$ falls outside of these limits. Experience dictates that this is simply a function of random sampling variability rather than some relationship between service times separated by ten jobs. We have set up our service-time model with independent service times, so we expect a flat sample autocorrelation function for service times. The spurious value can be ignored.

The high autocorrelation that typically exists in the time-sequenced stochastic data produced by a simulation makes the statistical analysis of the data a challenge. Specifically, if we wish to make an *interval* estimate of some steady-state statistic—for example, the average wait in a service node—we must be prepared to deal with the impact of autocorrelation on our ability to make an accurate estimate of the standard deviation. Most of so-called "classical" statistics relies on the assumption that the values sampled are drawn independently from a population. This is often not the case in discrete-event simulation, and appropriate measures must be taken in order to compute appropriate interval estimates.

Program acs. To implement Definition 4.4.6 as a one-pass algorithm for a fixed lag-value j involves nothing more than storing the values x_i, x_{i+j}, accumulating the x_i sum, and accumulating the x_i^2 and $x_i x_{i+j}$ "cosums." It is a greater challenge to construct a one-pass algorithm that will compute c_j for a *range* of lags $j = 1, 2, \ldots, k$. In addition to the accumulation of the x_i sum, the simultaneous computation of $c_0, c_1, \ldots, c_k$ involves storing the $k + 1$ consecutive values $x_i, x_{i+1}, \ldots, x_{i+k}$ and accumulating the $k + 1$ (lagged) $x_i x_{i+j}$ cosums for $j = 0, 1, 2, \ldots, k$. The $k + 1$ cosums can be stored as an array of length $k + 1$. A more interesting queue data structure is required to store the values $x_i, x_{i+1}, \ldots, x_{i+k}$. This queue has been implemented as a *circular* array in the program acs. A circular array is a natural choice here because the queue length is fixed at $k + 1$ and efficient access to *all* the elements in the queue, not just the head and tail, is required. In the following algorithm, the box indicates the rotating head of the circular queue. An array index p keeps track of the current location of the rotating head; the initial value is $p = 0$.

Algorithm 4.4.1   Program acs is based on the following algorithm. A circular queue is initially filled with $x_1, x_2, \ldots, x_k, x_{k+1}$, as illustrated by the boxed elements below. The lagged products $x_1 x_{1+j}$ are computed for all $j = 0, 1, \ldots, k$ thereby initializing the $k + 1$ cosums. Then the next data value is read into the (old) head of the queue location, p is incremented by 1 to define a new head of the queue location, the lagged products $x_2 x_{2+j}$ are computed for all $j = 0, 1, \ldots, k$, and the cosums are updated. This process is continued until all the data has been read and processed. (The case $n \bmod (k + 1) = 2$ is illustrated.)

$$
\begin{array}{llllllllll}
(i = k+1) & \boxed{x_1} & x_2 & x_3 & \cdots & x_{k-1} & x_k & x_{k+1} & (p = 0) \\
(i = k+2) & x_{k+2} & \boxed{x_2} & x_3 & \cdots & x_{k-1} & x_k & x_{k+1} & (p = 1) \\
(i = k+3) & x_{k+2} & x_{k+3} & \boxed{x_3} & \cdots & x_{k-1} & x_k & x_{k+1} & (p = 2) \\
& \vdots & \vdots & \vdots & \vdots & \vdots & \vdots & \vdots & \vdots \\
(i = 2k) & x_{k+2} & x_{k+3} & x_{k+4} & \cdots & x_{2k} & \boxed{x_k} & x_{k+1} & (p = k) \\
(i = 2k+1) & x_{k+2} & x_{k+3} & x_{k+4} & \cdots & x_{2k} & x_{2k+1} & \boxed{x_{k+1}} & (p = k+1) \\
(i = 2k+2) & \boxed{x_{k+2}} & x_{k+3} & x_{k+4} & \cdots & x_{2k} & x_{2k+1} & x_{2k+2} & (p = 0) \\
& \vdots & \vdots & \vdots & \vdots & \vdots & \vdots & \vdots & \vdots \\
(i = n) & x_{n-1} & x_n & \boxed{x_{n-k}} & \cdots & x_{n-4} & x_{n-3} & x_{n-2} & (p = 2)
\end{array}
$$

After the last data value, x_n, has been read, the associated lagged products computed, and the cosums updated, all that remains is to "empty" the queue. This can be accomplished by effectively reading k additional 0-valued data values. For more details, see program acs.

4.4.3 Exercises

4.4.1 Prove that the orthogonal distance from the point (u_i, v_i) to the line $a\,u + b\,v + c = 0$, where a, b, and c are constants, is in fact

$$d_i = \frac{|a\,u_i + b\,v_i + c|}{\sqrt{a^2 + b^2}}.$$

Hint: consider the squared distance $(u - u_i)^2 + (v - v_i)^2$ from (u_i, v_i) to a point (u, v) on the line and show that d_i^2 is the smallest possible value of this distance.

4.4.2 (a) If $u'_i = \alpha_u u_i + \beta_u$ and $v'_i = \alpha_v v_i + \beta_v$ for $i = 1, 2, \ldots, n$ and constants α_u, α_v, β_u, and β_v, how does the covariance of the u', v' data relate to the covariance of the u, v data? (b) Same question for the correlation coefficients. (c) Comment.

4.4.3 The orthogonal-distance regression derivation presented in this section treats both variables equally—there is no presumption that one variable is "independent" and the other is "dependent." Consider the more common regression approach, in which the equation of the regression line is $v = au + b$, consistent with a model that treats u as independent and v as dependent. That is, given the data (u_i, v_i) for $i = 1, 2, \ldots, n$ and the line defined by the equation $v = au + b$, the *conventional* (nonorthogonal) distance from the point (u_i, v_i) to the line is

$$\delta_i = |v_i - (au_i + b)|.$$

(a) What choice of the (a, b) parameters will minimize the conventional mean-square distance

$$\Delta = \frac{1}{n}\sum_{i=1}^{n}\delta_i^2 = \frac{1}{n}\sum_{i=1}^{n}(v_i - au_i - b)^2?$$

(b) Prove that the minimum value of Δ is $(1 - r^2)s_v^2$.

4.4.4 Prove Theorem 4.4.3.

4.4.5 To what extent are these two definitions of the autocovariance different?

$$c'_j = \frac{1}{n-j}\sum_{i=1}^{n-j}(x_i - \overline{x})(x_{i+j} - \overline{x}) \qquad \text{and} \qquad c_j = \left(\frac{1}{n-j}\sum_{i=1}^{n-j}x_i x_{i+j}\right) - \overline{x}^2$$

4.4.6 (a) Generate a figure like the one in Figure 4.4.6, but corresponding to the utilization 0.9. Do this for three different rngs streams. (b) Repeat for the utilization 0.7. (You can ignore the service autocorrelations.) (c) Comment.

4.4.7 If Definition 4.4.6 is used in conjunction with Definition 4.4.5, there is no *guarantee* that $|r_j| \leq 1$ for all $j = 1, 2, \ldots, k$. If it is important to guarantee that this inequality is true, then how should the two definitions be modified?

5

Next-Event Simulation

The three sections in this chapter all concern the *next-event* approach to discrete-event simulation. Section 5.1 defines the fundamental terminology used in next-event simulation: system state, events, simulation clock, event scheduling, and event list (which is also known as the *calendar*); and it provides an introduction to this fundamental approach as it applies to the simulation of a single-server service node with and without feedback. The algorithm associated with next-event simulation initializes the simulation clock (typically to time zero), event list (with an initial arrival, for example, in a queuing model), and system state, to begin the simulation. The simulation model continues to (1) remove the next event from the event list, (2) update the simulation clock to the time of the next event, (3) process the event, and (4) schedule the time of occurrence of any future events spawned by the event, until some terminal condition is satisfied.

Section 5.2 provides further illustrations of this approach in the simulation of a simple inventory system with delivery lag and of a multiple-server service node. The multiple-server service node provides an illustration of an event list that can have an arbitrarily large number of elements.

As the simulations in Sections 5.1 and 5.2 illustrate, an *event list* is an integral feature of the next-event approach. The data structures and algorithms that are used to manage the event list are crucial to the efficiency of a next-event simulation. Section 5.3 provides a sequence of examples associated with the management of an event list that begin with a naive and inefficient data structure and algorithm and iterate toward a more efficient scheme.

5.1 NEXT-EVENT SIMULATION

In this section, we will present a general *next-event* approach to building discrete-event simulation models. From this chapter on, this next-event approach will be the basis for all the discrete-event simulation models developed in this book.

The motivation for considering the next-event approach to discrete-event simulation is provided by considering the relative complexity of the effort required to extend the discrete-event simulation models in Section 3.1 to accommodate the slightly more sophisticated corresponding models in Section 3.3. For example, at the computational level, compare the simplicity of program ssq2 in Section 3.1 with the increased complexity of the extension to ssq2 that would be required to reproduce the results in Example 3.3.2; yet, the only increase in the complexity of the associated single-server service-node model is the addition of immediate feedback. Similarly, compare the simplicity of program sis2 in Section 3.1 with the increased complexity of the extension to sis2 that would be required to reproduce the results in Example 3.3.4; yet, in this case, the only increase in the complexity of the associated simple inventory system model is the addition of a delivery lag.

5.1.1 Definitions and Terminology

Programs ssq2 and sis2 and their corresponding extensions in Section 3.3 are valid and mean-ingful (albeit simple) discrete-event simulation programs, but they do not adapt easily to increased model complexity and they do not generalize well to other systems. From these observations, we see the need for a more general approach to discrete-event simulation—one that applies to queuing systems, inventory systems, and a variety of other systems. This more general approach—next-event simulation—is based on some important definitions and terminology: (1) system state, (2) events, (3) simulation clock, (4) event scheduling, and (5) event list (calendar).

System State

Definition 5.1.1   The *state* of a system is a complete characterization of the system at one instant—a comprehensive "snapshot" at a particular time. To the extent that the state of a system can be characterized by assigning values to variables, then *state variables* are used for this purpose.

To build a discrete-event simulation model via the next-event approach, the focus is on refining a description of the state of the system and its evolution in time. At the *conceptual*-model level, the state of a system exists only in the abstract and as a collection of possible answers to the following questions: What are the state variables, how are they interrelated, and how do they evolve in time? At the *specification* level, the state of the system exists as a collection of mathematical variables (the state variables), together with equations and logic describing how the state variables are interrelated and with an algorithm for computing their interaction and evolution in time. At the *computational* level, the state of the system exists as a collection of program variables that collectively characterize the system and are systematically updated as (simulated) time evolves.

Example 5.1.1
A natural way to describe the state of a single-server service node is to use the number of jobs in the service node as a state variable. As is demonstrated later in this section, by refining this system-state description, we can construct a next-event simulation model for a single-server service node with or without immediate feedback.

Example 5.1.2

Similarly, a natural way to describe the state of a simple inventory system is to use the current inventory level and the amount of inventory on order (if any) as state variables. As is demonstrated in the next section, by refining this system state description, we can construct a next-event simulation model of a simple inventory system with or without delivery lag.

Events

> **Definition 5.1.2**   An *event* is an occurrence that could change the state of the system. By definition, the state of the system cannot change except at an event time. Each event has an associated *event type*.

Example 5.1.3

For a single-server service-node model with or without immediate feedback, there are two types of events: the *arrival* of a job and the *completion of service* for a job. These two types of occurrence are events because they have the potential to change the state of the system. An arrival will always increase the number in the service node by one; if there is no feedback, a completion of service will always decrease the number in the service node by one. When there is feedback, a completion might decrease the number in the service node by one. In this case, there are two event types, because the "arrival" event type and the "completion of service" event type are not the same.

Example 5.1.4

For a simple inventory system with delivery lag, there are three event types: the occurrence of a *demand* instance, an *inventory review*, and the *arrival* of an inventory replenishment order. These are events, because they have the potential to change the state of the system: A demand will decrease the inventory level by one, an inventory review might lead to an increase in the amount of inventory on order, and the arrival of an inventory replenishment order will increase the inventory level and decrease the amount of inventory on order.

The *could* in Definition 5.1.2 is important; it is not necessary for an event to cause a change in the state of the system, as is illustrated in the following four examples: (1) events can be scheduled that statistically *sample*, but do not change, the state of a system; (2) for a single-server service node with immediate feedback, a job's completion of service will change the state of the system only if the job is not fed back; (3) for a single-server service node, an event may be scheduled at a prescribed time (e.g., 5 PM) to cut off the stream of arriving jobs to the node, which will not change the state of the system; and (4) for a simple inventory system with delivery lag, an inventory review will change the state of the system only if an order is placed.

Simulation Clock.   A discrete-event simulation model is dynamic; so, as the simulated system evolves, it is necessary to keep track of the current value of simulated time. In the implementation phase of a next-event simulation, the natural way keep track of simulated time is with a floating-point variable (typically named `t`, `time`, `tnow`, or `clock` in discrete-event simulation packages).

The two examples that follow the definition of the simulation clock highlight the inability of the discrete-event simulation approach to easily generalize or embellish models. The next-event framework overcomes this limitation.

> **Definition 5.1.3** The variable that represents the current value of simulated time in a next-event simulation model is called the *simulation clock*.

Example 5.1.5
The discrete-event simulation model that program ssq2 represents is heavily dependent on the job-processing order imposed by the FIFO queue discipline. Therefore, it is difficult to extend the model to account for immediate feedback, or for a finite service-node capacity, or for a priority queue discipline. In part, the reason for this difficulty is that there are effectively *two* simulation clocks: one coupled to the arrival events, the other coupled to the completion-of-service events. These two clocks are not synchronized, and so it is difficult to reason about the temporal order of events if arrivals are merged by feedback with completions of service.

Example 5.1.6
The discrete-event simulation model that program sis2 represents has only one type of event, inventory review, and events of this type occur deterministically at the beginning of each time interval. There is a simulation clock, but it is *integer-valued* and so is primitive at best. Because the simulation clock is integer-valued, we are essentially forced to ignore the individual demand instances that occur within each time interval. Instead, all the demands per time interval are aggregated into one random variable. This aggregation makes for a computationally efficient discrete-event-simulation program, but forces us in return to do some calculus to derive equations for the time-averaged holding and shortage levels. As was outlined in Section 3.3, when there is a delivery lag, the derivation of those equations is a significant task.

Event Scheduling.   In a discrete-event simulation model, it is necessary to use a *time-advance mechanism* to guarantee that events occur in the correct order—that is, to guarantee that the simulation clock never runs backward. The primary time-advance mechanism used in discrete-event simulation is known as *next-event* time advance; this mechanism is typically used in conjunction with *event scheduling*.

> **Definition 5.1.4** If event scheduling is used with a next-event time-advance mechanism as the basis for developing a discrete-event simulation model, the result is called a *next-event simulation model*.

To construct a next-event simulation model, three things must be done:

- construct a set of state variables that, together, provide a complete system description;
- identify the system's event types;

- construct a collection of algorithms that define the state changes that will take place when each type of event occurs.

The model is constructed so as to cause the (simulated) system to evolve in (simulated) time by executing the events in increasing order of their scheduled time of occurrence. Time does not flow continuously; instead, the simulation clock is advanced discontinuously from event time to event time. At the computational level, the simulation clock is frozen during the execution of each state-change algorithm so that each change of state, no matter how complex computationally, occurs instantaneously relative to the simulation clock.

Event List

> **Definition 5.1.5**   The data structure that represents the scheduled time of occurrence for the next possible event of each type is called the *event list* or *calendar*.

The event list is often, but not necessarily, represented as a priority queue sorted by the next scheduled time of occurrence for each event type.

5.1.2 Next-Event Simulation

> **Algorithm 5.1.1**   A next-event simulation model consists of the following four steps:
>
> - **Initialize.** The simulation clock is initialized (usually to zero). Then, by means of looking ahead, the first time of occurrence of each possible event type is found and scheduled; this process initializes the event list.
> - **Process current event.** The event list is scanned to locate the *most imminent* possible event. The simulation clock is then advanced to this event's scheduled time of occurrence, and the state of the system is updated to account for the occurrence of this event. This event is known as the "current" event.
> - **Schedule new events.** New events (if any) that might be spawned by the current event are placed on the event list (typically, in chronological order). The algorithm returns to the "Process current event" step.
> - **Terminate.** The process of advancing the simulation clock from one event time to the next continues until some terminal condition is satisfied. This terminal condition may be specified as a *pseudo*-event that only occurs once, at the end of the simulation, with the specification typically based on processing a fixed number of events, exceeding a fixed simulation clock time, or estimating an output measure to a prescribed precision.
>
> The next-event simulation model initializes once at the beginning of a simulation replication, then alternates between the second step (processing the current event) and third step (scheduling new events) until some termination criterion is encountered.

Because event times typically are random, the simulation clock runs *asynchronously*: Advances in the simulation clock occur only when an event occurs, which will almost certainly not synchronize with a fixed-advance system clock. Moreover, because state changes occur *only* at event times, periods of system inactivity are ignored by advancing the simulation clock from event time to event time. In a comparison to the alternative, which is a *fixed-increment* time-advance mechanism, there is a clear computational-efficiency advantage to asynchronous next-event processing.*

In the remainder of this section, next-event simulation will be illustrated by constructing a next-event model of a single-server service node. Additional illustrations are provided in the next section by constructing next-event simulation models of a simple inventory system with delivery lag and a multiple-server service node.

5.1.3 Single-Server Service Node

The state variable $l(t)$ provides a *complete* characterization of the state of a single-server service node, in the sense that

$$l(t) = 0 \iff q(t) = 0 \quad \text{and} \quad x(t) = 0$$

$$l(t) > 0 \iff q(t) = l(t) - 1 \quad \text{and} \quad x(t) = 1$$

where $l(t)$, $q(t)$, and $x(t)$ represent the number in the node, in the queue, and in service, respectively, at time $t > 0$. In words, if the number in the service node is known, then the number in the queue and the status (idle or busy) of the server are also known. Given that the state of the system is characterized by $l(t)$, we then ask, what events can cause $l(t)$ to change? The answer is that there are two such events: (1) an *arrival*, in which case $l(t)$ is increased by 1; and (2) a *completion of service*, in which case $l(t)$ is decreased by 1. Therefore, our conceptual model of a single-server service node consists of the state variable $l(t)$ and two associated event types: arrival and completion of service.

To turn this next-event conceptual model into a specification model, three additional assumptions must be made.

- The *initial state* $l(0)$ can have any nonnegative integer value. It is common, however, to assume that $l(0) = 0$, often referred to as the "empty and idle" state in reference to the initial queue condition and server status, respectively. Therefore, the first event must be an arrival.

- Although the *terminal state* can also have any nonnegative integer value, it is common to assume, as we will do, that the terminal state is also idle. Rather than specifying the number of jobs processed, our stopping criteria will be specified in terms of a time τ beyond which no new jobs can arrive. This assumption effectively "closes the door" at time τ but allows the system to continue operation until all jobs have been completely served. This would be the case, for instance, at an ice cream shop that closes at a particular

---

*Note that asynchronous next-event processing cannot be used if there is a need at the computational level for the simulation program to interact synchronously with some other process. For example, because of the need to interact with a person, so-called "real time" or "person-in-the-loop" simulation programs must use a fixed-increment time-advance mechanism. In this case, the underlying system model is usually based on a system of ordinary differential equations. In any case, fixed-increment time-advance simulation models are outside the scope of this book, but are included in some of the languages surveyed in Appendix A.

hour, but allows remaining customers to be served. Therefore, the last event must be a completion of service.*

• Some mechanism must be used to denote an event as *impossible*. One way to do this is to structure the event list so that it contains *possible* events only. This is particularly desirable if the number of event types is large. As an alternate, if the number of event types is not large, then the event list can be structured so that it contains both possible and impossible events—but with a numeric constant "∞" used for an event time to denote the impossibility of an event. For simplicity, this alternative event-list structure is used in Algorithm 5.1.2.

To complete the development of a specification model, the following notation is used. The next-event specification model is then sufficiently simple that we can write Algorithm 5.1.2 directly.

• The simulation clock (current time) is t.
• The terminal ("close the door") time is τ.
• The next scheduled arrival time is t_a.
• The next scheduled service completion time is t_c.
• The number in the node (state variable) is l.

The genius and allure of both discrete-event and next-event simulation is apparent, for example, in the generation of arrival times in Algorithm 5.1.2. The naive approach—generating and storing all arrivals prior to the execution of the simulation—is not necessary. Even if this naive approach were taken, the modeler would be beset by dual problems: memory consumption and not knowing how many arrivals to schedule. Next-event simulation simply primes the pump by scheduling the first arrival in the initialization phase, then schedules each subsequent arrival while processing the current arrival. Meanwhile, service completions weave their way into the event list at the appropriate moments in order to provide the appropriate sequencing of arrivals and service completions.

At the end of this section, we will discuss how to extend Algorithm 5.1.2 to account for several model extensions: immediate feedback, alternative queue disciplines, finite capacity, and random sampling.

Algorithm 5.1.2   This algorithm is a next-event simulation of a FIFO single-server service node with infinite capacity. The service node begins and ends in an empty and idle state. The algorithm presumes the existence of two functions, `GetArrival` and `GetService`, that return a random arrival time and service time, respectively.

```
l = 0;                        // initialize the system state
t = 0.0;                      // initialize the system clock
tₐ = GetArrival();            // initialize the event list
t_c = ∞;                      // initialize the event list
```

---

*The simulation will terminate at $t = \tau$ only if $l(\tau) = 0$. If, instead, $l(\tau) > 0$, then the simulation will terminate at $t > \tau$, because additional time will be required to complete service on the jobs in the service node.

```
while ((t_a < τ) or (l > 0)) {       // check for terminal condition
   t = min(t_a, t_c);                // scan the event list
   if (t == t_a) {                   // process an arrival
      l++;
      t_a = GetArrival();
      if (t_a > τ)
         t_a = ∞;
      if (l == 1)
         t_c = t + GetService();
   }
   else {                            // process a completion of service
      l--;
      if (l > 0)
         t_c = t + GetService();
      else
         t_c = ∞;
   }
}
```

If the service node is to be an $M/M/1$ queue (exponential interarrival and service times with a single server) with arrival rate 1.0 and service rate 1.25, for example, the two t_a = GetArrival() statements can be replaced with t_a = t + Exponential(1.0), and the two t_c = t + GetService() statements can be replaced with t_c = t + Exponential(0.8). The GetArrival and GetService functions can draw their values from a file (a "trace-driven" approach) or generate random variates to model these stochastic elements of the service node.[*]

Because there are just two event types, arrival and completion of service, the event list in Algorithm 5.1.2 contains at most two elements, t_a and t_c. Given that the event list is small and its size is bounded (by 2), there is no need for any special data structure to represent it. If the event list were larger, an array or structure would be a natural choice. The only drawback to storing the event list as t_a and t_c is the need to specify the arbitrary numeric constant "∞" to denote the impossibility of an event. In practice, ∞ can be any number that is *much* larger than the terminal time $τ$ ($100τ$ is used in program ssq3).

If the event list is large and its size is dynamic, then a dynamic data structure is required, with careful attention paid to its organization. This is necessary because the event list is scanned and updated each time an event occurs. Efficient algorithms for the insertion and deletion of events on the event list can affect the computational time required to execute the next-event simulation model. Henriksen (1983) reports that, for telecommunications-system models, the choice between an efficient and inefficient event-list processing algorithm can produce a five-fold difference in total processing time. Further discussion of data structures and algorithms associated with event lists is postponed to Section 5.3.

---

[*]The $M/M/1$ queue has analytic results available, so there is no need to simulate to estimate these known quantities.

Program ssq3. Program ssq3 is based on Algorithm 5.1.2. Note, in particular, the state variable number, which represents $l(t)$ (the number in the service node at time t) and the important time-management structure t, which contains

- the event list, t.arrival and t.completion (t_a and t_c from Algorithm 5.1.2),
- the simulation clock, t.current (t from Algorithm 5.1.2),
- the next event time, t.next ($\min(t_a, t_c)$ from Algorithm 5.1.2), and
- the last arrival time, t.last.

Event-list management is trivial. The event type (arrival or a completion of service) of the next event is determined by the statement t.next = Min(t.arrival, t.completion).

Note also that a statistics-gathering structure area is used to calculate the time-averaged number in the node, in the queue, and in service. These statistics are calculated exactly by accumulating time integrals via summation, as is valid because $l(\cdot)$, $q(\cdot)$, and $x(\cdot)$ are piecewise-constant functions and change value only at an event time (per Section 4.1). The structure area contains

- $\int_0^t l(s)\,ds$, evaluated as area.node,

- $\int_0^t q(s)\,ds$, evaluated as area.queue, and

- $\int_0^t x(s)\,ds$, evaluated as area.service.

Program ssq3 does not accumulate job-averaged statistics. Instead, the job-averaged statistics $\overline{w}$, $\overline{d}$, and $\overline{s}$ are computed from the time-averaged statistics $\overline{l}$, $\overline{q}$, and $\overline{x}$ via the equations in Theorem 1.2.1. The average interarrival time $\overline{r}$ is computed from the equation in Definition 1.2.4 by using the variable t.last. If it were not for the use of the assignment t.arrival = INFINITY to "close the door," $\overline{r}$ could be computed from the terminal value of t.arrival, thereby eliminating the need for t.last.

World Views and Synchronization. Programs ssq2 and ssq3 simulate exactly the same system. The programs work in different ways, however, with one clear consequence being that ssq2 naturally produces *job*-averaged statistics and ssq3 naturally produces *time*-averaged statistics. In the jargon of discrete-event simulation, the two programs are said to be based upon different *world views.** In particular, program ssq2 is based upon a *process-interaction* world view, program ssq3 upon an *event-scheduling* world view. Although other world views are sometimes advocated, process-interaction and event-scheduling are the two most common. Of these two, event-scheduling is the discrete-event simulation world view of choice in this and all the remaining chapters.

Because programs ssq2 and ssq3 simulate exactly the same system, these programs should be able to produce exactly the same output statistics. Getting them to do so, however, requires that

---

*A world view is the collection of concepts and views that guide the development of a simulation model. World views are also known as *conceptual frameworks*, *simulation strategies*, and *formalisms*.

both programs process exactly the same stochastic source of arriving jobs and associated service requirements. Because the arrival times a_i and service times s_i are ultimately produced by calls to Random, some thought is required to provide this synchronization. The random variates in program ssq2 are always generated in the alternating order $a_1, s_1, a_2, s_2, \ldots$; the order in which these random variates are generated in ssq3 cannot be known a priori. The best way to produce this synchronization is to use the library rngs, as is done in program ssq3. In Exercise 5.1.3, you are asked to modify program ssq2 to use the library rngs and, in that way, verify that the two programs can produce exactly the same output.

5.1.4 Model Extensions

We close this section by discussing how to modify program ssq3 to accommodate several important model extensions. For each of the four extensions, you are encouraged to consider what would be required to extend program ssq2 correspondingly.

Immediate Feedback.   Given the function GetFeedback from Section 3.3, we can modify program ssq3 to account for immediate feedback by just adding an if statement so that index and number are not changed if a feedback occurs following a completion of service, as illustrated.

```
else {                          // process a completion of service
   if (GetFeedback() == 0) {            // this statement is new
      index++;
      number--;
   }
```

For a job that is not fed back, the counter for the number of departed jobs (index) is incremented and the counter for the current number of jobs in the service node (number) is decremented. The simplicity of the immediate-feedback modification is a compelling example of how well next-event simulation models accommodate model extensions.

Alternative Queue Disciplines.   Program ssq3 can be modified to simulate *any* queue discipline. To do so, it is necessary to add a dynamic-queue data structure—for example, a singly linked list, where each list node contains the arrival time and service time for a job in the queue, as illustrated in Figure 5.1.1.*

Two supporting queue functions, Enqueue and Dequeue, are also needed to insert and delete jobs from the queue, respectively. Then

- use Enqueue each time an arrival event occurs and the server is busy;
- use Dequeue each time a completion-of-service event occurs and the queue is not empty.

---

*If this queue data structure is used, then service times are computed and stored at the time of arrival. In this way, each job's delay in the queue and wait in the service node can be computed at the time of entry into service, thereby eliminating the need to compute job-averaged statistics from time-averaged statistics.

Figure 5.1.1   Queue data structure.

The details of this important modification are left as an exercise. This modification will result in a program that can be tested for correctness by using a FIFO queue discipline and reproducing results from program ssq3.

Note, in particular, that this modification can be combined with the immediate-feedback modification illustrated previously. In this case, the arrival field in the linked list would hold the time of feedback for those jobs that are fed back. The resulting program would allow a priority-queue discipline to be used for fed-back jobs if (as is common) a priority assumption is appropriate.

Finite Service-Node Capacity.   Program ssq3 can be modified to account for a finite capacity by defining a constant CAPACITY, which represents the service-node capacity (one more than the queue capacity) and declaring an integer variable reject which counts rejected jobs. Then all that is required is a modification to the "process an arrival" portion of the program, as illustrated.

```
if (t.current == t.arrival) {                    // process an arrival
  if (number < CAPACITY) {
    number++;
    if (number == 1)
      t.completion = t.current + GetService();
  }
  else
    reject++;
  t.arrival       = GetArrival();
  if (t.arrival > STOP) {
    t.last        = t.current;
    t.arrival     = INFINITY;
  }
}
```

This code replaces the code in program ssq3 for processing an arrival. As with the immediate-feedback modification, again we see that the simplicity of this modification is a compelling example of how well next-event simulation models accommodate model extensions.

Random Sampling.   An important feature of program ssq3 is that its structure facilitates direct *sampling* of the current number in the service node or queue. This is easily accomplished by adding a sampling time element—say, t.sample—to the event list and constructing an associated algorithm to process the samples as they are acquired. Sampling times can then be scheduled *deterministically*, every δ time units, or *at random*, by generating sampling times with an *Exponential*(δ) random-variate inter-sample time. In either case, the details of this modification are left as an exercise.

5.1.5 Exercises

5.1.1   Consider a next-event simulation model of a three-server service node with a single queue and three servers. (*a*) What variable(s) are appropriate to describe the system state? (*b*) Define appropriate events for the simulation. Be specific: How does each event change the variable(s) you defined in part (*a*) that describe the system state? (*c*) What is the maximum length of the event list for the simulation? (Answer with and without considering the pseudo-event for termination of the replication.)

5.1.2   Consider a next-event simulation model of three single-server service nodes in series. (*a*) What variable(s) are appropriate to describe the system state? (*b*) Define appropriate events for the simulation. Be specific: How does each event change the variable(s) you defined in part (*a*) that describe the system state? (*c*) What is the maximum length of the event list for the simulation? (Answer with and without considering the pseudo-event for termination of the replication.)

5.1.3   (*a*) Use the library rngs to verify that programs ssq2 and ssq3 can produce *exactly* the same results. (*b*) Comment on the value of this as a consistency check for both programs.

5.1.4   Add a sampling capability to program ssq3. (*a*) With deterministic inter-sample time $\delta = 1.0$, sample the number in the service node and compare the average of these samples with the value for $\bar{l}$ computed by the program. (*b*) With average inter-sample time $\delta = 1.0$, sample *at random* the number in the service node and compare the average of these samples with the value for $\bar{l}$ computed by the program. (*c*) Comment.

5.1.5   Modify program ssq3 by adding a FIFO-queue data structure. Verify that this modified program and ssq3 produce *exactly* the same results.

5.1.6   As a continuation of Exercise 5.1.5, simulate a single-server service node for which the server uses a shortest-job-first priority-queue discipline that is based upon a knowledge of the service time for each job in the queue. (*a*) Generate a histogram of the wait in the node for the first 10000 jobs if interarrival times are *Exponential*(1.0) and service times are *Exponential*(0.8). (*b*) How does this histogram compare with the corresponding histogram generated when the queue discipline is FIFO? (*c*) Comment.

5.1.7   Modify program ssq3 to reproduce the feedback results obtained in Section 3.3.

5.1.8   Modify program ssq3 to account for a finite service-node capacity. (*a*) Find the proportion of rejected jobs for the capacities 1, 2, 3, 4, 5, and 6. (*b*) Repeat this experiment if the service time distribution is *Uniform*(1.0, 3.0). (*c*) Comment. (Use a large value of STOP.)

5.1.9   (*a*) Construct a next-event simulation model of a single-server machine shop. (*b*) Compare your program with the program ssms and verify that, with a proper use of the library rngs, the two programs can produce *exactly* the same output. (*c*) Comment on the value of this result as a consistency check for both programs.

5.1.10   An $M/M/1$ queue can be characterized by the following system-state change mechanism, where the system state is $l(t)$, the number of jobs in the node:

- The transition from state j to state $j + 1$ is exponential with rate $\lambda_1 > 0$ (the arrival rate) for $j = 0, 1, \ldots$.
- The transition from state j to state $j - 1$ is exponential with rate $\lambda_2 > 0$ (the service rate) for $j = 1, 2, \ldots$.

If the current state of the system is some positive integer j, what is the probability that the next transition will be to state $j + 1$?

5.2 NEXT-EVENT SIMULATION EXAMPLES

As a continuation of the discussion in the previous section, in this section, two next-event simulation models will be developed. The first is a next-event simulation model of a simple inventory system with delivery lag, the second is a next-event simulation model of a multiple-server service node.

5.2.1 A Simple Inventory System with Delivery Lag

To develop a next-event simulation model of a simple inventory system with delivery lag, we make two changes relative to the model on which program `sis2` is based. The first change is consistent with the discussion of delivery lag in Section 3.3. The second change is new and provides a more realistic demand model.

- There is a lag between the time of inventory review and the delivery of any inventory replenishment order that is placed at the time of review. This delivery lag is assumed to be a *Uniform*$(0, 1)$ random variable, independent of the size of the order. Consistent with this assumption, the delivery lag cannot be longer than a unit time interval; consequently, any order placed at the beginning of a time interval will arrive by the end of the time interval, before the next inventory review.
- The demands per time interval are no longer aggregated into one random variable and assumed to occur at a constant rate during the time interval. Instead, individual demand instances are assumed to occur *at random* throughout the simulated period of operation with an average rate of λ demand instances per time interval. In other words, each demand instance produces a demand for exactly one unit of inventory, and the inter-demand time is an *Exponential*$(1/\lambda)$ random variable.

Figure 5.2.1 shows the first six time intervals of a typical inventory-level time history, $l(t)$. Each demand instance causes the inventory level to decrease by one. Inventory review for the ith time interval occurs at $t = i - 1 = 0, 1, 2, \ldots$ with an inventory-replenishment order in the amount $o_{i-1} = S - l(i - 1)$ placed only if $l(i - 1) < s$. Following a delivery lag δ_i, the subsequent arrival of this order causes the inventory to experience an increase of o_{i-1} units at time $t = i - 1 + \delta_i$, as illustrated for $i = 3$ and $i = 6$ in Figure 5.2.1.

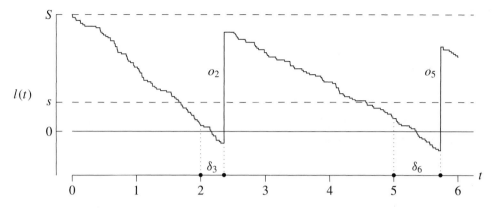

Figure 5.2.1   Inventory-level history.

Recall that, in program sis2, the aggregate demand in each time interval is generated as an *Equilikely*(10, 50) random variate. Although the aggregate demand in each time interval can be any value between 10 and 50, within each interval there is nothing random about the occurrence of the individual demands—the inter-demand time is constant. Thus, for example, if the random-variate aggregate demand in a particular interval is 25, then the inter-demand time throughout that interval is 0.04.

In contrast to the demand model in program sis2, it is more realistic to generate the inter-demand time as an *Exponential*$(1/\lambda)$ random variate. In this way, the demand is modeled as an arrival process (e.g., customers arriving at random to buy a car), with λ being the arrival rate *per time interval*. Thus, for example, if we want to generate demands with an average of 30 per time interval, then we would use $\lambda = 30$.

States.   To develop a next-event simulation model of this system at the specification level, the following notation is used.

- The simulation clock (current time) is t, and the terminal time is τ.
- At any time $t > 0$, the current inventory level is $l(t)$.
- At any time $t > 0$, the amount of inventory *on order* (if any) is $o(t)$.

In addition to $l(t)$, the new state variable $o(t)$ is necessary to keep track of an inventory-replenishment order that, because of a delivery lag, has not yet arrived. Together, $l(t)$ and $o(t)$ provide a complete state description of a simple-inventory system with delivery lag. Both $l(t)$ and $o(t)$ are integer-valued. Although t is real-valued, inventory reviews occur at integer values of t only. The terminal time τ corresponds to an inventory-review time, and so it is integer-valued.

We assume the initial state of the inventory system to be $l(0) = S$, $o(0) = 0$; that is, the initial inventory level is S, and the initial inventory replenishment order level is 0. Similarly, the terminal state is assumed to be $l(\tau) = S$, $o(\tau) = 0$, with the understanding that the ordering cost associated with increasing $l(t)$ to S at the end of the simulation (at $t = \tau$, with no delivery lag) should be included in the accumulated system statistics.

Events. Given that the state of the system is defined by $l(t)$ and $o(t)$, there are three types of events that can change the state of the system:

- a *demand* for an item at time t, in which case $l(t)$ will decrease by 1;
- an inventory *review* at (integer-valued) time t, in which case $o(t)$ will increase from 0 to $S - l(t)$, provided that $l(t) < s$; else $o(t)$ will remain 0;
- an *arrival* of an inventory-replenishment order at time t, in which case $l(t)$ will increase from its current level by $o(t)$ and then $o(t)$ will decrease to 0.

To complete the development of a specification model, the time variables t_d, t_r, and t_a are used to denote the next scheduled time for the three events inventory *demand*, inventory *review*, and inventory *arrival*, respectively. As in the previous section, ∞ is used to denote (schedule) an event that is not possible.

Algorithm 5.2.1   This algorithm is a next-event simulation of a simple inventory system with delivery lag. The algorithm presumes the existence of two functions, GetLag and GetDemand, that return a random value of delivery lag and the next demand time, respectively.

```
l = S;                                    // initialize inventory level
o = 0;                                    // initialize amount on order
t = 0.0;                                  // initialize simulation clock
t_d = GetDemand();                        // initialize the event list
t_r = t + 1.0;                            // initialize the event list
t_a = ∞;                                  // initialize the event list
while (t < τ) {
    t = min(t_d, t_r, t_a);                    // scan the event list
    if (t == t_d) {                   // process an inventory demand
        l--;
        t_d = GetDemand();
    }
    else if (t == t_r) {              // process an inventory review
        if (l < s) {
            o = S - l;
            δ = GetLag();
            t_a = t + δ;
        }
        t_r += 1.0;
    }
    else {                            // process an inventory arrival
        l += o;
        o = 0;
        t_a = ∞;
    }
}
```

Program sis3.   Program sis3 is an implementation of Algorithm 5.2.1. The event list consists of three elements: t.demand, t.review, and t.arrive, corresponding to t_d, t_r, and t_a, respectively. These are elements of the structure t. Similarly, the two state variables inventory and order correspond to $l(t)$ and $o(t)$. Also, the time-integrated holding and shortage integrals are sum.hold and sum.short.

5.2.2 A Multiple-Server Service Node

As another example of next-event simulation, we will now consider a *multiple*-server service node. The extension of this next-event simulation model to account for immediate feedback, or finite service-node capacity, or a priority-queue discipline is left as an exercise. This example serves three objectives.

- A multiple-server service node is one natural generalization of the single-server service node.
- A multiple-server service node has considerable practical and theoretical importance.
- In a next-event simulation model of a multiple-server service node, the size of the event list is dictated by the number of servers and, if this number is large, the data structure used to represent the event list is important.

Definition 5.2.1   A *multiple-server service node* consists of a single queue, if any, and two or more servers operating *in parallel* as illustrated in Figure 5.2.2.

At any instant in time, the state of each server will be either *busy* or *idle*, and the state of the queue will be either *empty* or *not empty*. If at least one server is idle, the queue must be empty. If the queue is not empty, then all the servers must be busy.

Jobs arrive at the node, generally at random, seeking service. When service is provided, the time involved generally is also random. At the completion of service, jobs depart. The service node operates as follows. As each job arrives, if all servers are busy, then the job enters the queue; else, an available server is selected and the job enters service. As each job departs from a server, if the queue is empty, then the server becomes idle; else, a job is selected from the queue

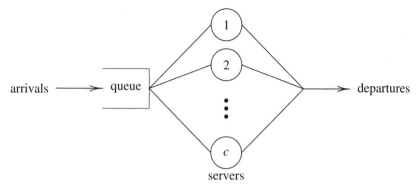

Figure 5.2.2   Multiple-server service-node system diagram.

to enter service at this server. Servers process jobs independently—they do not "team up" to process jobs more efficiently during periods of light traffic. This system configuration is popular, for example, at airport baggage check-in, banks, and roller coasters. Felt ropes or permanent dividers are often used to herd customers into queues. One advantage to this configuration, as opposed to a multiple-queue configuration, is that it is impossible to get stuck behind a customer with an unusually long service time due to a poor queue selection.

As in the single-server service-node model, control of the queue is determined by the *queue discipline*—the algorithm used when a job is selected from the queue to enter service (see Section 1.2). The queue discipline is typically FIFO.

Server-Selection Rule

Definition 5.2.2   A job can arrive to find two or more servers idle. In this case, the algorithm used to select an idle server is called the *server-selection rule*.

There are several possible server-selection rules. Of those listed below, the random, cyclic, and equity server-selection rules are designed to achieve an equal utilization of all servers. With the other two server-selection rules, typically, some servers will be more heavily utilized than others.

- Random selection—select at random from the idle servers.
- Selection in order—select server 1 if idle, else select server 2 if idle, etc.
- Cyclic selection—select the first available server beginning with the successor of the last server engaged (a circular search, if needed).
- Equity selection—select the server that has been idle longest *or* the idle server whose utilization is lowest.*
- Priority selection—choose the "best" idle server. This will require a specification from the modeler stating how "best" is evaluated.

For the purposes of mathematical analysis, multiple-server service nodes are frequently assumed to have *statistically identical, independent* servers. In this case, the server-selection rule has no effect on the average performance of the service node: Although the utilizations of the individual servers can be affected by the server-selection rule, if the servers are statistically identical and independent, then the *net* utilization of the node is not affected by the server-selection rule. Statistically identical servers are a convenient mathematical fiction; in a discrete-event simulation environment, if it is not appropriate, then there is no need to assume that the service times are statistically identical.

States.   In the queuing-theory literature, the parallel servers in a multiple-server service node are commonly called *service channels*. In the discussion that follows,

- the positive integer c will denote the number of servers (channels);
- the server index will be $s = 1, 2, \ldots, c$.

---

*There is an ambiguity in this server-selection rule, in that idle time can be measured from the most recent departure or from the beginning of the simulation. The modeler must specify which metric is appropriate.

As it does for a single-server node, the state variable $l(t)$ denotes the number of jobs in the service node at time t. For a multiple-server node with distinct servers, this single state variable does not provide a complete state description. If $l(t) \geq c$, then all servers are busy and $q(t) = l(t) - c$ jobs are in the queue. If $l(t) < c$, however, then, for a complete state description, we need to know which servers are busy and which are idle. Therefore, for $s = 1, 2, \ldots, c$, define

$$x_s(t) : \text{the number of jobs in service (0 or 1) by server } s \text{ at time } t,$$

or, equivalently, $x_s(t)$ is the state of server s at time t (with 0 denoting idle and 1 denoting busy). Finally, observe that

$$q(t) = l(t) - \sum_{s=1}^{c} x_s(t),$$

that is, the number of jobs in the queue at time t is the number of jobs in the service node at time t minus the number of servers busy at time t.

Events.   The $c + 1$ state variables $l(t), x_1(t), x_2(t), \ldots, x_c(t)$ provide a complete state description of a multiple-server service node. With a complete state description in hand, we then ask what types of events can cause the state variables to change. The answer is that, if the servers are distinct, then there are $c + 1$ event types—either an arrival to the service node or completion of service by one of the c servers. If an *arrival* occurs at time t, then $l(t)$ is incremented by 1. Then, if $l(t) \leq c$, an idle server s is selected and the job enters service at server s (and the appropriate completion of service is scheduled); else, all servers are busy and the job enters the queue. If a *completion of service* by server s occurs at time t, then $l(t)$ is decremented by 1. Then, if $l(t) \geq c$, a job is selected from the queue to enter service at server s; else, server s becomes idle.

   The additional assumptions needed to complete the development of the next-event simulation model at the specification level are consistent with those made for the single-server model in the previous section.

- The initial state of the multiple-server service node is empty and idle. Therefore, the first event must be an arrival.
- There is a terminal "close the door" time τ at which the arrival process is turned off, but the system continues operation until all jobs have been completed. Therefore, the terminal state of the multiple-server node is empty and idle, and the last event must be a completion of service.
- For simplicity, all servers are assumed to be independent and statistically identical. Moreover, equity selection is assumed to be the server-selection rule.

All of these assumptions can be relaxed.

Event List.   The event list for this next-event simulation model can be organized as an array of $c + 1$ event types indexed from 0 to c, as illustrated in Figure 5.2.3 for the case $c = 4$. The t field in each event structure is the scheduled time of next occurrence for that event; the x field is the current *activity status* of the event. The status field is used in this data structure as

0	t	x	arrival
1	t	x	completion of service by server 1
2	t	x	completion of service by server 2
3	t	x	completion of service by server 3
4	t	x	completion of service by server 4

Figure 5.2.3   Event-list data structure for multiple-server service node.

a superior alternative to the ∞ "impossibility flag" used in the model on which programs ssq3 and sis3 are based. For the 0th event type, x denotes whether the arrival process is on (1) or off (0). For the other event types, x denotes whether the corresponding server is busy (1) or idle (0).

An array data structure is appropriate for the event list, because the size of the event list cannot exceed $c + 1$. If c is large, however, it is preferable to use a variable-length data structure—for example, a linked list containing events sorted by time, so that the next (most imminent) event is always at the head of the list. Moreover, in this case, the event list should be partitioned into busy (event[e].x = 1) and idle (event[e].x = 0) sublists. This idea is discussed in more detail in the next section.

Program msq.   Program msq is an implementation of the next-event multiple-server service-node simulation model we have just developed.

- The state variable $l(t)$ is number.
- The state variables $x_1(t), x_2(t), \ldots, x_c(t)$ are incorporated into the event list.
- The time-integrated statistic $\int_0^t l(\theta)\,d\theta$ is area.
- The array named sum contains structures that are used to record, for each server, the sum of service times and the number served.
- The function NextEvent is used to search the event list to identify the index e of the next event.
- The function FindOne is used to search the event list to identify the index s of the available server that has been idle longest (because an equity-selection server-selection rule is used).

5.2.3 Exercises

5.2.1   Use program ddh in conjunction with program sis3 to construct a discrete-data histogram of the total demand per time interval. (Use 10 000 time intervals.) (*a*) Compare the result with the corresponding histogram for program sis2. (*b*) Comment on the difference.

5.2.2   (*a*) Modify program sis3 to account for a *Uniform*(0.5, 2.0) delivery lag. Assume that, if an order is placed at time t and if $o(t) > 0$, then the amount ordered will be $S - l(t) - o(t)$. (*b*) Discuss why you think your program is correct.

5.2.3   Modify program `sis3` so that the inventory review is no longer periodic but, instead, occurs after each demand instance. (This is *transaction reporting* as described in Section 1.3.) Assume that, when an order is placed, further review is stopped until the order arrives. This avoids the sequence of orders that otherwise would occur during the delivery lag. What impact does this modification have on the system statistics? Conjecture first, then simulate, using `STOP` equal to 10 000.0, to estimate steady-state statistics.

5.2.4   (*a*) Relative to program `msq`, provide a mathematical justification for the technique used to compute the average delay and the average number in the queue. (*b*) Does this technique require that the service node be idle at the beginning and end of the simulation for the computation of these statistics to be exact?

5.2.5   (*a*) Implement a "selection in order" server-selection rule for program `msq` and compute the statistics. (*b*) What impact does this have on the system performance statistics?

5.2.6   (*a*) Modify program `msq` so that the stopping criterion is based on "closing the door" after a fixed number of jobs have entered the service node. (*b*) Comment.

5.2.7   (*a*) Modify program `msq` to allow for feedback with probability β. What statistics are produced if $\beta = 0.1$? (*b*) At what value of β does the multiple-server service node saturate? (*c*) Provide a mathematical justification for why saturation occurs at this value of β.

5.2.8   Modify program `msq` to allow for a finite capacity: r jobs in the node at one time. (*a*) Draw a histogram of the time between lost jobs at the node. (*b*) Comment on the shape of this histogram.

5.2.9   Write a next-event simulation program that estimates the average time to complete the stochastic activity network given in Section 2.4. Compute the mean and variance of the time to complete the network.

5.3 EVENT-LIST MANAGEMENT

The next-event simulation models for the single-server service node and the simple inventory system from the previous two sections have such short event lists (two events for the single-server service node and three events for the simple inventory system) that their management does not require any special consideration. There are next-event simulations, however, that can have hundreds or even thousands of events on their event list simultaneously, and the efficient management of this list is crucial. The material in this section is based on a tutorial by Henriksen (1983) and Chapter 5 of Fishman (2001).

Although the discussion in this section is limited to managing event lists, practically all of the discussion applies equally well, for example, to the management of jobs in a single-server service node. A FIFO or LIFO queue discipline results in a trivial management of the jobs in the queue: jobs arriving when the server is busy are simply added to the tail (FIFO) or head (LIFO) of the queue. A "shortest processing time first" queue discipline (which is commonly advocated for minimizing the wait time in job shops), on the other hand, requires special data structures and algorithms for efficiently inserting jobs into the queue and deleting jobs from the queue.

5.3.1 Introduction

An event list is the data structure that contains a list of the events that are scheduled to occur in the future, along with any ancillary information associated with these events. The list, traditionally, is sorted by the scheduled time of occurrence, but, as indicated in the first example in this section, this is not a requirement. The event list is also known as the calendar, future events chain, sequencing set, future event set, and so on. The elements that compose an event list are known as future events, events, event notices, transactions, records, and so on. We will use the term *event notice* to describe these elements.

Why is efficient management of the event notices on the event list so important that it warrants an entire section? Many next-event simulation models expend more CPU time on managing the event list than on any other aspect (e.g., random-number generation, random-variate generation, processing of events, miscellaneous arithmetic operations, printing of reports) of the simulation.

Next-event simulations that require event-list management can be broken into four categories according to the following two boolean classifications:

- There is a either a *fixed* maximum or a *variable* maximum number of event notices on the event list. There are clear advantages to having the maximum number of events fixed in terms of memory allocation. All of the simulations seen thus far have had a fixed maximum number of event notices.
- The event-list management technique either is being devised for one specific model or is being developed for a general-purpose simulation language. If the focus is on a single model, then the scheduling aspects of that model can be exploited for efficiency. An event-list management technique designed for a general-purpose language must be robust, in the sense that it performs reasonably well for a variety of simulation models.

There are two critical operations in the management of the event notices that compose the event list. The first is the insertion (or *enqueue*) operation, whereby an event notice is placed on the event list. This operation is also referred to as "scheduling" the event. The second is the deletion (or *dequeue*) operation, whereby an event notice is removed from the event list. A deletion operation is performed to process the event (the more common case) or because a previously scheduled event needs to be canceled for some reason (the rare case). Insertion and deletion may occur at a prescribed position in the event list, or a search based on some criteria may need to be initiated first in order to locate the appropriate position. We will use the term *event-list management scheme* to refer to the data structures and associated algorithms corresponding to one particular technique of handling event-list insertions and deletions.

Of minor importance is a *change* operation, where a search for an existing event notice is followed by a change in some aspect of the event notice, such as changing its scheduled time of occurrence. Similarly, an *examine* operation searches for an existing event notice in order to examine its contents. A *count* operation is used to determine the number of event notices in the list. Due to their relative rarity in discrete-event simulation modeling and their similarity to insertion and deletion in principle, we will henceforth ignore the change, examine, and count operations and focus solely on the insertion and deletion operations.

5.3.2 Event-List Management Criteria

Three criteria that can be used to assess the effectiveness of the data structures and algorithms for an event-list management scheme are the following:

- **Speed:** The data structure and associated algorithms for inserting and deleting event notices should execute in minimal CPU time. Critical to achieving fast execution times is the efficient searching of the event list. The advantages of sophisticated data structures and algorithms for searching must be weighed against the associated extraneous overhead calculations (e.g., maintaining pointers for a list or heap operations) that they require. An effective general-purpose algorithm typically bounds the number of event notices searched for inserting or deleting.

- **Robustness:** Efficient event-list management should perform well for a wide range of scheduling scenarios. This criterion is much easier to achieve if the characteristics of one particular model can be exploited by the analyst. The designer of an event-list management scheme for a general-purpose language does not have this advantage.

- **Adaptability:** An effective event-list management scheme should be able to adapt its searching time to account both for the length of the event list and for the distribution of new events that are being scheduled. It is also advantageous for an event-list management scheme to be "parameter-free"—in the sense that the user should not be required to specify parameters that optimize the performance of the scheme. A "black box" approach to managing an event list is particularly appropriate for a general-purpose simulation language, whose users have a variety of sophistication levels.

Given these criteria, how do we know whether our event-list management scheme is effective? If this is for a single model, the only way to answer this question is by running several different schemes on the same model to see which executes fastest. It is typically not possible to prove that one particular scheme is superior to all other possible schemes, since a more clever analyst may exploit more of the structure associated with a specific model. It is, however, possible to show that one scheme likely dominates another in terms of execution time, by making several runs with different seeds and comparing execution times.

Testing event-list management schemes for a general-purpose language is much more difficult. In order to compare event-list management schemes, one must consider a representative test-bed of diverse simulation models on which to test various event-list management schemes. Average and worst-case performance in terms of the CPU time tests the speed, robustness, and adaptability of various event-list management schemes.

5.3.3 Example

We consider the timesharing computer system model presented by Henriksen (1983) to discuss event-list management schemes. In our usual manner, we start with the conceptual model before moving to the specification and computational models.

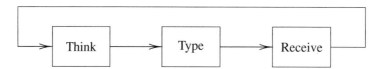

Figure 5.3.1   Timeshare system diagram.

Conceptual Model.   Consider a user at a timesharing computer system who endlessly cycles from (1) thinking, to (2) typing in a command, to (3) receiving the output from the command. The user does not take any breaks and never tires, and the workstation never fails. A system diagram that depicts this behavior is given in Figure 5.3.1.

This is the second time we have encountered what is known as a "closed" system. (See Section 3.3.3 for the first instance.) The "open" systems considered previously are the queuing models (jobs arrive, are processed, then depart) and inventory models (inventory arrives, is purchased, then departs). In the timesharing model, there is no arrival or departure. The three-step activity loops endlessly.

The times to perform the three operations, measured in seconds, are:

- The time to think requires *Uniform*(0, 10) seconds.

- There are an *Equilikely*(5, 15) number of keystrokes involved in typing in a command, and each keystroke requires *Uniform*(0.15, 0.35) seconds.

- The output from the command contains an *Equilikely*(50, 300) number of characters, each requiring (the ancient rate of) $1/120$ second to display at the workstation.

If we were to simulate just one user, this would be a rather trivial model, since there is only one event on the event list: the completion of the next activity (thinking, typing a keystroke, or receiving an output character). To make event-list management an issue, assume that the timesharing computer system is supporting *n* users at *n* terminals, each asynchronously cycling through the three-step process of thinking, typing, and receiving.

Many critiques of this model would be valid. Do all of the users actually think and type at the same rate? Does the distribution of thinking time actually "cut off" at ten seconds, as the *Uniform*(0, 10) thinking time implies? Are all of the *Uniform* and *Equilikely* distributions appropriate? Why doesn't the receive rate degrade when several users receive output simultaneously? Because the purpose of this model is to illustrate the management of the event list, we will forgo discussion about the reasonableness and accuracy of the model. The important topic of developing "input" models that accurately mimic the system of interest is addressed in Chapter 9.

Back-of-an-Envelope Calculations.   All of the distributions in this model are either *Uniform* or *Equilikely*, so it is worthwhile to do some preliminary calculations that might provide insight into model behavior prior to the simulation. The mean of a *Uniform* or *Equilikely* random variable is, not surprisingly, the average of their two parameters. Thus the expected length of each cycle for each user is

$$\left(\frac{0+10}{2}\right) + \left(\frac{5+15}{2}\right)\left(\frac{0.15+0.35}{2}\right) + \left(\frac{50+300}{2}\right)\left(\frac{1}{120}\right) = 5 + 10 \cdot 0.25 + 175 \cdot \frac{1}{120}$$

$$\cong 5 + 2.5 + 1.4583$$

$$= 8.9583$$

seconds. Thus, if one were to observe a user at a random instant in a system in steady state, the probabilities that the user will be thinking, typing, and receiving are

$$\frac{5}{8.9583} \cong 0.56, \qquad \frac{2.5}{8.9583} \cong 0.28, \quad \text{and} \quad \frac{1.4583}{8.9583} \cong 0.16,$$

respectively. These probabilities apply both to individual users and to the population of n users. At any particular point in simulated time, we could expect to see about 56% of the users thinking, 28% typing, and 16% receiving output.

Although it is clear from this analysis that the largest portion of a user's *simulated time* is spent thinking and the smallest portion of a user's *simulated time* is spent receiving, the opposite is true of the *number* of scheduled events. Each cycle has exactly one thinking event, an average of ten keystrokes, and an average of 175 characters received. Thus each cycle averages $1 + 10 + 175 = 186$ events. The expected fractions of events associated with thinking, typing a keystroke, and receiving an output character during a cycle are

$$\frac{1}{186} \cong 0.005, \qquad \frac{10}{186} \cong 0.054, \quad \text{and} \quad \frac{175}{186} \cong 0.941,$$

respectively. The vast majority of the events scheduled during the simulation will be receiving-a-character events. This observation will influence the probability distribution of the event times associated with event notices on the event list. This distribution can be exploited when designing an event-list management scheme.

Specification Model.   There are three events that compose the activity of a user on the time-sharing system:

- **(1)** complete thinking time;
- **(2)** complete a keystroke;
- **(3)** complete the display of a character.

For the second two events, some ancillary information must be stored as well: the number of keystrokes in the command, and the number of characters remaining to be displayed. Ancillary information of this kind is often called an "attribute" in general-purpose simulation languages. For all three events, an integer (1 for thinking, 2 for typing, 3 for receiving) is stored to denote the event type.

As with most next-event simulation models, the processing of each event triggers the scheduling of future events. For this particular model, we are fortunate: Each event triggers just one future event to schedule, the next activity for the user whose event is currently being processed.

Several data structures are capable of storing the event notices for this model. Two likely candidates are an array and a linked list. For simplicity, an array will be used to store the event notices. We can use an array here because we know in advance that there will always be n events on the event list, one event notice for the next activity for each user. We begin with this very simplistic (and grossly inefficient) data structure for the event list. We will subsequently refine this data structure, and eventually outline more sophisticated schemes. We will store the times of the events in an array of length n in an order associated with the n users and make a linear search of all elements of the array whenever the next event to be processed needs to be found. Thus, the deletion operation requires searching all n elements of the event list to find the event with the smallest event time; the insertion operation requires no searching, since the next event notice for a particular user simply overwrites the current event notice.

However, there is ancillary information that must be carried with each event notice. Instead of an array of n event times, the event list for this model should be organized as an array of n event structures. Each event structure consists of three fields: time, type, and info. The time field in the ith event structure stores the time of the event for the ith user ($i = 0, 1, \ldots, n - 1$). The type field stores the event type (1, 2, or 3). The info field stores the ancillary information associated with a keystroke or display of a character event: the number of keystrokes remaining (for a Type-2 event) or the number of characters remaining in the output (for a Type-3 event). The info field is not used for a thinking (Type-1) event, since no ancillary information is needed for that event type.

The initialization phase of the next-event simulation model schedules a complete thinking-time (Type 1) event for each of the n users on the system. The choice of having all users initially thinking is arbitrary.* After the initialization, the event list for a system with $n = 5$ users might look like the one presented in Figure 5.3.2. Each of the five *Uniform*(0, 10) completion-of-thinking event times is placed in the time field of the corresponding event structure. The value in the time field of the third structure is the smallest, indicating that the third user will be the first to stop thinking and begin typing, at time 1.305. All five of these events are completion-of-thinking events, as indicated by the type field in each event structure.

The remainder of the algorithm follows the standard next-event protocol: While the terminal condition has not been met, (1) scan the event list for the most imminent event, (2) update the simulation clock accordingly, (3) process the current event, and (4) schedule the subsequent event by placing the appropriate event notice on the event list.

time	9.803	3.507	1.305	2.155	8.243
type	1	1	1	1	1
info					
	0	1	2	3	4

Figure 5.3.2   Initial event list.

---

*All users begin thinking simultaneously at time 0, but will behave more independently after a few cycles as the timesharing system "warms up."

As the initial event (end of thinking for the third user at time 1.305) is deleted from the event list and processed, a number of keystrokes [an *Equilikely*(5, 15) random variate, which takes on the value 7 in this case—a slightly shorter-than-average command] and a time for the first keystroke [a *Uniform*(0.15, 0.35) random variate that takes on the value 0.301 in this case—a slightly longer-than-average keystroke time] are generated. The `time` field in the third event structure is incremented by 0.301, to $1.305 + 0.301 = 1.606$, and the number of keystrokes in this command is stored in the corresponding `info` field. Subsequent keystrokes for the third user decrement the integer in the corresponding `info` field. The condition of the event list after the processing of the first event is shown in Figure 5.3.3.

time	9.803	3.507	1.606	2.155	8.243
type	1	1	2	1	1
info			7		
	0	1	2	3	4

Figure 5.3.3   Updated event list.

To complete the development of the specification model, we use the following notation:

- The simulation clock is t, measured in seconds.
- The simulation terminates when the next scheduled event is τ seconds or more.

The algorithm is straightforward, as presented in Algorithm 5.3.1.

Algorithm 5.3.1   This algorithm is a next-event simulation of a think–type–receive time-sharing system with n concurrent users. The algorithm presumes the existence of four functions (GetThinkTime, GetKeystrokeTime, GetNumKeystrokes, and GetNumCharacters) that return the random time to think, time to enter a keystroke, number of keystrokes per command, and number of characters returned from a command, respectively. The function MinIndex returns the index of the most imminent event notice.

```
t = 0.0;                            // initialize system clock
for (i = 0; i < n; i++) {           // initialize event list
   event[i].time = GetThinkTime();
   event[i].type = 1;
}
while (t < τ) {                      // check for terminal condition
   j = MinIndex(event.time);        // find index of imminent event
   t = event[j].time;                             // update system clock
   if (event[j].type == 1) {   // process completion of thinking
      event[j].time = t + GetKeystrokeTime();
      event[j].type = 2;
      event[j].info = GetNumKeystrokes();
   }
```

```
    else if (event[j].type == 2) {
                                    // process keystroke completion
        event[j].info--;        // decrement # of keystrokes remaining
        if (event[j].info > 0)            // if more keystrokes remain
          event[j].time = t + GetKeystrokeTime();
        else {                               // else last keystroke
          event[j].time = t + 1.0 / 120.0;
          event[j].type = 3;
          event[j].info = GetNumCharacters();
        }
    }
    else if (event[j].type == 3) {  // process character received
      event[j].info--;        // decrement # of characters remaining
      if (event[j].info > 0)            // if more characters remain
        event[j].time = t + 1.0 / 120.0;
      else {                               // else last character
        event[j].time = t + GetThinkTime();
        event[j].type = 1;
      }
    }
}
```

Program ttr.   The think–type–receive specification model has been implemented in program
`ttr`, which prints the total number of events scheduled and the average number of event notices
searched for each deletion. For each value of n in the table below, the simulation was run three
times (with seeds 123456789, 987654321, and 555555555) for $\tau = 100$ seconds, and the averages
of the three replications are reported.

number of users n	expected number of events scheduled	average number of events scheduled	average number of event notices searched
5	10 381	9 902	5
10	20 763	20 678	10
50	103 814	101 669	50
100	207 628	201 949	100

The column headed "expected number of events scheduled" is generated as follows: The average
length of each cycle is 8.9583 seconds, so each user will go through an average of

$$\frac{100}{8.9583} \cong 11.16$$

cycles in $\tau = 100$ seconds. Also, the average number of events per cycle is 186, so we expect to see

$$(11.16) \cdot (186) \cdot n = 2076.3n$$

total events scheduled during the simulation. These values are reported in the second column of
the table. That the averages in the table from the three simulations are slightly lower than the

expected values is due to our arbitrary decision to begin each cycle by thinking, the longest event. In terms of event-list management, each deletion event (required to find the next event notice) requires an exhaustive search of the `time` field in all n event structures, so the average number of event notices searched for each deletion is simply n. The simplistic event-list management scheme used here sets the stage for more sophisticated schemes to follow.

5.3.4 An Improved Event-List Management Scheme

Our decision in the previous example to store event times unordered is a departure from the traditional convention in simulation languages, which is to order the event notices on the event list in ascending order of event times (i.e., the event list is maintained in chronological order). If we now switch to an ordered event list, a deletion requires no searching, and an insertion requires a search—just the opposite situation from the previous event-list management scheme. This switch will have no effect, time-wise, for the think–type–receive model: There is a deletion for every insertion during the simulation. Every deletion associated with the scheduling of an event pulls the first event notice from the head of the list. This subsection is focused, therefore, on efficient algorithms for inserting event notices into the event list.

There is good and bad news associated with the switch to an event list that is ordered by ascending event time. The good news is that the entire event list need not necessarily be searched every time an insertion operation is conducted. The bad news, however, is that arrays are no longer a natural choice for the data structure, as a result of the overhead associated with shifting event notices in the array when an event notice is placed at the beginning or middle of the list. A singly or doubly linked list is preferred because it offers the ability to insert items in the middle of the list easily. The overhead of maintaining pointers, however, dilutes the benefit of moving to an event list that is ordered by ascending event time. Also, direct access to the array is lost, and time-consuming element-by-element searches through the linked list are required.

A secondary benefit associated with switching to a linked list is that the maximum size of the list need not be specified in advance. This is of no consequence in our think–type–receive model, since there are always n events in the event list. In a general-purpose discrete-event simulation language, however, linked lists can expand until memory is exhausted.

Example 5.3.1
For the think–type–receive model with $n = 5$, for example, a singly linked list, linked from head (top) to tail (bottom), to store the elements of the event list corresponding to Figure 5.3.3 is shown in Figure 5.3.4. The three values stored on each event notice are the event time, event type, and ancillary information (seven keystrokes remaining in a command for the first element in the list). The event notices are ordered by event time. A deletion now involves no search, but a search is required for each insertion.

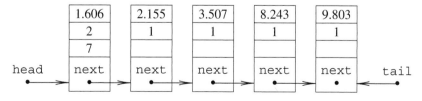

Figure 5.3.4   Event list as a linked list.

One question remains before implementing the new data structure and algorithm for the search. Should the list be searched from head to tail (top to bottom for a list with forward pointers) or tail to head (bottom to top for a list with backward pointers)? We begin by searching from tail to head and check our efficiency gains over the naive event-list management scheme presented in the previous subsection. The table below shows that the average number of events scheduled is identical to the previous event-list management scheme (as is expected because of the use of identical seeds) and that improvements in the average number of searches per insertion range from 18.8% ($n = 100$) to 23.4% ($n = 5$).

number of users n	average number of events scheduled	average number of event notices searched
5	9 902	3.83
10	20 678	8.11
50	101 669	40.55
100	201 949	81.19

These results are certainly not stunning. The improvement in search time is slight. What went wrong? The problem here is that we ignored our earlier back-of-an-envelope calculations. These calculations indicated that 94.1% of the events in the simulation would be the receipt of a character, which has a very short inter-event time. Thus, we should have searched the event list from head to tail; these short events were much more likely to have been inserted at or near the top of the list. We reprogrammed the search to go from head to tail, and the results are given in the table below.

number of users n	average number of events scheduled	average number of event notices searched
5	9 902	1.72
10	20 678	2.73
50	101 669	10.45
100	201 949	19.81

Confirming our calculations, the forward search performs far better than the backward search.* In this case, the more impressive savings in terms of the number of searches required over the exhaustive search ranges from 66% (for $n = 5$) to 80% (for $n = 100$).

---

*An interesting verification of the forward and backward searches can be made in this case, since using separate streams of random numbers assures an identical sequencing of events. For an identical event list of size n, the sum of the number of forward and backward searches equals n for an insertion at the top or bottom of the list. For an insertion in the middle, however, the sum equals $n + 1$ for identical lists. Therefore, the sum of the rightmost columns of the last two tables will always lie between n and $n + 1$, as it does in this case. The sums are 5.55, 10.84, 51.00, and 101.00 for $n = 5, 10, 50$, and 100. The sum tends to $n + 1$ for large n, since it is more likely to have an insertion in the middle of the list as n grows. Stated another way, when n is large, it is a near certainty that several users will be simultaneously receiving characters when an event notice is placed on the event list, meaning that the event is unlikely to be placed at the front of the list.

The think–type–receive model with $n = 100$ highlights our emphasis on efficient event-list management techniques. Even with the best of the three event-list management schemes employed so far (forward linear search of a singly linked list, averaging 19.81 searches per insertion), more time is spent on event-list management than the rest of the simulation operations (e.g., random-number generation, random-variate generation, processing events) combined.

5.3.5 More Advanced Event-List Management Schemes

The think–type–receive model represents a simple case for event-list management. First, the number of event notices on the event list remains constant throughout the simulation. Second, the fact that there are frequent short events (e.g., receiving a character) can be exploited in order to minimize the search time for an insertion using a forward search.

We now proceed to a discussion of the general case where (1) the number of event notices in the event list varies throughout the simulation, (2) the maximum length of the event list is not known in advance, and (3) the structure of the simulation model is unknown, so it cannot be exploited for optimizing an event-list management scheme.

In order to reduce the scope of the discussion, assume that a memory-allocation mechanism exists such that a memory location occupied by a deleted event notice may be immediately occupied by an event notice that is subsequently inserted into the event list. When memory space is released as soon as it becomes available in this fashion, the simulation will fail from lack of memory only when an insertion is made to an event list that exhausts the space allocated to it. Many general-purpose languages effectively place all entities (e.g., event notices in the event list, jobs waiting in a queue) in the simulation in a single partitioned list, in order to use memory in the most efficient manner. Data structures and algorithms associated with the allocation and deallocation of memory are detailed in Chapter 5 of Fishman (2001).

The following subsections briefly outline four event-list management schemes commonly used to insert event notices into an event list efficiently and to delete event notices from an event list in a general setting: multiple linked lists, binary search trees, heaps, and hybrid schemes.

Multiple Linked Lists.   One approach to reducing search time associated with insertions and deletions is to maintain multiple linear lists, each sorted by event time. Let k denote the number of such lists, and let n denote the number of event notices in all lists at one particular point in simulated time. Figure 5.3.5 shows $k = 2$ equal-length, singly linked lists for $n = 10$ initial think times in the think–type–receive model. An insertion can be made into either list. If the list sizes were not equal, choosing the shortest list minimizes the search time. The time savings associated with the insertion operation is offset by (1) the overhead associated with maintaining the multiple lists, and (2) a less efficient deletion operation. A deletion now requires a search of the top (head) event notices of the k lists, and the event notice with the smallest event time is deleted.

We close the discussion of multiple event lists with three issues that are important for minimizing CPU time for insertions and deletions:

- The decision about whether to fix the number (k) of lists throughout the simulation or to allow it to vary depends on many factors, including the largest value of n throughout the simulation, the distribution of the position of insertions in the event list, and how widely n varies throughout the simulation.

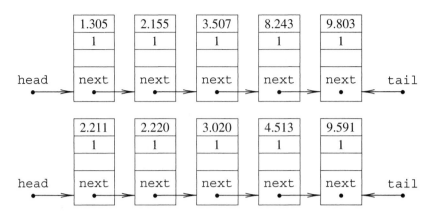

Figure 5.3.5   Multiple linked lists.

- If the number of lists, k, is allowed to vary throughout the simulation, the modeler must set appropriate thresholds to determine when lists are split (as n increases) and combined (as n decreases).
- The CPU time associated with inserting an event notice, deleting an event notice, combining lists, and splitting lists as functions of n and k should drive the optimization of this event-list management scheme.

The next two data structures, binary trees and heaps, are well-known data structures. Rather than developing the data structures and associated operations from scratch, we refer the reader to Carrano and Prichard (2002) for basic definitions, examples, and applications. Our discussion of these two data structures here will be rather general in nature.

Binary Trees.   We limit our discussion of trees to *binary trees*. A binary tree consists of n nodes connected by edges in a hierarchical fashion such that a *parent* node lies above, and is linked to, at most two *child* nodes. The parent–child relationship generalizes to the *ancestor–descendant* relationship in a fashion analogous to a family tree. A *subtree* in a binary tree consists of a node, along with all of the associated descendants. The top node in a binary tree is the only node in the tree without a parent and is called the *root*. A node with no children is called a *leaf*. The *height* of a binary tree is the number of nodes on the longest path from root to leaf. The *level* of a node is 1 if it is the root or 1 greater than the level of its parent if it is not the root. A binary tree of height h is *full* if every node at a level less than h has two children. Full trees have $n = 2^h - 1$ nodes. A binary tree of height h is *complete* if it is full down to level $h - 1$ and level h is filled from left to right. A full binary tree of height $h = 3$ with $n = 7$ nodes and a complete binary tree of height $h = 4$ with $n = 12$ nodes are displayed in Figure 5.3.6.

Nodes are often associated with a numeric value. In our setting, a node corresponds to an event notice, and the numeric value associated with the node is the event time. A *binary search tree* is a binary tree where the value associated with any node is greater than or equal to every value in its left subtree and less than or equal to every value in its right subtree. The "or equal to" portions of the previous sentence have been added to the standard definition of a binary search tree to allow for the possibility of equal event times.

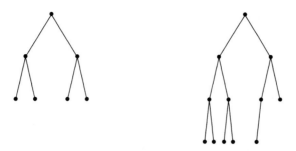

Figure 5.3.6   A full and a complete binary tree.

Example 5.3.2

There are many ways to implement a binary tree. Figure 5.3.7 shows a *pointer-based* complete-binary-tree representation of the ten initial events (from Figure 5.3.5) for a think–type–receive model with $n = 10$ users, where each user begins with a *Uniform*(0, 10) thinking activity. Each event notice has three fields (event time, event type, and event information), and each parent points to its child or children, if any. Every event time is greater than or equal to every event time in its left subtree and less than or

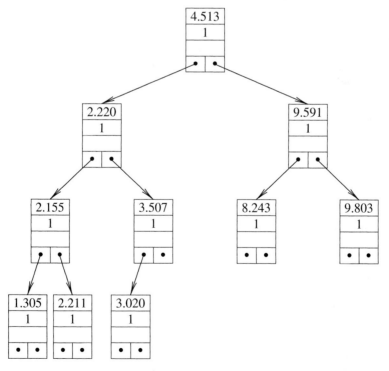

Figure 5.3.7   Binary search tree with pointers for the ten initial events in the think–type–receive model.

equal to every event time in its right subtree. Although there are many binary-search-tree configurations that could contain these particular event notices, the placement of the event notices in the *complete* binary search tree in Figure 5.3.7 is unique.

One advantage to binary search trees for storing event notices is that the "leftmost" leaf in the tree will always be the most imminent event. This makes a deletion operation fast, although it could require reconfiguring the tree after the deletion.

Insertions are faster than a linear search of a list, because fewer comparisons are necessary to find the appropriate insertion position. The key decision that remains for the scheme is whether the binary tree will be maintained as a complete tree (involving extra overhead associated with insertions and deletions) or allowed to evolve without the requirement that the tree remain complete. (The latter choice can result in an "imbalanced" tree, whose height increases over time, requiring more comparisons for insertions.) *Splay trees*, which require frequent rotations to maintain balance, have also performed well.

Heaps.   A heap is another data structure for storing event notices in order to minimize insertion and deletion times. In our setting, a heap is a complete binary tree with the following properties: (1) the event time associated with the root is less than or equal to the event time associated with each of its children, and (2) the root has heaps as subtrees. Figure 5.3.8 shows a heap associated

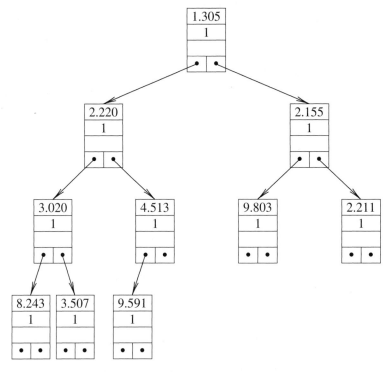

Figure 5.3.8   Heap with pointers for the ten initial events in the think–type–receive model.

with the first ten events in the think–type–receive model. This heap is not unique.* An obvious advantage to the heap data structure is that the most imminent event is at the root, making deletions fast. The heap property must be maintained, however, whenever an insertion or deletion operation is performed.

Hybrid Schemes.   The ideal event-list management scheme would perform well regardless of the size of the event list. Jones (1986) concludes that, when there are fewer than ten event notices on an event list, a singly linked list is optimal, because it avoids the overhead associated with more sophisticated schemes. Thus, to fully optimize an event-list management scheme, it might be necessary to have thresholds, similar to those that switch a thermostat on and off, that switch from one set of data structures and algorithms to another on the basis of the number of events on the list. It is important to avoid switching back and forth too often, however; typically, the switch requires processing-time overhead.

If a heap, for example, is used when the event list is long and a singly linked list is used when the event list is short, then appropriate thresholds should be determined that will cause a switch from one scheme to the other. As an illustration, when the number of event notices shrinks to $n = 5$ (e.g., n decreases from 6 to 5), the heap is converted to a singly linked list. Similarly, when the number of event notices grows to $n = 15$ (e.g., n increases from 14 to 15), the singly linked list is converted to a heap.

Henriksen's algorithm (Henriksen, 1983) provides adaptability to short and long event lists without alternating data structures via thresholds. Henriksen's algorithm uses a binary search tree and a doubly linked list simultaneously. This algorithm has been implemented in several simulation languages, including GPSS, SLX, and SLAM. At the conceptual level, Henriksen's algorithm employs two data structures:

- The event list is maintained as a single, linear, doubly linked list ordered by event times. The list is augmented by a dummy event notice on the left having the simulated time $-\infty$ and a dummy event notice on the right having the simulated time $+\infty$, to allow symmetry of treatment for all real event notices.
- A binary search tree with nodes associated with a *subset* of the event notices in the event list has nodes in the format shown in Figure 5.3.9. Leaf nodes have null values for left and right child pointers. The leftmost node in the tree has a null value for a pointer to the next-lower-time node. This binary search tree is degenerate at the beginning of the simulation (prior to scheduling initial events).

Pointer to next lower time tree node
Pointer to left child tree node
Pointer to right child tree node
Event time
Pointer to the event notice

Figure 5.3.9   Binary-search-tree-node format.

---

*If the event notices associated with times 2.220 and 2.211 in Figure 5.3.8, for example, were interchanged, the heap property would be retained.

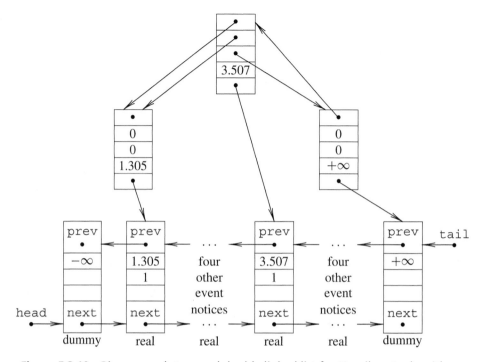

Figure 5.3.10 Binary search tree and doubly linked list for Henriksen's algorithm.

A three-node binary search tree associated with the ten initial events in the think–type–receive model is given in Figure 5.3.10. The binary search tree is given at the top of the figure, and the linear doubly linked list containing the ten initial event notices (plus the dummy event notices at both ends of the list) is shown at the bottom of the figure.

A deletion is an $O(1)$ operation, since the first non-dummy event notice in the event list is the most imminent event. To insert an event notice into the event list, the binary search tree is traversed in order to find the position of the event in the tree with the smallest event time greater than the event notice being inserted. A backward linear search of the doubly linked event list is initiated at the event notice to the immediate left of the event notice pointed to by the node found in the binary search tree. This backward linear search continues until either:

- the appropriate insertion position is found in l or fewer searches (Henriksen recommends $l = 4$), in which case the new event notice is linked into the event list at the appropriate position, or
- the appropriate insertion position is not found in l or fewer searches, in which case a "pull" operation is attempted. The pull operation begins by examining the pointer to the event in the binary search tree with the *next lower time* relative to the one found previously.

If this pointer is nonzero, its pointer is changed to the most recently examined event notice in the doubly linked list (i.e., the lth event encountered during the search), and the search continues for another l event notices as before. If the pointer is zero, there are no earlier binary tree nodes that can be updated, so the algorithm adds a new level to the tree. The new level is initialized by setting its leftmost leaf to point to the dummy notice on the right (event time $+\infty$) and setting all other new leaves to point to the dummy notice on the left (event time $-\infty$). The binary search tree is again searched as before.

Henriksen's algorithm works quite well at reducing the average search time for an insertion. Its only drawback seems to be that the maximum search time can be quite long. Other hybrid event-list management schemes might also have promise for reducing CPU times associated with insertions and deletions (e.g., Brown, 1988).

5.3.6 Exercises

5.3.1   (*a*) Modify program `ttr` so that the initial event for each user is the completion of the first character received as output, rather than the completion of thinking. Run the modified programs for $n = 5, 10, 50, 100$, and for initial seeds 123456789, 987654321, and 555555555. (*b*) Compare the average number of events for the three replications of the simulation relative to the results in Section 5.3.3. (*c*) Offer an explanation of why the observed average number of events goes up or down.

5.3.2   Modify program `ttr` to include an event list that is sorted by event time and is stored in a linked list. Verify the results for the forward search given in Example 5.3.1.

5.3.3   Assume that all events (thinking, typing a character, and receiving a character) in the think–type–receive model have deterministic durations of exactly $1/10$ second. Write a paragraph describing an event-list management scheme that requires *no* searching. Include the reason(s) that no searching is required.

5.3.4   Assume that all events (thinking, typing a character, and receiving a character) in the think–type–receive model have *Uniform*(0.4, 0.6) second durations. If you use a doubly-linked-list data structure to store the event list with events stored in chronological order, would it be wiser to begin an insertion operation with a search starting at the top (head) of the list or the bottom (tail) of the list? Justify your answer.

5.3.5   Assume that all events (thinking, typing a character, and receiving a character) in the think–type–receive model have *Exponential*(0.5) second durations. If you use a doubly-linked-list data structure to store the event list with events stored in chronological order, would it be wiser to begin an insertion operation with a search starting at the top (head) of the list or at the bottom (tail) of the list? Justify your answer.

5.3.6   The *verification* process from Algorithm 1.1.1 involves checking on whether a simulation model is working as expected. Program `ttr` prints the contents of the event list when the simulation reaches its terminal condition. What verification technique could be applied to this output to see if the program is executing as intended?

5.3.7 The *verification* process from Algorithm 1.1.1 involves checking on whether a simulation model is working as expected. Give a verification technique for comparing the think–type–receive model with (*a*) an unsorted event list with an exhaustive search for a deletion, and (*b*) an event list that is sorted by event time, with a backward search for an insertion.

6

Discrete Random Variables

This chapter begins a more thorough and methodical description of random variables, of their properties, of how they can be used to model the stochastic (random) components of a system of interest, and of the development of algorithms for generating the associated random variates for a Monte Carlo or discrete-event simulation model. This chapter is devoted entirely to describing and generating discrete random variables. The next chapter is devoted to describing and generating continuous random variables.

Section 6.1 defines a discrete random variable and introduces four popular models: the Bernoulli, binomial, Pascal, and Poisson distributions. Section 6.2 contains an approach to generating discrete random variables that is more general than the ad-hoc approaches given earlier for the *Equilikely*(a, b) variate and the *Geometric*(p) variate. Section 6.3 applies these variate-generation techniques to the simple inventory system. Section 6.4 contains a summary of the six discrete distributions encountered thus far. Finally, Section 6.5 departs slightly from the topic of discrete random variables and considers the related topic of the development of algorithms for shuffling and drawing random samples from a group of objects.

6.1 DISCRETE RANDOM VARIABLES

As illustrated in previous chapters, random variables, both discrete and continuous, appear naturally in discrete-event simulation models. Because of this, it is virtually impossible to build a valid discrete-event simulation model of a *system* without a good understanding of how to construct a valid random-variable model for each of the stochastic system *components*. In this chapter and the next, we will develop the mathematical and computational tools for building such stochastic models. Discrete random variables are considered in this chapter; continuous random variables are considered in the next chapter.

6.1.1 Discrete-Random-Variable Characteristics

The notation and development in this section largely follows the axiomatic approach to probability. As a convention, uppercase characters X, Y, ... are used to denote random variables (discrete or continuous), the corresponding lowercase characters x, y, ... are used to denote the specific values of X, Y, ..., and calligraphic characters $\mathcal{X}$, $\mathcal{Y}$, ... are used to denote the set of all possible values (often known as the *support* of the random variable). A variety of examples are used in this section to illustrate this notation.

Definition 6.1.1   The random variable X is *discrete* if and only if its set of possible values $\mathcal{X}$ is finite or, at most, countably infinite.

In a discrete-event simulation model, discrete random variables are often integers used for counting (e.g., the number of jobs in a queue or the amount of inventory demand). There is no inherent reason, however, why a discrete random variable has to be integer-valued.

Probability Density Function

Definition 6.1.2   A discrete random variable X is uniquely determined by its set of possible values $\mathcal{X}$ and associated *probability density function* (*pdf*), a real-valued function $f(\cdot)$ defined for each possible value $x \in \mathcal{X}$ as the probability that X assumes the value x:

$$f(x) = \Pr(X = x).$$

By definition, $x \in \mathcal{X}$ is a possible value of X if and only if $f(x) > 0$. In addition, $f(\cdot)$ is defined so that

$$\sum_x f(x) = 1,$$

where the sum is over all $x \in \mathcal{X}$.*

It is important to understand the distinction between a random variable, its set of possible values, and its pdf. The usual way to construct a *model* of a discrete random variable X is to first specify the set of possible values $\mathcal{X}$ and then, for each $x \in \mathcal{X}$, specify the corresponding probability $f(x)$. The following three examples are illustrations.

Example 6.1.1

If the random variable X is *Equilikely*(a, b), then $\mathcal{X}$ is the set of integers between a and b inclusive. Because $|\mathcal{X}| = b - a + 1$ and each possible value is equally likely, it follows that

$$f(x) = \frac{1}{b - a + 1} \qquad x = a, a + 1, \ldots, b.$$

As a specific example, if we were to roll one fair die and let the random variable X be the up face, then X would be *Equilikely*$(1, 6)$.

---

*The pdf of a discrete random variable is sometimes called a probability *mass* function (pmf) or probability function (pf); this terminology is not used in this book.

Example 6.1.2

Roll two fair dice. If the random variable X is the sum of the two up faces, then the set of possible values is $\mathcal{X} = \{x \mid x = 2, 3, \ldots, 12\}$, and, from the table in Example 2.3.1, the pdf of X is

$$f(x) = \frac{6 - |7 - x|}{36} \qquad x = 2, 3, \ldots, 12.$$

Although the sum of the up faces is the usual discrete random variable for games of chance that use two dice, see Exercise 6.1.2 for an alternative.

Example 6.1.3

Suppose a coin has p as its probability of a head and suppose we agree to toss this coin until the *first* tail occurs. If X is the number of heads (i.e., the number of tosses is $X + 1$), then $\mathcal{X} = \{x \mid x = 0, 1, 2, \ldots\}$, and the pdf is

$$f(x) = p^x(1 - p) \qquad x = 0, 1, 2, \ldots.$$

This random variable is said to be *Geometric*(p). (The coin is *fair* if $p = 0.5$.)

Because the set of possible values is *infinite*, for a *Geometric*(p) random variable, some math is required to verify that $\sum_x f(x) = 1$. Fortunately, this infinite series and other similar series can be evaluated by using the following properties of *geometric series*. If $p \neq 1$, then

$$1 + p + p^2 + p^3 + \cdots + p^x = \frac{1 - p^{x+1}}{1 - p} \qquad x = 0, 1, 2, \ldots;$$

and, if $|p| < 1$, then the following three *infinite series* converge to tractable quantities:

$$1 + p + p^2 + p^3 + p^4 + \cdots = \frac{1}{1 - p},$$

$$1 + 2p + 3p^2 + 4p^3 + \cdots = \frac{1}{(1 - p)^2},$$

$$1 + 2^2 p + 3^2 p^2 + 4^2 p^3 + \cdots = \frac{1 + p}{(1 - p)^3}.$$

Although the three infinite series converge for any $|p| < 1$, negative values of p have no meaning in Example 6.1.3. From the first of these infinite series, we have that

$$\sum_x f(x) = \sum_{x=0}^{\infty} p^x(1 - p) = (1 - p)(1 + p + p^2 + p^3 + p^4 + \cdots) = 1,$$

as required.

As the previous examples illustrate, discrete random variables have possible values that are generated by the outcomes of a random experiment. Therefore, a Monte Carlo simulation program can be used to generate these possible values consistent with their probability of occurrence—see, for example, program `galileo` in Section 2.3. Provided that the number of replications is large, a histogram of the values generated by replication should agree well with the random variable's pdf. Indeed, in the limit, as the number of replications becomes infinite, the discrete-data histogram and the discrete random variable's pdf should agree *exactly*, as illustrated in Section 4.2.

Cumulative Distribution Function

Definition 6.1.3 The *cumulative distribution function* (*cdf*) of the discrete random variable X is the real-valued function $F(\cdot)$ defined for each $x \in \mathcal{X}$ as

$$F(x) = \Pr(X \le x) = \sum_{t \le x} f(t)$$

where the sum is over all $t \in \mathcal{X}$ for which $t \le x$.

Example 6.1.4
If X is an *Equilikely*(a, b) random variable (Example 6.1.1), then the cdf is

$$F(x) = \sum_{t=a}^{x} 1/(b-a+1) = (x-a+1)/(b-a+1) \qquad x = a, a+1, \ldots, b.$$

Example 6.1.5
For the sum-of-two-dice random variable in Example 6.1.2, there is no simple equation for $F(x)$. That is of no real significance, however, because $|\mathcal{X}|$ is small enough that the cumulative pdf values are easily tabulated to yield the cdf. Figure 6.1.1 shows the pdf on the left and the cdf on the right (the vertical scale is different for the two figures). Any of the four styles for plotting a cdf given in Figure 4.2.8 is acceptable.

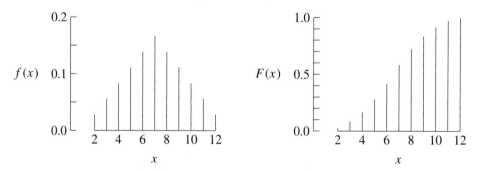

Figure 6.1.1 Two dice pdf and cdf.

Example 6.1.6
If X is a *Geometric*(p) random variable (Example 6.1.3), then the cdf is

$$F(x) = \sum_{t=0}^{x} p^t(1-p) = (1-p)(1 + p + p^2 + \cdots + p^x) = 1 - p^{x+1} \qquad x = 0, 1, 2, \ldots.$$

The cdf of a discrete random variable can always be generated from its corresponding pdf by recursion. If the possible values of X are the consecutive integers $x = a, a+1, \ldots, b$, for example, then

$$F(a) = f(a)$$

$$F(x) = F(x-1) + f(x) \qquad x = a+1, a+2, \ldots, b.$$

Similarly, a pdf can always be generated from its corresponding cdf by subtracting consecutive terms

$$f(a) = F(a)$$

$$f(x) = F(x) - F(x-1) \qquad x = a+1, a+2, \ldots, b.$$

Therefore, a model of a discrete random variable can be defined by specifying *either* the pdf *or* the cdf and then computing the other—there is no need to specify both.

As illustrated in Example 6.1.5, the cdf of a discrete random variable is strictly monotone increasing—if x_1 and x_2 are possible values of X with $x_1 < x_2$, then $F(x_1) < F(x_2)$. Moreover, since $F(x)$ is defined as a probability, $0 \le F(x) \le 1$. The monotonicity of $F(\cdot)$ is fundamentally important; in the next section, we will use it as the basis for a method to generate discrete random variates.

Mean and Standard Deviation.   Recall, from Definition 4.2.2, that, for a discrete-data sample, the sample mean $\bar{x}$ and standard deviation s can be computed from the discrete-data histogram as

$$\bar{x} = \sum_x x \hat{f}(x) \qquad \text{and} \qquad s = \sqrt{\sum_x (x - \bar{x})^2 \hat{f}(x)},$$

respectively. Moreover, the histogram relative frequencies $\hat{f}(x)$ converge to the corresponding probabilities $f(x)$ as the sample size becomes infinite. These two observations motivate the following definition.

Definition 6.1.4   The (population) *mean* μ and the corresponding (population) *standard deviation* σ are

$$\mu = \sum_x x f(x) \qquad \text{and} \qquad \sigma = \sqrt{\sum_x (x - \mu)^2 f(x)},$$

where the summations are over all $x \in \mathcal{X}$. The population *variance* is σ^2. An alternative, algebraically equivalent expression for σ is

$$\sigma = \sqrt{\left(\sum_x x^2 f(x) \right) - \mu^2}.$$

The population mean μ is a fixed constant, whereas the sample mean $\bar{x}$ is a random variable that varies from sample to sample. They are fundamentally different, although analogous, quantities. Since $\bar{x}$ is a random variable, it has a pdf and cdf that describe its distribution.

Example 6.1.7

If X is an *Equilikely* (a, b) random variable (Example 6.1.1), then

$$\mu = \frac{a+b}{2} \qquad \text{and} \qquad \sigma = \sqrt{\frac{(b-a+1)^2 - 1}{12}}.$$

The derivation of these equations is left as an exercise. In particular, if X is the result of rolling one fair die, then X is *Equilikely* $(1, 6)$, and so

$$\mu = 3.5 \qquad \text{and} \qquad \sigma = \sqrt{\frac{35}{12}} \cong 1.708,$$

illustrating the fact that μ need not necessarily be a member of $\mathcal{X}$. The geometric interpretation of the mean is the horizontal center of mass of the pdf.

Example 6.1.8

If X is the sum-of-two-dice random variable (Example 6.1.2), then

$$\mu = \sum_{x=2}^{12} x f(x) = \cdots = 7 \qquad \text{and} \qquad \sigma^2 = \sum_{x=2}^{12} (x - \mu)^2 f(x) = \cdots = 35/6.$$

Therefore, the population standard deviation is $\sigma = \sqrt{35/6} \cong 2.415$.

Example 6.1.9

If X is a *Geometric* (p) random variable (Example 6.1.3), then

$$\mu = \sum_{x=0}^{\infty} x f(x) = \sum_{x=1}^{\infty} x p^x (1 - p) = \cdots = \frac{p}{1 - p}$$

using the infinite series following Example 6.1.3, and

$$\sigma^2 = \left(\sum_{x=0}^{\infty} x^2 f(x) \right) - \mu^2 = \left(\sum_{x=1}^{\infty} x^2 p^x (1 - p) \right) - \frac{p^2}{(1 - p)^2} = \cdots = \frac{p}{(1 - p)^2}$$

so that $\sigma = \sqrt{p}/(1 - p)$. The derivation of these equations is left as an exercise. In particular, tossing a fair $(p = 0.5)$ coin until the first tail occurs generates a *Geometric* (0.5) random variable (X is the number of heads) with

$$\mu = 1 \qquad \text{and} \qquad \sigma = \sqrt{2} \cong 1.414.$$

The population mean and the population variance are two special cases of a more general notion known as "expected value."

Expected Value

> **Definition 6.1.5** The mean of a random variable (discrete or continuous) is also known as its *expected value*. It is conventional to denote the expected value as $E[\,\cdot\,]$. That is, the expected value of the discrete random variable X is
>
> $$E[X] = \sum_x x f(x) = \mu$$
>
> where the summation is over all $x \in \mathcal{X}$.*

If we were to use a Monte Carlo simulation to generate a large random variate sample $x_1, x_2, \ldots, x_n$ corresponding to the random variable X and then calculate the sample mean $\overline{x}$, we would expect to find that $\overline{x} \to E[X] = \mu$ as $n \to \infty$. Thus, the expected value of X is really the "expected average"; in that sense, the term "expected value" is potentially misleading. The expected value (the mean) is not necessarily the *most likely* possible value [which is the *mode*, the element in $\mathcal{X}$ corresponding to the largest value of $f(x)$].

Example 6.1.10
If a fair coin is tossed until the first tail appears, then the *most likely* number of heads is 0 and the *expected* number of heads is 1 (see Example 6.1.9). In this case, 0 occurs with probability 1/2 and 1 occurs with probability 1/4. Thus the most likely value (the mode) is twice as likely as the expected value (the mean). On the other hand, for some random variables, the mean and mode will be the same. For example, if X is the sum-of-two-dice random variable, then the expected value and the most likely value are both 7 (see Examples 6.1.5 and 6.1.8).

> **Definition 6.1.6** If $h(\cdot)$ is a function defined for all possible values of X, then as x takes on all possible values in $\mathcal{X}$, the equation $y = h(x)$ defines the set $\mathcal{Y}$ of possible values for a *new* random variable $Y = h(X)$. The expected value of Y is
>
> $$E[Y] = E[h(X)] = \sum_x h(x) f(x),$$
>
> where the sum is over all $x \in \mathcal{X}$.†

Example 6.1.11
If $y = (x - \mu)^2$ with $\mu = E[X]$, then from Definitions 6.1.4 and 6.1.6,

$$E[Y] = E\left[(X - \mu)^2\right] = \sum_x (x - \mu)^2 f(x) = \sigma^2.$$

---

*The expected value could fail to exist if there are infinitely many possible values.

†Definition 6.1.6 can be established as a theorem. When presented as a definition, it is sometimes called the "law of the unconscious statistician." If the set of possible values $\mathcal{X}$ is infinite, then $E[Y]$ could fail to exist.

This equation indicates that the variance σ^2 is the expected value of the squared difference about the mean. Similarly, if $y = x^2 - \mu^2$, then

$$E[Y] = E\left[X^2 - \mu^2\right] = \sum_x (x^2 - \mu^2) f(x) = \left(\sum_x x^2 f(x)\right) - \mu^2 = \sigma^2,$$

and so

$$\sigma^2 = E[X^2] - E[X]^2.$$

This last equation demonstrates that the two operations $E[\,\cdot\,]$ and $(\cdot)^2$ do *not* commute; the expected value of X^2 is not equal to the square of $E[X]$. Indeed, $E[X^2] \geq E[X]^2$, with equality if and only if X is not really random at all, i.e., $\sigma^2 = 0$, often known as a random variable with a *degenerate* distribution.

Example 6.1.12
If $Y = aX + b$ for constants a and b, then

$$E[Y] = E[aX + b] = \sum_x (ax + b) f(x) = a\left(\sum_x x f(x)\right) + b = a E[X] + b.$$

In particular, suppose that X is the number of heads before the first tail and that you are playing a game with a fair coin, where you win \$2 for every head. Let the random variable Y be the amount you win. The possible values of Y are defined by

$$y = h(x) = 2x \qquad x = 0, 1, 2, \ldots$$

and your *expected winnings* (for each play of the game) are

$$E[Y] = E[2X] = 2E[X] = 2.$$

If you play this game repeatedly and pay more than \$2 per game to do so, then, in the long run, you can expect to lose money.

6.1.2 Discrete-Random-Variable Models

Let X be any discrete random *variable*. In the next section, we will consider a unified algorithmic approach to generating possible values of X. The values so generated are random *variates*. The distinction between a random variable (discrete or continuous) and a corresponding random variate is subtle, but important. The former is an abstract, but well-defined, mathematical function that maps the outcome of an experiment to a real number (per Definition 6.1.2); the latter is an algorithmically generated possible value (realization) of the former (per Definition 2.3.2). For example, the functions `Equilikely` (Definition 2.3.4) and `Geometric` (Definition 3.1.2) generate random *variates* corresponding to *Equilikely*(a, b) and *Geometric*(p) random *variables*, respectively.

Bernoulli Random Variable

Example 6.1.13

The discrete random *variable* X having the possible values $\mathcal{X} = \{0, 1\}$ is said to be *Bernoulli*(p) if $X = 1$ with probability p and $X = 0$ otherwise (i.e., with probability $1 - p$). In effect, X is a Boolean random variable with 1 as `true` and 0 as `false`. The pdf for a *Bernoulli*(p) random variable is $f(x) = p^x (1 - p)^{1-x}$ for $x \in \mathcal{X}$. The corresponding cdf is $F(x) = (1 - p)^{1-x}$ for $x \in \mathcal{X}$. The mean is $\mu = 0 \cdot (1 - p) + 1 \cdot p = p$, and the variance is $\sigma^2 = (0 - p)^2 (1 - p) + (1 - p)^2 p = p(1 - p)$. Therefore, the standard deviation is $\sigma = \sqrt{p(1 - p)}$. We can generate a corresponding *Bernoulli*(p) random *variate* as follows.

```
if (Random() < 1.0 - p)
    return 0;
else
    return 1;
```

As illustrated in Section 2.3, no matter how sophisticated or computationally complex, a Monte Carlo simulation that uses n replications to estimate an (unknown) probability p is equivalent to generating an *iid* sequence of n *Bernoulli*(p) random variates.

Example 6.1.14

A popular state lottery game, *Pick-3*, requires players to pick a 3-digit number from the 1000 numbers between 000 and 999. It costs \$1 to play the game. If the 3-digit number picked by a player matches the 3-digit number chosen, at random, by the state, then the player wins \$500, minus the original \$1 investment, for a net yield of +\$499. Otherwise, the player's yield is $-\$1$. (See Exercise 6.1.7 for another way to play this game.) Let the discrete random variable X represent the result of playing the game with the convention that $X = 1$ denotes a win and $X = 0$ denotes a loss. Then X is a *Bernoulli*(p) random variable with $p = 1/1000$. In addition, let the discrete random variable $Y = h(X)$ be the player's yield, where

$$h(x) = \begin{cases} -1 & x = 0 \\ 499 & x = 1. \end{cases}$$

From Definition 6.1.6, the player's *expected* yield is

$$E[Y] = \sum_{x=0}^{1} h(x) f(x) = h(0)(1 - p) + h(1)p = -1 \cdot \frac{999}{1000} + 499 \cdot \frac{1}{1000} = -0.5.$$

In this case, Y has just two possible values—one is 999 times more likely than the other, and neither is the expected value. Even though the support values for Y are far apart (-1 and 499), the value of $E[Y]$ shows that playing *Pick-3* is the equivalent of a voluntary 50-cent donation to the state for every dollar bet.

Because it has only two possible values, a *Bernoulli* (p) random variable may seem to have limited applicability. That is not the case, however, because this simple random variable can be used to construct more sophisticated stochastic models, as illustrated by the following examples.

Binomial Random Variable

Example 6.1.15

In the spirit of Example 6.1.3, suppose a coin has p as its probability of tossing a head and suppose we toss this coin n times. Let X be the number of heads; in this case X is said to be a *Binomial* (n, p) random variable. The set of possible values is $\mathcal{X} = \{0, 1, 2, \ldots, n\}$, and the associated pdf is

$$f(x) = \binom{n}{x} p^x (1 - p)^{n-x} \qquad x = 0, 1, 2, \ldots, n.$$

That is, $p^x (1 - p)^{n-x}$ is the probability of x heads and $n - x$ tails, and the binomial coefficient accounts for the number of different sequences in which these heads and tails can occur. Equivalently, n tosses of the coin generate an *iid* sequence $X_1, X_2, \ldots, X_n$ of *Bernoulli* (p) random variables ($X_i = 1$ corresponds to a head on the ith toss), and

$$X = X_1 + X_2 + \cdots + X_n.$$

It might appear intuitive that the pdf in Example 6.1.15 is correct, but building a discrete-random-variable model requires us to confirm that, in fact, $f(x) > 0$ for all $x \in \mathcal{X}$ (which is obvious in this case) and that $\sum_x f(x) = 1$ (which is not obvious). To verify that the pdf sum is 1, we can use the *binomial equation*

$$(a + b)^n = \sum_{x=0}^{n} \binom{n}{x} a^x b^{n-x}.$$

In the particular case where $a = p$ and $b = 1 - p$,

$$1 = (1)^n = \left(p + (1 - p)\right)^n = \sum_{x=0}^{n} \binom{n}{x} p^x (1 - p)^{n-x},$$

which is equivalent to $f(0) + f(1) + \cdots + f(n) = 1$, as was desired. To compute the mean of a binomial random variable:

$$\mu = E[X] = \sum_{x=0}^{n} x f(x)$$

$$= \sum_{x=0}^{n} x \binom{n}{x} p^x (1 - p)^{n-x}$$

$$= \sum_{x=0}^{n} x \frac{n!}{x!\,(n-x)!} p^x (1-p)^{n-x}$$

$$= \sum_{x=1}^{n} x \frac{n!}{x!\,(n-x)!} p^x (1-p)^{n-x}$$

$$= \sum_{x=1}^{n} \frac{n!}{(x-1)!\,(n-x)!} p^x (1-p)^{n-x}$$

$$= np \sum_{x=1}^{n} \frac{(n-1)!}{(x-1)!\,(n-x)!} p^{x-1} (1-p)^{n-x}.$$

To evaluate the last sum, let $m = n - 1$ and $t = x - 1$, so that $m - t = n - x$. Then, from the binomial equation,

$$\mu = np \sum_{t=0}^{m} \frac{m!}{t!\,(m-t)!} p^t (1-p)^{m-t} = np \big(p + (1-p)\big)^m = np(1)^m = np.$$

Hence, the mean of a *Binomial*(n, p) random variable is $\mu = np$. In a similar way, it can be shown that the variance is

$$\sigma^2 = E[X^2] - \mu^2 = \left(\sum_{x=0}^{n} x^2 f(x) \right) - \mu^2 = \cdots = np(1-p),$$

and so the standard deviation of a *Binomial*(n, p) random variable is $\sigma = \sqrt{np(1-p)}$.

Pascal Random Variable

Example 6.1.16

As a second example of using *Bernoulli*(p) random variables to build a more sophisticated stochastic model, suppose a coin has p as its probability of a head and suppose we toss this coin until the nth tail occurs. If X is the number of heads (i.e., the number of tosses is $X + n$), then X is said to be a *Pascal*(n, p) random variable. The set of possible values is $\mathcal{X} = \{0, 1, 2, \ldots\}$, and the associated pdf is

$$f(x) = \binom{n+x-1}{x} p^x (1-p)^n \qquad x = 0, 1, 2, \ldots.$$

In other words, $p^x (1-p)^n$ is the probability of x heads and n tails, and the binomial coefficient accounts for the number of different sequences in which these $n + x$ heads and tails can occur, given that the last coin toss must be a tail.

It might appear intuitive that the *Pascal* (n, p) pdf equation is correct, but it is necessary to prove that the infinite pdf sum does, in fact, converge to 1. The proof of this property is based on another (negative-exponent) version of the *binomial equation* — for any positive integer n,

$$(1 - p)^{-n} = 1 + \binom{n}{1}p + \binom{n+1}{2}p^2 + \cdots + \binom{n+x-1}{x}p^x + \cdots,$$

provided that $|p| < 1$. (See Example 6.1.3 for $n = 1$ and $n = 2$ versions of this equation.) By using the negative-exponent binomial equation, we see that[*]

$$\sum_{x=0}^{\infty}\binom{n+x-1}{x}p^x(1-p)^n = (1-p)^n\sum_{x=0}^{\infty}\binom{n+x-1}{x}p^x = (1-p)^n(1-p)^{-n} = 1,$$

which confirms that $f(0) + f(1) + f(2) + \cdots = 1$. Moreover, in a similar way, it can be shown that the mean is

$$\mu = E[X] = \sum_{x=0}^{\infty}xf(x) = \cdots = \frac{np}{1-p}$$

and that the variance is

$$\sigma^2 = E[X^2] - \mu^2 = \left(\sum_{x=0}^{\infty}x^2 f(x)\right) - \mu^2 = \cdots = \frac{np}{(1-p)^2}$$

[and so the standard deviation is $\sigma = \sqrt{np}/(1-p)$]. The details of this derivation are left as an exercise.

Example 6.1.17
As a third example of using *Bernoulli* (p) random variables to build a more sophisticated stochastic model, a *Geometric* (p) random variable is a special case of a *Pascal* (n, p) random variable when $n = 1$. If instead $n > 1$ and if $X_1, X_2, \ldots, X_n$ is an *iid* sequence of n *Geometric* (p) random variables, then the sum

$$X = X_1 + X_2 + \cdots + X_n$$

is a *Pascal* (n, p) random variable. For example, if $n = 4$ and if p is large, then a typical head/tail sequence might look like

$$\underbrace{hhhhhht}_{X_1 = 6}\ \underbrace{hhhhhhhhht}_{X_2 = 9}\ \underbrace{hhhht}_{X_3 = 4}\ \underbrace{hhhhhhht}_{X_4 = 7},$$

---

[*]Because the negative-exponent binomial equation is applicable to the mathematical analysis of its characteristics, a *Pascal* (n, p) random variable is also known as a *negative binomial* random variable. Also, some authors define X to be the toss number of the nth tail, which alters both $f(x)$ and $\mathcal{X}$.

where X_1, X_2, X_3, X_4 count the number of heads in each $h \ldots ht$ sequence and, in this case,

$$X = X_1 + X_2 + X_3 + X_4 = 26.$$

The number of heads in each $h \ldots ht$ *Bernoulli*(p) sequence is an independent realization of a *Geometric*(p) random variable. In this way, we see that a *Pascal*(n, p) random variable is the sum of n *iid Geometric*(p) random variables. From Example 6.1.9, we know that, if X is *Geometric*(p), then $\mu = p/(1-p)$ and $\sigma = \sqrt{p}/(1-p)$.

Poisson Random Variable. A *Poisson*(μ) random variable is a limiting case of a *Binomial*$(n, \mu/n)$ random variable. Let X be a *Binomial*(n, p) random variable with $p = \mu/n$. Fix the values of μ and x, and consider what happens in the limit as $n \to \infty$. The pdf of X is

$$f(x) = \frac{n!}{x!(n-x)!} \left(\frac{\mu}{n}\right)^x \left(1 - \frac{\mu}{n}\right)^{n-x} = \frac{\mu^x}{x!} \left(\frac{n!}{(n-x)!(n-\mu)^x}\right) \left(1 - \frac{\mu}{n}\right)^n$$

for $x = 0, 1, \ldots, n$. It can be shown that

$$\lim_{n \to \infty} \left(\frac{n!}{(n-x)!(n-\mu)^x}\right) = 1 \qquad \text{and} \qquad \lim_{n \to \infty} \left(1 - \frac{\mu}{n}\right)^n = \exp(-\mu),$$

and so

$$\lim_{n \to \infty} f(x) = \frac{\mu^x \exp(-\mu)}{x!}.$$

This limiting case is the motivation for defining a *Poisson*(μ) random variable. For large values of n, *Binomial*$(n, \mu/n)$ and *Poisson*(μ) random variables are virtually identical, particularly if μ is small. As an exercise, you are asked to prove that the parameter μ in the definition of the *Poisson*(μ) pdf

$$f(x) = \frac{\mu^x \exp(-\mu)}{x!} \qquad\qquad x = 0, 1, 2, \ldots$$

is in fact the mean and that the standard deviation is $\sigma = \sqrt{\mu}$.

6.1.3 Exercises

6.1.1   (a) Simulate rolling a pair of fair dice 360 times with five different seeds, and generate five histograms of the resulting sum of the two up faces. Compare the histogram mean, standard deviation, and relative frequencies with the corresponding population mean, standard deviation, and pdf. (b) Repeat for 3600, 36 000, and 360 000 replications. (c) Comment.

6.1.2   Repeat the previous exercise, *except* that the random variable of interest is the absolute value of the *difference* between the two up faces.

6.1.3   Derive the equations for μ and σ in Example 6.1.7. (See Exercise 6.1.5.)

6.1.4   Prove that $\sum_x (x - \mu)^2 f(x) = \left(\sum_x x^2 f(x)\right) - \mu^2$.

6.1.5 Let X be a discrete random variable with possible values $x = 1, 2, \ldots, n$. (*a*) If the pdf of X is $f(x) = \alpha x$, then what is α (as a function of n)? (*b*) What are the cdf, mean, and standard deviation of X? *Hint*:

$$\sum_{x=1}^{n} x = \frac{n(n+1)}{2} \qquad \sum_{x=1}^{n} x^2 = \frac{n(n+1)(2n+1)}{6} \qquad \sum_{x=1}^{n} x^3 = \frac{n^2(n+1)^2}{4}.$$

6.1.6 Fill in the $= \cdots =$'s in Example 6.1.9.

6.1.7 As an alternative to Example 6.1.14, another way to play *Pick-3* is for the player to opt for a win *in any order*. For example, if the player's number is 123, then the player will win (the same amount) if the state draws any of 123, 132, 231, 213, 321, or 312. Because this is an easier win, the pay-off is suitably smaller—namely, \$80 (for a net yield of +\$79). (*a*) What is the player's expected yield (per game) if this variation of the game is played? (Assume the player is bright enough to pick a 3-digit number with three different digits.) (*b*) Construct a Monte Carlo simulation to supply "convincing numerical evidence" of the correctness of your solution.

6.1.8 An urn is initially filled with one amber ball and one black ball. Each time a ball is drawn, at random, if it is a black ball then it *and* another black ball are put back in the urn. Let X be the number of random draws required to find the amber ball. (*a*) What is $E[X]$? (*b*) Construct a Monte Carlo simulation to estimate $E[X]$ based on 1000, 10 000, 100 000, and 1 000 000 replications. (*c*) Comment.

6.1.9 The location of an interval of fixed length $r > 0$ is selected at random on the real number line. Let X be the number of integers within the interval. Find the mean and standard deviation of X as a function of (r, p), where $p = r - \lfloor r \rfloor$.

6.1.10 If X is a *Poisson*(μ) random variable, prove that the mean of X is the parameter μ and that the standard deviation is $\sqrt{\mu}$.

6.2 GENERATING DISCRETE RANDOM VARIATES

In this section, we will consider the development of correct, exact algorithms for the generation of discrete random variates. We begin with an important definition, an associated theorem, and an algorithm.

6.2.1 Inverse Distribution Function

> **Definition 6.2.1** Let X be a discrete random variable with cdf $F(\cdot)$. The *inverse distribution function* (*idf*) of X is the function $F^*: (0, 1) \to \mathcal{X}$ defined for all $u \in (0, 1)$ as
>
> $$F^*(u) = \min_x \{x : u < F(x)\}$$
>
> where the minimum is over all possible values $x \in \mathcal{X}$. This means that, if $F^*(u) = x$, then x is the smallest possible value of X for which $F(x)$ is greater than u.*

---

*The rather unusual definition of the inverse [and the avoidance of the notation $F^{-1}(u)$] is to account for the fact that $F(x)$ is not 1–1. A specific value of the idf is also called a *fractile*, *quantile*, or *percentile* of the distribution of X. For example, $F^*(0.95)$ is often called the 95th percentile of the distribution of X.

Example 6.2.1

Figure 6.2.1 provides a graphical illustration of the idf for a discrete random variable with $\mathcal{X} = \{a, a+1, \ldots, b\}$ for two common ways of plotting the same cdf. The value of $x \in \mathcal{X}$ corresponding to a given u, in this case $u = 0.45$, is found by extending the dashed line horizontally until it strikes one of the cdf "spikes" in the left-hand cdf or one of the "risers" on the right-hand cdf. The value of $X \in \mathcal{X}$ corresponding to that cdf spike defines the value of $x = F^*(u)$.

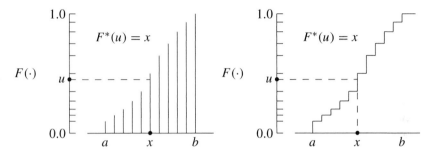

Figure 6.2.1 Generating discrete random variables.

The following theorem is based on Definition 6.2.1 and on the insight provided by the previous example. The significance of the theorem is that it provides an important algorithmic characterization of the idf.

Theorem 6.2.1   Let X be an integer-valued random variable with $\mathcal{X} = \{a, a+1, \ldots, b\}$, where b may be ∞, and let $F(\cdot)$ be the cdf of X. For any $u \in (0, 1)$, if $u < F(a)$, then $F^*(u) = a$; else, $F^*(u) = x$ where $x \in \mathcal{X}$ is the (unique) possible value of X for which $F(x - 1) \le u < F(x)$.

Although Theorem 6.2.1 (and others that follow in this section) are written for *integer-valued* discrete random variables, they are easily extended to the more general case.

Algorithm 6.2.1   Let X be an integer-valued random variable with $\mathcal{X} = \{a, a+1, \ldots, b\}$, where b may be ∞, and let $F(\cdot)$ be the cdf of X. For any $u \in (0, 1)$, the following linear search algorithm defines the idf $F^*(u)$:

```
x = a;
while (F(x) <= u)
    x++;
return x;                                          // x is F*(u)
```

Because $u < 1.0$ and $F(x) \to 1.0$ as x increases, the loop in Algorithm 6.2.1 is guaranteed to terminate. Note, however, that the search is linear and begins at the smallest possible value of X; hence, *average-case* efficiency can be a real problem if $\mu = E[X]$ is large relative to a. More specifically, in repeated applications of this algorithm using values of u generated by calls to Random, the average number of passes through the while loop will be $\mu - a$. To see this, let u be a *Uniform*$(0, 1)$ random variate, and let X be the discrete random variable corresponding to the random variate x generated by Algorithm 6.2.1. If the discrete random variable Y represents the number of while loop passes, then $Y = X - a$. Therefore, from Definition 6.1.6, $E[Y] = E[X - a] = E[X] - a = \mu - a$.

Algorithm 6.2.1 can be made more efficient by starting the search at a better (more likely) point. The best (most likely) starting point is the mode.* We then have the following more efficient version of Algorithm 6.2.1.

Algorithm 6.2.2   Let X be an integer-valued random variable with $\mathcal{X} = \{a, a + 1, \ldots, b\}$, where b may be ∞, and let $F(\cdot)$ be the cdf of X. For any $u \in (0, 1)$, the following linear search algorithm defines the idf $F^*(u)$:

```
x = mode;                          // initialize with the mode of X
if  (F(x) <= u)
    while  (F(x) <= u)
        x++;
else if  (F(a) <= u)
    while  (F(x - 1) > u)
        x--;
else
    x = a;
return x;                                             // x is F*(u)
```

Although Algorithm 6.2.2 is still a linear search, generally it is more efficient than Algorithm 6.2.1, perhaps significantly so, unless the mode of X is a. If $|\mathcal{X}|$ is very large, even more efficiency may be needed. In this case, a *binary* search should be considered. (See Exercise 6.2.9.)

Idf Examples.   In some important cases, the idf $F^*(u)$ can be expressed explicitly by using Theorem 6.2.1 to solve the equation $F(x) = u$ for x. The following three examples are illustrations.

Example 6.2.2
If X is *Bernoulli*(p) and $F(x) = u$, then $x = 0$ if and only if $0 < u < 1 - p$; and $x = 1$ otherwise. Therefore

$$F^*(u) = \begin{cases} 0 & 0 < u < 1 - p \\ 1 & 1 - p \le u < 1. \end{cases}$$

---

*The mode of X is the value of $x \in \mathcal{X}$ for which $f(x)$ is largest. For many discrete random variables, but *not* all, $\lfloor \mu \rfloor$ is an essentially equivalent choice.

Example 6.2.3

If X is *Equilikely* (a, b), then

$$F(x) = \frac{x - a + 1}{b - a + 1} \qquad x = a, a + 1, \ldots, b.$$

Therefore, provided $u \geq F(a)$,

$$F(x - 1) \leq u < F(x) \iff \frac{(x - 1) - a + 1}{b - a + 1} \leq u < \frac{x - a + 1}{b - a + 1}$$

$$\iff x - a \leq (b - a + 1)u < x - a + 1$$

$$\iff x \leq a + (b - a + 1)u < x + 1.$$

Thus, for $F(a) \leq u < 1$, the idf is

$$F^*(u) = a + \lfloor (b - a + 1)u \rfloor.$$

Moreover, if $0 < u < F(a) = 1/(b - a + 1)$, then $0 < (b - a + 1)u < 1$, and so $F^*(u) = a$, as required. Therefore, the idf equation is valid for all $u \in (0, 1)$.

Example 6.2.4

If X is *Geometric* (p), then

$$F(x) = 1 - p^{x+1} \qquad x = 0, 1, 2, \ldots.$$

Therefore, provided $u \geq F(0)$,

$$F(x - 1) \leq u < F(x) \iff 1 - p^x \leq u < 1 - p^{x+1}$$

$$\iff -p^x \leq u - 1 < -p^{x+1}$$

$$\iff p^x \geq 1 - u > p^{x+1}$$

$$\iff x \ln(p) \geq \ln(1 - u) > (x + 1) \ln(p)$$

$$\iff x \leq \frac{\ln(1 - u)}{\ln(p)} < x + 1.$$

Thus, for $F(0) \leq u < 1$ the idf is

$$F^*(u) = \left\lfloor \frac{\ln(1 - u)}{\ln(p)} \right\rfloor.$$

Moreover, if $0 < u < F(0) = 1 - p$, then $p < 1 - u < 1$, and so $F^*(u) = 0$, as required. Therefore, the idf equation is valid for all $u \in (0, 1)$.

6.2.2 Random-Variate Generation by Inversion

The following theorem is of fundamental importance in random-variate generation applications. Because of the importance of this theorem, a detailed proof is presented. The proof makes use of a definition and two results, listed here for reference.

- Two random variables X_1 and X_2 with corresponding pdfs $f_1(\cdot)$ and $f_2(\cdot)$ defined on $\mathcal{X}_1$ and $\mathcal{X}_2$, respectively, are *identically distributed* if and only if they have a common set of possible values $\mathcal{X}_1 = \mathcal{X}_2$ and, for all x in this common set, $f_1(x) = f_2(x)$.
- The idf F^*: $(0, 1) \to \mathcal{X}$ maps the interval $(0, 1)$ *onto* $\mathcal{X}$.
- If U is *Uniform*$(0, 1)$ and $0 \le \alpha < \beta \le 1$, then $\Pr(\alpha \le U < \beta) = \beta - \alpha$.*

Theorem 6.2.2   (Probability integral transformation) If X is a discrete random variable with idf $F^*(\cdot)$, and the continuous random variable U is *Uniform*$(0, 1)$, and Z is the discrete random variable defined by $Z = F^*(U)$, then Z and X are identically distributed.

Proof. Let $\mathcal{X}$, $\mathcal{Z}$ be the set of possible values for X, Z respectively. If $x \in \mathcal{X}$ then, because F^* maps $(0, 1)$ onto $\mathcal{X}$, there exists $u \in (0, 1)$ such that $F^*(u) = x$. From the definition of Z it follows that $x \in \mathcal{Z}$ and so $\mathcal{X} \subseteq \mathcal{Z}$. Similarly, if $z \in \mathcal{Z}$, then, from the definition of Z, there exists $u \in (0, 1)$ such that $F^*(u) = z$. Because F^*: $(0, 1) \to \mathcal{X}$, it follows that $z \in \mathcal{X}$, and so $\mathcal{Z} \subseteq \mathcal{X}$. This proves that $\mathcal{X} = \mathcal{Z}$. Now, let $\mathcal{X} = \mathcal{Z} = \{a, a + 1, \ldots, b\}$ be the common set of possible values. To prove that X and Z are identically distributed, we must show that $\Pr(Z = z) = f(z)$ for any z in this set of possible values, where $f(\cdot)$ is the pdf of X. From the definition of Z and $F^*(\cdot)$ and Theorem 6.2.1, if $z = a$, then

$$\Pr(Z = a) = \Pr\left(U < F(a)\right) = F(a) = f(a).$$

Moreover, if $z \in \mathcal{Z}$ with $z \neq a$, then

$$\Pr(Z = z) = \Pr\left(F(z - 1) \le U < F(z)\right) = F(z) - F(z - 1) = f(z),$$

which proves that Z and X have the same pdf and so are identically distributed.   □

Theorem 6.2.2 is the basis for the following algorithm by which *any* discrete random variable can be generated with just *one* call to Random. In Section 7.2, we will see that there is an analogous theorem and algorithm for continuous random variables. This elegant algorithm is known as the *inversion* method for random-variate generation.

Algorithm 6.2.3 If X is a discrete random variable with idf $F^*(\cdot)$, then a corresponding discrete random variate x can be generated as follows

```
u = Random();
return F*(u);
```

---

*See Example 7.1.2 in the next chapter for more discussion of this result.

Inversion Examples.   Random-variate generation by inversion has a clear intuitive appeal and always works. It is *the* method of choice for discrete-random-variate generation, provided that the idf can be manipulated into a form that is mathematically tractable for algorithmic implementation. We begin with an example for an arbitrary discrete distribution.

Example 6.2.5
Consider the discrete random variable X having the pdf

$$f(x) = \begin{cases} 0.1 & x = 2 \\ 0.3 & x = 3 \\ 0.6 & x = 6. \end{cases}$$

The cdf for X is plotted in two different formats in Figure 6.2.2.

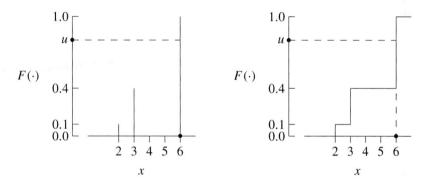

Figure 6.2.2   Inversion for an arbitrary discrete random variable.

For a random number u (which strikes the vertical axes in Figure 6.2.2, for example, at $u = 0.803$), $F^*(u)$ assumes the value 2, 3, or 6, depending on the intersection point of the associated horizontal line with the cdf. The reader is encouraged to reflect on the geometry in Figure 6.2.2 to appreciate the fact that the algorithm

```
u = Random();
if (u < 0.1)
    return 2;
else if (u < 0.4)
    return 3;
else
    return 6;
```

returns 2 with probability 0.1, returns 3 with probability 0.3, and returns 6 with probability 0.6, which corresponds to the pdf of X. The inversion method for generating discrete random variates always carries this intuitive geometric interpretation. Problems arise, however, when $|\mathcal{X}|$ is large or countably infinite. The average execution time for this particular example can be minimized by checking the ranges for u associated with $x = 6$ (the mode) first, then $x = 3$, then $x = 2$. This way, 60% of the invocations of this algorithm require the evaluation of just one condition.

Random-variate-generation algorithms for popular discrete probability models that include parameters follow.

Example 6.2.6
This algorithm uses inversion and the idf expression from Example 6.2.2 to generate a *Bernoulli*(p) random variate (see Example 6.1.13):

```
u = Random();
if (u < 1 - p)
    return 0;
else
    return 1;
```

Example 6.2.7
This algorithm uses inversion and the idf expression from Example 6.2.3 to generate an *Equilikely*(a, b) random variate:

```
u = Random();
return a + (long) (u * (b - a + 1));
```

This is the algorithm used to define the function `Equilikely`(a, b) (for which, see Definition 2.3.4).

Example 6.2.8
This algorithm uses inversion and the idf expression from Example 6.2.4 to generate a *Geometric*(p) random variate:

```
u = Random();
return (long) (log(1.0 - u) / log(p));
```

This is the algorithm used to define the function `Geometric`(p) (for which, see Definition 3.1.2).

Examples 6.2.6, 6.2.7, and 6.2.8 illustrate random-variate generation by inversion for three of the six parametric discrete-random-variable models presented in the previous section. For the other three models, inversion is not so easily applied.

Example 6.2.9
If X is a *Binomial*(n, p) random variable, the cdf is

$$F(x) = \sum_{t=0}^{x} \binom{n}{t} p^t (1 - p)^{n-t} \qquad x = 0, 1, 2, \ldots, n.$$

It can be shown that this sum is the complement of an *incomplete beta function* (see Appendix D)

$$F(x) = \begin{cases} 1 - I(x + 1, n - x, p) & x = 0, 1, \ldots, n - 1 \\ 1 & x = n. \end{cases}$$

Except for special cases, an incomplete beta function cannot be inverted algebraically to form a "closed form" expression for the idf. Therefore, inversion is not easily applied to the generation of *Binomial*(n, p) random variates.

Algorithm Design Criteria. As suggested by Example 6.2.9, for many discrete random *variables*, the design of a correct, exact, and efficient algorithm to generate corresponding random *variates* is significantly more complex than just a straightforward implementation of Algorithm 6.2.3. Here are some important, generally accepted algorithm-design criteria; with minor modifications, these criteria apply equally well to continuous-random-variate-generation algorithms also.

- *Portability*—a random-variate-generation algorithm should be implementable in a high-level language in such a way that it ports easily to any common contemporary computing environment. Hardware-dependent implementations should be rejected, even if the cost to do so is an increase in execution time.
- *Exactness*—assuming a perfect random-number generator, the histogram that results from generating random variates from an *exact* algorithm converges to the corresponding random-variable pdf as the number of random variates generated goes to infinity. Example 6.2.10, presented later, is an illustration of a random-variate generator that is approximate, not exact.
- *Robustness*—the performance of a random-variate-generation algorithm should be insensitive to small changes in the random-variable-model parameters [e.g., n and p for a *Binomial*(n, p) model] and work properly for all reasonable values of the parameters.
- *Efficiency*—although this criterion is commonly overrated, a random-variate-generation algorithm should be both time and memory efficient. Execution time often consists of two parts. *Set-up* time occurs once (for example, to compute and store a cdf array); *marginal* execution time occurs each time a random variate is generated. Ideally both times are small. If not, an implementation-dependent algorithm judgment must be made about which is more important.
- *Clarity*—if all other things are equal, a random-variate-generation algorithm that is easy to understand and implement is always preferred. For some people, this is the most important criterion; however, for random-variate-generation specialists, portability, exactness, robustness, and efficiency tend to be most important.
- *Synchronization*—a random-variate-generation algorithm is synchronized if exactly one call to Random is required for each random variate generated. This property and monotonicity are often needed to implement certain *variance reduction techniques*—for example, the *common random numbers technique*, illustrated in Example 3.1.7.
- *Monotonicity*—a random-variate-generation algorithm is monotone if it is synchronized and, like Algorithm 6.2.3, the transformation from u to x is monotone increasing (or monotone decreasing).

Although some of these criteria will be satisfied automatically if inversion is used, others (e.g., efficiency) may not. Generally, however, like the algorithms in Examples 6.2.6, 6.2.7, and 6.2.8, the best random-variate-generation algorithm is based on Algorithm 6.2.3.

Example 6.2.10

In practice, an algorithm that uses inversion, but is only approximate, could be satisfactory, provided that the approximation is sufficiently good. As an illustration, suppose we want to generate *Binomial* (10, 0.4) random variates. To the 0.*ddd* precision indicated, the corresponding pdf is

x	0	1	2	3	4	5	6	7	8	9	10
$f(x)$	0.006	0.040	0.121	0.215	0.251	0.201	0.111	0.042	0.011	0.002	0.000

To be consistent with these *approximate* pdf values, random variates can be generated by filling a 1000-element integer-valued array $a[\cdot]$ with 6 zeros, 40 ones, 121 twos, etc. Then the algorithm

```
j = Equilikely(0, 999);
return a[j];
```

can be used each time a *Binomial* (10, 0.4) random variate is needed. The approximate nature of this algorithm is evident. A 10, for example, will never be generated, even though $f(10) = \binom{10}{10}(0.4)^{10}(0.6)^0 = 1/9\,765\,625$. It is "approximately exact," however, with an accuracy that may be acceptable in some applications. Moreover, the algorithm is portable, robust, clear, synchronized, and monotone. Marginal execution time efficiency is good. Set-up time efficiency is a potential problem, however, as is memory efficiency. For example, if 0.*ddddd* precision is desired, then it would be necessary to use a 100 000-element array.

Example 6.2.11

The algorithm in Example 6.2.10 for generating *Binomial* (10, 0.4) random variates is inferior to an *exact* algorithm based on filling an 11-element floating-point array with cdf values and then using Algorithm 6.2.2 with $x = 4$ (the mode) to initialize the search.

In general, inversion can be used to generate *Binomial* (n, p) random variates by computing a floating-point array of $n + 1$ cdf values and then using Algorithm 6.2.2 with $x = \lfloor np \rfloor$ to initialize the search. The capability provided by the library rvms (see Appendix D) can be used in this case to compute the cdf array by calling the cdf function cdfBinomial (n, p, x) for $x = 0, 1, \ldots, n$. Because this approach is inversion, it is in many respects ideal. The only drawback to this approach is some inefficiency, in the sense of set-up time (to compute the cdf array) and memory (to store the cdf array), particularly if n is large.*

---

*As discussed in the next section, if n is large, the size of the cdf array can be *truncated* significantly and thereby partially compensate for the inefficiency in memory and set-up time.

Example 6.2.12

The need to store a cdf array can be eliminated completely, at the expense of increased marginal execution time. This can be done by using Algorithm 6.2.2 with the cdf capability provided by the library rvms to compute cdf values only as needed. Indeed, in the library rvms, this is how the idf function idfBinomial (n, p, u) is evaluated. Given that rvms provides this capability, $Binomial(n, p)$ random variates can be generated by inversion as

```
u = Random();
return idfBinomial(n, p, u);          // use the library rvms
```

With appropriate modifications, the approach in Examples 6.2.11 and 6.2.12 can be used to generate $Poisson(\mu)$ and $Pascal(n, p)$ random variates. Therefore, inversion can be used to generate random variates for all six of the parametric random variable models presented in the previous section. For $Equilikely(a, b)$, $Bernoulli(p)$, and $Geometric(p)$ random variates, inversion is essentially ideal. For $Binomial(n, p)$, $Pascal(n, p)$, and $Poisson(\mu)$ random variates, however, time and memory efficiency can be a problem if inversion is used. In part, this justifies the development of alternative generation algorithms for $Binomial(n, p)$, $Pascal(n, p)$, and $Poisson(\mu)$ random variates.

6.2.3 Alternative Random-Variate-Generation Algorithms

Binomial Random Variates. As an alternative to inversion, a $Binomial(n, p)$ random variate can be generated by summing an *iid* sequence of $Bernoulli(p)$ random variates. (See Example 6.1.15.)[*]

Example 6.2.13

This algorithm uses a $Bernoulli(p)$ random-variate generator to generate a $Binomial(n, p)$ random variate:

```
x = 0;
for (i = 0; i < n; i++)
    x += Bernoulli(p);
return x;
```

This algorithm is portable, exact, robust, and clear. It is not synchronized or monotone, and the $O(n)$ marginal execution time complexity can be a problem if n is large.

Poisson Random Variates. A $Poisson(\mu)$ random variable is the limiting case of a $Binomial(n, \mu/n)$ random variable as $n \to \infty$. Therefore, if n is large, then one of these random variates can be generated as an approximation to the other. Unfortunately, because the algorithm in Example 6.2.13 is $O(n)$, if n is large, we must look for other ways to generate a Poisson(μ)

---

[*]Random-variate generation via the summing of an *iid* sequence of more elementary random variates is known as a *convolution method*.

random variate. The *Poisson*(μ) cdf $F(\cdot)$ is equal to the complement of an *incomplete gamma function* (see Appendix D):

$$F(x) = 1 - P(x + 1, \mu) \qquad\qquad x = 0, 1, 2 \dots.$$

Except for special cases, an incomplete gamma function cannot be inverted to form an idf. Therefore, inversion can be used to generate a *Poisson*(μ) random variate, but the cdf is not simple enough to avoid the need to use a numerical approach like the ones in Examples 6.2.11 or 6.2.12.

There are standard algorithms for generating a *Poisson*(μ) random variate that do not rely on either inversion or a "large n" version of the algorithm in Example 6.2.13. The following is an example.

Example 6.2.14
This algorithm generates a *Poisson*(μ) random variate:

```
a = 0.0;
x = 0;
while (a < μ) {
    a += Exponential(1.0);
    x++;
}
return x - 1;
```

This algorithm is portable, exact, and robust. It is neither synchronized nor monotone, and marginal execution-time efficiency can be a problem if μ is large, because the expected number of passes through the `while` loop is $\mu + 1$. Although the algorithm is obscure at this point, clarity will be provided in Section 7.3.

Pascal Random Variates.   Like a *Binomial*(n, p) cdf, a *Pascal*(n, p) cdf contains an *incomplete beta function* (for which, see Appendix D). Specifically, a *Pascal*(n, p) cdf is

$$F(x) = 1 - I(x + 1, n, p) \qquad\qquad x = 0, 1, 2, \dots.$$

Except for special cases, an incomplete beta function cannot be inverted algebraically to form a closed-form idf. Therefore, inversion can be used to generate a *Pascal*(n, p) random variate, but the cdf is not simple enough to avoid the need to use a numerical approach like the ones in Examples 6.2.11 or 6.2.12.

As an alternative to inversion, recall (from Section 6.1) that the random variable X is *Pascal*(n, p) if and only if

$$X = X_1 + X_2 + \cdots + X_n,$$

where $X_1, X_2, \dots, X_n$ is an *iid Geometric*(p) sequence. Therefore, a *Pascal*(n, p) random variate can be generated by summing an *iid* sequence of n *Geometric*(p) random variates.

Example 6.2.15

This algorithm uses a *Geometric*(p) random variate generator to generate a *Pascal*(n, p) random variate:

```
x = 0;
for (i = 0; i < n; i++)
    x += Geometric(p);
return x;
```

This algorithm is portable, exact, robust, and clear. It is neither synchronized nor monotone, and the $O(n)$ marginal execution-time complexity can be a problem if n is large.

Library rvgs.   See Appendix E for the library rvgs (Random Variate GeneratorS). This library consists of six functions for generating discrete random variates and seven functions for generating continuous random variates (for which, see Chapter 7). The six discrete-random-variate generators in the library are the following:

- long Bernoulli(double p) —returns 1 with probability p or 0 otherwise;
- long Binomial(long n, double p) —returns a *Binomial*(n, p) random variate;
- long Equilikely(long a, long b) —returns an *Equilikely*(a, b) random variate;
- long Geometric(double p) —returns a *Geometric*(p) random variate;
- long Pascal(long n, double p) —returns a *Pascal*(n, p) random variate;
- long Poisson(double μ) —returns a *Poisson*(μ) random variate.

These random-variate generators feature minimal set-up times (in some cases, at the expense of potentially large marginal execution times).

Library rvms.   The *Bernoulli*(p), *Equilikely*(a, b), and *Geometric*(p) generators in rvgs use inversion and in that sense are ideal. The other three generators do not use inversion. If that is a problem, then, as an alternative to the library rvgs, the idf functions in the library rvms (Random Variable ModelS, Appendix D) can be used to generate *Binomial*(n, p), *Pascal*(n, p), and *Poisson*(μ) random variates by inversion, as illustrated in Example 6.2.12.

Because the idf functions in the library rvms were designed for accuracy at the possible expense of marginal execution-time inefficiency, use of this approach is generally not recommended when many observations need to be generated. Instead, in that case, set up an array of cdf values and use inversion (Algorithm 6.2.2), as illustrated in Example 6.2.11. This approach is considered in more detail in the next section.

6.2.4 Exercises

6.2.1   Prove that $F^*(\cdot)$ is a monotone increasing function [i.e., prove that, if $0 < u_1 \le u_2 < 1$, then $F^*(u_1) \le F^*(u_2)$].

6.2.2   If inversion is used to generate a *Geometric*(p) random variate x, use your knowledge of the largest and smallest possible values of Random to compute (as a function of p) the largest and smallest possible value of x that can be generated. (See also Exercise 3.1.3.)

6.2.3   Find the pdf associated with the random-variate-generation algorithm

```
u = Random();
return ⌈3.0 + 2.0 * u²⌉;
```

6.2.4   (a) Generate a $Poisson(9)$ random-variate sample of size $1\,000\,000$, using the appropriate generator function in the library $\mathtt{rvgs}$, and form a histogram of the results. (b) Compare the resulting relative frequencies with the corresponding $Poisson(9)$ pdf, using the appropriate pdf function in the library $\mathtt{rvms}$. (c) Comment on the value of this process as a test of correctness for the two functions used.

6.2.5   (a) If X is a discrete random variable with possible values $x = 1, 2, \ldots, n$ and pdf $f(x) = \alpha x$, find an equation for the idf $F^*(u)$ as a function of n. (See Exercise 6.1.5.) (b) Construct an inversion function that will generate a value of X with one call to $\mathtt{Random}$. (c) How would you convince yourself that this random-variate generator function is correct?

6.2.6   Let X be a discrete random variable with cdf $F(\cdot)$ and idf $F^*(\cdot)$. Prove or disprove the following: (a) If $u \in (0, 1)$, then $F(F^*(u)) = u$. (b) If $x \in \mathcal{X}$, then $F^*(F(x)) = x$.

6.2.7   (a) Implement Algorithm 6.2.1 for a $Poisson(\mu)$ random variable and use Monte Carlo simulation to verify that the expected number of passes through the $\mathtt{while}$ loop is μ. Use $\mu = 1, 5, 10, 15, 20, 25$. (b) Repeat with Algorithm 6.2.2. (c) Comment. [Use the function $\mathtt{cdfPoisson}$ in the library $\mathtt{rvms}$ to generate the $Poisson(\mu)$ cdf values.]

6.2.8   Design and then implement a consistent "invalid input" error-trapping mechanism for the functions in the library $\mathtt{rvgs}$. Write a defense of your design.

6.2.9   Suppose X is a discrete random variable with cdf $F(\cdot)$ and possible values $\mathcal{X} = \{a, a + 1, \ldots, b\}$, with both a and b finite. (a) Construct a binary search algorithm that will return $F^*(u)$ for any $u \in (0, 1)$. (b) Present convincing numerical evidence that your algorithm is correct when used to generate $Binomial(100, 0.2)$ random variates, and compare its efficiency with that of Algorithms 6.2.1 and 6.2.2. (c) Comment.

6.2.10   (a) As an extension of Exercise 6.1.8, what is the idf of X? (b) Provide convincing numerical evidence that this idf is correct.

6.2.11   Two integers X_1, X_2 are drawn at random, without replacement, from the set $\{1, 2, \ldots, n\}$ with $n \geq 2$. Let $X = |X_1 - X_2|$. (a) What are the possible values of X? (b) What are the pdf, cdf, and idf of X? (c) Construct an inversion function that will generate a value of X with just one call to $\mathtt{Random}$. (d) What did you do to convince yourself that this random-variate generator function is correct?

6.2.12   Same as the previous exercise, except that the draw is with replacement.

6.3 DISCRETE-RANDOM-VARIABLE APPLICATIONS

The purpose of this section is to demonstrate several discrete-random-variable applications using the capabilities provided by the discrete-random-variate generators in the library $\mathtt{rvgs}$ and the pdf, cdf, and idf functions in the library $\mathtt{rvms}$. We begin by considering alternatives to the inventory-demand models used in programs $\mathtt{sis2}$ and $\mathtt{sis3}$.

6.3.1 Alternative Inventory-Demand Models

Example 6.3.1

The inventory-demand model in program `sis2` is that the demand per time interval is generated as an *Equilikely*(10, 50) random variate d. In this case, the mean is 30, the standard deviation is $\sqrt{140} \cong 11.8$, and the demand pdf is flat, as illustrated in Figure 6.3.1.

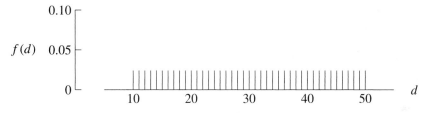

Figure 6.3.1 *Equilikely*(10, 50)-demand-model pdf.

As discussed in Example 3.1.5, this is not a very realistic model.* Therefore, we consider alternative models, one of which is that there are, say, 100 instances per time interval when a demand for 1 unit might occur. For each of these instances, the probability that the demand *will* occur is, say, 0.3, independent of what happens at the other demand instances. The inventory demand per time interval is then the sum of 100 independent *Bernoulli*(0.3) random variates or, equivalently, it is a *Binomial*(100, 0.3) random variate. In this case, the function `GetDemand` in the program `sis2` should be

```
long GetDemand(void)
{
    return (Binomial(100, 0.3));
}
```

The resulting random demand per time interval will have the mean 30, the standard deviation $\sqrt{21} \cong 4.6$, and a pdf approximately symmetric about the mean, as illustrated in Figure 6.3.2.

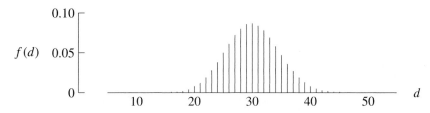

Figure 6.3.2 *Binomial*(100, 0.3)-demand-model pdf.

*Discussion of matching models to data collected from the system of interest is delayed until Chapter 9.

Example 6.3.2

If we believe that the *Binomial*(100, 0.3) model in the previous example could be close to reality, then a *Poisson*(30) model should also be considered. For this model, the random variate returned by the function GetDemand in program sis2 would be

```
return (Poisson(30.0));
```

The resulting random-variate demand per time interval will have mean 30, standard deviation $\sqrt{30} \cong 5.5$, and the pdf illustrated in Figure 6.3.3. Although similar to the *Binomial*(100, 0.3) pdf in Example 6.3.1, the *Poisson*(30) pdf has slightly "heavier" tails, as is consistent with the larger standard deviation.

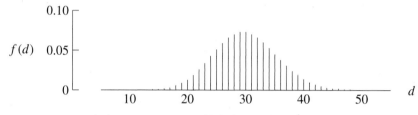

Figure 6.3.3   *Poisson(30)-demand-model pdf.*

For reasons discussed in Section 7.3, a traditional model for inventory systems is what was used in Example 6.3.2—the discrete demand is a *Poisson*(λ) random variable with the parameter λ matched to the expected amount of demand per time interval (the demand rate). Indeed, this is the inventory-demand model used in program sis3, with $\lambda = 30$.

Example 6.3.3

Yet another potential model for the inventory demand is that there are, say, 50 instances per time interval at which a *Geometric*(p) inventory demand will occur with p equal to, say, 0.375, and that the amount of demand that actually occurs at each of these instances is independent of the amount of demand that occurs at the other instances. The demand per time interval is then the sum of 50 independent *Geometric*(0.375) random variates or, equivalently, it is a *Pascal*(50, 0.375) random variate. The random variate returned by the function GetDemand in program sis2 would then be

```
return (Pascal(50, 0.375));
```

producing a random-variate demand per time interval having mean 30, standard deviation $\sqrt{48} \cong 6.9$, and the pdf illustrated in Figure 6.3.4.

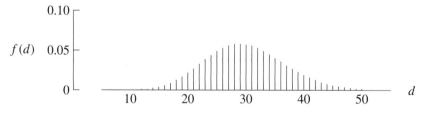

Figure 6.3.4   *Pascal(50, 0.375)-demand-model pdf.*

Consistent with the increase in the standard deviation from 5.5 to 6.9, this pdf has slightly heavier tails than the *Poisson*(30) pdf.

Example 6.3.4

As an extension of the previous example, one might argue that the inventory-demand model should allow for the number of demand instances per time interval to be an independent discrete random variable also—say, *Poisson*(50). The function GetDemand in program sis2 would then be*

```
long GetDemand(void)
{
   long instances = Poisson(50.0);
                              // must truncate to avoid 0
   return (Pascal(instances, 0.375));
}
```

With this extension, the mean of the resulting random-variate demand per time interval will remain 30. As illustrated in Figure 6.3.5, the pdf will become further dispersed about the mean, as is consistent with the increase of the standard deviation to $\sqrt{66} \cong 8.1$.

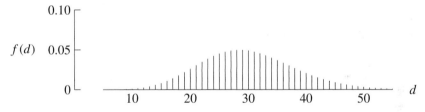

Figure 6.3.5   Compound-demand-model pdf.

To compute the pdf of the "compound" (or "stochastic parameters") random variable corresponding to the random variates generated by the function GetDemand in Example 6.3.4, let the discrete random variables D and I denote the demand *amount* and number of demand *instances* per time interval, respectively. From the law of total probability, it follows that

$$f(d) = \Pr(D = d) = \sum_{i=0}^{\infty} \Pr(I = i)\, \Pr(D = d \mid I = i) \qquad d = 0, 1, 2, \ldots.$$

For any value of d, the probability $f(d)$ can be evaluated by using the pdf capability in the library rvms. To do so, however, the infinite sum over possible demand instances must be *truncated* to a finite range, say $0 < a \leq i \leq b$. The details of how to determine

---

*There is a potential problem with this function—see Exercise 6.3.7.

a and *b* are discussed later in this section. Provided *a* and *b* are selected appropriately, the following algorithm can then be used to compute $f(d)$:

```
double sum = 0.0;
for (i = a; i <= b; i++)                  // use the library rvms
    sum += pdfPoisson(50.0, i) * pdfPascal(i, 0.375, d);
return sum;                               // sum is f(d)
```

These compound random variables are of interest in a branch of statistics known as *Bayesian statistics.*

Any of the inventory-demand models in the previous examples can be used in program sis2 as a replacement for the *Equilikely*(*a*, *b*) model. In most applications, this will result in a model that more accurately reflects the system. Moreover, with some minor modification, a more accurate inventory-demand model can be used in program sis3. The result is program sis4.

Program sis4.   Program sis4 is based on the next-event simulation program sis3, but it has a more realistic inventory-demand model, one that allows for a random amount of demand at each demand instance. As in program sis3, demand instances are assumed to occur at random throughout the period of operation, with an average rate of λ instances per time interval, so that the inter-demand time is an *Exponential*$(1/\lambda)$ random variate. Unlike the model on which program sis3 is based, however, these demand instances correspond to times where a demand *may* occur. Whether a demand *actually* occurs at these demand instances is random with probability *p*. Moreover, to allow for the possibility of more than one unit of inventory demand at a demand instance, the demand amount is assumed to be a *Geometric*(*p*) random variate. Because the expected value of a *Geometric*(*p*) random variable is $p/(1 - p)$ and the expected number of demand instances per time interval is λ, the expected demand per time interval is $\lambda p/(1 - p)$.

Example 6.3.5

In terms of the automobile-dealership example considered previously, the inventory-demand model on which program sis4 is based corresponds to an average of λ customers per week that visit the dealership with the *potential* to buy one or more automobiles. Each customer will, independently, not buy an automobile with probability $1 - p$, or they will buy one automobile with probability $(1 - p)p$, or they will buy two automobiles with probability $(1 - p)p^2$, or three with probability $(1 - p)p^3$, etc. The parameter values used in program sis4 are $\lambda = 120.0$ and $p = 0.2$, which correspond to an expected value of 30.0 automobiles purchased per week. For future reference, note that

$$30.0 = \frac{\lambda p}{1 - p} = \lambda \sum_{x=0}^{\infty} x(1 - p)p^x = \underbrace{\lambda(1 - p)p}_{19.200} + \underbrace{2\lambda(1 - p)p^2}_{7.680} + \underbrace{3\lambda(1 - p)p^3}_{2.304} + \cdots.$$

Therefore, on average, of the 120 customers that visit per week,

- $\lambda(1 - p) = 96.0$ do not buy anything;
- $\lambda(1 - p)p = 19.200$ buy one automobile;

- $\lambda (1 - p)p^2 = 3.840$ buy two automobiles;
- $\lambda (1 - p)p^3 = 0.768$ buy three automobiles;
- etc.

In the remainder of this section, we show how to truncate a discrete random variable. Truncation makes the demand model more realistic by limiting the number of automobiles a customer buys, for example, to 0, 1, or 2. (See Exercise 6.3.2.)

6.3.2 Truncation

Let X be a discrete random variable with possible values $\mathcal{X} = \{x \mid x = 0, 1, 2, \ldots\}$ and cdf $F(x) = \Pr(X \leq x)$. A modeler might be interested in formulating a pdf that effectively restricts the possible values to a finite range of integers $a \leq x \leq b$, with $a \geq 0$ and $b < \infty$.* If $a > 0$, then the probability that X is strictly less than a is

$$\alpha = \Pr(X < a) = \Pr(X \leq a - 1) = F(a - 1).$$

Similarly, the probability that X is strictly greater than b is

$$\beta = \Pr(X > b) = 1 - \Pr(X \leq b) = 1 - F(b).$$

In general, then

$$\Pr(a \leq X \leq b) = \Pr(X \leq b) - \Pr(X < a) = F(b) - F(a - 1).$$

A value of X outside the range $a \leq X \leq b$ is essentially impossible if and only if $F(b) \cong 1.0$ and $F(a - 1) \cong 0.0$. There are two cases to consider.

- If a and b are specified, then the cdf of X can be used to compute the left-tail and right-tail probabilities —

$$\alpha = \Pr(X < a) = F(a - 1) \qquad \text{and} \qquad \beta = \Pr(X > b) = 1 - F(b),$$

  respectively. The cdf transformation from possible values to probabilities is exact.
- If, instead, the (positive) left-tail and right-tail probabilities α and β are specified, then the idf of X can be used to compute the possible values

$$a = F^*(\alpha) \qquad \text{and} \qquad b = F^*(1 - \beta),$$

  respectively. Because X is a discrete random variable, this idf transformation from probabilities to possible values is not exact. Instead a and b provide only *bounds*, in the sense that $\Pr(X < a) \leq \alpha$ and $\Pr(X > b) < \beta$.

---

*In the case of the automobile-dealership example, $a = 0$ and $b = 45$ implies that there is no left-hand truncation and no week contains more than 45 customer demands.

Example 6.3.6

For the *Poisson*(50) random variable I in the discussion following Example 6.3.4, it was necessary to specify a, b so that $\Pr(a \le I \le b) \cong 1.0$ with negligible error. By experimentation, it was found that, in this case, "negligible error" could be interpreted as $\alpha = \beta = 10^{-6}$. The idf capability in the random-variate models library rvms was then used to compute

```
a = idfPoisson(50.0, α);                    // α = 10⁻⁶
b = idfPoisson(50.0, 1.0 - β);              // β = 10⁻⁶
```

to produce the results $a = 20$, $b = 87$. Consistent with the bounds produced by the (α, β) to (a, b) conversion, $\Pr(I < 20) = \mathtt{cdfPoisson(50.0, 19)} \cong 0.48 \times 10^{-6} < \alpha$ and $\Pr(I > 87) = 1.0 - \mathtt{cdfPoisson(50.0, 87)} \cong 0.75 \times 10^{-6} < \beta$.

   To facilitate the evaluation of the compound pdf values in Example 6.3.4, the possible values of a *Poisson*(50) random variable were truncated to the integers 20 through 87, inclusive. As demonstrated in Example 6.3.6, this truncation is so slight as to be of no practical significance. In other words, because a *Poisson*(50) pdf is effectively zero for all integers less than 20 or greater than 87, the truncated random variable and the untruncated random variable are essentially identically distributed. In some simulation applications, truncation is more significant. There are two reasons for this:

- *Efficiency* — when discrete random variates are generated via inversion and the idf is not available as a simple algebraic expression, then a cdf search technique like Algorithm 6.2.2 must be used. To facilitate this search, cdf values are usually stored in an array. Because the range of possible values defines the size of the array, traditional computer science space/time considerations dictate that the range of possible values should be as small as possible (assuming that the model remains realistic).
- *Realism* — some discrete random variates can take on arbitrarily large values (at least in theory). If you have created a discrete random variable model having mean 30 and standard deviation 10, do you really want your random variate generator to (surprise!) return the value 100?

When it is significant, truncation must be done correctly, because the result is a *new* random variable. The four examples that follow are illustrations.

Incorrect Truncation

Example 6.3.7

As in Example 6.3.2, suppose we want to use a *Poisson*(30) demand model in program sis2 but, to avoid extreme values, we want to truncate the demand to the range $20 \le d \le 40$. In this case, we might be tempted to generate the demand with

```
d = Poisson(30.0);
if (d < 20)
   d = 20;
```

```
if (d > 40)
   d = 40;
return d;
```

Because the pdf values below 20 and above 40 "fold" back to 20 and 40, respectively, as the resulting pdf spikes at $d = 20$ and $d = 40$ shown in Figure 6.3.6 indicate, this is *not* a correct truncation for most applications.

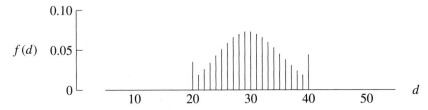

Figure 6.3.6 *Poisson*(30)-demand-model pdf incorrectly truncated.

Truncation by cdf Modification

Example 6.3.8

As in Example 6.3.7, we want to generate *Poisson*(30) demands truncated to the range $20 \le d \le 40$. This time, however, we want to do it correctly. Before truncation, the *Poisson*(30) pdf is

$$f(d) = \frac{30^d \exp(-30)}{d!} \qquad d = 0, 1, 2, \ldots$$

with

$$\Pr(20 \le D \le 40) = F(40) - F(19) = \sum_{d=20}^{40} f(d) \cong 0.945817.$$

The truncation correction is to compute a new truncated random variable D_t with pdf $f_t(d)$. This is done by increasing the corresponding value of $f(d)$ via division by the factor $F(40) - F(19)$ as

$$f_t(d) = \frac{f(d)}{F(40) - F(19)} \qquad d = 20, 21, \ldots, 40.$$

The corresponding truncated cdf is

$$F_t(d) = \sum_{t=20}^{d} f_t(t) = \frac{F(d) - F(19)}{F(40) - F(19)} \qquad d = 20, 21, \ldots, 40.$$

A random variate correctly truncated to the range $20 \le d \le 40$ can now be generated by inversion, by using the truncated cdf $F_t(\cdot)$ and Algorithm 6.2.2 as follows:

```
u = Random();
d = 30;
```

```
if  (F_t(d) <= u)
   while  (F_t(d) <= u)
      d++;
else if  (F_t(20) <= u)
   while  (F_t(d - 1) > u)
      d--;
else
   d = 20;
return d;
```

Compared to the mean and standard deviation of D, which are $\mu = 30.0$ and $\sigma = \sqrt{30} \cong 5.477$, the mean and standard deviation of D_t are

$$\mu_t = \sum_{d=20}^{40} d\, f_t(d) \cong 29.841 \qquad \text{and} \qquad \sigma_t = \sqrt{\sum_{d=20}^{40} (d - \mu_t)^2 f_t(d)} \cong 4.720.$$

The pdf of the correctly truncated $Poisson(30)$ random variable in Example 6.3.8 is illustrated in Figure 6.3.7. For reference, the incorrectly truncated pdf from Example 6.3.7 and the untruncated pdf from Example 6.3.2 are also shown.

In general, if it is desired to truncate the integer-valued discrete random variable X to the set of possible values $\mathcal{X}_t = \{a, a + 1, \ldots, b\} \subset \mathcal{X}$, thereby forming the new discrete random variable

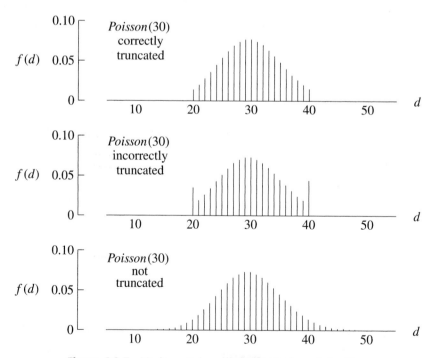

Figure 6.3.7   Various $Poisson(30)$-demand-model pdfs.

X_t, then the pdf and cdf of X_t are defined as

$$f_t(x) = \frac{f(x)}{F(b) - F(a-1)} \qquad x \in \mathcal{X}_t$$

and

$$F_t(x) = \frac{F(x) - F(a-1)}{F(b) - F(a-1)} \qquad x \in \mathcal{X}_t,$$

where $f(\cdot)$ and $F(\cdot)$ are the pdf and cdf of X. Random values of X_t can then be generated via inversion and Algorithm 6.2.2 with $F_t(\cdot)$, as in Example 6.3.8. [The equations for $f_t(x)$ and $F_t(x)$ presume that $a - 1$ is a possible value of X.]

There may be *rare* occasions where "incorrect truncation" is appropriate in a discrete-event simulation model. Decisions concerning the style of truncation are driven by the fit between the model and the real-world system.

As an alternative to truncation by cdf modification, we can use truncation by *constrained inversion*, as illustrated next. Provided that the idf of X can be easily applied, this is the truncation approach of choice.

Truncation by Constrained Inversion.　　If $F^*(\cdot)$ is the idf of X, then the following algorithm can be used to generate the random variate X_t correctly truncated to the range $a \le x \le b$:

```
α = F(a - 1);            // assumes that a - 1 is a possible value of X
β = 1.0 - F(b);
u = Uniform(α, 1.0 - β);
x = F*(u);
return x;
```

The key here is that u is *constrained* to a subrange $(\alpha, 1 - \beta) \subset (0, 1)$ in such a way that correct truncation is automatically enforced prior to the idf inversion. This is another illustration of the elegance of random-variate generation by inversion.

Example 6.3.9

The cdf and idf capabilities in the library rvms can be used to generate a *Poisson*(30) random demand, correctly truncated to the range $20 \le d \le 40$:

```
α = cdfPoisson(30.0, 19);                         // set-up
β = 1.0 - cdfPoisson(30.0, 40);                   // set-up
u = Uniform(α, 1.0 - β);
d = idfPoisson(30.0, u);
return d;
```

This algorithm should be implemented so that α and β are static variables that are computed once only.

Truncation by Acceptance–Rejection

Example 6.3.10

As an alternative to the techniques in Example 6.3.8 and 6.3.9, correct truncation can also be achieved, at the potential expense of some extra calls to the function Poisson(30), by using *acceptance–rejection* as follows:

```
d = Poisson(30.0);
while ((d < 20) or (d > 40))
    d = Poisson(30.0);
return d;
```

Although easily remembered and easily programmed, acceptance–rejection is not synchronized or monotone, even if the untruncated generator has these properties. Generally, the cdf modification technique in Example 6.3.8 or the constrained inversion technique in Example 6.3.9 is preferable.

6.3.3 Exercises

6.3.1   (*a*) Suppose you wish to use inversion to generate a *Binomial*(100, 0.1) random variate X *truncated* to the subrange $x = 4, 5, \ldots, 16$. How would you do it? Work through the details. (*b*) What is the value of the mean and standard deviation of the resulting truncated random variate?

6.3.2   The function GetDemand in program sis4 can return demand amounts outside the range $0, 1, 2$. (*a*) What is the largest demand amount that this function can return? In some applications, integers outside the range $0, 1, 2$ might not be meaningful, no matter how unlikely. (*b*) Modify GetDemand so that the value returned is correctly truncated to the range $0, 1, 2$. Do not use acceptance–rejection. (*c*) With truncation, what is the resulting average demand per time interval, and how does that compare to the average with no truncation?

6.3.3   Prove that the "truncation by constrained inversion" algorithm is correct. Also, draw a figure that illustrates the geometry behind the "truncation by constrained inversion" algorithm.

6.3.4   Prove that the mean of the compound demand in Example 6.3.4 is 30 and that the variance is 66.

6.3.5   (*a*) Do Exercise 6.2.10. (*b*) In addition, prove that the expected value of X is infinitely large (i.e., $E[X] = \infty$.) (*c*) Comment on the potential paradox that using inversion makes just *one* call to Random is somehow equivalent to a direct Monte Carlo simulation of this random experiment that requires, on average, an *infinite* number of calls to Random. *Hint*: the pdf of X is

$$f(x) = \frac{1}{x(x+1)} = \frac{1}{x} - \frac{1}{x+1} \qquad x = 1, 2, 3, \ldots.$$

6.3.6   (*a*) Implement the function GetDemand in Example 6.3.4 in such a way that you can use Monte Carlo simulation to estimate the expected number of calls to Random per call to GetDemand. (*b*) Construct an alternative version of GetDemand that requires exactly *one* call to Random per call to GetDemand and yet is effectively equivalent to the original function in the sense that the alternative function produces a random variate whose pdf matches the pdf illustrated in Example 6.3.4. (*c*) Provide convincing numerical evidence of correctness.

6.3.7   (*a*) Is it necessary to modify the function `GetDemand` in Example 6.3.4 so that `instances` must be positive? (*b*) If so, why and how would you do it? If not necessary, why?

6.4 DISCRETE-RANDOM-VARIABLE MODELS

For convenience, the characteristic properties of the following six discrete-random-variable models are summarized in this section: *Equilikely*(*a*, *b*), *Bernoulli*(*p*), *Geometric*(*p*), *Pascal*(*n*, *p*), *Binomial*(*n*, *p*), and *Poisson*(*μ*). For more details about these models, see Sections 6.1 and 6.2; for supporting software, see the random-variable-models library `rvms` in Appendix D and the random-variate-generators library `rvgs` in Appendix E.

6.4.1 Equilikely

Definition 6.4.1   A discrete random variable X is *Equilikely*(*a*, *b*) if and only if

- the parameters a, b are integers with $a < b$,
- the possible values of X are $\mathcal{X} = \{a, a+1, \ldots, b\}$,
- the pdf of X is

$$f(x) = \frac{1}{b - a + 1} \qquad x = a, a+1, \ldots, b,$$

  as illustrated in Figure 6.4.1,
- the cdf of X is

$$F(x) = \frac{x - a + 1}{b - a + 1} \qquad x = a, a+1, \ldots, b,$$

- the idf of X is

$$F^*(u) = a + \lfloor (b - a + 1)u \rfloor \qquad 0 < u < 1,$$

- the mean of X is

$$\mu = \frac{a + b}{2},$$

- the standard deviation of X is

$$\sigma = \sqrt{\frac{(b - a + 1)^2 - 1}{12}}.$$

Figure 6.4.1   *Equilikely*(a, b) pdf.

An *Equilikely*(a, b) random variable is used to model those situations in which a discrete random variable is restricted to the integers between a and b inclusive and all values in this range are equally likely. A typical application will involve a model derived from a statement such as "... an element is selected *at random* from a finite set ...". An *Equilikely* random variable is also known as a *discrete uniform*, *DU*, or *rectangular* random variable—terminology that is not used in this book.

6.4.2 Bernoulli

Definition 6.4.2 A discrete random variable X is *Bernoulli*(p) if and only if

- the real-valued parameter p satisfies $0 < p < 1$,
- the possible values of X are $\mathcal{X} = \{0, 1\}$,
- the pdf of X is

$$f(x) = p^x(1 - p)^{1-x} \qquad x = 0, 1,$$

  as illustrated in Figure 6.4.2 for $p = 0.6$,
- the cdf of X is

$$F(x) = (1 - p)^{1-x} \qquad x = 0, 1,$$

- the idf of X is

$$F^*(u) = \begin{cases} 0 & 0 < u < 1 - p \\ 1 & 1 - p \leq u < 1, \end{cases}$$

- the mean of X is

$$\mu = p,$$

- the standard deviation of X is

$$\sigma = \sqrt{p(1 - p)}.$$

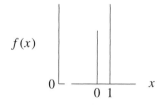

$f(x)$

0 $0 \quad 1$ x

Figure 6.4.2 *Bernoulli*(0.6) pdf.

A *Bernoulli*(*p*) random variable is used to model the Boolean situation, where only two outcomes are possible—success or failure, true or false, 1 or 0, and so on. The parameter *p* determines the probability of the two possible outcomes with the convention that

$$p = \Pr(\text{success}) = \Pr(X = 1)$$

and

$$1 - p = \Pr(\text{failure}) = \Pr(X = 0).$$

The random variables X_1, X_2, X_3, ..., define an *iid* sequence of so-called *Bernoulli*(*p*) *trials* if and only if each X_i is *Bernoulli*(*p*) and each is statistically independent of all the others. The repeated tossing of a coin (biased or not) is the classic example of an *iid* sequence—a head is equally likely on each toss (*p* does not change), and the coin has no memory of the previous outcomes (independence).

6.4.3 Geometric

Definition 6.4.3   A discrete random variable X is *Geometric*(*p*) if and only if
- the real-valued parameter p satisfies $0 < p < 1$,
- the possible values of X are $\mathcal{X} = \{0, 1, 2, \ldots\}$,
- the pdf of X is
$$f(x) = p^x(1 - p) \qquad\qquad x = 0, 1, 2, \ldots,$$
as illustrated in Figure 6.4.3 for $p = 0.75$,
- the cdf of X is
$$F(x) = 1 - p^{x+1} \qquad\qquad x = 0, 1, 2, \ldots,$$
- the idf of X is
$$F^*(u) = \left\lfloor \frac{\ln(1 - u)}{\ln(p)} \right\rfloor \qquad\qquad 0 < u < 1,$$
- the mean of X is
$$\mu = \frac{p}{1 - p},$$
- the standard deviation of X is
$$\sigma = \frac{\sqrt{p}}{1 - p}.$$

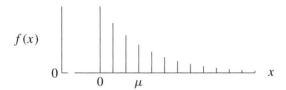

Figure 6.4.3   *Geometric*(0.75) pdf.

A *Geometric*(*p*) random variable is conventionally used to model the number of "successes" (1's) before the first "failure" (0) in a sequence of independent *Bernoulli*(*p*) trials. For example, if a coin has *p* as the probability of a head (success), then *X* counts the number of heads before the first tail (failure). Equivalently, a *Geometric*(*p*) random variable can be interpreted as a model for random sampling from a urn containing balls, a fraction *p* of which are 1's with the remainder 0's. A *Geometric*(*p*) random variable counts the number of 1's that are drawn, with replacement, before the first 0. If *X* is *Geometric*(*p*), then *X* has the *memoryless* property—for any nonnegative integer *x'*,

$$\Pr(X \ge x + x' \mid X \ge x') = \Pr(X \ge x) = p^x \qquad x = 0, 1, 2, \ldots,$$

independent of *x'*. An intuitive interpretation of this property is that a string of $x' - 1$ consecutive 1's is followed by a sequence of 0's and 1's that is probabilistically the same as a brand new sequence.

6.4.4 Pascal

Definition 6.4.4 A discrete random variable *X* is *Pascal*(*n*, *p*) if and only if

- the parameter *n* is a positive integer,
- the real-valued parameter *p* satisfies $0 < p < 1$,
- the possible values of *X* are $\mathcal{X} = \{0, 1, 2, \ldots\}$,
- the pdf of *X* is

$$f(x) = \binom{n + x - 1}{x} p^x (1 - p)^n \qquad x = 0, 1, 2, \ldots,$$

  as illustrated in Figure 6.4.4 for $(n, p) = (5, 2/7)$,
- the cdf of *X* contains the incomplete beta function (see Appendix D)

$$F(x) = 1 - I(x + 1, n, p) \qquad x = 0, 1, 2, \ldots,$$

- except for special cases, the idf of *X* must be found by numerical inversion,
- the mean of *X* is

$$\mu = \frac{np}{1 - p},$$

- the standard deviation of *X* is

$$\sigma = \frac{\sqrt{np}}{1 - p}.$$

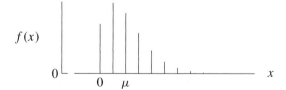

Figure 6.4.4 *Pascal*(5, 2/7) pdf.

A *Pascal*(*n*, *p*) random variable is the number of 1's before the nth 0 in an *iid* sequence of independent *Bernoulli*(*p*) trials. Therefore, a *Pascal*(1, *p*) random variable and a *Geometric*(*p*) random variable are equivalent. Moreover, X is a *Pascal*(*n*, *p*) random variable if and only if

$$X = X_1 + X_2 + \cdots + X_n,$$

where $X_1, X_2, \ldots, X_n$ is an *iid* sequence of *Geometric*(*p*) random variables. In terms of the urn model presented in Section 6.4.3, a *Pascal*(*n*, *p*) random variable counts the number of 1's that are drawn, with replacement, before the nth 0. A *Pascal*(*n*, *p*) random variable is also called a *negative binomial*—terminology that is not used in this book.

6.4.5 Binomial

Definition 6.4.5   A discrete random variable X is *Binomial*(*n*, *p*) if and only if

- the parameter n is a positive integer,
- the real-valued parameter p satisfies $0 < p < 1$,
- the possible values of X are $\mathcal{X} = \{0, 1, 2, \ldots, n\}$,
- the pdf of X is

$$f(x) = \binom{n}{x} p^x (1 - p)^{n-x} \qquad x = 0, 1, \ldots, n,$$

as illustrated in Figure 6.4.5 for $(n, p) = (10, 0.3)$,
- the cdf of X contains the incomplete beta function (see Appendix D)

$$F(x) = \begin{cases} 1 - I(x + 1, n - x, p) & x = 0, 1, \ldots, n - 1 \\ 1 & x = n, \end{cases}$$

- except for special cases, the idf of X must be computed by numerical inversion,
- the mean of X is

$$\mu = np,$$

- the standard deviation of X is

$$\sigma = \sqrt{np(1 - p)}.$$

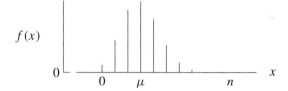

Figure 6.4.5   *Binomial*(10, 0.3) pdf.

A *Binomial*(n, p) random variable is the number of 1's in a sequence of n independent *Bernoulli*(p) trials. Therefore a *Binomial*(1, p) random variable and a *Bernoulli*(p) random variable are the same. Equivalently, X is a *Binomial*(n, p) random variable if and only if

$$X = X_1 + X_2 + \cdots + X_n,$$

where $X_1, X_2, \ldots, X_n$ is an *iid* sequence of *Bernoulli*(p) random variables. For example, if a coin has p as the probability of a head (success), then X counts the number of heads in a sequence of n tosses. In terms of the urn model, a *Binomial*(n, p) random variable counts the number of 1's that will be drawn, with replacement, if exactly n balls are drawn. Because drawing x 1's out of n is equivalent to drawing n − x 0's, it follows that X is *Binomial*(n, p) if and only if n − X is a *Binomial*(n, 1 − p) random variable.

6.4.6 Poisson

Definition 6.4.6 A discrete random variable X is *Poisson*(μ) if and only if

- the real-valued parameter μ satisfies $\mu > 0$,
- the possible values of X are $\mathcal{X} = \{0, 1, 2, \ldots\}$,
- the pdf of X is

$$f(x) = \frac{\mu^x \exp(-\mu)}{x!} \qquad x = 0, 1, 2 \ldots,$$

as illustrated in Figure 6.4.6 for $\mu = 5$,
- the cdf of X contains the incomplete gamma function (see Appendix D)

$$F(x) = 1 - P(x + 1, \mu) \qquad x = 0, 1, 2 \ldots,$$

- except for special cases, the idf of X must be computed by numerical inversion,
- the mean of X is μ,
- the standard deviation of X is

$$\sigma = \sqrt{\mu}.$$

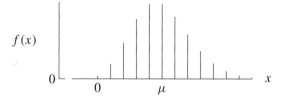

Figure 6.4.6 *Poisson*(5) pdf.

A *Poisson*(μ) random variable is a limiting case of a *Binomial* random variable—that is, let X be a *Binomial*$(n, \mu/n)$ random variable $(p = \mu/n)$, fix the values of μ and x, and consider what happens in the limit as $n \to \infty$. The pdf of X is

$$f(x) = \frac{n!}{x!(n-x)!} \left(\frac{\mu}{n}\right)^x \left(1 - \frac{\mu}{n}\right)^{n-x}$$

$$= \frac{\mu^x}{x!} \left(\frac{n!}{(n-x)!(n-\mu)^x}\right) \left(1 - \frac{\mu}{n}\right)^n.$$

It can be shown that

$$\lim_{n\to\infty} \left(\frac{n!}{(n-x)!(n-\mu)^x}\right) = 1 \qquad \text{and} \qquad \lim_{n\to\infty} \left(1 - \frac{\mu}{n}\right)^n = \exp(-\mu).$$

Therefore

$$\lim_{n\to\infty} f(x) = \frac{\mu^x \exp(-\mu)}{x!}$$

and so, for large values of n, *Binomial*$(n, \mu/n)$ and *Poisson*(μ) random variables are virtually identical, particularly if μ is small.

6.4.7 Summary

Effective discrete-event-simulation modeling requires that the modeler be familiar with several parametric distributions that can be used to mimic the stochastic elements of the model. In order to choose the proper distribution, it is important to know

- how these distributions arise;
- the support, $\mathcal{X}$;
- the mean, μ;
- the variance, σ^2; and
- the shape of the pdf.

Thus, when a modeling situation arises, the modeler has a wide array of options for selecting a stochastic model.

It is also important to know how these distributions relate to one another. Figure 6.4.7 summarizes relationships among the six distributions considered in this text. Listed in each oval are the name, parameter(s), and support of each distribution. The solid arrows connecting the ovals denote special cases [e.g., the *Bernoulli*(p) distribution is a special case of the *Binomial*(n, p) distribution when $n = 1$] and transformations [e.g., the sum (convolution) of n independent and identically

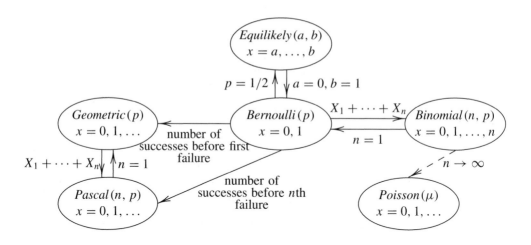

Figure 6.4.7 Relationships among discrete distributions.

distributed *Bernoulli*(p) random variables has a *Binomial*(n, p) distribution]. The dashed line connecting the *Binomial*(n, p) and *Poisson*(μ) distributions indicates that the limiting distribution of a binomial random variable as $n \to \infty$ is a Poisson distribution.

There are internal characteristics associated with these distributions that are not shown in Figure 6.4.7. One such characteristic is that the sum of independent Poisson random variables also has a Poisson distribution. Another is that the sum of independent binomial random variables with identical parameters p also has a binomial distribution.

The mean and variance of a random variable are special cases of what are called *moments*. The table below summarizes the first four moments of the distribution of the six discrete random variables surveyed in this chapter. The first moment, the mean $\mu = E[X]$, and the second moment about the mean, the variance $\sigma^2 = E[(X - \mu)^2]$, have been defined earlier. The skewness and kurtosis are the third and fourth standardized centralized moments about the mean, defined by

$$E\left[\left(\frac{X - \mu}{\sigma}\right)^3\right] \qquad \text{and} \qquad E\left[\left(\frac{X - \mu}{\sigma}\right)^4\right].$$

The skewness is a measure of the symmetry of a distribution. A symmetric pdf has skewness zero. A positive skewness typically indicates that the distribution is "leaning" to the left, and a negative skewness typically indicates that the distribution is "leaning" to the right. The geometric distribution, for example, has a positive skewness for all values of its parameter p. The kurtosis is a measure of the peakedness and tail behavior of a distribution. In addition to being measures associated with particular distributions, the higher-order moments are occasionally used to differentiate between the parametric models.

Distribution	Mean	Variance	Skewness	Kurtosis
$Equilikely\,(a, b)$	$\dfrac{a + b}{2}$	$\dfrac{(b - a + 1)^2 - 1}{12}$	0	$\dfrac{3}{5}\left[3 - \dfrac{4}{(b - a)(b - a + 2)}\right]$
$Bernoulli\,(p)$	p	$p(1 - p)$	$\dfrac{1 - 2p}{\sqrt{p(1 - p)}}$	$\dfrac{1}{p(1 - p)} - 3$
$Geometric\,(p)$	$\dfrac{p}{1 - p}$	$\dfrac{p}{(1 - p)^2}$	$\dfrac{1 + p}{\sqrt{1 - p}}$	$\dfrac{p^2 + 7p + 1}{p}$
$Pascal\,(n, p)$	$\dfrac{np}{1 - p}$	$\dfrac{np}{(1 - p)^2}$	$\dfrac{1 + p}{\sqrt{n(1 - p)}}$	$3 + \dfrac{6}{n} + \dfrac{(1 - p)^2}{np}$
$Binomial\,(n, p)$	np	$np(1 - p)$	$\dfrac{1 - 2p}{\sqrt{np(1 - p)}}$	$3 - \dfrac{6}{n} + \dfrac{1}{np(1 - p)}$
$Poisson\,(\mu)$	μ	μ	$\dfrac{1}{\sqrt{\mu}}$	$3 + \dfrac{1}{\mu}$

Although discussion is limited here to just six discrete distributions, there are many other parametric distributions capable of modeling discrete distributions. For a complete list of common discrete and continuous parametric distributions, including pdf, cdf, idf, and moments, we recommend the compact work of Evans, Hastings, and Peacock (2000) or the encyclopedic works of Johnson, Kotz, and Kemp (1993) and Johnson, Kotz, and Balakrishnan (1994, 1995, 1997).

We conclude this section with a brief discussion of the techniques and software associated with the evaluation of the pdf, cdf, and idf of the six discrete distributions surveyed in this chapter.

6.4.8 Pdf, Cdf and Idf Evaluation

Library rvms. Pdfs, cdfs, and idfs for all six of the discrete random variable models in this section can be evaluated by using the functions in the library rvms, listed in Appendix D. For example, if X is $Binomial\,(n, p)$, then, for any $x = 0, 1, 2, \ldots, n$,

```
pdf = pdfBinomial (n, p, x) ;                                     // f(x)
cdf = cdfBinomial (n, p, x) ;                                     // F(x)
```

and for any $0.0 < u < 1.0$

```
idf = idfBinomial (n, p, u) ;                                     // F*(u)
```

This library also has functions to evaluate pdfs, cdfs, and idfs for all the continuous random variables cataloged in Chapter 7.

Alternative, Recursive Approaches for Calculating pdf Values. Although this approach is not used in the library rvms to evaluate pdfs and cdfs, if X is a discrete random variable, then pdf values (and cdf values, by summation) can usually be easily generated recursively. In particular,

- if X is *Geometric*(p), then

$$f(0) = 1 - p$$

$$f(x) = pf(x-1) \qquad x = 1, 2, 3, \ldots;$$

- if X is *Pascal*(n, p), then

$$f(0) = (1-p)^n$$

$$f(x) = \frac{(n+x-1)p}{x} f(x-1) \qquad x = 1, 2, 3, \ldots;$$

- if X is *Binomial*(n, p), then

$$f(0) = (1-p)^n$$

$$f(x) = \frac{(n-x+1)p}{x(1-p)} f(x-1) \qquad x = 1, 2, 3, \ldots, n;$$

- if X is *Poisson*(μ), then

$$f(0) = \exp(-\mu)$$

$$f(x) = \frac{\mu}{x} f(x-1) \qquad x = 1, 2, 3, \ldots.$$

6.4.9 Exercises

6.4.1   (*a*) Simulate the tossing of a fair coin 10 times, and record the number of heads. (*b*) Repeat this experiment 1000 times and generate a discrete-data histogram of the results. (*c*) Verify numerically that the relative frequency of the number of heads is approximately equal to the pdf of a *Binomial*(10, 0.5) random variable.

6.4.2   Prove that a *Geometric*(p) random variable has the memoryless property.

6.4.3   Derive the *Geometric*(p), *Pascal*(n, p), *Binomial*(n, p), and *Poisson*(μ) recursive pdf equations in Section 6.4.8.

6.4.4   Use the *Binomial*(n, p) recursive pdf equations to implement the functions pdfBinomial(n,p,x) and cdfBinomial(n,p,x). Compare your implementations, in terms of both accuracy and speed, with the corresponding functions in the library rvms. Use $n = 10, 100, 1000$, with $\mu = 5$ and $p = \mu/n$. The comparison can be restricted to possible values for x within the range $\mu \pm 3\sigma$.

6.4.5   Verify numerically that the pdf of a *Binomial*(25, 0.04) random variable is virtually identical to the pdf of a *Poisson*(μ) random variable for an appropriate value of μ. Evaluate these pdfs in two ways: by using the appropriate pdf functions in the library rvms, and by using the *Binomial*(n, p) recursive pdf equations.

6.4.6   Prove or disprove: If X is a *Pascal*(n, p) random variable, then the cdf of X is $F(x) = I(n, x+1, 1-p)$.

6.4.7   Prove or disprove: If X is a *Binomial*(n, p) random variable, then the cdf of X is $F(x) = I(n-x, x+1, 1-p)$.

6.4.8   Prove that, if X is *Binomial* (n, p), then

$$\Pr(X > 0) = \sum_{x=0}^{n-1} p(1 - p)^x.$$

What is the probability interpretation of this equation?

6.4.9   (*a*) If you play *Pick-3* (as in Example 6.1.14) once a day for 365 consecutive days, what is the probability that you will be ahead at the end of this period? (*b*) What is your expected yield (winnings) at the end of this period?

6.4.10   Let X be a *Geometric* (p) random variable. As an alternative to the definition used in this book, some authors define the random variable $Y = X + 1$ to be *Geometric* (p). (*a*) What are the pdf, cdf, idf, mean, and standard deviation of Y? (*b*) Does this random variable Y also have the memoryless property?

6.5 RANDOM SAMPLING AND SHUFFLING

The topic of primary importance in this section is *random sampling* — that is, we will focus on algorithms that use a random-number generator to sample, at random, from a fixed collection (set) of objects. We begin, however, with the related topic of *random shuffling*.

6.5.1 Random Shuffling

Definition 6.5.1   Exactly $m!$ different permutations $(a_0, a_1, \ldots, a_{m-1})$ can be formed from a finite set $\mathcal{A}$ of $m = |\mathcal{A}|$ distinct elements. A *random shuffle generator* is an algorithm that will produce any one of these $m!$ permutations in such a way that all are equally likely.

Example 6.5.1
If $\mathcal{A} = \{0, 1, 2\}$ then the $3! = 6$ different possible permutations of $\mathcal{A}$ are

$$(0, 1, 2) \quad (0, 2, 1) \quad (1, 0, 2) \quad (1, 2, 0) \quad (2, 0, 1) \quad (2, 1, 0).$$

A random shuffle generator can produce any of these six possible permutations, each with equal probability $1/6$.

Algorithm 6.5.1   The intuitive way to generate a random shuffle of $\mathcal{A}$ is to draw the first element a_0 at random from $\mathcal{A}$. Then draw the second element a_1 at random from the set $\mathcal{A} - \{a_0\}$ and the third element a_2 at random from the set $\mathcal{A} - \{a_0, a_1\}$, etc. The following *in place* algorithm does that, provided the elements of the set $\mathcal{A}$ are stored (in any order) in the array $a[0], a[1], \ldots, a[m - 1]$.

```
for (i = 0; i < m - 1; i++) {
   j = Equilikely(i, m - 1);
   hold = a[j];
```

```
    a[j] = a[i];                              // swap a[i] and a[j]
    a[i] = hold;
}
```

Algorithm 6.5.1 is an excellent example of the elegance of simplicity. Figure 6.5.1 (corresponding to $m = 9$) illustrates the two indices i and j [j is an *Equilikely*$(i, m - 1)$ random variate] and the state of the $a[\cdot]$ array (initially filled with the integers 0 through 8 in sorted order, although any order is acceptable) for the first three passes through the `for` loop.

prior to $a[0], a[3]$ swap   prior to $a[1], a[6]$ swap   prior to $a[2], a[4]$ swap

Figure 6.5.1   Permutation-generation algorithm.

This algorithm is ideal for shuffling a deck of cards ($m = 52$ for a standard deck)—always a useful simulation skill. Moreover, as discussed later in this section, a minor modification to this algorithm makes it suitable for random sampling, without replacement.

6.5.2 Random Sampling

We now turn to the topic of primary importance in this section: algorithms for random sampling. To provide a common basis for comparison of the algorithms, the following notation and terminology is used.

- We are given a *population* $\mathcal{P}$ of $m = |\mathcal{P}|$ elements $a_0, a_1, \ldots, a_{m-1}$. Typically, m is large, perhaps so large that the population must be stored in secondary memory as a disk file, for example. If m is not large, then the population could be stored in primary memory as an array $a[0], a[1], \ldots, a[m - 1]$ or a linked list. In either case, regardless of whether m is large, the population is a *list*: $\mathcal{P}$ is ordered, so that there is a first element a_0, a second element a_1, etc.
- We wish to obtain a random *sample* $\mathcal{S}$ of $n = |\mathcal{S}|$ elements $x_0, x_1, \ldots, x_{n-1}$ from $\mathcal{P}$. Like $\mathcal{P}$, the sample $\mathcal{S}$ is also a *list*. Typically, n is small relative to m, but not necessarily so. If n is small, then the sample can be stored in primary memory as the array $x[0], x[1], \ldots, x[n - 1]$, with the data type of $x[\cdot]$ being the same as the data type of the population elements. For most algorithms, however, the use of an array is not a fundamental restriction; a linked list could be used instead, for example, or the sample could be written to secondary memory as a disk file if n is large.
- The algorithms use two generic functions `Get(&z, ` $\mathcal{L}$`, j)` and `Put(z, ` $\mathcal{L}$`, j)` where the list $\mathcal{L}$ could be either $\mathcal{P}$ or $\mathcal{S}$, and

        `Get(&z, ` $\mathcal{L}$`, j)` returns the value of the jth element in the list $\mathcal{L}$ as z;

Put $(z,\ \mathcal{L},\ j)$ assigns the value of z to the jth element in the list $\mathcal{L}$.

Both $\mathcal{P}$ and $\mathcal{S}$ can be accessed by Get and Put. The use of these generic functions allows the algorithms to be presented in a form that is essentially independent of how the lists $\mathcal{P}$ and $\mathcal{S}$ are actually stored.

- In some random-sampling applications, it is important to preserve the order of the population in the sample; in other applications, *order preservation* is not important.
- A random-sampling algorithm may be *sequential*, in which case $\mathcal{P}$ is traversed once, in order, and elements are selected at random to form $\mathcal{S}$. A sequential algorithm is necessarily used when random access to $\mathcal{P}$ is not possible. In contrast, a *nonsequential* random-sampling algorithm is based on the assumption that random access to $\mathcal{P}$ is possible and reasonably efficient. Note that it is access to $\mathcal{P}$, not to $\mathcal{S}$, that determines whether a random-sampling algorithm must be sequential or may be nonsequential.
- A random-sampling algorithm may use sampling *with replacement*, in which case the sample can contain multiple instances of the same population element. If so, then n could be larger than m. Instead, if sampling *without replacement* is used, then the sample cannot contain multiple instances of the same population element, and so $n \leq m$. Sampling without replacement is the usual case in practice. For the special (trivial) case of sampling without replacement when $n = m$, $\mathcal{P} \equiv \mathcal{S}$.

Relative to random-sampling algorithms, the phrase "at random" can be interpreted in at least two different (but closely related) ways: (i) each element of $\mathcal{P}$ is equally likely to be an element of $\mathcal{S}$ or (ii) each possible sample of size n is equally likely to be selected. Because these two interpretations of randomness are not equivalent, it is important to recognize what kind of samples a particular random-sampling algorithm actually produces.

Nonsequential Sampling

Algorithm 6.5.2   This $O(n)$ algorithm provides nonsequential random sampling with replacement. The value of m must be known.

```
for (i = 0; i < n; i++) {
   j = Equilikely(0, m - 1);
   Get(&z, P, j);                              // random access
   Put(z, S, i);
}
```

Because sampling is with replacement, n can be larger than m in Algorithm 6.5.2. If the elements of $\mathcal{P}$ are distinct, the number of possible samples is m^n, and all samples are equally likely to be generated. The order of $\mathcal{P}$ is not preserved in $\mathcal{S}$. Algorithm 6.5.2 should be compared with Algorithm 6.5.3, which is the *without*-replacement analog. For both algorithms, the $O(n)$ complexity is based on the number of random variates generated and ignores, perhaps unrealistically, the complexity of access to $\mathcal{P}$ and $\mathcal{S}$.

Example 6.5.2

In most nonsequential random-sampling applications, the population and sample are stored as arrays $a[0], a[1], \ldots, a[m - 1]$ and $x[0], x[1], \ldots, x[n - 1]$, respectively, in which case Algorithm 6.5.2 is equivalent to

```
for (i = 0; i < n; i++) {
    j = Equilikely(0, m - 1);
    x[i] = a[j];
}
```

Algorithm 6.5.3   This $O(n)$ algorithm provides nonsequential random sampling without replacement. The value of m must be known a priori, and $n \le m$.

```
for (i = 0; i < n; i++) {
    j = Equilikely(i, m - 1);          // note, i not 1
    Get(&z, P, j);                      // random access
    Put(z, S, i);
    Get(&x, P, i);                      // sequential access
    Put(z, P, i);                       // sequential access
    Put(x, P, j);                       // random access
}
```

If the elements of $\mathcal{P}$ are distinct, the number of possible samples is $m(m - 1) \ldots (m - n + 1) = m!/(m - n)!$, and all samples are equally likely to be generated. The order of $\mathcal{P}$ is not preserved in $\mathcal{S}$. Also, the order that exists in $\mathcal{P}$ is destroyed by this sampling algorithm. If this effect is undesirable, use a copy of $\mathcal{P}$ instead.

Example 6.5.3

In most nonsequential random-sampling applications, the population and sample are stored as arrays $a[0], a[1], \ldots, a[m - 1]$ and $x[0], x[1], \ldots, x[n - 1]$, respectively. In this case, Algorithm 6.5.3 is equivalent to

```
for (i = 0; i < n; i++) {
    j = Equilikely(i, m - 1);
    x[i] = a[j];
    a[j] = a[i];
    a[i] = x[i];
}
```

In this form, it is clear that Algorithm 6.5.3 is a simple extension of Algorithm 6.5.1.

Sequential Sampling.   The next three algorithms provide *sequential* sampling. For each algorithm, the basic idea is the same—traverse $\mathcal{P}$ once, in order, and select elements to put in $\mathcal{S}$. The selection or nonselection of elements is random with probability p and is effected via the generic statements

```
Get(&z, P, j);
if (Bernoulli(p))
    Put(z, S, i);
```

For Algorithm 6.5.4, p is independent of i and j. For Algorithms 6.5.5 and 6.5.6, however, the probability of selection is adaptively conditioned on the values of i, j, and the number of elements previously selected.

Algorithm 6.5.4   This $O(m)$ algorithm provides sequential random sampling without replacement. Each element of P can be selected, independently, with probability p.

```
i = 0;                                    // i indexes the sample
j = 0;                                    // j indexes the population
while ( more data in P ) {
    Get(&z, P, j);                        // sequential access
    j++;
    if (Bernoulli(p)) {
        Put(z, S, i);
        i++;
    }
}
```

Although $p = n/m$ is a logical choice for the probability of selection, Algorithm 6.5.4 does not make use of either m or n explicitly. Note also that Algorithm 6.5.4 is not consistent with the idea of a specified sample size. Although the *expected* size of the sample is mp, the *actual* size is a *Binomial*(m, p) random variable. That is, no matter how p is chosen (with $0 < p < 1$), there is no way to specify the exact size of the sample. It can range from 0 to m. The order of P is preserved in S.

Example 6.5.4

In many sequential-random-sampling applications, the population is stored in secondary memory as a sequential file, and the sample is stored in primary memory as an array $x[0], x[1], \ldots, x[n-1]$, in which case Algorithm 6.5.4 is equivalent to

```
i = 0;
while ( more data in P ) {
    z = GetData();
    if (Bernoulli(p)) {
        x[i] = z;
        i++;
    }
}
```

The fact that Algorithm 6.5.4 does not make use of m can be an advantage in some discrete-event simulation and real-time data-acquisition applications. In these applications, the objective may be to sample, at random, say 1% of the population elements, independent of the population size. Algorithm 6.5.4 with $p = 0.01$ would provide this ability, but at the expense of not being able to specify the sample size exactly. In contrast, Algorithm 6.5.5 provides the ability to specify the sample size, provided that m is known, and Algorithm 6.5.6 provides this ability even if m is unknown.

Algorithm 6.5.5 This $O(m)$ algorithm provides sequential random sampling of n sample values from $\mathcal{P}$ without replacement, provided that $m = |\mathcal{P}|$ is known a priori.

```
i = 0;                                          // i indexes the sample
j = 0;                                          // j indexes the population
while (i < n) {
   Get(&z, P, j);                               // sequential access
   p = (n - i) / (m - j);
   j++;
   if (Bernoulli(p)) {
      Put(z, S, i);
      i++;
   }
}
```

The key to the correctness of Algorithm 6.5.5 is that the population element a_j is selected with a probability $(n - i)/(m - j)$ that is conditioned on the number of sample elements left to be selected and the number of population elements left to be considered. Access to the population list is sequential (as is access to the sample list). Because sampling is without replacement, $n \leq m$. If the elements of $\mathcal{P}$ are distinct, then the number of possible samples is the binomial coefficient $m!/(m - n)!\,n!$ (see Appendix D), and all samples are equally likely to be generated. The order of $\mathcal{P}$ is preserved in $\mathcal{S}$.

Example 6.5.5

In many sequential-random-sampling applications, the population is stored in secondary memory as a sequential file, and the sample is stored in primary memory as an array $x[0], x[1], \ldots, x[n - 1]$, in which case Algorithm 6.5.5 is equivalent to

```
i = 0;
j = 0;
while (i < n) {
   z = GetData();
   p = (n - i) / (m - j);
   j++;
   if (Bernoulli(p)) {
```

```
      x[i] = z;
      i++;
   }
}
```

Algorithm 6.5.6 This $O(m)$ algorithm provides sequential random sampling of n sample values from $\mathcal{P}$ without replacement, even if $m = |\mathcal{P}|$ is not known a priori.

```
for (i = 0; i < n; i++) {
   Get(&z, P, i);                           // sequential access
   Put(z, S, i);
}
j = n;
while ( more data in P ) {
   Get(&z, P, j);                           // sequential access
   j++;
   p = n / j;
   if (Bernoulli(p)) {
      i = Equilikely(0, n - 1);
      Put(z, S, i);
   }
}
```

Algorithm 6.5.6 is based on initializing the sample with the first n elements of the population. Then, for each of the (unknown number of) additional population elements, with conditional probability $n/(j + 1)$ the population element a_j (for $j \geq n$) overwrites an existing randomly selected element in the sample. Access to the population list is sequential. Access to the sample list is *not* sequential, however, and so the order in $\mathcal{P}$ is not preserved in S. Because sampling is without replacement, $n \leq m$. If the elements of $\mathcal{P}$ are distinct, then the number of possible samples is the binomial coefficient $m!/(m - n)!\, n!$ (see Appendix D), and all samples are equally likely to be generated. There is an important caveat, however—see Example 6.5.8.

Example 6.5.6

In many sequential-random-sampling applications, the population is stored in secondary memory as a sequential file, and the sample is stored in primary memory as an array $x[0], x[1], \ldots, x[n - 1]$, in which case Algorithm 6.5.6 is equivalent to

```
for (i = 0; i < n; i++) {
   z = GetData();
   x[i] = z;
}
j = n;
```

```
while ( more data in P ) {
   z = GetData();
   j++;
   p = n / j;
   if (Bernoulli(p)) {
     i = Equilikely(0, n - 1);
     x[i] = z;
   }
}
```

Algorithm Differences. Although Algorithms 6.5.5 and 6.5.6 have some similarities, there are two important differences.

- Algorithm 6.5.5 requires knowledge of the population size m, and Algorithm 6.5.6 does not. This makes Algorithm 6.5.6 valuable in those discrete-event simulation and real-time data-acquisition applications where m is not known a priori, particularly if the sample size is sufficiently small that the sample can be stored in primary memory as an array.
- Beyond this obvious difference, there is a more subtle difference related to the number of possible samples when *order* is considered. Algorithm 6.5.5 preserves the order present in the population, Algorithm 6.5.6 does not. This difference is illustrated by the following two examples.

Example 6.5.7
If Algorithm 6.5.5 is used to select samples of size $n = 3$ from the population list $(0, 1, 2, 3, 4)$, then $m = 5$ and, because the sampling is sequential and the order of the population list is preserved in the samples, we find that exactly

$$\binom{5}{3} = \frac{5!}{2!\,3!} = 10$$

ordered samples are possible, which are:

$(0, 1, 2)$ $(0, 1, 3)$ $(0, 1, 4)$ $(0, 2, 3)$ $(0, 2, 4)$ $(0, 3, 4)$ $(1, 2, 3)$ $(1, 2, 4)$ $(1, 3, 4)$ $(2, 3, 4)$.

Each of these (ordered) samples will occur with equal probability $1/10$.

Example 6.5.8
If Algorithm 6.5.6 is used, then 13 samples are possible, as illustrated

$(0, 1, 2)$ $(0, 1, 3)$ $(0, 1, 4)$ $(0, 3, 2)$ $(0, 4, 2)$ $\begin{matrix}(0, 3, 4)\\(0, 4, 3)\end{matrix}$ $(3, 1, 2)$ $(4, 1, 2)$ $\begin{matrix}(3, 1, 4)\ (3, 4, 2)\\(4, 1, 3)\ (4, 3, 2)\end{matrix}$

These samples are *not* equally likely, however. Instead, each of the six samples that are pairwise alike except for permutation will occur with probability $1/20$. The other seven samples will occur with probability $1/10$. If permutations are combined (for example, by sorting and then combining like results) then, as desired, each of the resulting 10 samples will have equal probability $1/10$. Because order in $\mathcal{P}$ is not preserved in $\mathcal{S}$ by Algorithm 6.5.6, some ordered samples, like $(1, 3, 4)$, cannot occur except in permuted order, like $(3, 1, 4)$ and $(4, 1, 3)$. For this reason, Algorithm 6.5.6 produces the correct number of equally likely samples only if all alike-but-for-permutation samples are combined. This potential need for postprocessing is the (small) price paid for not knowing the population size a priori.

6.5.3 Urn Models

Many discrete stochastic models are based on random sampling. We close this section with four examples. Three of these discrete stochastic models were considered earlier in this chapter; the other is new:

- *Binomial* (n, p);
- *Hypergeometric* (n, a, b);
- *Geometric* (p);
- *Pascal* (n, p).

All four of these models can be motivated by drawing, at random, from a conceptual urn initially filled with a amber balls and b black balls.

Binomial

Example 6.5.9

An urn contains a amber balls and b black balls. A total of n balls are drawn from the urn, *with* replacement. Let x be the number of amber balls drawn. A Monte Carlo simulation of this random experiment is easily constructed. Indeed, one obvious way to construct this simulation is to use Algorithm 6.5.2, as implemented in Example 6.5.2, applied to a population array of length $m = a + b$ with a 1's (amber balls) and b 0's (black balls). Then x is the number of 1's in a random sample of size n. The use of Algorithm 6.5.2 is overkill in this case, however, because an equivalent simulation can be constructed without using either a population or sample data structure. Let $p = a/(a + b)$, and use the $O(n)$ algorithm

```
x = 0;
for (i = 0; i < n; i++)
    x += Bernoulli(p);
return x;
```

The discrete random variate x so generated is *Binomial* (n, p). The associated pdf of X is

$$f(x) = \binom{n}{x} p^x (1 - p)^{n-x} \qquad x = 0, 1, 2, \ldots, n.$$

Hypergeometric

Example 6.5.10
As a variation of Example 6.5.9, suppose that the draw from the urn is *without* replacement (and so $n \leq m$). Although Algorithm 6.5.3 could be used, as implemented in Example 6.5.3, it is better to use a properly modified version of the algorithm in Example 6.5.9 instead:

```
m = a + b;
x = 0;
for (i = 0; i < n; i++) {
    p = (a - x) / (m - i);
    x += Bernoulli(p);
}
return x;
```

The discrete random variate x so generated is said to be *Hypergeometric*(n, a, b). The associated pdf of X is

$$f(x) = \frac{\binom{a}{x}\binom{b}{n-x}}{\binom{a+b}{n}} \qquad x = \max\{0, n-b\}, \ldots, \min\{a, n\}.$$

The lower limit in the range of possible values of X accounts for the possibility that the sample size could exceed the number of black balls ($n > b$). If $n > b$, then at least $n - b$ amber balls must be drawn. Similarly, the upper limit accounts for the possibility that the sample size could exceed the number of amber balls ($n > a$). If $n > a$, then at most a amber balls can be drawn. In applications where n is less than or equal to both a and b, the range of possible values is $x = 0, 1, 2, \ldots, n$.

Geometric

Example 6.5.11
As another variation of Example 6.5.9, suppose that the draw is from the urn *with* replacement, but that we draw only until the first black ball is obtained. Let x be the number of amber balls drawn. With $p = a/(a + b)$, the following algorithm simulates this random experiment

```
x = 0;
while (Bernoulli(p))
    x++;
return x;
```

The discrete random variate x so generated is *Geometric*(p). The associated pdf of X is

$$f(x) = p^x(1 - p) \qquad x = 0, 1, 2, \ldots.$$

This stochastic model is commonly used in reliability studies, in which case p is usually close to 1.0 and x counts the number of successes before the first failure. This algorithm for generating geometric variates is inferior to the inversion algorithm presented in Example 6.2.8. The algorithm presented here is (1) not synchronized, (2) not monotone, and (3) inefficient. The expected number of passes through the `while` loop is

$$1 + \frac{p}{1 - p}.$$

If p is close to 1, the execution time for this algorithm can be quite high.

Pascal

Example 6.5.12
As an extension of Example 6.5.11, suppose that we draw *with* replacement until the nth black ball is obtained. Let x be the number of amber balls drawn (so that a total of $n + x$ balls are drawn, the last of which is black). With $p = a/(a + b)$, the following algorithm simulates this random experiment.

```
x = 0;
for (i = 0; i < n; i++)
    while (Bernoulli(p))
        x++;
return x;
```

The discrete random variate x so generated is *Pascal*(n, p). The associated pdf is

$$f(x) = \binom{n + x - 1}{x} p^x (1 - p)^n \qquad x = 0, 1, 2, \ldots.$$

This stochastic model is commonly used in reliability studies, in which case p is usually close to 1.0 and x counts the number of successes before the nth failure.

Further reading on random sampling and shuffling can be found in the textbooks by Nijenhuis and Wilf (1978) and Wilf (1989) or in the journal articles by McLeod and Bellhouse (1983) and Vitter (1984).

6.5.4 Exercises

6.5.1   Two cards are drawn in sequence from a well-shuffled ordinary deck of 52 cards. (*a*) If the draw is without replacement, use Monte Carlo simulation to verify that the second card will be higher in rank than the first with probability 8/17. (*b*) Estimate this probability if the draw is with replacement.

6.5.2   Which of the four discrete stochastic models presented in this section characterizes the actual (random-variate) sample size when Algorithm 6.5.4 is used?

6.5.3   (*a*) For $a = 30$, $b = 20$, and $n = 10$, use Monte Carlo simulation to verify the correctness of the probability equations in Examples 6.5.9 and 6.5.10. (*b*) Verify that, in both cases, the mean of X is $na/(a + b)$. (*c*) Are the standard deviations the same?

6.5.4   Three people toss a fair coin in sequence until the first occurrence of a head. Use Monte Carlo simulation to estimate (*a*) the probability that the first person to toss will eventually win the game by getting the first head, (*b*) the second person's probability of winning, and (*c*) the third person's probability of winning. Also, (*d*) compare your estimates with the analytic solutions.

6.5.5   How would the algorithm and pdf equation need to be modified if the random draw in Example 6.5.12 is *without* replacement?

6.5.6   An urn is initially filled with one amber ball and one black ball. Each time a black ball is drawn, it is replaced *and* another black ball is also placed in the urn. Let X be the number of random draws required to find the amber ball. (*a*) What is the pdf of X for $x = 1, 2, 3, \ldots$? (*b*) Use Monte Carlo simulation to estimate the pdf and the expected value of X.

6.5.7   Same as the previous exercise, except that there is a reservoir of just eight black balls to add to the urn. Comment.

7
Continuous Random Variables

In this chapter, the focus shifts from discrete random variables to continuous random variables. The first four sections in this chapter have titles identical to those of the previous chapter, but with "discrete" replaced by "continuous." As in the previous chapter, we proceed with a more thorough and methodical description of continuous random variables, of their properties, of how they can be used to model the stochastic (random) components of a system of interest, and of the development of algorithms for generating the associated random variates for a Monte Carlo or discrete-event simulation model.

Section 7.1 defines a continuous random variable and introduces eight popular models: the uniform, exponential, standard normal, normal, lognormal, Erlang, chisquare, and student distributions. Section 7.2 contains an approach to generating continuous random variables that is more general than the ad-hoc approaches given earlier for the *Uniform*(a, b) and *Exponential*(μ) variates. Section 7.3 applies these variate-generation techniques to the generation of arrival processes and service times. Section 7.4 contains a summary of the eight continuous distributions encountered thus far. Section 7.5 considers the topic of the modeling and generation of arrivals to a system when the arrival rate varies with time. A model known as a *nonstationary Poisson process* is introduced. Finally, Section 7.6 reintroduces a variate-generation technique known as *acceptance–rejection*, which was used for generating truncated random variables in the previous chapter.

7.1 CONTINUOUS RANDOM VARIABLES

The formal theory of continuous random variables closely parallels the corresponding theory of discrete random variables. In many instances, little more is involved than replacing sums with integrals and differences with derivatives. With this in mind, we will cover the corresponding topics in this chapter more rapidly than in the previous chapter, pausing to elaborate only when the "continuous" theory is significantly different from the "discrete" theory.

7.1.1 Continuous-Random-Variable Characteristics

> **Definition 7.1.1**   The random variable X is *continuous* if and only if the cdf of X, $F(x) = P(X \leq x)$ is a *continuous* function.*

Continuous random variables are real-valued measurements, such as size, weight, distance, elapsed time, volume, and density. In practice, it is common to assume that $\mathcal{X}$ is an open interval (a, b), where a may be $-\infty$ and b may be ∞.

Probability Density Function

> **Definition 7.1.2**   As in the discrete case, the continuous random variable X is uniquely determined by its set of possible values $\mathcal{X}$ and its corresponding *probability density function* (pdf), which is a real-valued function $f(\cdot)$ defined for all $x \in \mathcal{X}$ in such a way that
>
> $$\int_a^b f(x)\,dx = \Pr(a \leq X \leq b)$$
>
> for any interval $(a, b) \subseteq \mathcal{X}$. By definition, $x \in \mathcal{X}$ is a possible value of X if and only if $f(x) > 0$. In addition, $f(\cdot)$ is defined so that
>
> $$\int_x f(x)\,dx = 1$$
>
> where the integration is over all $x \in \mathcal{X}$.

Example 7.1.1
If the random variable X is *Uniform*(a, b), then the possible values of X are all the real numbers between a and b [i.e., $\mathcal{X} = (a, b)$]. Because the length of this interval is $b - a$ and all values in this interval are equally likely, it follows that

$$f(x) = \frac{1}{b - a} \qquad a < x < b.$$

If X is a discrete random variable, then it makes sense to evaluate a "point" probability, such as $\Pr(X = 1)$. However, if X is a continuous random variable, then *all* "point" probabilities are zero. In the continuous case, probability is defined by an *area* under the pdf curve, and so $\Pr(X = x) = 0$ for *any* $x \in \mathcal{X}$. For the same reason, endpoint inclusion is not an issue—that is, if

---

*Thus, the set of possible values in $\mathcal{X}$ is continuous, *not* discrete—see Definition 6.1.1.

X is a continuous random variable and if $[a, b] \subseteq \mathcal{X}$, then all of the probabilities $\Pr(a \leq X \leq b)$, $\Pr(a < X \leq b)$, $\Pr(a \leq X < b)$, $\Pr(a < X < b)$ are equal to $\int_a^b f(x)\,dx$.

Cumulative Distribution Function

> **Definition 7.1.3**   The *cumulative distribution function* (cdf) of the continuous random variable X is the continuous real-valued function $F(\cdot)$ defined for each $x \in \mathcal{R}$ as
>
> $$F(x) = \Pr(X \leq x) = \int_{t \leq x} f(t)\,dt.$$

Example 7.1.2
If the random variable X is *Uniform*(a, b) then $\mathcal{X} = (a, b)$, and the cdf is

$$F(x) = \int_a^x \frac{1}{b - a}\,dt = \frac{x - a}{b - a} \qquad a < x < b.$$

In the important special case where U is *Uniform*$(0, 1)$, the cdf is

$$F(u) = \Pr(U \leq u) = u \qquad 0 < u < 1.$$

Figure 7.1.1 illustrates the relation between a general pdf on the left and the associated cdf on the right. The two vertical axes are not drawn to the same scale. The shaded area in the graph of the pdf represents $F(x_0)$ for some arbitrary x-value x_0. As in the discrete case, the cdf is strictly monotone increasing—if x_1 and x_2 are possible values of X with $x_1 < x_2$ and all the real numbers between x_1 and x_2 are also possible values, then $F(x_1) < F(x_2)$. Moreover $F(\cdot)$ is bounded between 0.0 and 1.0.

As is illustrated in Figure 7.1.1 and as is consistent with Definition 7.1.3, the cdf can be obtained from the pdf by integration. Conversely, the pdf can be obtained from the cdf by differentiation, as

$$f(x) = \frac{d}{dx} F(x) \qquad x \in \mathcal{X}.$$

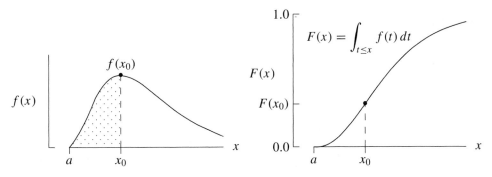

Figure 7.1.1   The pdf and cdf of a continuous random variable X.

Thus, as in the discrete case, a continuous random variable model can be constructed by specifying $\mathcal{X}$ and either the pdf or the cdf. One can be determined from the other.

Example 7.1.3

In Section 3.1, an *Exponential*(μ) random variable X was defined by the transformation $X = -\mu \ln(1 - U)$, where U is *Uniform*$(0, 1)$. To derive the cdf of X from this definition, recognize that, for any $x > 0$,

$$
\begin{aligned}
F(x) &= \Pr(X \le x) \\
&= \Pr\left(-\mu \ln(1 - U) \le x\right) \\
&= \Pr\left(1 - U \ge \exp(-x/\mu)\right) \\
&= \Pr\left(U \le 1 - \exp(-x/\mu)\right) \\
&= 1 - \exp(-x/\mu),
\end{aligned}
$$

where the last equation follows from Example 7.1.2. Therefore, the pdf of X is

$$
f(x) = \frac{d}{dx} F(x) = \frac{d}{dx}\left(1 - \exp(-x/\mu)\right) = \frac{1}{\mu} \exp(-x/\mu) \qquad x > 0.
$$

Mean and Standard Deviation

Definition 7.1.4   The *mean* μ of the continuous random variable X is

$$
\mu = \int_x x f(x)\, dx
$$

and the corresponding *standard deviation* σ is

$$
\sigma = \sqrt{\int_x (x - \mu)^2 f(x)\, dx} = \sqrt{\left(\int_x x^2 f(x)\, dx\right) - \mu^2}
$$

where the integration is over all possible values $x \in \mathcal{X}$. The *variance* is σ^2.*

Example 7.1.4

If X is *Uniform*(a, b), then it can be shown that

$$
\mu = \frac{a + b}{2} \qquad \text{and} \qquad \sigma = \frac{b - a}{\sqrt{12}}.
$$

This derivation is left as an exercise.

---

*Compare Definition 7.1.4 to Definition 4.3.3.

Example 7.1.5

If X is *Exponential*(μ), then from the pdf equation in Example 7.1.3,

$$\int_x xf(x)\,dx = \int_0^\infty \frac{x}{\mu}\exp(-x/\mu)\,dx = \mu \int_0^\infty t\,\exp(-t)\,dt = \cdots = \mu$$

(via integration by parts). So the parameter μ is, in fact, the mean. Moreover, the variance is

$$\sigma^2 = \left(\int_0^\infty \frac{x^2}{\mu}\exp(-x/\mu)\,dx\right) - \mu^2 = \left(\mu^2 \int_0^\infty t^2\exp(-t)\,dt\right) - \mu^2 = \cdots = \mu^2.$$

For this particular distribution, the parameter μ is also the standard deviation.

Expected Value

Definition 7.1.5   As in the discrete case, the mean of a continuous random variable is also known as the *expected value*, and the expected-value operator is denoted as $E[\,\cdot\,]$. In particular, the mean of X is the expected value of X,

$$\mu = E[X] = \int_x xf(x)\,dx,$$

and the variance is the expected value of $(X - \mu)^2$,

$$\sigma^2 = E\left[(X - \mu)^2\right] = \int_x (x - \mu)^2 f(x)\,dx.$$

In general, if $g(X)$ is some function of the random variable X [i.e., $Y = g(X)$], then the expected value of the random variable Y is

$$E[Y] = E[g(X)] = \int_x g(x)f(x)\,dx.$$

In all three of these equations, the integration is over all $x \in \mathcal{X}$.

Example 7.1.6

Consider a circle of fixed radius r and a fixed point Q on the circumference of the circle. Suppose a second point P is selected *at random* on the circumference of the circle, and let the random variable Y be the distance of the line segment (chord) joining points P and Q. To determine the expected value of Y, let Θ be the random angle of point P as measured counterclockwise from point Q. It can be shown that the relationship between Y and Θ is $Y = 2r\sin(\Theta/2)$, as illustrated in Figure 7.1.2.

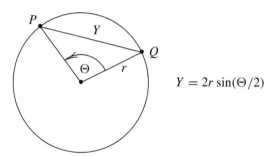

Figure 7.1.2   A random chord.

Therefore, if we interpret the phrase "*P* is selected at random" to mean that the angle Θ is *Uniform*$(0, 2\pi)$, then the pdf of Θ is $f(\theta) = 1/2\pi$ for $0 < \theta < 2\pi$, and the expected length of the chord, $E[Y]$, is

$$E[Y] = \int_0^{2\pi} 2r\sin(\theta/2)f(\theta)\,d\theta = \int_0^{2\pi} \frac{2r\sin(\theta/2)}{2\pi}\,d\theta = \cdots = \frac{4r}{\pi}.$$

Because some possible values of Y are more likely than others, Y is *not* a *Uniform*$(0, 2r)$ random variable; if it were, the expected value of Y would be r. (See Exercise 7.1.6 for an alternative approach to this example.)

Example 7.1.7
As in the discrete case, if X is a continuous random variable and Y is the continuous random variable $Y = aX + b$, for constants a and b, then

$$E[Y] = E[aX + b] = aE[X] + b.$$

7.1.2 Continuous-Random-Variable Models

In addition to the *Uniform*(a, b) and *Exponential*(μ) continuous random variables already discussed, in the remainder of this section we will consider several other parametric models, beginning with the so-called *Normal*$(0, 1)$ or *standard normal* continuous random variable. All students of elementary statistics will recognize this as a special case of the ubiquitous "bell-shaped curve" used by many (appropriately or not) to model virtually all things random.

Standard Normal Random Variable

Definition 7.1.6   The continuous random variable Z is said to be *Normal*$(0, 1)$ if and only if the set of all possible values is $\mathcal{Z} = (-\infty, \infty)$ and the pdf is

$$f(z) = \frac{1}{\sqrt{2\pi}}\exp(-z^2/2) \qquad -\infty < z < \infty,$$

as illustrated in Figure 7.1.3.

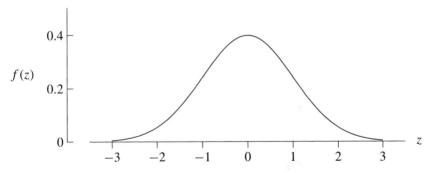

Figure 7.1.3   Standard normal pdf.

If Z is a *Normal* $(0, 1)$ random variable, then Z is "standardized" (see Example 4.1.3)—that is, the mean is

$$\mu = \int_{-\infty}^{\infty} z f(z)\, dz = \frac{1}{\sqrt{2\pi}} \int_{-\infty}^{\infty} z \exp(-z^2/2)\, dz = \cdots = 0$$

and the variance is

$$\sigma^2 = \int_{-\infty}^{\infty} (z - \mu)^2 f(z)\, dz = \frac{1}{\sqrt{2\pi}} \int_{-\infty}^{\infty} z^2 \exp(-z^2/2)\, dz = \cdots = 1.$$

The $\mu = 0$ derivation is easy, the $\sigma^2 = 1$ derivation is not. Both are left as an exercise. The standard deviation is $\sigma = 1$.

If Z is a *Normal* $(0, 1)$ random variable then the corresponding cdf is

$$F(z) = \int_{-\infty}^{z} f(t)\, dt = \Phi(z) \qquad\qquad -\infty < z < \infty,$$

where the special function $\Phi(\cdot)$ is defined as

$$\Phi(z) = \frac{1}{\sqrt{2\pi}} \int_{-\infty}^{z} \exp(-t^2/2)\, dt \qquad\qquad -\infty < z < \infty.$$

As a special function, $\Phi(\cdot)$ cannot be evaluated in "closed form." It can, however, be written as a function of an incomplete gamma function $P(a, x)$, as detailed in Appendix D, and evaluated numerically. Specifically, it can be shown that, if $z \geq 0$, then

$$\Phi(z) = \frac{1 + P(1/2, z^2/2)}{2} \qquad\qquad z \geq 0.$$

For $z < 0$, $\Phi(z)$ can be evaluated by using the identity $\Phi(-z) = 1 - \Phi(z)$. The special function $\Phi(z)$ is available in the library rvms as the function cdfNormal(0.0, 1.0, z).

One compelling reason for the popularity of the expected-value notation is that it applies equally well to both discrete and continuous random variables. For example, suppose that X is a random variable (discrete or continuous) with mean μ and standard deviation σ. For constants a and b, define the new random variable $X' = aX + b$ with mean μ' and standard deviation σ'. As an extension of Example 7.1.7,

$$\mu' = E[X'] = E[aX + b] = aE[X] + b = a\mu + b$$

and

$$(\sigma')^2 = E[(X' - \mu')^2] = E[(aX - a\mu)^2] = a^2 E[(X - \mu)^2] = a^2\sigma^2.$$

Therefore

$$\mu' = a\mu + b \qquad \text{and} \qquad \sigma' = |a|\sigma.$$

(See Section 4.1 for analogous sample-statistics equations.)

Example 7.1.8

Suppose that Z is a random variable with mean 0 and standard deviation 1 (often known as a *standardized random variable*) and that we wish to construct a new random variable X with *specified* mean μ and standard deviation σ. The usual way to do this is to define $X = \sigma Z + \mu$. Then, because $E[Z] = 0$ and $E[Z^2] = 1$, the mean and variance of X are

$$E[X] = \sigma E[Z] + \mu = \mu \qquad \text{and} \qquad E[(X - \mu)^2] = E[\sigma^2 Z^2] = \sigma^2 E[Z^2] = \sigma^2,$$

respectively, as desired. The standard deviation of X is σ. This example is the basis for the following definition.

Normal Random Variable

Definition 7.1.7   The continuous random variable X is $Normal(\mu, \sigma)$ if and only if

$$X = \sigma Z + \mu,$$

where $\sigma > 0$ and the random variable Z is $Normal(0, 1)$.

From Example 7.1.7, it follows that, if X is a $Normal(\mu, \sigma)$ random variable, then the mean of X is μ and the standard deviation is σ —that is, for this random-variable model, the mean and standard deviation are *explicit* parameters. The point here is that a $Normal(\mu, \sigma)$ random variable is constructed from a $Normal(0, 1)$ random variable by "shifting" the mean from 0 to μ via the addition of μ and by "scaling" the standard deviation from 1 to σ via multiplication by σ.

The cdf of a $Normal(\mu, \sigma)$ random variable is

$$F(x) = \Pr(X \le x) = \Pr(\sigma Z + \mu \le x) = \Pr\left(Z \le (x - \mu)/\sigma\right) \qquad -\infty < x < \infty,$$

and so

$$F(x) = \Phi\left(\frac{x-\mu}{\sigma}\right) \qquad -\infty < x < \infty,$$

where the special function $\Phi(\cdot)$ is the cdf of a *Normal*(0, 1) random variable. Because the pdf of a *Normal*(0, 1) random variable is

$$\frac{d}{dz}\Phi(z) = \frac{1}{\sqrt{2\pi}}\exp(-z^2/2) \qquad -\infty < z < \infty,$$

the associated pdf of a *Normal*(μ, σ) random variable is

$$f(x) = \frac{dF(x)}{dx} = \frac{d}{dx}\Phi\left(\frac{x-\mu}{\sigma}\right) = \cdots = \frac{1}{\sigma\sqrt{2\pi}}\exp\left(-(x-\mu)^2/2\sigma^2\right) \qquad -\infty < x < \infty,$$

as illustrated in Figure 7.1.4.

The shift parameter μ and the scale parameter σ allow the normal distribution to serve a wide variety of mathematical modeling applications, such as the weights of newborn babies, measurements on industrial products, agricultural yields, and IQ scores. For all choices of the parameters μ and σ, the pdf of a normal random variable has the bell shape shown in Figure 7.1.4.

In a field of study known as "classical statistics," observations are assumed to be drawn from populations having a *Normal*(μ, σ) distribution, which allows for conclusions to be drawn concerning μ and σ. An important result in statistical theory is known as the *central limit theorem*, which states that the limiting distribution of sums of independent and identically distributed random variables tends to the normal distribution as the number of random variables in the sum goes to infinity (Hogg, McKean, and Craig, 2005).

We summarize here a few facts about normal random variables:

- A *Normal*(μ, σ) random variable is also commonly called a *Gaussian* random variable.
- The 68–95–99.73 rule for *Normal*(μ, σ) random variables indicates that the area under the pdf within 1, 2, and 3 standard deviations of the mean μ is approximately 0.68, 0.95, and 0.9973.
- The pdf has inflection points at $\mu \pm \sigma$.

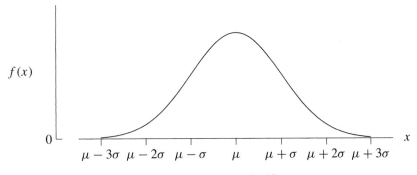

Figure 7.1.4 Normal pdf.

- Our choice of using $Normal(\mu, \sigma)$ to denote a normal random variable differs significantly from the more common $N(\mu, \sigma^2)$. We use the nonstandard notation so that the parameters in our software are consistent with the presentation in the text.

Finally, a warning about the $Normal(\mu, \sigma)$ distribution: The support is $\mathcal{X} = \{x \mid -\infty < x < \infty\}$, so the $Normal(\mu, \sigma)$ distribution is *not* an appropriate choice for modeling certain aspects (e.g., time between arrivals or service times in a queue) of a discrete-event simulation model, since it could potentially move the simulation clock backward, with potentially disastrous consequences. Thus the $Normal(\mu, \sigma)$ distribution must often be manipulated in a manner to produce only positive values. One such manipulation is the lognormal distribution, presented next.

Lognormal Random Variable

Definition 7.1.8   The continuous random variable X is $Lognormal(a, b)$ if and only if

$$X = \exp(a + bZ),$$

where the random variable Z is $Normal(0, 1)$ and $b > 0$.

Like a $Normal(\mu, \sigma)$ random variable, a $Lognormal(a, b)$ random variable is also based on transforming a $Normal(0, 1)$ random variable. The transformation in this case is nonlinear, and the result is therefore more difficult to analyze. However, the nonlinear transformation provides an important benefit; a $Lognormal(a, b)$ random variable is inherently positive and therefore potentially well suited to modeling naturally positive quantities, such as service times.

The cdf of a $Lognormal(a, b)$ random variable is

$$F(x) = \Pr(X \le x) = \Pr\left(\exp(a + bZ) \le x\right) = \Pr\left(a + bZ \le \ln(x)\right) \qquad x > 0,$$

and so

$$F(x) = \Pr\left(Z \le (\ln(x) - a)/b\right) = \Phi\left(\frac{\ln(x) - a}{b}\right) \qquad x > 0,$$

where the special function $\Phi(\cdot)$ is the cdf of a $Normal(0, 1)$ random variable. Differentiation by x then establishes that the pdf of a $Lognormal(a, b)$ random variable is

$$f(x) = \frac{dF(x)}{dx} = \cdots = \frac{1}{bx\sqrt{2\pi}} \exp\left(-(\ln(x) - a)^2/2b^2\right) \qquad x > 0,$$

as illustrated in Figure 7.1.5 for $(a, b) = (-0.5, 1.0)$.

It can be verified that the mean and standard deviation of a $Lognormal(a, b)$ random variable are

$$\mu = \exp\left(a + b^2/2\right) \qquad \text{and} \qquad \sigma = \exp\left(a + b^2/2\right)\sqrt{\exp(b^2) - 1}.$$

For the pdf illustrated, $\mu = 1.0$, and $\sigma = \sqrt{e - 1} \cong 1.31$.

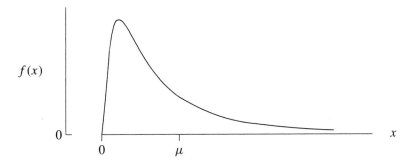

Figure 7.1.5   Lognormal pdf.

A *Uniform*(a, b) random variable is the continuous analog of a discrete *Equilikely*(a, b) random variable. Similarly, an *Exponential*(μ) random variable is the continuous analog of a *Geometric*(p) random variable. Given that relationship, and because a *Pascal*(n, p) random variable is the sum of n *iid Geometric*(p) random variables, it is natural to assume that the sum of n *iid Exponential*(μ) random variables defines a commonly used continuous random variable model. That is the case.

Erlang Random Variable

Definition 7.1.9   The continuous random variable X is *Erlang*(n, b) if and only if

$$X = X_1 + X_2 + \cdots + X_n,$$

where $X_1, X_2, \ldots, X_n$ is an *iid* sequence of *Exponential*(b) random variables.

Although the details of the derivation are beyond the scope of this book, from Definition 7.1.9, it can be shown that the pdf of an *Erlang*(n, b) random variable is

$$f(x) = \frac{1}{b(n-1)!}(x/b)^{n-1}\exp(-x/b) \qquad x > 0,$$

as illustrated in Figure 7.1.6 for $(n, b) = (3, 1.0)$. In this case, $\mu = 3.0$ and $\sigma \cong 1.732$.

The corresponding cdf is an incomplete gamma function (see Appendix D):

$$F(x) = \int_0^x f(t)\, dt = 1 - \left[\exp\left(-\frac{x}{b}\right)\right]\left(\sum_{i=0}^{n-1}\frac{(x/b)^i}{i!}\right) = P(n, x/b) \qquad x > 0.$$

The associated mean and standard deviation are

$$\mu = nb \qquad \text{and} \qquad \sigma = \sqrt{n}\, b.$$

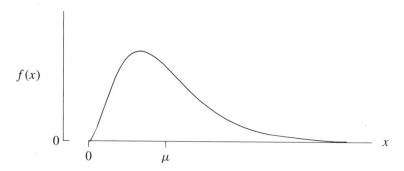

Figure 7.1.6 Erlang pdf.

Chisquare and Student Random Variables. In addition to *Erlang* (n, b), *Exponential* (μ), *Lognormal* (a, b), *Normal* (μ, σ), and *Uniform* (a, b) random variables, two other continuous random variables commonly used in simulation applications are *Chisquare* (n) and *Student* (n). These two random variables are defined in the next section and summarized in Section 7.4. Although they are less likely to be used for parametric modeling in a discrete-event simulation model, having integer parameters, they are often used for statistical inference.

7.1.3 Exercises

7.1.1   The possible values of the continuous random variable X are $a < x < b$. (*a*) If the pdf of X is $f(x) = \alpha(b - x)$, what is α in terms of a and b? (*b*) Derive equations for the cdf as a function of x, in terms of a and b. (*c*) Derive equations for the mean and standard deviation in terms of a and b.

7.1.2   (*a*) Derive the equations for μ and σ in Example 7.1.4, and compare them to the corresponding equations in Example 6.1.7. (*b*) If σ_u^2 and σ_e^2 are the variances of a *Uniform* (a, b) and an *Equilikely* (a, b) random variable, respectively, show that $\sigma_e^2 > \sigma_u^2$ and that $\sigma_e^2 / \sigma_u^2 \to 1$ as $(b - a) \to \infty$.

7.1.3   If X is a continuous random variable with pdf $f(\cdot)$, mean μ, and variance σ^2, prove that

$$\int_x x^2 f(x)\, dx = \mu^2 + \sigma^2,$$

where the integration is over all possible values of X.

7.1.4   If X is an *Exponential* (μ) random variable and $Y = \lfloor X \rfloor$, prove that Y is a *Geometric* (p) random variable with a value of p that depends on μ.

7.1.5   (*a*) If $X = U/(1 - U)$, with U a *Uniform* $(0, 1)$ random variable, then what is the set of possible values $\mathcal{X}$? (*b*) What is the cdf of X? (*c*) What is the pdf of X?

7.1.6   (*a*) If $Y = 2r \sin(\Theta/2)$, where the random variable Θ is *Uniform* $(0, 2\pi)$ and $r > 0$, prove that the pdf of Y is

$$f(y) = \frac{2}{\pi \sqrt{4r^2 - y^2}} \qquad 0 < y < 2r.$$

(b) Why is there a 2 in the numerator? (c) As an alternative to the derivation in Example 7.1.6, use this pdf to prove that $E[Y] = 4r/\pi$. (d) What is the standard deviation of Y?

7.1.7   (a) If U is $Uniform(0, 1)$, if Θ is $Uniform(-\pi/2, \pi/2)$, if U and Θ are independent, and if $V = U + \cos(\Theta)$, prove that the possible values of V are $0 < v < 2$ and that the pdf of V is

$$f(v) = \begin{cases} 1 - (2/\pi)\arccos(v) & 0 < v < 1 \\ (2/\pi)\arccos(v-1) & 1 \le v < 2. \end{cases}$$

(b) How does this relate to Example 4.3.1?

7.1.8   If X is a $Normal(\mu, \sigma)$ random variable, what is the numerical value of $\Pr(|X - \mu| \le k\sigma)$ for $k = 1, 2, 3, 4$?

7.1.9   (a) If $X = \exp(U)$, with U a $Uniform(a, b)$ random variable, then what is $\mathcal{X}$? (b) What is the cdf of X? (c) What is the pdf of X? (d) What is the mean of X?

7.1.10   Show that the inflection points on the pdf for a $Normal(\mu, \sigma)$ random variable occur at $x = \mu \pm \sigma$.

7.1.11   Let X be a random variable with finite mean μ and finite variance σ^2. Find the mean and variance of the *standardized* random variable $Y = (X - \mu)/\sigma$.

7.2 GENERATING CONTINUOUS RANDOM VARIATES

This section is the continuous-random-variate analog of Section 6.2. As in Section 6.2, the emphasis is on the development of inversion algorithms for random-variate generation.

7.2.1 Inverse Distribution Function

Definition 7.2.1   Let X be a continuous random variable with cdf $F(\cdot)$ and set of possible values $\mathcal{X}$. The *inverse distribution function* (*idf*) of X is the function $F^{-1}: (0, 1) \to \mathcal{X}$, defined for all $u \in (0, 1)$ as

$$F^{-1}(u) = x$$

where $x \in \mathcal{X}$ is the unique possible value for which $F(x) = u$.

Because the cdf is strictly monotone increasing, there is a one-to-one correspondence between possible values $x \in \mathcal{X}$ and cdf values $u = F(x) \in (0, 1)$.* Therefore, unlike the idf for a discrete random variable, which is a pseudo-inverse, the idf for a continuous random variable is a true inverse, as illustrated in Figure 7.2.1.

Because the idf is a true inverse, it is sometimes possible to express the idf in "closed form" by solving the equation $F(x) = u$ for x, using common algebraic techniques. Generally, this is much

---

*The condition $f(x) > 0$ for all $x \in \mathcal{X}$ is sufficient to guarantee that $F(\cdot)$ is strictly monotone increasing—see Definition 7.1.3.

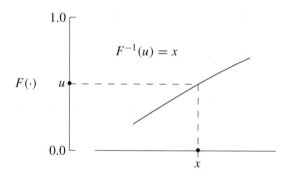

Figure 7.2.1 Generating continuous random variates.

easier to do in the continuous case than in the discrete case, because no inequality manipulations are required.

Example 7.2.1
If X is $Uniform(a, b)$, then $F(x) = (x - a)/(b - a)$ for $a < x < b$. Solving the equation $u = F(x)$ for x yields the idf

$$x = F^{-1}(u) = a + (b - a)u \qquad 0 < u < 1.$$

Example 7.2.2
If X is $Exponential(\mu)$, then $F(x) = 1 - \exp(-x/\mu)$ for $x > 0$. Solving the equation $u = F(x)$ for x yields the idf

$$x = F^{-1}(u) = -\mu \ln(1 - u) \qquad 0 < u < 1.$$

Example 7.2.3
If X is a continuous random variable with possible values $0 < x < b$ and pdf $f(x) = 2x/b^2$, then the cdf is $F(x) = (x/b)^2$. The equation $u = F(x) = (x/b)^2$ can be solved for x to establish that the idf of X is $F^{-1}(u) = b\sqrt{u}$ for $0 < u < 1$.

7.2.2 Random-Variate Generation by Inversion

As in the discrete case, the idf of a continuous random variable can be used to generate corresponding random variates via inversion. The theoretical basis for this is the following theorem, which is known as the *probability integral transformation*. The proof is left as an exercise.

Theorem 7.2.1   If X is a continuous random variable with idf $F^{-1}(\cdot)$, if the continuous random variable U is $Uniform(0, 1)$, and if Z is the continuous random variable defined by $Z = F^{-1}(U)$, then Z and X are identically distributed.*

---

*Compare to Theorem 6.2.2.

Algorithm 7.2.1   If X is a continuous random variable with idf $F^{-1}(\cdot)$, then a corresponding continuous random variate can be generated by inversion as follows.

```
u = Random();
return F⁻¹(u);
```

Example 7.2.4
This algorithm uses inversion and Example 7.2.1 to generate a *Uniform*(a, b) random variate:

```
u = Random();
return a + (b - a) * u;
```

This is the algorithm used to define the function Uniform(a, b). (See Definition 2.3.3.)

Example 7.2.5
This algorithm uses inversion and Example 7.2.2 to generate an *Exponential*(μ) random variate:

```
u = Random();
return −μ * log(1 - u);
```

This is the algorithm used to define the function Exponential(μ). (See Definition 3.1.1.)

Note that the random variable U is *Uniform*$(0, 1)$ if and only if the random variable $1 - U$ is also *Uniform*$(0, 1)$. Therefore, it is also valid to generate an *Exponential*(μ) random variate as follows, thereby saving the essentially negligible cost of a subtraction

```
u = Random();
return −μ * log(u);
```

Because this algorithm has reverse monotonicity (e.g., large values of u correspond to small values of x and conversely), for esthetic reasons we prefer the algorithm in Example 7.2.5.

The random-variate-generation algorithms in Examples 7.2.4 and 7.2.5 are essentially ideal: Both algorithms are portable, exact, robust, efficient, clear, synchronized, and monotone.

Generally it is not possible to solve for a continuous-random-variable idf explicitly by algebraic techniques. There are, however, two other options that might be available.

- Use a function that accurately *approximates* $F^{-1}(\cdot)$.
- Find the idf by solving the equation $u = F(x)$ *numerically*.

We will see examples of both approaches in this section.

Approximate Inversion. If Z is a *Normal*$(0, 1)$ random variable, then the cdf is the special function $\Phi(\cdot)$ and the idf $\Phi^{-1}(\cdot)$ cannot be evaluated in closed form. However, as demonstrated by Odeh and Evans (1974), the idf can be *approximated* with reasonable efficiency and essentially negligible error as the ratio of two fourth-degree polynomials: For any $u \in (0, 1)$, a *Normal*$(0, 1)$ idf approximation is $\Phi^{-1}(u) \cong \Phi_a^{-1}(u)$, where

$$
\Phi_a^{-1}(u) = \begin{cases} -t + p(t)/q(t) & 0.0 < u < 0.5 \\ t - p(t)/q(t) & 0.5 \leq u < 1.0, \end{cases}
$$

$$
t = \begin{cases} \sqrt{-2\ln(u)} & 0.0 < u < 0.5 \\ \sqrt{-2\ln(1-u)} & 0.5 \leq u < 1.0, \end{cases}
$$

and

$$
p(t) = a_0 + a_1 t + \cdots + a_4 t^4
$$

$$
q(t) = b_0 + b_1 t + \cdots + b_4 t^4.
$$

If the ten polynomial coefficients are chosen carefully, then this approximation is accurate with an absolute error less than 10^{-9} for all $0.0 < u < 1.0$. (See Exercise 7.2.8.)*

Example 7.2.6

Given that the $\Phi_a^{-1}(u)$ approximation to $\Phi^{-1}(u)$ is essentially exact and can be evaluated with reasonable efficiency, inversion can be used to generate *Normal*$(0, 1)$ random variates, as illustrated:

```
u = Random();
return Φₐ⁻¹(u);
```

This algorithm is portable, essentially exact, robust, reasonably efficient, synchronized, and monotone, at the expense of clarity.

Alternative Methods. For historical perspective, we mention two other *Normal*$(0, 1)$ random-variate-generation algorithms, although we prefer the inversion algorithm in Example 7.2.6.

* One algorithm is based on the fact that, if $U_1, U_2, \ldots, U_{12}$ is an *iid* sequence of *Uniform*$(0, 1)$ random variables, then the random variable

$$
Z = U_1 + U_2 + \cdots + U_{12} - 6
$$

is approximately *Normal*$(0, 1)$. The mean of this random variable is exactly 0.0, the standard deviation is exactly 1.0, and the pdf is symmetric about 0. The set of possible values is $-6.0 < z < 6.0$, however, and so it is clear that some approximation is involved. The justification for this algorithm is provided by the central limit theorem—see Section 8.1. This algorithm is portable, robust, relatively efficient, and clear; it is not exact, synchronized, or monotone. We do not recommended this algorithm, since it is not exact.

---

*The values of the ten constants are listed in the file `rvgs.c`. The numerical-analysis details of how these coefficients were originally calculated are beyond the scope of this book.

- The other algorithm, due to Box and Muller (1958), is based on the fact that, if U_1, U_2 are independent *Uniform* (0, 1) random variables, then

$$Z_1 = \sqrt{-2\ln(U_1)}\,\cos(2\pi U_2) \qquad \text{and} \qquad Z_2 = \sqrt{-2\ln(U_1)}\,\sin(2\pi U_2)$$

will be independent *Normal* (0, 1) random variables. If only one normal variate is needed, the second normal variate must be stored for a subsequent call or discarded. The algorithm is portable, exact, robust, and relatively efficient. It is not clear (unless the underlying theory is understood), it is not monotone, and it is synchronized only in pair-wise fashion. The polar method (Marsaglia and Bray, 1964) improves the computational efficiency.

Normal Random Variates.   Consistent with Definitions 7.1.7 and 7.1.8, random variates corresponding to *Normal* (μ, σ) and *Lognormal* (a, b) random variables can be generated by using a *Normal* (0, 1) generator. This illustrates the importance of having a good *Normal* (0, 1) random-variate generator and helps explain why a significant amount of past research has been devoted to the development of such generators.

Example 7.2.7
This algorithm generates a *Normal* (μ, σ) random variate.

```
z = Normal(0.0, 1.0);
return  μ + σ * z;                        // see definition 7.1.7
```

Lognormal Random Variates

Example 7.2.8
This algorithm generates a *Lognormal* (a, b) random variate.

```
z = Normal(0.0, 1.0);
return exp(a + b * z);                    // see definition 7.1.8
```

The algorithm in Example 7.2.6 generates a *Normal* (0, 1) random variate by approximate inversion. Provided that this algorithm is used to generate the *Normal* (0, 1) random variate, the algorithms in Examples 7.2.7 and 7.2.8 use inversion also and, therefore, are essentially ideal.

Numerical Inversion.   *Numerical* inversion provides another way to generate continuous random variates when $F(x)$ can't be inverted algebraically [i.e., when $F^{-1}(u)$ can't be expressed in closed form]. The equation $u = F(x)$ can be solved for x iteratively. Several iterative algorithms can be used. Of these, *Newton's method* provides a good compromise between rate of convergence and robustness. Given $u \in (0, 1)$, to derive Newton's method for inverting a cdf, let t be numerically close to the value of x for which $u = F(x)$. If $F(\cdot)$ is expanded in a Taylor's series about the point t, then

$$F(x) = F(t) + F'(t)(x - t) + \frac{1}{2!}F''(t)(x - t)^2 + \cdots$$

where $F'(t)$ and $F''(t)$ are the first and second derivative of the cdf evaluated at t. Because the first derivative of the cdf is the pdf, $F'(t) = f(t)$. Moreover, if $|x - t|$ is sufficiently small, then the $(x - t)^2$ and higher order terms can be ignored, and so

$$u = F(x) \cong F(t) + f(t)(x - t).$$

Solving the equation $u \cong F(t) + f(t)(x - t)$ for x yields

$$x \cong t + \frac{u - F(t)}{f(t)}.$$

With initial guess t_0, this last equation defines an *iterative* numerical algorithm of the form

$$t_{i+1} = t_i + \frac{u - F(t_i)}{f(t_i)} \qquad\qquad i = 0, 1, 2, \ldots$$

that solves the equation $u = F(x)$ for x in the sense that $t_i \to x$ as $i \to \infty$, as illustrated in Figure 7.2.2.

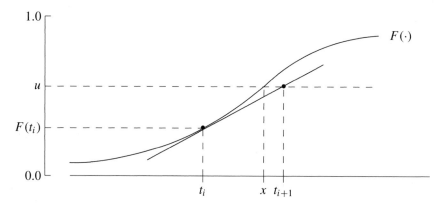

Figure 7.2.2   Numerical inversion of the cdf.

There are two remaining issues relative to Newton's method: the choice of an initial value t_0, and the test for convergence.

- Generally, as in the discrete case, the best choice for the initial value is the mode. For most of the continuous-random-variable models considered in this book, $t_0 = \mu$ is an essentially equivalent choice.
- Generally, for a given (user-supplied, tiny) convergence parameter $\epsilon > 0$, the condition $|t_{i+1} - t_i| < \epsilon$ is a reasonable test for convergence.

With the initial value and convergence issues resolved in this way, Newton's method for numerical inversion is summarized by the following algorithm.

> **Algorithm 7.2.2**   Given $u \in (0, 1)$, the pdf $f(\cdot)$, the cdf $F(\cdot)$, and a convergence parameter $\epsilon > 0$, this algorithm will solve for $x = F^{-1}(u)$:
>
> ```
> x = μ;                                              // μ is E[X]
> do {
>     t = x;
>     x = t + (u - F(t)) / f(t);
> } while (|x − t| > ε);
> return x;                                           // x is F⁻¹(u)
> ```

If Algorithm 7.2.2 is used to compute the idf for an inherently positive random variable, and if u is small, then a negative value of x could occur early in the iterative process. A negative value of x will cause t to be negative on the next iteration, and then $F(t)$ and $f(t)$ will be undefined. The following modification of Algorithm 7.2.2 can be used to avoid this problem.

```
x = μ;                                              // μ is E[X]
do {
    t = x;
    x = t + (u - F(t)) / f(t);
    if (x <= 0.0)
        x = 0.5 * t;
} while (|x − t| > ε);
return x;                                           // x is F⁻¹(u)
```

Algorithms 7.2.1 and 7.2.2 together provide a general-purpose inversion approach to continuous-random-variate generation. For example, the *Erlang*(n, b) idf capability provided in the library rvms, which is based on Algorithm 7.2.2, can be used with Algorithm 7.2.1 as

```
u = Random();
return idfErlang(n, b, u);
```

7.2.3 Alternative Random-Variate-Generation Algorithms

Properties relating parametric distributions to one another can be used to develop variate-generation algorithms for some common parametric distributions. We exploited these special properties in the last chapter to devise variate-generation algorithms for discrete random variables [e.g., a *Binomial*(n, p) random variable is the sum of n independent and identically distributed *Bernoulli*(p) random variables]. We consider the Erlang, chisquare, and student distributions here.

Erlang Random Variates.   As an alternative to inversion, an *Erlang*(n, b) random variate can be generated by summing n independent and identically distributed *Exponential*(b) random variates (see Definition 7.1.9), as follows:

```
x = 0.0;
for (i = 0; i < n;  i++)
   x += Exponential(b);
return x;
```

This algorithm is portable, exact, robust, and clear. Because it is $O(n)$, if n is large, then the algorithm is not efficient. For any n, it is not synchronized or monotone. To increase computational efficiency, this algorithm can be modified as follows:

```
t = 1.0;
for (i = 0; i < n;  i++)
   t *= (1.0 - Random());
return -b * log(t);
```

If n is large, the time saved by this modification is potentially significant, because only one `log()` evaluation is required rather than n of them. Moreover, the product of n random numbers, $\prod_{i=1}^{n} U_i$, has the same distribution as $\prod_{i=1}^{n}(1 - U_i)$, so still more processing time (i.e., n subtractions) can be saved by replacing the statement in the `for` loop with `t *= Random();`.

Both algorithms presented here are inherently $O(n)$, however, and are thus not efficient if n is large. If floating-point arithmetic could be done with infinite precision, these two algorithms would produce identical output. See Exercise 7.2.2.

Chisquare Random Variates.   If n is an even positive integer, then an *Erlang*$(n/2, 2)$ random variable is equivalent to a *Chisquare*(n) random variable. As an alternative characterization, if n is even or odd then X is a *Chisquare*(n) random variable if and only if $X = Z_1^2 + Z_2^2 + \cdots + Z_n^2$, where $Z_1, Z_2, \ldots, Z_n$ is an *iid* sequence of *Normal*$(0, 1)$ random variables. This alternative characterization can be used with a *Normal*$(0, 1)$ random-variate generator to generate a *Chisquare*(n) random variate as follows:

```
x = 0.0;
for (i = 0; i < n; i++) {
   z = Normal(0.0, 1.0);
   x += (z * z);
}
return x;
```

This algorithm is portable, exact, robust, and clear. Because it is $O(n)$, if n is large, then the algorithm is not efficient. For any $n > 1$, it is not synchronized or monotone.

Student Random Variates.   The random variable X is *Student*(n) if and only if $X = Z/\sqrt{V/n}$, where Z is *Normal*$(0, 1)$, V is *Chisquare*(n), and Z, V are independent.* Therefore, the following

---

*This distribution was discovered by W.S. Gossett when he was working for an Irish brewery. The brewery did not want its competitors to know that it was using statistical methods, so it required that he use the pseudonym "Student" to publish his work. Most textbooks refer to this distribution as either the Student's-*t* or just the *t* distribution. This distribution is the basis for one of the cornerstone statistical tests—the two-sample *t*-test for comparing the means of two populations (Hogg, McKean, and Craig, 2005).

algorithm uses a *Normal*(0, 1) random-variate generator and a *Chisquare*(n) random-variate generator to generate a *Student*(n) random variate as

```
z = Normal(0.0, 1.0);
v = Chisquare(n);
return z / sqrt(v / n);
```

This algorithm is exact, robust, portable, and clear. Because it is $O(n)$, if n is large, then the algorithm is not efficient. For any n, it is not synchronized or monotone.

7.2.4 Testing for Correctness

Any continuous random-variate-generation algorithm should be tested for correctness. One quick method to test for *incorrectness* is to calculate the sample mean and sample standard deviation for a large sample size. Since $\bar{x} \to \mu$ and $s \to \sigma$ as $n \to \infty$, the sample mean and standard deviation should be "reasonably close" to the associated population mean and standard deviation. The meaning of "reasonably close" will be investigated further in the next two chapters, but, for now, we can say that, if the sample mean and sample standard deviations are not closing in on the appropriate theoretical values, the algorithm is incorrect. The convergence described here is a necessary, but not a sufficient, condition for the correctness of the algorithm. The reason that we can't conclude that the algorithm is correct is that two different distributions can have identical mean and variance [e.g., a *Uniform*$(-\sqrt{3}, \sqrt{3})$ and a *Normal*(0, 1) random variable]. Two empirical tests for correctness are based on the pdf $f(x)$ and the cdf $F(x)$ which, unlike the mean and standard deviation, completely characterize the distribution.

Comparing the Histogram and *f(x)*.   A natural way to test the correctness of a random-variate-generation algorithm at the computational level is to use the algorithm to generate a sample of n random variates and, as in Section 4.3, construct a k-bin continuous-data histogram with bin width δ. If $\mathcal{X}$ is the set of possible values, $\hat{f}(x)$ is the histogram density, and $f(x)$ is the pdf, then, provided that the algorithm is correct, for all $x \in \mathcal{X}$, we should find that $\hat{f}(x) \to f(x)$ as $n \to \infty$ and $\delta \to 0$. Of course, in practice, these limits cannot be achieved. Instead, we use a large but finite value of n and a small but nonzero value of δ.

Given that n is finite and δ is not zero, perfect agreement between $\hat{f}(x)$ and $f(x)$ will not be achieved. In the discrete case, this lack of perfect agreement is due to natural sampling variability only. In the continuous case, the *quantization error* associated with binning the sample is an additional factor that will contribute to the lack of agreement. In other words, if we let $\mathcal{B} = [m - \delta/2, m + \delta/2)$ be a small histogram bin with width δ and midpoint m, and if $f(x)$ has a Taylor's series expansion at $x = m$, we can write

$$f(x) = f(m) + f'(m)(x - m) + \frac{1}{2!}f''(m)(x - m)^2 + \frac{1}{3!}f'''(m)(x - m)^3 + \cdots,$$

from which it follows that

$$\Pr(X \in \mathcal{B}) = \int_{\mathcal{B}} f(x)\, dx = \cdots = f(m)\delta + \frac{1}{24}f''(m)\delta^3 + \cdots.$$

If terms of order δ^4 or higher are ignored, then, for all $x \in \mathcal{B}$, the histogram density is

$$\hat{f}(x) = \frac{1}{\delta} \Pr(X \in \mathcal{B}) \cong f(m) + \frac{1}{24} f''(m)\delta^2.$$

Therefore [unless $f''(m) = 0$], depending on the sign of $f''(m)$, there is a positive or negative *bias* between the (experimental) density of the histogram bin and the (theoretical) pdf evaluated at the bin midpoint. This bias can be significant if the curvature of the pdf is large at the bin midpoint.

Example 7.2.9
If X is a continuous random variable with pdf

$$f(x) = \frac{2}{(x+1)^3} \qquad x > 0,$$

then the cdf X is

$$F(x) = \int_0^x f(t)\, dt = 1 - \frac{1}{(x+1)^2} \qquad x > 0;$$

and, by solving the equation $u = F(x)$ for x, we find that the idf is

$$F^{-1}(u) = \frac{1}{\sqrt{1-u}} - 1 \qquad 0 < u < 1.$$

In this case, the pdf curvature is very large close to $x = 0$. Therefore, if inversion is used to generate random variates corresponding to X, and if the correctness of the inversion algorithm is tested by constructing a histogram, unless δ is very small we expect that the histogram will not match the pdf well for the bins close to $x = 0$. In particular, if a histogram bin width of $\delta = 0.5$ is used then, in the limit as $n \to \infty$, for the first six histogram bins, $\hat{f}(x)$ and $f(m)$ are (with $d.dddd$ precision)

m	0.25	0.75	1.25	1.75	2.25	2.75
$\hat{f}(x)$	1.1111	0.3889	0.1800	0.0978	0.0590	0.0383
$f(m)$	1.0240	0.3732	0.1756	0.0962	0.0583	0.0379

The "curvature bias" is evident in the first few bins. Indeed, for the first bin ($m = 0.25$) the curvature bias is

$$\frac{1}{24} f''(m)\delta^2 = 0.08192,$$

which explains essentially all the difference between $\hat{f}(x)$ and $f(m)$.

Comparing the Empirical cdf and F(x). Comparing the empirical cdf, introduced in Section 4.3, with the population cdf $F(x)$ is a second method for assessing the correctness of a variate-generation algorithm. The quantization error associated with binning is eliminated. For large samples, the empirical cdf should converge to $F(x)$ [i.e., $\hat{F}(x) \to F(x)$ as $n \to \infty$, if the variate-generation algorithm is valid].

Library rvgs. All of the continuous-random-variate generators presented in this section are provided in the library rvgs as the following seven functions:

- double Chisquare(long n)—returns a *Chisquare*(n) random variate;
- double Erlang(long n, double b)—returns an *Erlang*(n, b) random variate;
- double Exponential(double μ)—returns an *Exponential*(μ) random variate;
- double Lognormal(double a, double b)—returns a *Lognormal*(a, b) random variate;
- double Normal(double μ, double σ)—returns a *Normal*(μ, σ) random variate;
- double Student(long n)—returns a *Student*(n) random variate;
- double Uniform(double a, double b)—returns a *Uniform*(a, b) random variate.

7.2.5 Exercises

7.2.1   Prove Theorem 7.2.1.

7.2.2   To what extent will the two *Erlang*(n, b) random-variate-generation algorithms in Section 7.2.3 produce the same output? Why?

7.2.3   What are the largest and smallest possible values that can be returned by the function Normal$(0, 1)$?

7.2.4   A continuous random variable X is *Weibull*(a, b) if the real-valued parameters a, b are positive, the possible values of X are $x > 0$, and the cdf is

$$F(x) = 1 - \exp\left(-(bx)^a\right) \qquad x > 0.$$

What are the pdf and idf?

7.2.5   A continuous random variable X is *Logistic*(a, b) if the real-valued parameter a is unconstrained, the real-valued parameter b is positive, the possible values of X are $-\infty < x < \infty$, and the idf is

$$F^{-1}(u) = a - b \ln\left(\frac{1-u}{u}\right) \qquad 0 < u < 1.$$

What are the pdf and cdf?

7.2.6   A continuous random variable X is *Pareto*(a, b) if the real-valued parameters a, b are positive, the possible values of X are $x > a$, and the pdf is

$$f(x) = ba^b / x^{b+1} \qquad x > a.$$

(a) Find the mean and standard deviation of X, paying careful attention to any restrictions that must be imposed on a and b. (b) Construct a function that uses inversion to generate values of X. (c) Present convincing numerical evidence that this random-variate generator is correct in the case $a = 1.0$, $b = 2.0$. (d) Comment.

7.2.7   (a) Given a circle of radius r centered at the origin of a conventional (x, y) coordinate system, construct an algorithm that will generate points *uniformly*, at random, interior to the circle. This algorithm cannot use more than two calls to Random per (x, y) point generated. (b) Explain clearly the theory behind the correctness of your algorithm. (Guessing this algorithm, even if correct, is not enough.)

7.2.8   (*a*) If $\Phi_a^{-1}(\cdot)$ is the *Normal*(0, 1) idf approximation as implemented in the library rvgs by the function Normal, and $\Phi(\cdot)$ is the *Normal*(0, 1) cdf as implemented in the library rvms by the function cdfNormal, use Monte Carlo simulation to *estimate* the smallest value of $\epsilon > 0$ such that $|\Phi(\Phi_a^{-1}(u)) - u| < \epsilon$ for all $u \in (0, 1)$. (Use the libraries rngs, rvgs, and rvms directly; do not cut-and-paste from rvgs.c and rvms.c to build the Monte Carlo simulation program.) (*b*) Comment on the value of this process as a cdf/idf consistency check.

7.2.9   (*a*) Give the *theoretical* largest and smallest standard normal random variates that can be generated by the Box–Muller method when it is using a random-number generator with $m = 2^{31} - 1$. (*b*) Compute the *actual* largest and smallest standard normal random variates that can be generated by the Box–Muller method when $a = 48\,271$, $m = 2^{31} - 1$, and the random numbers U_1 and U_2 are generated by consecutive calls to Random, using a single stream.

7.3 CONTINUOUS-RANDOM-VARIABLE APPLICATIONS

The purpose of this section is to demonstrate several continuous-random-variable applications, using the capabilities provided by the continuous-random-variate generators in the library rvgs and the pdf, cdf, and idf functions in the library rvms. We begin by developing an after-the-fact justification for some of the assumptions used in the discrete-event simulation models developed earlier in this book.

7.3.1 Arrival-Process Models

Recall that the usual convention is to model *interarrival* times as a sequence of independent positive random variables $R_1, R_2, R_3, \ldots$ and then construct the corresponding *arrival* times as a sequence of random variables $A_1, A_2, A_3, \ldots$ defined by

$$A_0 = 0$$

$$A_i = A_{i-1} + R_i \qquad i = 1, 2, \ldots.$$

By induction, for all i (with $A_0 = 0$),

$$A_i = R_1 + R_2 + \cdots + R_i \qquad i = 1, 2, \ldots.$$

The arrival times are ordered, so $0 = A_0 < A_1 < A_2 < A_3 < \cdots$, as illustrated:

Example 7.3.1

This approach is used in programs ssq2 and ssq3 to generate job arrivals with the additional assumption that the random variables $R_1, R_2, R_3, \ldots$ are *Exponential*$(1/\lambda)$.

In both programs, the arrival rate is equal to $\lambda = 0.5$ jobs per unit time. Similarly, this approach and an *Exponential* $(1/\lambda)$-inter-demand random-variable model is used in programs sis3 and sis4 to generate demand instances. The demand rate corresponds to an average of $\lambda = 30.0$ *actual* demands per time interval in program sis3 and to $\lambda = 120.0$ *potential* demands per time interval in program sis4.

Definition 7.3.1 If R_1, R_2, R_3, ... is an *iid* sequence of random positive interarrival times with common mean $1/\lambda > 0$, then the corresponding random sequence of arrival times A_1, A_2, A_3, ... is a *stationary arrival process* with *rate* λ.[*]

The "units" of the mean interarrival time $1/\lambda$ are time per arrival. Thus, the arrival rate λ has the reciprocal units arrivals per unit time. For example, if the average interarrival *time* is 0.1 minutes, then the arrival *rate* is $\lambda = 10.0$ arrivals per minute. Although it is traditional to specify stationary arrival processes in terms of λ, it is usually more convenient for discrete-event simulation software to use $1/\lambda$ for the purpose of generating the arrival process, as is done in programs ssq2, ssq3, sis3, and sis4.

If the arrival rate λ *varies* with time, then the arrival process is *nonstationary*. Truly stationary arrival processes are a convenient fiction.[†] Despite that, stationary arrival processes (*i*) are important theoretically, (*ii*) sometimes provide a satisfactory approximation over a short time interval, and (*iii*) must be understood before one attempts to study nonstationary arrival processes. Nonstationary arrival processes are considered in Section 7.5.

Stationary Poisson Arrival Process. Although the interarrival times R_1, R_2, R_3, ... can be any type of positive random variable, in the absence of information to the contrary, both theory and practice support the hypothesis that it is often appropriate to assume that the interarrival times are *Exponential* $(1/\lambda)$, as is done in programs ssq2, ssq3, sis3, and sis4.

Definition 7.3.2 If R_1, R_2, R_3, ... is an *iid* sequence of *Exponential* $(1/\lambda)$ interarrival times, then the corresponding sequence A_1, A_2, A_3, ... of arrival times is a stationary *Poisson* arrival process with rate λ. Equivalently, for $i = 1, 2, 3, ...$ the arrival time A_i is an *Erlang* $(i, 1/\lambda)$ random variable.

Algorithm 7.3.1 Given $\lambda > 0$ and $t > 0$, this algorithm generates a realization of a stationary Poisson arrival process with arrival rate λ over the time interval $(0, t)$:

```
a₀ = 0.0;                                    // a convention
n = 0;
```

---

[*]A stationary arrival process is also known as a *renewal* process, which models the sequence of *event* times $A_1, A_2, A_3, \ldots$. (In our case, events are arrivals.) A stationary arrival process is also called a *homogeneous* arrival process.

[†]Consider, for example, the rate at which jobs arrive to a fileserver or the rate at which customers arrive to a fast-food restaurant. In both cases, the arrival rate will vary dramatically during a typical 24-hour day of operation.

```
while (a_n < t) {
    a_{n+1} = a_n + Exponential(1 / λ);
    n++;
}
return a_1, a_2, ..., a_{n-1};
```

Random Arrivals.　We now turn to several fundamental theoretical results that demonstrate the interrelation between *Uniform, Exponential,* and *Poisson* random variables. These results will help explain why arrival processes are commonly assumed to be stationary Poisson processes. In the discussion that follows, (i) $t > 0$ defines a fixed time interval $(0, t)$, (ii) n represents the number of arrivals in the interval $(0, t)$, and (iii) $r > 0$ is the length of a small subinterval located *at random* interior to $(0, t)$.

Correspondingly, $\lambda = n/t$ is the arrival rate, $p = r/t$ is the probability that a particular arrival will be in the subinterval, and $np = nr/t = \lambda r$ is the expected number of arrivals in the subinterval. This notation and the discussion to follow are consistent with the stochastic experiment illustrated in Example 4.3.6.

Theorem 7.3.1　Let $A_1, A_2, \ldots, A_n$ be *iid Uniform*$(0, t)$ random variables, and let the discrete random variable X be the number of A_i that fall in a fixed subinterval of length $r = pt$ interior to $(0, t)$. Then X is a *Binomial*(n, p) random variable.*

Proof.　Each A_i is in the subinterval of length r with probability $p = r/t$. For each i, define

$$X_i = \begin{cases} 1 & \text{if } A_i \text{ is in the subinterval} \\ 0 & \text{otherwise.} \end{cases}$$

Because $X_1, X_2, \ldots, X_n$ is an *iid* sequence of *Bernoulli*(p) random variables and because the number of A_i that fall in the subinterval is $X = X_1 + X_2 + \cdots + X_n$, it follows that X is a *Binomial*(n, p) random variable.　　　　　　　　　　　　　　□

Random Arrivals Produce Poisson Counts.　Because $p = \lambda r/n$, X is a *Binomial*$(n, \lambda r/n)$ random variable. As discussed in Chapter 6, if n is large and $\lambda r/n = r/t$ is small, then X will be indistinguishable from a *Poisson*(λr) random variable. Therefore, the previous theorem can be restated as follows.

Theorem 7.3.2　Let $A_1, A_2, \ldots, A_n$ be *iid Uniform*$(0, t)$ random variables, and let the discrete random variable X be the number of A_i that fall in a fixed subinterval of length r interior to $(0, t)$. If n is large and r/t is small, then X is indistinguishable from a *Poisson*(λr) random variable with $\lambda = n/t$.

---

*As in Example 4.3.6, think of the A_i as *unsorted* arrival times.

Example 7.3.2

As in Example 4.2.2, suppose $n = 2000$ *Uniform*$(0, t)$ random variates are generated and are tallied into a continuous-data histogram with 1000 bins of size $r = t/1000$. If the resulting 1000 bin counts are then tallied into a discrete-data histogram, then, in this case, $\lambda r = (n/t)(t/1000) = 2$ and, consistent with Theorem 7.3.2, the resulting discrete-data-histogram relative frequencies will agree with the pdf of a *Poisson*(2) random variable.

To paraphrase Theorem 7.3.2, if many arrivals occur *at random* with a rate of λ, then the number of arrivals X that will occur in an interval of length r is a *Poisson*(λr) random variable. Therefore, the probability of x arrivals in any interval of length r is

$$\Pr(X = x) = \frac{(\lambda r)^x \exp(-\lambda r)}{x!} \qquad x = 0, 1, 2, \ldots.$$

In particular, $\Pr(X = 0) \doteq \exp(-\lambda r)$ is the probability of *no* arrivals in an interval of length r. Correspondingly, the probability of *at least one* arrival in the interval is

$$\Pr(X > 0) = 1 - \Pr(X = 0) = 1 - \exp(-\lambda r).$$

For a fixed arrival rate λ, the probability of at least one arrival increases with increasing interval length r.

Random Arrivals Produce Exponential Interarrivals. If the continuous random variable R represents the time between consecutive arrivals (the interarrival time), then the possible values of R are $r > 0$. What is the cdf of R?

Consider an arrival time A_i selected at random and a corresponding interval of length r, beginning at A_i, as illustrated below.

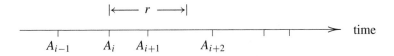

The random variable $R = A_{i+1} - A_i$ will be less than r, as illustrated, if and only if there is at least one arrival in this interval. Therefore, the cdf of R is

$$\Pr(R \le r) = \Pr(\text{at least one arrival in this interval}) = 1 - \exp(-\lambda r) \qquad r > 0.$$

From this cdf equation, we see that R is an *Exponential*$(1/\lambda)$ random variable. (See Definition 7.4.2.) This proves the following important theorem.

Theorem 7.3.3 If arrivals occur at random (i.e., as a stationary Poisson process) with rate λ, then the corresponding interarrival times form an *iid* sequence of *Exponential*$(1/\lambda)$ random variables.*

---

*Theorem 7.3.3 is the justification for Definition 7.3.2.

Consistent with the previous discussion, if all that is known about an arrival process is that arrivals occur at random with a constant rate λ, then the function `GetArrival` in programs `ssq2` and `ssq3` is appropriate. Similarly, if all that is known about a demand process is that demand instances occur at random with a constant rate λ, then the function `GetDemand` in programs `sis3` and `sis4` is appropriate.

Are there occasions for which a stationary Poisson arrival process is an appropriate model? Over short time intervals, such a model could provide a good approximation. Also, a result given in Fishman (2001, page 464) states that the superposition of many general stationary arrival processes (each not necessarily Poisson) is a stationary Poisson arrival process.

The following example is based on the observation that, if arrivals occur at random with rate $\lambda = 1$ (which is often referred to as a *unit* Poisson process), then the number of arrivals X in an interval of length μ will be a $Poisson(\mu)$ random variate. (The case $X = 4$ is illustrated below.)

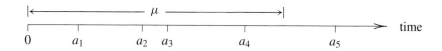

Example 7.3.3
An algorithm that generates a $Poisson(\mu)$ random variate is

```
a = 0.0;
x = 0;
while (a < μ) {
    a += Exponential(1.0);
    x++;
}
return x - 1;
```

Example 7.3.3 provides an explanation for the choice of the $Poisson(\mu)$ random-variate generator presented in Example 6.2.14. The algorithm can be made a bit more efficient by replacing the calls to `Exponential(1.0)` with corresponding calls to `Random`. The details are left as an exercise.

Summary.   Given a fixed time interval $(0, t)$, there are two ways of generating a realization of a stationary Poisson arrival process with rate λ over this time interval.

- Generate the number of arrivals as a $Poisson(\lambda t)$ random-variate n. Then generate a $Uniform(0, t)$ random variate sample of size n, and *sort* this sample to form the arrival process $0 < a_1 < a_2 < a_3 < \cdots < a_n$, as in Example 4.3.6.
- Use Algorithm 7.3.1 to generate $Exponential(1/\lambda t)$ interarrival times.

Although these two approaches are statistically equivalent, they are certainly *not* computationally equivalent. Because of the need to temporarily store the entire n-point sample and then sort it, there is no computational justification for the use of the first approach, particularly if n is large. Instead, the second approach, which is based on Theorem 7.3.3, is always preferred. Indeed, from a computational perspective, the second approach represents a triumph of theory over brute-force computing.

It is important to remember that the *mode* of the exponential distribution is 0, which means that a stationary Poisson arrival process will exhibit a clustering of arrivals. The top axis below shows the clustering associated with the arrival times of a stationary Poisson arrival process with $\lambda = 1$; the bottom axis shows a stationary arrival process with *Erlang* $(4, 1/4)$ interarrival times. The mean time between arrivals for both processes is 1. The top axis has many short interarrival times, which results in clusters of arrivals. The arrival pattern associated with a stationary Poisson arrival process is often referred to as "random" arrivals.

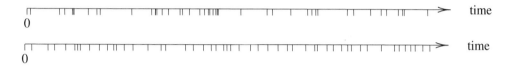

The stationary Poisson arrival process generalizes to (*i*) a stationary arrival process when the exponential time between arrivals is replaced by any continuous random variable with positive support, and (*ii*) a nonstationary Poisson process when the arrival rate λ varies with time.

This concludes our discussion of *arrival* processes. We now turn to a discussion of *service* processes.

7.3.2 Service-Process Models

For a service-node-simulation model, a default service-process model is not well defined; there are only application-dependent guidelines.

- *Uniform* (a, b) service times are rarely ever an appropriate model. Service times seldom "cut off" at minimum and maximum values a and b. Service-time models typically have "tails" on their pdfs.
- Because service times are inherently positive, they can *not* be *Normal* (μ, σ) unless this random variable is truncated to positive values only.
- Probability models such as the *Lognormal* (a, b) distribution generate inherently positive random variates, making them candidates for modeling service times.
- When service times consist of the sum of n *iid* *Exponential* (b) subtask times, the *Erlang* (n, b) model is appropriate.

Program ssq4. Program ssq4 is based on program ssq3, but with a more realistic, *Erlang* $(5, 0.3)$ service-time model. The corresponding service rate is 2/3. As in program ssq3, program ssq4 uses *Exponential* (2) random-variate interarrivals. The corresponding arrival rate is 1/2.

Example 7.3.4

The arrival rate is $\lambda = 0.5$ for both program `ssq3` and program `ssq4`. For both programs, the service rate is $\nu = 2/3 \cong 0.667$. As illustrated in Figure 7.3.1, however, the distribution of service times for the two programs is very different.

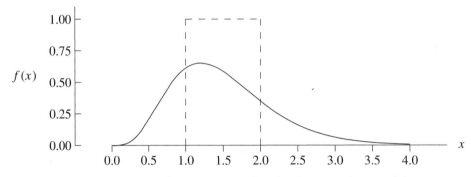

Figure 7.3.1   *Erlang*(5, 0.3) and *Uniform*(1, 2) service-time models.

The solid line is the *Erlang* (5, 0.3) service time pdf in `ssq4`; the dashed line represents the *Uniform* (1, 2) pdf in `ssq3`. Both have mean 1.5. Because of this difference in the service-time distributions, one would expect a difference in some (all?) of the steady-state service-node statistics. As an exercise, you are asked to *first* conjecture how the steady-state statistics will differ for these two programs and to then investigate the correctness of your conjecture via discrete-event simulation.*

Erlang Service Times.   The *Erlang* (n, b) service-time model is appropriate when service processes can be naturally decomposed into a series (tandem) of independent "subprocesses" as illustrated in Figure 7.3.2.

    If there are n service subprocesses (the case $n = 3$ is illustrated) then the total service time will be the sum of the service times for the subprocesses. If, in addition, the time to perform each service subprocess is independent of the other times, then a random-variate service time can be generated by generating a service time for each subprocess and then summing the subprocess-service-time variates. In the particular case that each of the n service subprocesses is *Exponential* (b), the total service time will be *Erlang* (n, b), and the service rate will be $1/nb$.

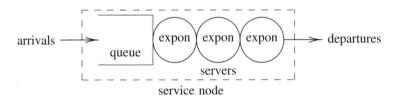

Figure 7.3.2   Three-subprocess service time.

---

*Which statistics will change, and in what way?

Truncation. Let X be a continuous random variable with the set of possible values $\mathcal{X}$ and the cdf $F(x) = \Pr(X \leq x)$. Consider an interval $(a, b) \subset \mathcal{X}$, and suppose that we wish to restrict (truncate) the possible values of X to this interval. Truncation in this continuous-variable context is similar to, but simpler than, truncation in the discrete-variable context. (See Section 6.3.) By definition, the probability that X is less than or equal to a is

$$\Pr(X \leq a) = F(a).$$

Similarly, the probability that X is greater than or equal to b is

$$\Pr(X \geq b) = 1 - \Pr(X < b) = 1 - F(b).$$

In general, then,
$$\Pr(a < X < b) = \Pr(X < b) - \Pr(X \leq a) = F(b) - F(a).$$

Relative to truncation, there are two cases to consider.

- If the left-truncation value a and the right-truncation value b are specified, then the cdf of X can be used to calculate the left-tail and right-tail truncation probabilities as

    $$\alpha = \Pr(X \leq a) = F(a) \qquad \text{and} \qquad \beta = \Pr(X \geq b) = 1 - F(b),$$

    respectively.
- If instead α and β are specified, then the idf of X can be used to calculate the left-tail and right-tail possible truncation values as

    $$a = F^{-1}(\alpha) \qquad \text{and} \qquad b = F^{-1}(1 - \beta),$$

    respectively.

In either case, for a continuous random variable, the cdf transformation from possible values to probabilities or the idf transformation from probabilities to possible values is exact.

Example 7.3.5
As a continuation of Example 7.3.4, suppose we want to use a *Normal*(1.5, 2.0) random variable to model service times. Because service times cannot be negative, it is necessary to truncate the left tail of this random variable to nonnegative service times. Moreover, to prevent extremely large service times, suppose we choose to truncate the right tail to values less than 4.0. Therefore $a = 0.0$, $b = 4.0$, and the cdf capability in the library rvms can be used to calculate the corresponding truncation probabilities α and β as

```
α = cdfNormal(1.5, 2.0, a);                    // a is 0.0
β = 1.0 − cdfNormal(1.5, 2.0, b);              // b is 4.0
```

The result is $\alpha = 0.2266$ and $\beta = 0.1056$. Note that the resulting *truncated Normal* (1.5, 1.0) random variable has mean 1.85 (not 1.5) and standard deviation 1.07 (not 2.0). The mean increases, since more probability was lopped off the left-hand tail than off the right-hand tail. The standard deviation, on the other hand, decreases, since the most extreme values have been eliminated.

Constrained Inversion.   Once the left-tail and right-tail truncation probabilities α and β have been identified, either by specification or by computation, then the corresponding truncated random variate can be generated by constrained inversion as

```
u = Uniform(α, 1.0 - β);
return F⁻¹(u);
```

Example 7.3.6
The idf capability in the library `rvms` can be used to generate the truncated *Normal* (1.5, 2.0) random variate in Example 7.3.5 as

```
α = 0.2266274;
β = 0.1056498;
u = Uniform(α, 1.0 - β);
return idfNormal(1.5, 2.0, u);
```

The geometry for the generation of a $Normal(1.5, 2.0)$ random variate truncated on the left at $a = 0.0$ and on the right at $b = 4.0$ is shown in Figure 7.3.3. The random number in the figure is $u = 0.7090$, which corresponds to the truncated-normal variate $x = 2.601$.

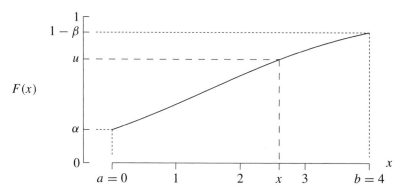

Figure 7.3.3 Truncated-normal constrained inversion.

Triangular Random Variable.   Although it is not one of the standard continuous random-variate models summarized in the next section, the three-parameter *Triangular*(a, b, c) model is commonly used in those situations where all that is known is the range of possible values along with the most likely possible value (the mode). In this case, it is reasonable to assume that the pdf of the random variable has the shape illustrated in Figure 7.3.4.*

Formally, the continuous random variable X is *Triangular*(a, b, c) if and only if the real-valued parameters a, b, and c satisfy $a < c < b$, the possible values of X are $\mathcal{X} = \{x \mid a < x < b\}$, and the pdf of X is

---

*See Exercise 7.3.9 for the special cases $c = a$ or $c = b$.

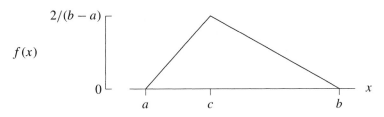

Figure 7.3.4   Triangular pdf.

$$
f(x) = \begin{cases} \dfrac{2(x-a)}{(b-a)(c-a)} & a < x \leq c \\[2mm] \dfrac{2(b-x)}{(b-a)(b-c)} & c < x < b. \end{cases}
$$

When modeling service times with a known finite range of possible values, the *Triangular* (a, b, c) random variable should be considered as a viable alternative to truncating one of the more traditional service-time models—for example, an *Erlang* (n, b) or *Lognormal* (a, b) random-variable model. Another scenario where the triangular distribution is appropriate is the case in which no data is available. One of the authors of this text was modeling the construction of the Space Station and, of course, no data existed for the various construction operations. The construction times were modeled by interviewing an expert at NASA who was able to make informed guesses about the shortest construction time a, the longest construction time b, and the most likely construction time c.

It follows from the definition of a *Triangular* (a, b, c) random variable that the mean and standard deviation are

$$
\mu = \frac{a+b+c}{3} \qquad \text{and} \qquad \sigma = \frac{\sqrt{(a-b)^2 + (a-c)^2 + (b-c)^2}}{6}.
$$

Moreover, integrating the pdf yields the cdf

$$
F(x) = \begin{cases} \dfrac{(x-a)^2}{(b-a)(c-a)} & a < x \leq c \\[2mm] 1 - \dfrac{(b-x)^2}{(b-a)(b-c)} & c < x < b, \end{cases}
$$

and solving the equation $u = F(x)$ for x yields the idf as

$$
F^{-1}(u) = \begin{cases} a + \sqrt{(b-a)(c-a)u} & 0 < u \leq \dfrac{c-a}{b-a} \\[2mm] b - \sqrt{(b-a)(b-c)(1-u)} & \dfrac{c-a}{b-a} < u < 1. \end{cases}
$$

All of these derivations are left as an exercise.

7.3.3 Exercises

7.3.1   Modify the function `GetService` in program `ssq4` so that service times greater than 3.0 are not possible. Do this in two ways: (*a*) using acceptance–rejection; (*b*) using truncation by constrained inversion. (*c*) Comment.

7.3.2   (*a*) Investigate the issue concerning steady-state statistics discussed in Example 7.3.4. Specifically, compare the steady-state statistics produced by programs `ssq3` and `ssq4`. (*b*) Which of these statistics are different, and why?

7.3.3   Repeat Exercises 7.3.1 and 7.3.2, but using instead the truncated-random-variable model from Example 7.3.5.

7.3.4   (*a*) Prove that, if Y is a *Normal*(μ, σ) random variable X truncated to the interval (a, b), then

$$E[Y] = \mu - \left(\frac{f(b) - f(a)}{F(b) - F(a)} \right) \sigma^2,$$

where $f(\cdot)$ and $F(\cdot)$ are the pdf and cdf of X (not Y). (*b*) How could you check that this result is correct?

7.3.5   (*a*) Derive the equations for the mean and standard deviation of a *Triangular*(a, b, c) random variable. (*b*) Similarly, derive the equations for the cdf and idf.

7.3.6   (*a*) Use Monte Carlo simulation to demonstrate that the following algorithm is computationally more efficient than the algorithm in Example 7.3.3.

```
t = exp(μ);
x = 0;
while (t > 1.0) {
    t *= 1.0 - Random();
    x++;
}
return x - 1;
```

(*b*) Demonstrate also that the two algorithms produce identical random variates for identical streams of random numbers. (*c*) There is, however, a computational problem with the more efficient algorithm if μ is large. What is that problem?

7.3.7   (*a*) Starting from the cdf equation for a *Triangular*(a, b, c) random variable, derive the corresponding idf equation. (*b*) Implement an algorithm that will generate a *Triangular*(a, b, c) random variate with just one call to `Random`. (*c*) In the specific case $(a, b, c) = (1.0, 7.0, 3.0)$, use Monte Carlo simulation to generate a continuous-data histogram that provides convincing evidence that your random-variate generator is correct.

7.3.8   (*a*) There is an alternative to the use of a truncated *Normal*$(1.5, 2.0)$ random variable in Example 7.3.5. If a *Normal*$(\mu, 2.0)$ random variable is used instead as the basis for truncation, use the capabilities in the library `rvms` to determine what value of the parameter μ is required to yield the service rate $\nu = 2/3$. (*b*) Discuss your approach to solving for μ. (*c*) This choice of μ forces the truncated random variable to have the mean 1.5. What are the corresponding values of α and β? (*d*) What is the standard deviation of the resulting truncated random variable? *Hint*: See Exercise 7.3.4.

7.3.9   (*a*) Relative to the definition of a *Triangular*(a, b, c) random variable, verify that the parameter constraint $a < c < b$ can be relaxed to $a \leq c \leq b$, if done correctly. (*b*) Specifically, what are the pdf, cdf, and idf equations if $a = c < b$? (*c*) Same question if $a < b = c$.

7.4 CONTINUOUS-RANDOM-VARIABLE MODELS

For convenience, properties of the following eight continuous-random-variable models are summarized in this section: *Uniform*(a, b), *Exponential*(μ), *Erlang*(n, b), *Normal*$(0, 1)$ *Normal*(μ, σ), *Lognormal*(a, b), *Chisquare*(n), and *Student*(n). For more details about these models, see Sections 7.1 and 7.2. For supporting software, see the random-variable-models library `rvms` in Appendix D and the random-variate-generators library `rvgs` in Appendix E.

7.4.1 Uniform

Definition 7.4.1 A continuous random variable X is *Uniform*(a, b) if and only if

- the real-valued *location* (see Section 7.4.9) parameters a, b satisfy $a < b$,
- the possible values of X are $\mathcal{X} = \{x \mid a < x < b\}$,
- the pdf of X is

$$f(x) = \frac{1}{b - a} \qquad a < x < b,$$

as illustrated in Figure 7.4.1,
- the cdf of X is

$$F(x) = \frac{x - a}{b - a} \qquad a < x < b,$$

- the idf is

$$F^{-1}(u) = a + (b - a)u \qquad 0 < u < 1,$$

- the mean of X is

$$\mu = \frac{a + b}{2},$$

- the standard deviation of X is

$$\sigma = \frac{b - a}{\sqrt{12}}.$$

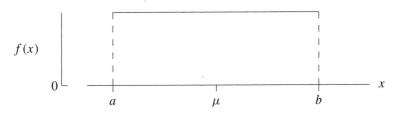

Figure 7.4.1 *Uniform*(a, b) pdf.

A *Uniform*(a, b) random variable is used to model situations in which a continuous random variable is restricted to a subrange (a, b) of the real line and all values in this subrange are equally likely. Thus a *Uniform*(a, b) random variable is the continuous analog of the discrete *Equilikely*(a, b) random variable. A typical *Uniform*(a, b) application will involve a model derived from a statement like "... the service time is chosen *at random* from the time interval (a, b)." The special case $a = 0$ and $b = 1$ is particularly important, because virtually all random-number generators are designed to generate a sequence of *Uniform*$(0, 1)$ random variates.

7.4.2 Exponential

Definition 7.4.2   A continuous random variable X is *Exponential*(μ) if and only if

- the real-valued *scale* parameter μ satisfies $\mu > 0$,
- the possible values of X are $\mathcal{X} = \{x \mid x > 0\}$,
- the pdf of X is

$$f(x) = \frac{1}{\mu} \exp(-x/\mu) \qquad x > 0,$$

as illustrated in Figure 7.4.2,
- the cdf of X is

$$F(x) = 1 - \exp(-x/\mu) \qquad x > 0,$$

- the idf is

$$F^{-1}(u) = -\mu \ln(1 - u) \qquad 0 < u < 1,$$

- the mean of X is μ,
- the standard deviation of X is $\sigma = \mu$.

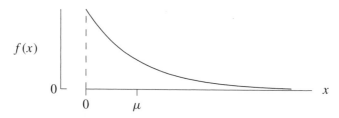

Figure 7.4.2   *Exponential*(μ) pdf.

Exponential random variables are frequently used to model the interarrival distribution associated with arrival processes. *Exponential* random variables are also used to model the service-time distribution for individual service nodes in a network of queues and for component lifetimes in system-reliability analysis. If X is *Exponential* (μ), then X has the important *memoryless* property: for any $x' > 0$,

$$\Pr\left(X \geq x + x' \mid X \geq x'\right) = \Pr(X \geq x) = \exp(-x/\mu) \qquad x > 0,$$

independent of x'. If a light bulb, for example, has an exponential lifetime, then a used bulb that has been has successfully functioning for x' hours has a remaining-lifetime pdf identical to that of a brand-new bulb. Graphically, any right-hand tail of an exponential pdf that is rescaled so that it integrates to one is identical (except for location) to the original pdf. An *Exponential* (μ) random variable is the continuous analog of the discrete *Geometric* (p) random variable. Just as the normal distribution plays an important role in classical statistics, because of the central limit theorem, the exponential distribution plays an important role in stochastic processes (particularly reliability and queuing theory), since it is the only continuous distribution with the memoryless property. (The only discrete distribution with the memoryless property is the geometric distribution.)

7.4.3 Erlang

Definition 7.4.3   A continuous random variable X is *Erlang* (n, b) if and only if

- the *shape* parameter n is a positive integer,
- the real-valued *scale* parameter b satisfies $b > 0$,
- the possible values of X are $\mathcal{X} = \{x \mid x > 0\}$,
- the pdf of X is

$$f(x) = \frac{1}{b\,(n-1)!}(x/b)^{n-1}\exp(-x/b) \qquad x > 0,$$

  as illustrated in Figure 7.4.3 for $n = 3$ and $b = 1$,
- the cdf of X is the incomplete gamma function (see Appendix D)

$$F(x) = P(n, x/b) \qquad x > 0,$$

- except for special cases, the idf of X must be found by numerical inversion,
- the mean of X is

$$\mu = nb,$$

- the standard deviation of X is

$$\sigma = \sqrt{n}\,b.$$

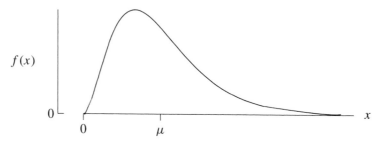

Figure 7.4.3 *Erlang(n, b)* pdf.

Erlang random variables are frequently used to model random service times, particularly when the service is defined by a series of independent subprocesses. That is consistent with the fact that a continuous random variable X is *Erlang* (n, b) if and only if

$$X = X_1 + X_2 + \cdots + X_n,$$

where $X_1, X_2, \ldots, X_n$ is an *iid* sequence of *Exponential* (b) random variables. An *Erlang* $(1, b)$ random variable and an *Exponential* (b) random variable are equivalent. An *Erlang* (n, b) random variable is the continuous analog of the discrete *Pascal* (n, p) random variable.

7.4.4 Standard Normal

Definition 7.4.4   A continuous random variable Z is *Normal* $(0, 1)$ (*standard normal*) if and only if

* the possible values of Z are $\mathcal{Z} = \{z \mid -\infty < z < \infty\}$,
* the pdf of Z is

$$f(z) = \phi(x) = \frac{1}{\sqrt{2\pi}} \exp\left(-z^2/2\right) \qquad -\infty < z < \infty,$$

  as illustrated in Figure 7.4.4,
* the cdf of Z is the following special function $\Phi(\cdot)$:

$$F(z) = \Phi(z) = \frac{1}{\sqrt{2\pi}} \int_{-\infty}^{z} \exp(-t^2/2)\, dt \qquad -\infty < z < \infty,$$

* the idf of Z is

$$F^{-1}(u) = \Phi^{-1}(u) \qquad 0 < u < 1,$$

  where $\Phi^{-1}(\cdot)$ is defined by an algorithm (see Section 7.2) that computes the numerical inverse of the equation $u = \Phi(z)$,
* the mean of Z is $\mu = 0$,
* the standard deviation of Z is $\sigma = 1$.

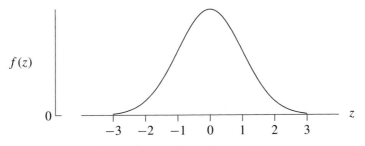

Figure 7.4.4   *Normal*(0, 1) pdf.

A *Normal*(0, 1) random variable is a special case of a *Normal*(μ, σ) random variable. The letter Z for a *Normal*(0, 1) random variable, and the symbols ϕ and Φ for its pdf and cdf, are used because of the centrality of the normal distribution in probability and statistics. The pdf has inflection points at ± 1. For both the *Normal*(0, 1) and *Normal*(μ, σ) models, the areas under the pdf within 1, 2, and 3 standard deviations from the mean are approximately 0.68, 0.95, and 0.997, respectively.

7.4.5 Normal

Definition 7.4.5   A continuous random variable X is *Normal*(μ, σ) if and only if

- the real-valued *location* parameter μ can have any value,
- the real-valued *scale* parameter σ satisfies $\sigma > 0$,
- the possible values of X are $\mathcal{X} = \{x \mid -\infty < x < \infty\}$,
- the pdf of X is

$$f(x) = \frac{1}{\sigma\sqrt{2\pi}} \exp\left(-(x - \mu)^2/2\sigma^2\right) \qquad -\infty < x < \infty,$$

  as illustrated in Figure 7.4.5,
- the cdf of X is defined by the *Normal*(0, 1) cdf as

$$F(x) = \Phi\left(\frac{x - \mu}{\sigma}\right) \qquad -\infty < x < \infty,$$

- the idf of X is defined by the *Normal*(0, 1) idf as

$$F^{-1}(u) = \mu + \sigma \, \Phi^{-1}(u) \qquad 0 < u < 1,$$

- the mean of X is μ,
- the standard deviation of X is σ.

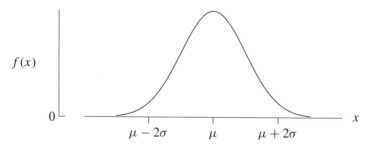

Figure 7.4.5   *Normal*(μ, σ) pdf.

The random variable X is *Normal*(μ, σ) if and only if

$$X = \mu + \sigma Z$$

where the random variable Z is *Normal*$(0, 1)$. Elementary statistics textbooks usually state this result as: If X is *Normal*(μ, σ) then $Z = (X - \mu)/\sigma$ is *Normal*$(0, 1)$. The pdf of the *Normal*(μ, σ) has inflection points at $\mu \pm \sigma$. The primary importance of a *Normal*(μ, σ) random variable in simulation is its role as the *asymptotic* form of the sum of other random variables—that is, if $X_1, X_2, \ldots, X_n$ is an *iid* sequence of random variables, each having mean μ and standard deviation σ, and if

$$X = X_1 + X_2 + \cdots + X_n,$$

then the random variable X has mean $n\mu$ and standard deviation $\sqrt{n}\,\sigma$. In addition, X is asymptotically *Normal*$(n\mu, \sqrt{n}\,\sigma)$ as $n \to \infty$. This result is one form of the central limit theorem.

7.4.6 Lognormal

Definition 7.4.6   A continuous random variable X is *Lognormal*(a, b) if and only if

- the real-valued parameter a can have any value [exp(a), the median, is a scale parameter],
- the real-valued *shape* parameter b satisfies $b > 0$,
- the possible values of X are $\mathcal{X} = \{x \mid x > 0\}$,
- the pdf of X is

$$f(x) = \frac{1}{bx\sqrt{2\pi}} \exp\left(-(\ln(x) - a)^2 / 2b^2\right) \qquad x > 0,$$

as illustrated in Figure 7.4.6 for $(a, b) = (-0.5, 1)$,
- the cdf of X is defined by the *Normal*$(0, 1)$ cdf as

$$F(x) = \Phi\left(\frac{\ln(x) - a}{b}\right) \qquad x > 0,$$

- the idf of X is defined by the $Normal(0, 1)$ idf as

$$F^{-1}(u) = \exp\left(a + b\,\Phi^{-1}(u)\right) \qquad\qquad 0 < u < 1,$$

- the mean of X is

$$\mu = \exp\left(a + b^2/2\right),$$

- the standard deviation of X is

$$\sigma = \exp\left(a + b^2/2\right)\sqrt{\exp(b^2) - 1}.$$

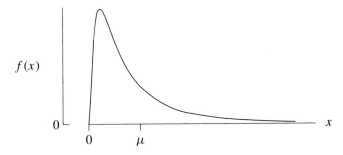

Figure 7.4.6 *Lognormal(a, b)* pdf.

The random variable X is $Lognormal(a, b)$ if and only if

$$X = \exp(a + bZ),$$

where the random variable Z is $Normal(0, 1)$. The *Lognormal* distribution is sometimes used as an alternate to the *Erlang* distribution as a model of, for example, a service time.

7.4.7 Chisquare

Definition 7.4.7 A continuous random variable X is *Chisquare*(n) if and only if

- the *shape* parameter n is a positive integer, known as the "degrees of freedom,"
- the possible values of X are $\mathcal{X} = \{x \mid x > 0\}$,
- the pdf of X is

$$f(x) = \frac{1}{2\Gamma(n/2)}(x/2)^{n/2-1}\exp(-x/2) \qquad\qquad x > 0,$$

as illustrated in Figure 7.4.7 for $n = 5$,

- the cdf of X is the incomplete gamma function (see Appendix D)

$$F(x) = P(n/2, x/2) \qquad x > 0,$$

- except for special cases, the idf of X must be found by numerical inversion,
- the mean of X is

$$\mu = n,$$

- the standard deviation of X is

$$\sigma = \sqrt{2n}.$$

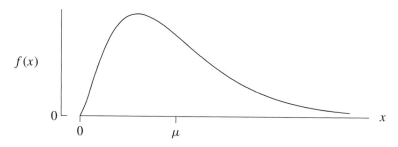

$f(x)$

Figure 7.4.7 *Chisquare(n) pdf.*

The random variable X is *Chisquare*(n) if and only if

$$X = Z_1^2 + Z_2^2 + \cdots + Z_n^2$$

where $Z_1, Z_2, \ldots, Z_n$ is an *iid* sequence of *Normal*$(0, 1)$ random variables. The *Chisquare* random variable is commonly used in statistical "goodness-of-fit" tests. (See Section 9.2.) The discrete nature of the parameter n precludes the use of this distribution for modeling stochastic elements of a system in most applications. If n is an even number, then a *Chisquare*(n) random variable and a *Erlang*$(n/2, 2)$ random variable are equivalent.

7.4.8 Student

Definition 7.4.8   A continuous random variable X is *Student*(n) if and only if

- the shape parameter n is a positive integer, known as the "degrees of freedom,"
- the possible values of X are $\mathcal{X} = \{x \mid -\infty < x < \infty\}$,
- the pdf of X is

$$f(x) = \frac{1}{\sqrt{n}\, B(1/2,\, n/2)} (1 + x^2/n)^{-(n+1)/2} \qquad -\infty < x < \infty,$$

as illustrated in Figure 7.4.8 for $n = 10$, where $B(\cdot, \cdot)$ is the incomplete beta function defined in Appendix D,

- if $x \geq 0$, then the cdf of X is a function of the incomplete beta function

$$F(x) = \frac{1 + I\left(1/2,\, n/2,\, n/(n + x^2)\right)}{2} \qquad x \geq 0,$$

and, if $x < 0$, then $F(x) = 1 - F(-x)$,
- the idf of X must be found by numerical inversion,
- the mean of X is $\mu = 0$ for $n = 2, 3, 4 \ldots$,
- the standard deviation of X is

$$\sigma = \sqrt{\frac{n}{n - 2}} \qquad n = 3, 4, 5 \ldots.$$

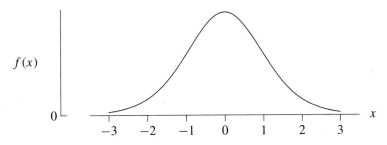

Figure 7.4.8 *Student(n)* pdf.

A continuous random variable X is *Student*(n) if and only if

$$X = \frac{Z}{\sqrt{V/n}},$$

where Z is a *Normal*$(0, 1)$ random variable, V is a *Chisquare*(n) random variable, and Z, V are independent. The primary use of a *Student*(n) random variable in simulation is in interval estimation. The pdf is symmetric and bell-shaped, with slightly heavier tails than the pdf of a standard normal random variable. The limiting distribution (Hogg, McKean, and Craig, 2005, Chapter 4) of a *Student*(n) distribution as $n \to \infty$ is a *Normal*$(0, 1)$ distribution.

7.4.9 Summary

Effective discrete-event-simulation modeling requires that the modeler be familiar with several continuous parametric distributions that can be used to mimic the stochastic elements of the model. In order to choose the proper distribution, it is important to know

- how these distributions arise and how they are related to one another;
- the support, $\mathcal{X}$;
- the mean, μ;
- the variance, σ^2; and
- the shape of the pdf.

If the situation calls for a continuous model (e.g., interarrival times), a modeler can quickly examine the shape of the histogram associated with a data set to rule in and rule out certain distributions on the basis of the shapes of their pdfs.

Parameters.   Parameters in a distribution allow modeling of such diverse applications as machine-failure time, patient postsurgery survival time, and customer interarrival time, by a single distribution [for example, the *Lognormal* (a, b) distribution]. In our brief overview of each of the eight continuous distributions presented in this chapter, we have used the adjectives *scale*, *shape*, and *location* to classify parameters. We now consider the meaning behind each of these terms.

Location parameters are used to shift the distribution to the left or right along the x-axis. If c_1 and c_2 are two values of a location parameter for a distribution with cdf $F(x; c)$, then there exists a real constant α such that $F(x; c_1) = F(\alpha + x; c_2)$. A familiar example of a location parameter is the mean μ of the *Normal* (μ, σ) distribution. This parameter simply translates the bell-shaped curve to the left and right for various values of μ. Location parameters are also known as "shift" parameters.

Scale parameters are used to expand or contract the x-axis by a factor of α. If λ_1 and λ_2 are two values for a scale parameter for a distribution with cdf $F(x; \lambda)$, then there exists a real constant α such that $F(\alpha x; \lambda_1) = F(x; \lambda_2)$. A familiar example of a scale parameter is μ in the *Exponential* (μ) distribution. The pdf always has the same shape, and the units on the x-axis are determined by the value of μ.

Shape parameters are appropriately named, since they affect the shape of the pdf. A familiar example of a shape parameter is n in the *Erlang* (n, b) distribution. The pdf assumes different shapes for various values of n.

In summary, location parameters *translate* distributions along the x-axis; scale parameters *expand* or *contract* the scale for distributions; and all other parameters are shape parameters.

Relationships Among Distributions.   It is also important to know how these distributions relate to one another. Figure 7.4.9 highlights some relationships among the eight distributions summarized in this section, plus the *Uniform* $(0, 1)$ distribution. Listed in each oval are the name, parameter(s), and support of a distribution. Each distribution, including the standard normal distribution, is designated by the generic random variable X. The solid arrows connecting the ovals denote special cases [e.g., the *Normal* $(0, 1)$ distribution is a special case of the *Normal* (μ, σ) distribution when $\mu = 0$ and $\sigma = 1$] and transformations [e.g., the sum (convolution) of n independent and identically distributed *Exponential* (μ) random variables has an *Erlang* (n, μ) distribution]. The dashed line between the *Student* (n) and *Normal* $(0, 1)$ distributions, for example, indicates that the limiting distribution of a *Student* (n) random variable as $n \to \infty$ has a standard normal distribution.

There are internal characteristics associated with these distributions that are not shown in Figure 7.4.9. One such characteristic is that the sum of independent normal random variables also

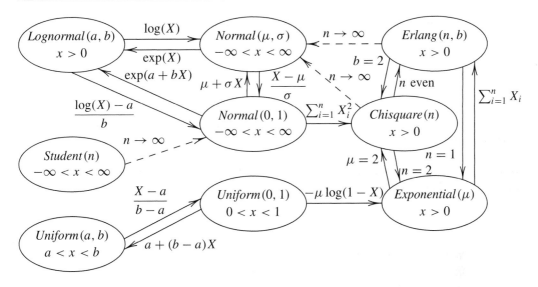

Figure 7.4.9   Relationships among continuous distributions.

has a normal distribution. Another is that the sum of independent chi-square random variables also has chi-square distribution.

There could be an arrow drawn from the *Uniform*(0, 1) distribution to every other distribution, in theory, because of the probability integral transformation (Theorem 7.2.1). Only the transformations to the *Uniform*(a, b) and *Exponential*(μ) distributions are included in the figure, however, as these are the only distributions with monotone, synchronized variate-generation algorithms that do not require numerical methods.

The alert reader will notice that the sequence of arrows running from the *Uniform*(0, 1) distribution to the *Exponential*(2) distribution to the *Chisquare*(2) distribution to the *Normal*(0, 1) distribution in Figure 7.4.9 is related to the Box–Muller algorithm for generating normal variates given in Section 7.2.

Moments.   As in the case of discrete random variables, the table below summarizes the first four moments of the distribution of the continuous random variables surveyed in this chapter. earlier. To save space, the *Lognormal*(a, b) distribution uses $\omega = \exp(b^2)$. As before the mean, variance, skewness, and kurtosis are defined as

$$\mu = E[X], \quad \sigma^2 = E\left[(X - \mu)^2\right], \quad E\left[\left(\frac{X - \mu}{\sigma}\right)^3\right], \quad \text{and} \quad E\left[\left(\frac{X - \mu}{\sigma}\right)^4\right].$$

The ith moment for the *Student*(n) distribution is defined only for $n > i$, for $i = 1, 2, 3, 4$. As with discrete distributions, the skewness is a measure of the symmetry of a distribution. The *Exponential*(μ) distribution, for example, has a positive skewness for all values of its parameter μ. The kurtosis is a measure of the peakedness and tail behavior of a distribution. The *Student*(n) distribution, for example, with its heavier tails than a standard normal distribution, has a kurtosis that converges to 3 from above as $n \to \infty$.

Distribution	Mean	Variance	Skewness	Kurtosis
$Uniform(a, b)$	$\dfrac{a+b}{2}$	$\dfrac{(b-a)^2}{12}$	0	$\dfrac{9}{5}$
$Exponential(\mu)$	μ	μ^2	2	9
$Erlang(n, b)$	nb	nb^2	$\dfrac{2}{\sqrt{n}}$	$3+\dfrac{6}{n}$
$Normal(0, 1)$	0	1	0	3
$Normal(\mu, \sigma)$	μ	σ^2	0	3
$Lognormal(a, b)$	$e^{a+b^2/2}$	$e^{2a}\omega(\omega-1)$	$(\omega+2)\sqrt{\omega-1}$	$\omega^4+2\omega^3+3\omega^2-3$
$Chisquare(n)$	n	$2n$	$\dfrac{2^{3/2}}{\sqrt{n}}$	$3+\dfrac{12}{n}$
$Student(n)$	0	$\dfrac{n}{n-2}$	0	$\dfrac{3(n-2)}{n-4}$

Although discussion here is limited to just the continuous distributions introduced in this chapter, there are many other parametric distributions capable of modeling continuous random variables. For a complete list of common discrete and continuous parametric distributions, including pdf, cdf, idf, and moments, we recommend the compact work of Evans, Hastings, and Peacock (2000) or the encyclopedic works of Johnson, Kotz, and Kemp (1993) and Johnson, Kotz, and Balakrishnan (1994, 1995, 1997).

7.4.10 Pdf, Cdf, and Idf Evaluation

Pdfs, cdfs, and idfs for all eight of the continuous-random-variable models in this section can be evaluated by using the functions in the library `rvms`. (See Appendix D.) For example, if X is $Erlang(n, b)$, then, for any $x > 0$,

```
pdf = pdfErlang(n, b, x);                              // f(x)
cdf = cdfErlang(n, b, x);,                             // F(x)
```

and, for any $0.0 < u < 1.0$,

```
idf = idfErlang(n, b, u);                              // F⁻¹(u)
```

This library also has functions to evaluate pdfs, cdfs, and idfs for all the discrete random variables in Chapter 6.

7.4.11 Exercises

7.4.1   Prove that the cdf of an $Erlang(n, b)$ random variable is

$$F(x) = P(n, x/b),$$

where $P(\cdot, \cdot)$ is the incomplete gamma function.

7.4.2   Prove that, if $n > 0$ is an even integer, then a *Chisquare*(n) random variable and a *Erlang*$(n/2, 2)$ random variable are equivalent.

7.4.3   Prove that an *Exponential*(μ) random variable has the "memoryless" property.

7.4.4   From the *Normal*$(0, 1)$ cdf definition—namely,

$$\Phi(z) = \frac{1}{\sqrt{2\pi}} \int_{-\infty}^{z} \exp(-t^2/2)\, dt \qquad -\infty < z < \infty,$$

prove that $\Phi(-z) + \Phi(z) = 1$.

7.4.5   Let X be a random variable having finite mean μ and standard deviation σ. Prove that the *standardized* random variable

$$\frac{X - \mu}{\sigma}$$

has mean 0 and standard deviation 1.

7.4.6   (*a*) Find the mean, variance, skewness, and kurtosis of

$$Z = U_1 + U_2 + \cdots + U_{12} - 6,$$

where $U_1, U_2, \ldots, U_{12}$ is an *independent and identically distributed* sequence of *Uniform*$(0, 1)$ random variables. (*b*) Generate 10 000 variates by using this approach (see Section 7.2), and provide convincing numerical evidence that the variates so generated are approximately normally distributed.

7.5 NONSTATIONARY POISSON PROCESSES

In this section, we will develop two algorithms for simulating a *nonstationary* Poisson process—that is, rather than events that occur at a constant rate of occurrence λ, we now consider processes whose time-varying rate is governed by $\lambda(t)$. We introduce nonstationary Poisson processes by developing one incorrect and two correct algorithms for generating a realization of simulated event times.

7.5.1 An Incorrect Algorithm

For reference, recall that we can simulate a *stationary* Poisson process with rate λ for the time interval $0 \le t < \tau$ by using Algorithm 7.3.1, reproduced here for convenience:

```
a₀ = 0.0;
n = 0;
while (aₙ < τ) {
    aₙ₊₁ = aₙ + Exponential(1 / λ);
    n++;
}
return a₁, a₂, ..., aₙ₋₁;
```

This algorithm generates a realization of a stationary Poisson process with event times $a_1, a_2,$ $a_3, \ldots$ and constant event rate λ. Given this algorithm, it might be conjectured that, if the event rate varies with time, then the following minor modification will correctly generate a *nonstationary* Poisson process with event rate function $\lambda(t)$:

```
a₀ = 0.0;
n = 0;
while (aₙ < τ) {
    aₙ₊₁ = aₙ + Exponential(1 / λ(aₙ));
    n++;
}
return a₁, a₂, ..., aₙ₋₁;
```

Unfortunately, this naive modification of the correct stationary-Poisson-process algorithm is *not* correct. The algorithm considers the value of the event-rate function $\lambda(t)$ only at the time of the previous event a_n and, hence, ignores the evolution of $\lambda(t)$ after time a_n. In general, the incorrectness of this algorithm will manifest itself as an "inertia error" that can be characterized, for a rate function $\lambda(t)$ that is monotone on (a_n, a_{n+1}), as follows:

- if $\lambda(a_n) < \lambda(a_{n+1})$, then $a_{n+1} - a_n$ will tend to be too large;
- if $\lambda(a_n) > \lambda(a_{n+1})$, then $a_{n+1} - a_n$ will tend to be too small.

If $\lambda(t)$ varies slowly with t, then the inertia error will be uniformly small, and this incorrect algorithm could produce an acceptable approximation to what is desired. Of course, this is a weak justification for using an algorithm *known* to be incorrect, particularly in light of the existence of two *correct* algorithms for simulating a nonstationary Poisson process.

Example 7.5.1

A nonstationary Poisson process with the piecewise-constant event-rate function

$$\lambda(t) = \begin{cases} 1 & 0 \le t < 15 \\ 2 & 15 \le t < 35 \\ 1 & 35 \le t < 50 \end{cases}$$

was simulated by using the incorrect algorithm with $\tau = 50$. Note that the expected number of events in the interval $0 \le t < 15$ is 15; in the interval $15 \le t < 35$, the expected number is $2 \times (35 - 15) = 40$; and 15 events are expected in the interval $35 \le t < 50$. Therefore, the expected number of events in the interval $0 \le t < 50$ is 70. The nonstationary Poisson process was replicated 10 000 times, so that the expected number of events over all replications was 700 000, and a time-dependent relative-frequency histogram was generated by partitioning the interval $0 \le t \le 50$ into 50 bins with bin size 1.0, as illustrated in Figure 7.5.1.

The number of events was accumulated for each bin and then divided by the number of replications (10 000) to provide an *estimate* $\hat{\lambda}(t)$ of the expected rate at time t where t corresponds to the midpoint of the bin.* As suggested in the previous discussion,

---

*If the bin width had not been 1.0, to compute $\hat{\lambda}(t)$, it would have been necessary to divide by the bin width as well.

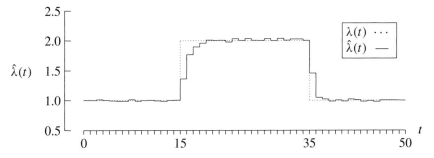

Figure 7.5.1   Incorrect nonstationary-Poisson-process generation.

the inertia error causes $\hat{\lambda}(t)$ to underestimate $\lambda(t)$ immediately to the right of the step-increase at $t = 15$ and to overestimate $\lambda(t)$ immediately to the right of the step-decrease at $t = 35$. Although this error is accentuated by the discontinuous nature of $\lambda(t)$, some error would remain even if the discontinuities in $\lambda(t)$ were removed. You are asked to investigate this issue further in Exercise 7.5.2.

Correct Algorithms.   Two correct algorithms for generating a realization of a nonstationary Poisson process will be presented in the two subsequent subsections: thinning and inversion. Neither algorithm dominates the other in terms of performance—thinning works for a wider range of event-rate functions $\lambda(t)$, but inversion is synchronized and typically is more efficient computationally. The existence of these two exact algorithms is sufficient reason to never use the incorrect algorithm in practice.

7.5.2 Thinning Method

The first correct algorithm for generating a nonstationary Poisson process is based upon the existence of an *upper bound* (majorizing constant) on the event rate λ_{max}, such that $\lambda(t) \le \lambda_{max}$ for $0 \le t < \tau$. As we will see, the efficiency of the algorithm is dependent upon whether λ_{max} is a tight bound. The algorithm is based upon generating a stationary Poisson process with rate λ_{max}, then correcting for this too-high event rate and making the process nonstationary by occasionally thinning (discarding) some of the event times. The following algorithm, known as the *thinning method*, is due to Lewis and Shedler (1979). (As demonstrated in the next section, the thinning method is actually just a clever use of the acceptance–rejection method for random-variate generation.)

Algorithm 7.5.1   Provided that $\lambda(t) \le \lambda_{max}$ for $0 \le t < \tau$, this algorithm generates a nonstationary Poisson process with event rate function $\lambda(t)$:

```
a₀ = 0.0;
n = 0;
while (aₙ < τ) {
    s = aₙ;
```

```
   do {                                             // thinning loop
      s = s + Exponential(1 / λmax);     // possible next event time
      u = Uniform(0, λmax);
   } while ((u > λ(s)) and (s < τ));                // thinning criteria
   a_{n+1} = s;
   n++;
}
return a_1, a_2, ..., a_{n-1};
```

The key to the correctness of the thinning method is the rate correction defined by the do-while loop, as illustrated geometrically in Figure 7.5.2. If it were not for this loop, the process generated would be stationary with event rate λ_{max}. When the event-rate function $\lambda(t)$ is low (high), more (fewer) "trial" event times s are thinned out, as expected. Because occasional multiple "thinning" passes through the do-while loop are necessary to make the process nonstationary, efficiency suffers if $\lambda(s)$ differs significantly from λ_{max}.

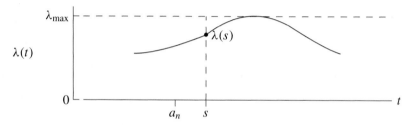

Figure 7.5.2   Thinning algorithm geometry.

Example 7.5.2
Algorithm 7.5.1 was used with $\lambda_{max} = 2.0$ to repeat the experiment in Example 7.5.1. In this case, the nonstationary Poisson process is correctly simulated, as illustrated in Figure 7.5.3.

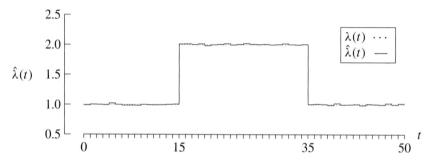

Figure 7.5.3   Thinning implementation results.

Relative to the incorrect algorithm, there is increased complexity with Algorithm 7.5.1. In particular, as compared with the time to compute the time-dependent relative frequency histogram in Example 7.5.1, the thinning-method time was approximately 2.2 times greater.

So long as λ_{max} is a tight upper bound to $\lambda(t)$, so that thinning is infrequent, Algorithm 7.5.1 is relatively efficient and, primarily because it is easy to implement, popular. At least two random variates are required per event, however, and, even if `rngs` is used with separate streams allocated to the `Exponential` and `Uniform` calls, synchronization is a problem, because multiple passes through the thinning loop can occur. The inversion method, presented next, eliminates the need for multiple calls to `Random`.

7.5.3 Inversion Method

The second correct algorithm is based upon the inverse-transformation method of random-variate generation. (See Sections 6.2 and 7.2.) Because of this, and in contrast to the thinning method, this second algorithm has the desirable property that only one call to `Random` is required to generate each new event time. The algorithm is based upon the *cumulative* event-rate function, defined as

$$\Lambda(t) = \int_0^t \lambda(s)\, ds \qquad 0 \le t < \tau,$$

which represents the expected number of events in the interval $[0, t)$. Note that $\lambda(\cdot)$ is analogous to the pdf of a continuous random variable and that $\Lambda(\cdot)$ is analogous to the cdf.

We assume that $\lambda(t) > 0$ for all $t \in [0, \tau)$, except perhaps at a finite number of times where $\lambda(t)$ may be 0. Therefore, $\Lambda(t)$ is a strictly monotone increasing function with positive slope. The event rate function can be found by differentiation:

$$\frac{d\Lambda(t)}{dt} = \lambda(t) \qquad 0 \le t < \tau.$$

Because $\Lambda(\cdot)$ is a strictly monotone increasing function, there exists an *inverse* function $\Lambda^{-1}(\cdot)$, analogous to the idf of a continuous random variable. From these observations, it can be shown that the following algorithm is correct.

Algorithm 7.5.2   If $\Lambda^{-1}(\cdot)$ is the inverse of the function $\Lambda(\cdot)$ defined by

$$\Lambda(t) = \int_0^t \lambda(s)\, ds \qquad 0 \le t < \tau,$$

then the following algorithm (due to a result by Çinlar, 1975) generates a nonstationary Poisson process having event-rate function $\lambda(t)$ for $0 \le t < \tau$:

```
a0 = 0.0;                      // initialize nonstationary Poisson process
y0 = 0.0;                      // initialize unit Poisson process
```

```
n = 0;                                    // initialize event counter
while (aₙ < τ) {
    yₙ₊₁ = yₙ + Exponential(1.0);
    aₙ₊₁ = Λ⁻¹(yₙ₊₁);
    n++;
}
return a₁, a₂, ..., aₙ₋₁;
```

Algorithm 7.5.2 works by generating a sequence of event times $y_1, y_2, y_3, \dots$ corresponding to a *stationary* "unit" Poisson process having rate 1.* The inverse function $\Lambda^{-1}(\cdot)$ is then used to transform the unit Poisson values $y_1, y_2, y_3, \dots$ into a sequence of event times $a_1, a_2, a_3, \dots$ corresponding to a *nonstationary* Poisson process with event-rate function $\lambda(t)$, as illustrated in Figure 7.5.4.

The extent to which Algorithm 7.5.2 is useful is determined by the extent to which the inverse function $\Lambda^{-1}(\cdot)$ can be evaluated efficiently. As illustrated in the following example, if the event-rate function $\lambda(t)$ is assumed to be piecewise constant, then $\Lambda(t)$ will be piecewise linear, and the evaluation of $\Lambda^{-1}(\cdot)$ is easy.

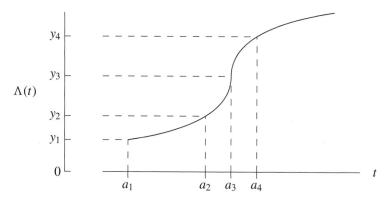

Figure 7.5.4   Inversion-algorithm geometry.

Example 7.5.3

If $\lambda(t)$ is the piecewise-constant event-rate function in Examples 7.5.1 and 7.5.2, then it follows by integration that

$$\Lambda(t) = \begin{cases} t & 0 \le t < 15 \\ 2t - 15 & 15 \le t < 35 \\ t + 20 & 35 \le t < 50. \end{cases}$$

---

*From the results in Section 7.3, this is statistically equivalent to generating n points *at random* in the interval $0 < y < \Lambda(\tau)$, where $\Lambda(\tau)$ is the expected number of events in the interval $[0, \tau)$.

Thus, by solving the equation $y = \Lambda(t)$ for t, we find that

$$\Lambda^{-1}(y) = \begin{cases} y & 0 \le y < 15 \\ (y+15)/2 & 15 \le y < 55 \\ y-20 & 55 \le y < 70. \end{cases}$$

As illustrated in Figure 7.5.5, the nonstationary Poisson process can then be correctly simulated by inversion via Algorithm 7.5.2.

In contrast to the timing results in Example 7.5.2 for the thinning method, the time required to generate this time-dependent relative-frequency histogram by inversion is essentially identical to the time required to generate the incorrect results in Example 7.5.1. The event-rate function in this particular example was ideal, in the sense that the cumulative event-rate function $\Lambda(t)$ can easily be inverted in closed form. If this is not the case, the modeler is left with three options: (i) use thinning if a majorizing constant λ_{max} can be found; (ii) use numerical methods to invert $\Lambda(t)$; or (iii) use an approximate algorithm.

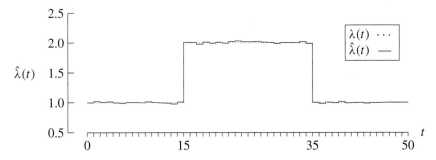

Figure 7.5.5   Inversion-implementation results.

Next-Event-Simulation Orientation.   The three algorithms for generating a nonstationary Poisson process (the incorrect algorithm, thinning, and inversion) all return event times $a_1, a_2, \ldots,$ a_{n-1}, violating next-event-simulation practice. All three, however, can easily be adapted for a next-event simulation. With inversion, for example, an arrival event being processed at time t should schedule the next arrival at time

$$\Lambda^{-1}\big(\Lambda(t) + Exponential(1.0)\big),$$

if this time does not exceed the termination time τ. This procedure would be appropriate for generating the next event time a_3, given the current event time $t = a_2$, for example, in Figure 7.5.4. The geometry associated with this instance traces a path from a_2 to y_2 to y_3 to a_3 in Figure 7.5.4.

What happens in a next-event simulation when the time of a scheduled event exceeds τ? If the interval $0 \le t < \tau$ represents a single cycle that cuts off at time τ (e.g., a restaurant that opens at 11:00 AM and closes at 9:00 PM), then the event to be scheduled after time τ is simply discarded (not scheduled), and the arrival stream is cut off. Alternatively, the interval $0 \leq t < \tau$ could represent one of many cycles in a repeating cyclical process, such as a hospital emergency room (where $\tau = 24$ hours). Furthermore, there could be different functions $\Lambda_1(t), \Lambda_2(t), \ldots$ for each cycle (e.g., $\Lambda_1(t)$ for Sunday, $\Lambda_2(t)$ for Monday and so on). In this case, when

$$\Lambda_1^{-1}\big(\Lambda_1(t) + Exponential(1.0)\big)$$

exceeds τ, the residual amount of the unit stationary Poisson process,

$$\Lambda_1(t) + Exponential(1.0) - \Lambda_1(\tau),$$

is used as an argument in $\Lambda_2(t)$ to generate the first event time in the second cycle (e.g., the first patient to arrive to the emergency room on Monday).

Piecewise-Constant Versus Piecewise-Linear.   In most applications, a piecewise-constant event-rate function $\lambda(t)$ is not realistic. Instead, it is more likely that $\lambda(t)$ will vary continuously with t, avoiding discontinuous jumps like those at $t = 15$ and $t = 35$ in Examples 7.5.1, 7.5.2, and 7.5.3. On the other hand, because accurate estimates of $\lambda(t)$ are difficult to obtain (because of the large amount of data required), it is hard to justify "high-order" smooth polynomial approximations to $\lambda(t)$. Therefore, as a compromise, we will examine the case where $\lambda(t)$ is a piecewise-linear function (spline). That is, we will specify $\lambda(t)$ at a sequence of (t_j, λ_j) points (spline knots), then use linear interpolation to "fill in" other values of $\lambda(t)$ as necessary.

7.5.4 Piecewise-Linear Time-Dependent Event-Rate Functions

As discussed in the previous paragraph, it is common to assume that the event-rate function $\lambda(t)$ is piecewise linear—that is, we assume that $\lambda(t)$ is specified as a piecewise-linear function defined for $0 \leq t < \tau$ by the $k + 1$ *knot pairs* (t_j, λ_j) for $j = 0, 1, \ldots, k$, with the understanding that $0 = t_0 < t_1 < \cdots < t_k = \tau$ and $\lambda_j \geq 0$ for $j = 0, 1, \ldots, k$, as illustrated for the case $k = 6$ in Figure 7.5.6.

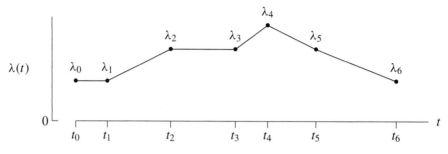

Figure 7.5.6 Piecewise-linear event-rate function.

Algorithm 7.5.3   Given a piecewise-linear, time-dependent event-rate function $\lambda(t)$ specified by the $k + 1$ knot pairs (t_j, λ_j) for $j = 0, 1, \ldots, k$, then the mathematical construction of the piecewise-linear function $\lambda(t)$, of the corresponding piecewise-quadratic cumulative event-rate function $\Lambda(t)$, and of the inverse function $\Lambda^{-1}(u)$ involves four steps.

- Define the k *slopes* of the piecewise-linear segments of $\lambda(t)$:

$$s_j = \frac{\lambda_{j+1} - \lambda_j}{t_{j+1} - t_j} \qquad j = 0, 1, \ldots, k - 1.$$

- Define the $k + 1$ *cumulative event rates* $0 = \Lambda_0 \le \Lambda_1 \le \cdots \le \Lambda_k$ as

$$\Lambda_j = \int_0^{t_j} \lambda(t) \, dt \qquad j = 1, 2, \ldots, k.$$

With $\Lambda_0 = 0$, the cumulative event rates can be calculated recursively as

$$\Lambda_j = \Lambda_{j-1} + \frac{1}{2}(\lambda_j + \lambda_{j-1})(t_j - t_{j-1}) \qquad j = 1, 2, \ldots, k.$$

- For each subinterval $t_j \le t < t_{j+1}$, if $s_j \ne 0$, then

$$\lambda(t) = \lambda_j + s_j(t - t_j)$$

is a linear function of $t - t_j$. Instead, if $s_j = 0$, then $\lambda(t)$ is a constant function. Similarly, if $s_j \ne 0$, then

$$\Lambda(t) = \Lambda_j + \lambda_j(t - t_j) + \frac{1}{2}s_j(t - t_j)^2$$

is a quadratic function of $t - t_j$, but, if $s_j = 0$, then $\Lambda(t)$ is a linear function of $t - t_j$.
- For each subinterval $\Lambda_j \le y < \Lambda_{j+1}$, if $s_j \ne 0$, then the quadratic equation $y = \Lambda(t)$ can be solved for t to yield the inverse function

$$\Lambda^{-1}(y) = t_j + \frac{2(y - \Lambda_j)}{\lambda_j + \sqrt{\lambda_j^2 + 2s_j(y - \Lambda_j)}}.$$

Instead, if $s_j = 0$, then the linear equation $y = \Lambda(t)$ can be solved for t to yield the inverse function

$$\Lambda^{-1}(y) = t_j + \frac{y - \Lambda_j}{\lambda_j}.$$

The mathematical derivation of these four steps is left as an exercise.

The following algorithm is just Algorithm 7.5.2 with an additional index j to keep track of which branch of the $\Lambda^{-1}(\cdot)$ function should be used. As the event times increase, the index j of the current $\Lambda_j < y \le \Lambda_{j+1}$ subinterval is updated accordingly, thereby eliminating the need to search for the index of the subinterval in which y_n lies.

Algorithm 7.5.4   If $\lambda(t)$ is a piecewise-linear function with components defined to be consistent with Algorithm 7.5.3, then the following inversion algorithm generates a nonstationary Poisson process with event-rate function $\lambda(t)$ for $0 \le t < \tau$:

```
a₀ = 0.0;
y₀ = 0.0;
n = 0;
j = 0;
while (aₙ < τ) {
    yₙ₊₁ = yₙ + Exponential(1.0);
    while ((Λⱼ₊₁ < yₙ₊₁) and (j < k))
      j++;
    aₙ₊₁ = Λ⁻¹(yₙ₊₁);                                    // Λⱼ < yₙ₊₁ ≤ Λⱼ₊₁
    n++;
}
return a₁, a₂, ..., aₙ₋₁;
```

In Algorithm 7.5.3, the slopes s_j and cumulative event rates Λ_j can be generated once, external to Algorithm 7.5.4; or these parameters can be generated internal to Algorithm 7.5.4, as needed, each time j is incremented. In either case, the time complexity of Algorithm 7.5.4 is essentially independent of the number of knots. When $\lambda(t)$ is piecewise linear, Algorithm 7.5.4 is the preferred way to generate a nonstationary Poisson process.

7.5.5 Exercises

7.5.1   Work through the details of Algorithm 7.5.3. Be sure to address the issue of how we know to always use the $+$ sign when solving the $y = \Lambda(t)$ quadratic equation, and verify that, as $s_j \to 0$, the $s_j \ne 0$ version of the $\Lambda^{-1}(y)$ equation degenerates into the $s_j = 0$ version.

7.5.2   (a) Use the event-rate function

$$\lambda(t) = \begin{cases} 1 & 0 \le t < 10 \\ t/10 & 10 \le t < 20 \\ 2 & 20 \le t < 30 \\ (50 - t)/10 & 30 \le t < 40 \\ 1 & 40 \le t < 50 \end{cases}$$

and the incorrect nonstationary Poisson process generation algorithm to produce a figure similar to the one in Example 7.5.1. (b) Comment.

7.5.3   (a) Use the event-rate function in Exercise 7.5.2 and Algorithm 7.5.1 to produce a figure similar to the one in Example 7.5.2. (b) Comment.

7.5.4   (a) Use the event-rate function in Exercise 7.5.2 and Algorithm 7.5.4 to produce a figure similar to the one in Example 7.5.3. (b) Comment.

7.5.5   Prove that, if $\lambda(t)$ is constant with value λ_{max}, then Algorithm 7.5.1 reduces to Algorithm 7.3.1.

7.5.6   Prove that, if $\lambda(t)$ is constant with value λ, then Algorithm 7.5.2 reduces to Algorithm 7.3.1.

7.5.7   Use Algorithm 7.5.4 with $\tau = 2000$ to construct a finite-horizon simulation of an initially idle single-server service node with a nonstationary Poisson arrival process. Assume that the arrival-rate function $\lambda(t)$ is the piecewise-linear function illustrated and that the service-time distribution is *Erlang* (4, 0.25).

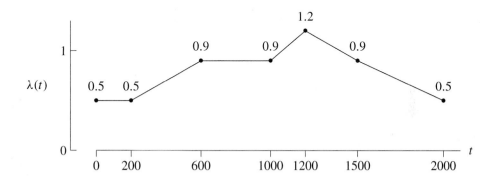

(*a*) The simulation should be replicated 64 times to estimate the mean of the instantaneous (snapshot) number in the node at each of the $\lambda(t)$ time knots (other than $t_0 = 0$). (*b*) Comment. (*c*) If you were to approximate the nonstationary Poisson arrival process with an "equivalent" stationary arrival process with constant rate $\bar{\lambda}$, what would the numerical value of $\bar{\lambda}$ be?

7.5.8   Draw $\Lambda(t)$ associated with $\lambda(t)$ from Figure 7.5.6. Show the geometry associated with Algorithm 7.5.4 on your graph.

7.5.9   Construct an algorithm for generating a realization of a nonstationary Poisson process with a piecewise-constant event-rate function $\lambda(t)$.

7.5.10   Thinning can be performed in a more general setting than that presented in this section. A *majorizing function* $\lambda_{max}(t)$ can replace the majorizing constant λ_{max}. Reformulate the thinning algorithm in its more general setting.

7.6 ACCEPTANCE–REJECTION

The thinning method was presented in the previous section as a technique for generating a realization of a nonstationary Poisson arrival process. The thinning method is based on a more general technique, acceptance–rejection, commonly used in nonuniform random-variate generation. We reintroduce the acceptance–rejection technique in this section, and illustrate its use by developing an efficient algorithm for generating a *Gamma* (*a*, *b*) random variate.

7.6.1 Background

Definition 7.6.1   *Acceptance–rejection* is a technique for generating nonuniform random variates. The technique is most commonly applied to generating *continuous* random variates, as follows. Let X be a continuous random variable with an associated set of possible values $\mathcal{X}$ and pdf $f(x)$ defined for all $x \in \mathcal{X}$. Choose a *majorizing* pdf $g(x)$ with the same set of possible values $x \in \mathcal{X}$ and an associated real-valued constant $c > 1$ such that

- $f(x) \le c\,g(x)$ for all $x \in \mathcal{X}$;
- the idf $G^{-1}(\cdot)$ associated with the majorizing pdf can be evaluated *efficiently*.

The basic idea behind acceptance–rejection (von Neumann, 1951) is to use the inverse-transformation method to generate a possible value $x \in \mathcal{X}$ consistent with the majorizing pdf $g(\cdot)$ and then either accept or reject x on the basis of how well $c\,g(x)$ (the dashed line in Figure 7.6.1) approximates $f(x)$ (the solid line in Figure 7.6.1).

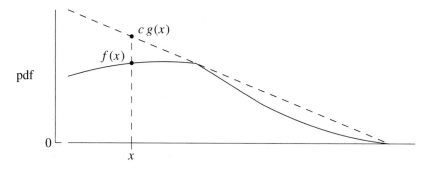

Figure 7.6.1   Acceptance–rejection geometry.

Algorithm 7.6.1   Given the two pdfs $f(\cdot)$ and $g(\cdot)$ with common domain $\mathcal{X}$, given a constant $c > 1$ such that $f(x) \le c\,g(x)$ for all $x \in \mathcal{X}$, and given the idf $G^{-1}(\cdot)$ corresponding to $g(\cdot)$, then the following acceptance–rejection algorithm generates a random variate X with pdf $f(\cdot)$

```
do {
    u = Random();
    x = G⁻¹(u);
    v = Random();
} while (c * g(x) * v > f(x));              // rejection criterion
return x;
```

Algorithm Correctness.   To prove that Algorithm 7.6.1 is correct, define the *acceptance region*

$$\mathcal{A} = \left\{ (t, v) \,\middle|\, v \le \frac{f(t)}{c\, g(t)} \right\} \subset \mathcal{X} \times (0, 1)$$

and then write Algorithm 7.6.1 in the equivalent form

```
do {
    u = Random();
    t = G⁻¹(u);                                        // generate T
    v = Random();                                      // generate V
} while ((t, v) ∉ A);
x = t;                                                 // generate X
return x;
```

This formulation helps make it clear that Algorithm 7.6.1 works by repeatedly generating samples (t, v) of the bivariate random variable (T, V) until a sample falls in the acceptance region. When this occurs, $x = t$ is a sample of the univariate random variable X.

Proof.   What we must prove is that $\Pr(X \le x) = F(x)$ for all $x \in \mathcal{X}$, where

$$F(x) = \int_{t \le x} f(t)\, dt.$$

From the definition of conditional probability,

$$\Pr(X \le x) = \Pr\left(T \le x \mid (T, V) \in \mathcal{A}\right) = \frac{\Pr\left((T \le x) \text{ and } (T, V) \in \mathcal{A}\right)}{\Pr\left((T, V) \in \mathcal{A}\right)},$$

where the numerator in the last expression is

$$\Pr\left((T \le x) \text{ and } (T, V) \in \mathcal{A}\right) = \int_{t \le x} \Pr\left((T, V) \in \mathcal{A} \mid T = t\right) g(t)\, dt$$

and the denominator $\Pr\left((T, V) \in \mathcal{A}\right)$ is this same integral integrated over all $t \in \mathcal{X}$. The correctness of Algorithm 7.6.1 then follows from the fact that V is $Uniform\,(0, 1)$ and so

$$\Pr\left((T, V) \in \mathcal{A} \mid T = t\right) = \Pr\left(V \le \frac{f(t)}{c\, g(t)}\right) = \frac{f(t)}{c\, g(t)}.$$

In other words,

$$\Pr\left((T \le x) \text{ and } (T, V) \in \mathcal{A}\right) = \int_{t \le x} \frac{f(t)}{c\, g(t)} g(t)\, dt = \frac{F(x)}{c},$$

and $\Pr\left((T, V) \in \mathcal{A}\right) = 1/c$; so $\Pr(X \le x) = F(x)$, as desired.   □

Note that, in addition to calling Random twice, each do-while loop iteration in Algorithm 7.6.1 involves

- one evaluation of $G^{-1}(u)$,
- one evaluation of $f(x)$, and
- one evaluation of $g(x)$.

Therefore, the extent to which Algorithm 7.6.1 is efficient is determined, in part, by the efficiency of these three function evaluations. In addition, the expected number of loop iterations is also important, and this number is determined by how closely $c\,g(x)$ bounds $f(x).$*

The design of a good acceptance–rejection algorithm, customized to a specific pdf $f(\cdot)$, is usually a mathematically oriented exercise. Important trade-off choices must be made between the expected number of do-while loop iterations and the computational simplicity of $g(\cdot)$ and $G^{-1}(\cdot)$ function evaluations.

Example 7.6.1
Suppose that the pdf $f(x)$ (the solid line in Figure 7.6.2 defined on $0 < x < w$) is a smoothed histogram of many service times. Suppose also that $f(x)$ can be fit well with a *triangular* majorizing pdf $c\,g(x)$, also defined on $0 < x < w$ with parameters (h, w). [The dashed line is $c\,g(x)$, not $g(x)$.]

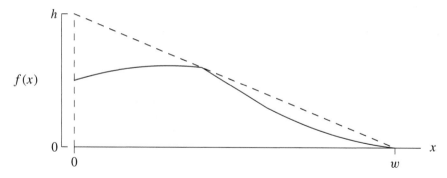

Figure 7.6.2   Service-time pdf with triangular majorizing function.

Appropriate values for (h, w) can be identified by inspection. The triangular majorizing pdf can then be written as

$$c\,g(x) = (1 - x/w)h \qquad 0 < x < w,$$

and the constant c can be calculated from the pdf requirement $\int_0^w g(x)\,dx = 1$. In other words,

$$c = c\int_0^w g(x)\,dx = \int_0^w c\,g(x)\,dx = \int_0^w (1 - x/w)h\,dx = \cdots = \frac{1}{2}hw,$$

---

*See Exercise 7.6.1.

and so we find that, in this case, a majorizing pdf is

$$g(x) = \left(1 - \frac{x}{w}\right)\frac{2}{w} \qquad 0 < x < w.$$

The associated majorizing cdf is

$$G(x) = \int_0^x g(s)\, ds = \frac{2}{w}\int_0^x \left(1 - \frac{s}{w}\right) ds = \cdots = \frac{2x}{w}\left(1 - \frac{x}{2w}\right) \qquad 0 < x < w.$$

The attractive feature of this triangular majorizing pdf is that the associated cdf is quadratic, and so the idf can be found via the quadratic formula: We can write $u = G(x)$ and solve for u, to find that the idf of the majorizing pdf is

$$x = G^{-1}(u) = w\left(1 - \sqrt{1 - u}\right) \qquad 0 < u < 1.$$

The resulting acceptance–rejection algorithm is

```
do {
    u = Random();
    x = w * (1.0 - sqrt(1.0 - u));
    v = Random();
} while (h * (1.0 - x / w) * v > f(x));
return x;
```

Example 7.6.2
As an alternative to the triangular majorizing pdf in Example 7.6.1, suppose that we choose an *Exponential*(a) majorizing pdf,

$$g(x) = \frac{1}{a}\exp(-x/a) \qquad 0 < x < \infty,$$

where the parameters (a, c) are selected so that $f(x) \le c\, g(x)$ for all $x > 0$. Then

$$G^{-1}(u) = -a\ln(1 - u) \qquad 0 < u < 1,$$

and the resulting acceptance–rejection algorithm is

```
do {
    x = Exponential(a);
    v = Random();
} while ((c / a) * exp(-x / a) * v > f(x));
return x;
```

Relative to Examples 7.6.1 and 7.6.2, the choice of whether to use the triangle or exponential majorizing pdf is determined by the shape of $f(\cdot)$ and the resulting value of c. By necessity, c is always greater than 1; the closer c is to 1, the better, since the expected number of passes through the do-while loop (rejections) is smaller. For more insight, see Exercise 7.6.1.

7.6.2 Gamma Random Variates

We now use the acceptance–rejection technique to develop an efficient algorithm for generating *Gamma* (a, b) random variates. We begin with the following definition.

Definition 7.6.2   A continuous random variable X is *Gamma* (a, b) if and only if

- the real-valued shape parameter a satisfies $a > 0$,
- the real-valued scale parameter b satisfies $b > 0$,
- the possible values of X are $\mathcal{X} = \{x \mid x > 0\}$,
- the pdf of X is

$$f(x) = \frac{1}{b\Gamma(a)} (x/b)^{a-1} \exp(-x/b) \qquad x > 0,$$

- the cdf of X is the incomplete gamma function

$$F(x) = P(a, x/b) \qquad x > 0,$$

- the mean of X is

$$\mu = ab,$$

- the standard deviation of X is

$$\sigma = b\sqrt{a}.$$

As shown in Figure 7.6.3, a *Gamma* (a, b) random variable is an important generalization of three common continuous random variables encountered previously.

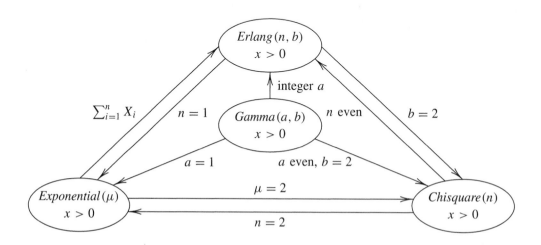

Figure 7.6.3   Special cases of the gamma distribution.

In specific,

- A *Gamma*(1, *b*) random variable is an *Exponential*(*b*) random variable.
- If *n* is a positive integer, then a *Gamma*(*n*, *b*) random variable is an *Erlang*(*n*, *b*) random variable.
- If *n* is a positive integer, then a *Gamma*(*n*/2, 2) random variable is a *Chisquare*(*n*) random variable.

Except for special cases (in particular, $a = 1$), because the cdf of a *Gamma*(*a*, *b*) random variable is an incomplete gamma function, there is no hope of finding a closed-form equation for the idf $F^{-1}(\cdot)$. Therefore, the inverse-transformation method is not directly applicable. However, as demonstrated in the remainder of this section, the acceptance–rejection technique is applicable. The reason for this is based, in part, on the following theorem, which effectively reduces the problem of generating a two-parameter gamma random variate to generating a one-parameter gamma random variate. The proof is left as an exercise.

Theorem 7.6.1   The random variable X is *Gamma*(*a*, 1) if and only if the random variable $Y = bX$ is *Gamma*(*a*, *b*), for all $a > 0$ and $b > 0$.

The significance of Theorem 7.6.1 is that it is sufficient to develop an efficient algorithm to generate a *Gamma*(*a*, 1) random variate; a *Gamma*(*a*, *b*) random variate can then be generated via multiplication by *b*. Therefore, in the discussion that follows, we assume that the pdf to be majorized is the *Gamma*(*a*, 1) pdf,

$$f(x) = \frac{1}{\Gamma(a)} x^{a-1} \exp(-x) \qquad x > 0.$$

As illustrated in Figure 7.6.4, the x^{a-1} term in $f(x)$ causes the shape of the pdf to be fundamentally different in the two cases $a > 1$ and $a < 1$. The pdf for $a < 1$ has a vertical asymptote at $x = 0$. Because of this, it should be no surprise to find that two different majorizing pdfs are required, one for each case.

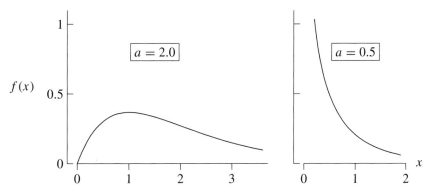

Figure 7.6.4   *Gamma*(2.0, 1) and *Gamma*(0.5, 1) pdfs.

Cheng's Algorithm.   We begin with the more common case, $a > 1$, and develop a majorizing pdf by using an approach developed by Cheng (1977). The majorizing cdf that Cheng used is

$$G(x) = \frac{x^t}{x^t + a^t} \qquad\qquad x > 0,$$

which has the associated pdf

$$g(x) = \frac{dG(x)}{dx} = \frac{ta^t x^{t-1}}{(x^t + a^t)^2} \qquad\qquad x > 0$$

and idf

$$G^{-1}(u) = a\left(\frac{u}{1-u}\right)^{1/t} \qquad\qquad 0 < u < 1.$$

In these three equations, the real-valued parameter t is positive, but otherwise arbitrary.

Note that $G(x)$ is, in fact, a valid cdf for a positive, continuous random variable—that is, $G(x)$ is a continuous function that increases monotonically from the value 0 at $x = 0$ to the value 1 as $x \to \infty$. This cdf was chosen, in part, so that the idf could be easily evaluated.

The key to Cheng's approach is a clever choice of t and the following theorem. The proof is left as a (challenging) exercise.

Theorem 7.6.2   If $f(x)$ is the pdf of a *Gamma*$(a, 1)$ random variable, if $g(x)$ is the majorizing pdf given previously, and if $t = \sqrt{2a - 1}$ with $a > 1$, then, for all $x > 0$,

$$\frac{f(x)}{g(x)} = \frac{x^{a-t}(x^t + a^t)^2 \exp(-x)}{ta^t \Gamma(a)} \leq \frac{f(a)}{g(a)} = \frac{4a^a \exp(-a)}{t\Gamma(a)}.$$

In other words, the largest possible value of the ratio $f(x)/g(x)$ occurs at $x = a$.

Note that, if $t = \sqrt{2a - 1}$ with $a > 1$, then $1 < t < a$. In addition, if

$$c = \frac{f(a)}{g(a)} = \frac{4a^a \exp(-a)}{\sqrt{2a - 1}\,\Gamma(a)},$$

then it can be verified numerically that we have the following table

a	1	1.5	2	3	5	10	20	50	∞
c	1.47	1.31	1.25	1.20	1.17	1.15	1.14	1.13	1.13

Because c is the largest possible value of $f(x)/g(x)$, it follows that

$$f(x) \leq c\, g(x) \qquad\qquad x > 0.$$

The fact that c is only slightly greater than 1.0 for all $a > 1$ is important, because it means that $g(x)$ is a good majorizing pdf (i.e., few rejections) for $f(x)$ for all values of $a > 1$.*

If $f(x)$ is the pdf of a *Gamma* $(a, 1)$ random variable and $g(x)$ is the pdf given previously, then Cheng's acceptance–rejection algorithm for generating a *Gamma* $(a, 1)$ random variate is

```
c = f(a) / g(a);
t = sqrt(2.0 * a - 1.0);
do {
    u = Random();
    s = log(u / (1.0 - u));
    x = a * exp(s / t);
    v = Random();
} while (c * g(x) * v > f(x));
return x;
```

Efficiency Improvement.   Most acceptance–rejection methods can be made more efficient by analyzing the computational details of the implementation in an attempt to eliminate unnecessary and redundant calculations. The following example is an illustration of this.

Example 7.6.3

The $c\,g(x)\,v > f(x)$ test at the bottom of the do-while loop in Cheng's *Gamma* $(a, 1)$-variate-generation algorithm can be simplified considerably by observing that, if

$$x = G^{-1}(u) = a \left(\frac{u}{1 - u} \right)^{1/t},$$

then

$$x^t = \frac{a^t u}{1 - u} \qquad \text{and} \qquad x^t + a^t = \frac{a^t}{1 - u}.$$

The inequality test can then be written as

$$c\,g(x)\,v > f(x) \iff \left(\frac{4a^a \exp(-a)}{t\Gamma(a)} \right) \left(\frac{ta^t x^{t-1}}{(x^t + a^t)^2} \right) v > \frac{1}{\Gamma(a)} x^{a-1} \exp(-x)$$

$$\iff 4va^a a^t x^t \exp(-a) > x^a (x^t + a^t)^2 \exp(-x)$$

$$\iff 4va^a a^t \left(\frac{a^t u}{1 - u} \right) \exp(-a) > a^a \left(\frac{u}{1 - u} \right)^{a/t} \left(\frac{a^t}{(1 - u)} \right)^2 \exp(-x)$$

$$\iff 4vu(1 - u) > \left(\frac{u}{1 - u} \right)^{a/t} \exp(a - x).$$

---

*Cheng's algorithm becomes increasingly more efficient as $a \to \infty$.

All this symbol manipulation yields the following simple yet efficient acceptance–rejection algorithm. The correctness of this algorithm can be verified (for particular values of $a > 1$ and b) by generating, say, 100 000 values and comparing the resulting continuous-data histogram or empirical cdf with the corresponding $Gamma\,(a, b)$ pdf or cdf, respectively.

Algorithm 7.6.2 If $a > 1$, then, for any $b > 0$, the following algorithm generates a $Gamma$ (a, b) random variate without calculating c directly:

```
t = sqrt(2.0 * a - 1.0);
do {
    u = Random();
    s = log(u / (1.0 - u));
    x = a * exp(s / t);
    v = Random();
} while (4.0 * v * u * (1.0 - u) > exp(a - x + a * s / t));
x = b * x;
return x;
```

Ahrens and Dieter's Algorithm.   We now turn to the case $0 < a \le 1$ and develop a majorizing pdf using an approach developed by Ahrens and Dieter (1974). The development begins with the observation that, if $a \le 1$ and if we define

$$c\,g(x) = \begin{cases} x^{a-1}/\Gamma(a) & 0 < x \le 1 \\ \exp(-x)/\Gamma(a) & 1 < x < \infty, \end{cases}$$

then, by inspection,

$$f(x) = \frac{x^{a-1}\exp(-x)}{\Gamma(a)} \le c\,g(x) \qquad 0 < x < \infty.$$

To compute the constant c, write

$$c = c \int_0^\infty g(x)\,dx = \int_0^1 \frac{x^{a-1}}{\Gamma(a)}dx + \int_1^\infty \frac{\exp(-x)}{\Gamma(a)}dx = \cdots = \frac{1}{t\,\Gamma(a+1)},$$

where the parameter $t < 1$ is

$$t = \frac{e}{e+a} \qquad (e = \exp(1) = 2.71828\ldots).$$

It can be verified numerically that $c > 1$ for all $0 < a \le 1$; indeed, we have the following representative table:

a	0	1/32	1/16	1/8	1/4	1/2	1
c	1.00	1.03	1.06	1.11	1.20	1.34	1.37

Therefore, the majorizing function is

$$g(x) = \begin{cases} at x^{a-1} & 0 < x \le 1 \\ at \exp(-x) & 1 < x < \infty, \end{cases}$$

which has the associated cdf

$$G(x) = \begin{cases} t x^{a} & 0 < x \le 1 \\ 1 - at \exp(-x) & 1 < x < \infty. \end{cases}$$

The idf can be expressed in closed form as

$$G^{-1}(u) = \begin{cases} (u/t)^{1/a} & 0 < u \le t \\ -\ln\big((1-u)/at\big) & t < u < 1. \end{cases}$$

As in Cheng's algorithm, the choice of majorizing pdf is dictated, in part, by the simplicity of $G^{-1}(u)$. Moreover, because c is uniformly close to 1 for all $0 < a \le 1$, Ahrens and Dieter's algorithm is efficient, becoming more so as $a \to 0$.

Finally, we observe that the $c\,g(x)v > f(x)$ inequality in Ahrens and Dieter's algorithm can be written (depending on whether $x < 1$ or $x > 1$) as

$$c\,g(x)\,v > f(x) \iff \begin{cases} v > \exp(-x) & 0 < x \le 1 \\ v > x^{a-1} & 1 < x < \infty, \end{cases}$$

which is summarized by the following algorithm.

Algorithm 7.6.3   If $0 < a \le 1$, then, for any $b > 0$, the following algorithm generates a *Gamma*(a, b) random variate:

```
t = e / (e + a);                        // e = exp(1) ≅ 2.718281828459...
do {
   u = Random();
   if (u ≤ t) {                         // 0 < x ≤ 1
      x = exp(log(u / t) / a);
      s = exp(-x);
   }
   else {                               // 1 < x < ∞
      x = -log((1.0 - u) / (a * t));
      s = exp((a - 1.0) * log(x));
   }
   v = Random();
} while (v > s);
x = b * x;
return x;
```

This completes our discussion of the algorithms for generating a *Gamma*(*a*, *b*) random variate. The problem was initially reduced to generating a *Gamma*(*a*, 1) random variate. If *a* = 1 (the exponential case), then an *Exponential*(1) variate is generated. If *a* > 1, then Cheng's algorithm (Algorithm 7.6.2) is executed. If *a* < 1, then Ahrens and Dieter's algorithm (Algorithm 7.6.3) is executed. Finally, in all three cases, the *Gamma*(*a*, 1) random variate is multiplied by *b* to produce a *Gamma*(*a*, *b*) random variate. We end this section with a review of the performance of the two acceptance–rejection algorithms.

Performance

Example 7.6.4

Algorithms 7.6.2 and 7.6.3 were implemented and compared with the corresponding generators in the library `rvgs`. The times (in seconds) required to generate 1000 random observations are as illustrated for selected values of *a*:

a	0.5	1.0	2.0	4.0	8.0	16.0	32.0	64.0
algorithms 7.6.2 and 7.6.3	2.8	2.7	2.8	2.7	2.6	2.6	2.5	2.5
library rvgs	1.5	0.7	1.6	3.2	6.2	12.3	24.5	48.8

The second row of the table corresponds to Algorithms 7.6.2 and 7.6.3. As expected, the times are essentially independent of *a*. The third row used functions from the library `rvgs`, as follows: for *a* = 0.5, *Chisquare*(1); for *a* = 1.0, *Exponential*(1.0); for all other values of *a*, *Erlang*(*a*, 1.0). Because the algorithm used in `rvgs` to generate an *Erlang*(*a*, *b*) random variate is $O(a)$, it should be avoided if *a* is larger than, say, 4; instead, use Algorithm 7.6.2.

Further Reading.   This section has provided an elementary introduction to the acceptance–rejection technique for generating random variates, along with a fairly sophisticated application [i.e., generating a *Gamma*(*a*, *b*) random variate]. We recommend Devroye's (1986) text for a more thorough treatment.

7.6.3 Exercises

7.6.1   Let the discrete random variable *R*, with possible value *r* = 0, 1, 2, . . . , count the number of rejections (per return) by Algorithm 7.6.1. (*a*) Find the pdf of *R*. (*b*) Prove that $E[R] = c - 1$.

7.6.2   Work through the details of Examples 7.6.1 and 7.6.2. How do you know to choose the '−' sign when using the quadratic equation to determine $G^{-1}(u)$?

7.6.3   Prove Theorem 7.6.1.

7.6.4   Prove Theorem 7.6.2. *Hint:* Consider the first derivative of the ratio $f(x)/g(x)$, and verify that this first derivative is zero at some point $x > 0$ if and only if $s(x) = 0$, where

$$s(x) = (a - x - t)a^t + (a - x + t)x^t \qquad x > 0.$$

Then show that $s(x) = 0$ if and only if $x = a$.

7.6.5   Implement the *Gamma*(*a*, *b*) generator defined by Algorithms 7.6.2 and 7.6.3. Present convincing numerical evidence that your implementation is correct in the three cases (*a*, *b*) = (0.5, 1.0), (2.0, 1.0), and (10.0, 1.5).

8

Output Analysis

This chapter contains a more thorough presentation of techniques that can be applied to the output produced by a Monte Carlo or discrete-event simulation. In the case of discrete-event simulation models, these techniques are complicated by the presence of serial (auto)correlation—which means that standard statistical techniques, when based on the assumption of independence, must be modified to account for the autocorrelation.

Section 8.1 contains a graphical demonstration of the fundamentally important central limit theorem. It also contains an introduction to *interval estimation*, or *confidence intervals* (as they are more commonly known). These intervals give an indication of the accuracy or precision associated with a *point estimate* of a performance measure (such as a mean, variance, or fractile). Section 8.2 applies the concepts associated with interval estimation to Monte Carlo simulation. The purpose is to augment the point estimate with an interval that can be taken to contain the true value of the parameter of interest with high "confidence." Section 8.3 introduces the concepts of finite-horizon and infinite-horizon statistics as two classifications of statistics associated with a system. The $M/G/1$ queue is used to illustrate the difference between the two types of statistics. Section 8.4 introduces "batch means," a popular technique for reducing the effect of autocorrelation in analyzing simulation output. Section 8.5 provides insight into the mathematics of traditional queueing theory as applied to the steady-state statistics associated with the single-server service node.

Important topics in output analysis not addressed here include: (i) tactical issues associated with the design of a simulation experiment (e.g., run length, number of runs, adjusting for initialization bias), considered by Whitt (2006); (ii) statistical issues associated with selecting the best system, considered by Kim and Nelson (2006); (iii) metamodeling and response surface methodology, considered by Barton (2006); (iv) random-search techniques, considered by Andradóttir (2006); and (v) metaheuristics, considered by Ólafsson (2006).

8.1 INTERVAL ESTIMATION

This chapter is concerned with simulation applications based on the fundamentally important *central limit theorem* and with its relation to interval estimation. Because discrete-event simulation models produce statistical estimates (means, relative frequencies, etc.) that are inherently uncertain, some technique must be used to quantify that uncertainty; many such techniques are based on the central limit theorem.

8.1.1 Central Limit Theorem

Although a rigorous proof of the central limit theorem is beyond the scope of this book (see Hogg, McKean, and Craig, 2005, for a proof), we can use Monte Carlo simulation to demonstrate the theorem's correctness and, by so doing, gain insight. We begin with a statement of the theorem.

Theorem 8.1.1 If $X_1, X_2, \ldots, X_n$ are *iid* random variables having finite common mean μ and finite common standard deviation σ, and if $\overline{X}$ is the *mean* of these random variables

$$\overline{X} = \frac{1}{n} \sum_{i=1}^{n} X_i,$$

then the mean of $\overline{X}$ is μ and the standard deviation of $\overline{X}$ is $\sigma/\sqrt{n}$. Furthermore, $\overline{X}$ approaches a *Normal*$(\mu, \sigma/\sqrt{n})$ random variable as $n \to \infty$.

Sample Mean Distribution

Example 8.1.1
Suppose we choose one of the random variate generators in the library `rvgs`—it does not matter which one—and use it to generate a sequence of random-variate samples, each of fixed sample size $n > 1$. As each of the n-point samples indexed $j = 1, 2, \ldots$ is generated, the corresponding sample mean $\overline{x}_j$ and sample standard deviation s_j can be calculated from Algorithm 4.1.1 (Welford's Algorithm), as illustrated:

$$\underbrace{x_1, x_2, \ldots, x_n}_{\overline{x}_1, s_1}, \underbrace{x_{n+1}, x_{n+2}, \ldots, x_{2n}}_{\overline{x}_2, s_2}, \underbrace{x_{2n+1}, x_{2n+2}, \ldots, x_{3n}}_{\overline{x}_3, s_3}, x_{3n+1}, \ldots$$

A continuous-data histogram can then be created by using program `cdh`, as illustrated in Figure 8.1.1.

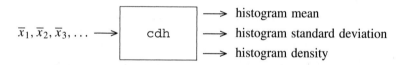

Figure 8.1.1   Calculating a histogram of the sample means.

Let μ and σ denote the theoretical (population) mean and standard deviation of the random variates. As is consistent with Theorem 8.1.1, the resulting *sample mean histogram* will have the following three properties:

- Independent of n, the histogram mean will be approximately μ.
- Independent of n, the histogram standard deviation will be approximately $\sigma/\sqrt{n}$.
- The histogram approaches the pdf of a *Normal*$(\mu, \sigma/\sqrt{n})$ random variable as $n \to \infty$.

Example 8.1.2

To illustrate Example 8.1.1 in more detail, the library `rvgs` was used to generate 10 000 *Exponential*(μ) random-variate samples of size $n = 9$ and 10 000 samples of size $n = 36$ (using the default `rngs` stream and initial seed in both cases). These histograms are shown in Figure 8.1.2.

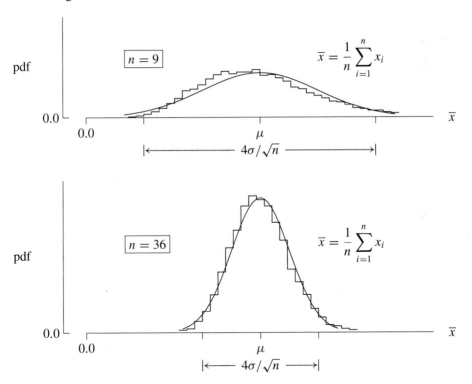

Figure 8.1.2 Histograms of sample means.

As is consistent with the central limit theorem, for both cases, the histogram mean and standard deviation are approximately μ and $\sigma/\sqrt{n}$ respectively. (Because *Exponential*(μ) random variate samples were used, the population standard deviation is $\sigma = \mu$ in both cases.) Moreover, the histogram density corresponding to the 36-point sample means (the stairstep curve) is matched almost exactly by the pdf of a *Normal*$(\mu, \sigma/\sqrt{n})$ random variable (the continuous curve). This agreement illustrates that, for *Exponential*(μ) samples, $n = 36$ is large enough for the sample mean to be approximately a *Normal*($\mu, \sigma/\sqrt{n}$) random variate. Similarly, the histogram density corresponding to the 9-point sample means matches relatively well, but with a skew to the left. This skew is caused by the highly asymmetric shape of the *Exponential*(μ) pdf and the fact that, in this case, $n = 9$ is not a large enough sample—the sample-mean histogram retains a vestige of the asymmetric population pdf.

As stated in Example 8.1.1 and illustrated in Example 8.1.2, the standard deviation of the sample-mean histogram is $\sigma/\sqrt{n}$. Moreover, in the sense of Chebyshev's inequality, this standard deviation defines an *interval* within which essentially all the sample means lie. Because the existence of this interval is so important as the theoretical basis for the development of an interval-estimation technique, for emphasis we provide the following two related summary points.

- Essentially all (about 95%) of the sample means lie inside an interval of width of $4\sigma/\sqrt{n}$ and centered about μ.
- Because $\sigma/\sqrt{n} \to 0$ as $n \to \infty$, if n is large, then essentially all of the sample means will be arbitrarily close to μ.

The random-variate samples in Examples 8.1.1 and 8.1.2 are drawn from a population with a fixed pdf. As illustrated in Example 8.1.2, the accuracy of the *Normal*($\mu, \sigma/\sqrt{n}$) pdf approximation of the distribution of $\overline{X}$ is dependent on the shape of this population pdf. If the samples are drawn from a population with a highly asymmetric pdf (like the *Exponential*(μ) pdf for example) then n may need to be as large as 30 or more for the *Normal*($\mu, \sigma/\sqrt{n}$) pdf to be a good fit to the histogram density. If, instead, the samples are drawn from a population with a pdf symmetric about the mean [like the *Uniform*(a, b) pdf, for example], then values of n as small as 10 or less often produce a satisfactory fit.

Standardized Sample-Mean Distribution

Example 8.1.3
As an extension of Example 8.1.1, there is another way to characterize the central limit theorem. Alternatively to computing the sample means $\overline{x}_1, \overline{x}_2, \overline{x}_3, \ldots$, we could standardize each sample mean, by subtracting μ and dividing the result by $\sigma/\sqrt{n}$, to form the *standardized* sample means $z_1, z_2, z_3, \ldots$, defined by

$$z_j = \frac{\overline{x}_j - \mu}{\sigma/\sqrt{n}} \qquad j = 1, 2, 3, \ldots.$$

We can then generate a continuous-data histogram for the standardized sample means by using program cdh, as illustrated in Figure 8.1.3.

The resulting *standardized sample-mean histogram* will have the following three properties:

- Independent of n, the histogram mean will be approximately 0.
- Independent of n, the histogram standard deviation will be approximately 1.
- Provided n is sufficiently large, the histogram density will approximate the pdf of a *Normal*(0, 1) random variable.

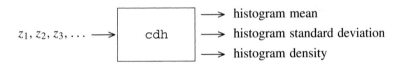

Figure 8.1.3   Calculating a histogram of the standardized sample means.

Example 8.1.4

To illustrate Example 8.1.3 in more detail, the sample means from Example 8.1.2 were standardized to form the two histograms illustrated in Figure 8.1.4.

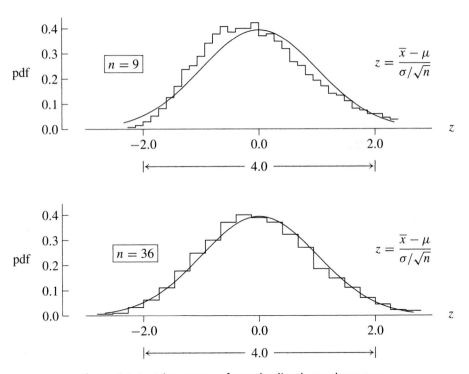

Figure 8.1.4   Histograms of standardized sample means.

Consistent with Example 8.1.3, in both cases, the histogram mean and standard deviation are approximately 0.0 and 1.0 respectively. Moreover, the histogram density corresponding to the 36-point samples (the stairstep curve) matches the pdf of a *Normal*(0, 1) random variable (the continuous curve) almost exactly. Similarly, the histogram density corresponding to the 9-point samples also matches, but not quite as well, because it retains a vestige of the population's asymmetric pdf.

t-Statistic Distribution.   The next issue to consider is the effect of replacing the *population* standard deviation σ with the *sample* standard deviation s_j in the equation

$$z_j = \frac{\overline{x}_j - \mu}{\sigma/\sqrt{n}} \qquad j = 1, 2, 3, \ldots;$$

but, first, we give a definition and some discussion.

Definition 8.1.1   Each sample mean $\overline{x}_j$ is a *point estimate* of μ, each sample variance s_j^2 is a *point estimate* of σ^2, and each sample standard deviation s_j is a *point estimate* of σ.

In the context of Definition 8.1.1 and the discussion in Section 4.1, the following points address the issue of *bias* as it relates to the sample mean and sample variance.

- The sample mean is an *unbiased* point estimate of μ—that is, the mean of the sample means $\overline{x}_1, \overline{x}_2, \overline{x}_3, \ldots$ will converge to μ as the number of sample means is increased. Indeed, this is guaranteed by the central limit theorem.
- The sample variance is a *biased* point estimate of σ^2. Specifically, the mean of the sample variances $s_1^2, s_2^2, s_3^2, \ldots$ will converge to $(n-1)\sigma^2/n$, not σ^2. To remove this $(n-1)/n$ bias, it is conventional to multiply the sample variance by the *bias correction* factor $n/(n-1)$. The result is an *unbiased* point estimate of σ^2.*

The square root of the unbiased point estimate of σ^2 is a point estimate of σ. The corresponding point estimate of $\sigma/\sqrt{n}$ is then

$$\frac{\left(\sqrt{\dfrac{n}{n-1}} \right) s_j}{\sqrt{n}} = \frac{s_j}{\sqrt{n-1}}$$

The following two examples illustrate what it is that changes in Examples 8.1.3 and 8.1.4 if $\sigma/\sqrt{n}$ is replaced with $s_j/\sqrt{n-1}$ in the definition of z_j.

Example 8.1.5

As an extension of Example 8.1.3, instead of computing z_j for each n-point sample, we can calculate the t-statistic

$$t_j = \frac{\overline{x}_j - \mu}{s_j/\sqrt{n-1}} \qquad j = 1, 2, 3, \ldots$$

and then generate a continuous-data histogram by using program cdh, as illustrated in Figure 8.1.5.

---

*If the "divide by $n-1$" definition of the sample variance had been used in Section 4.1 then multiplication by the $n/(n-1)$ bias correction factor would not be necessary.

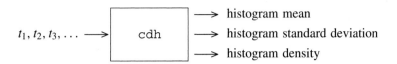

Figure 8.1.5   Calculating a histogram of sample *t*-statistics.

The resulting *t-statistic histogram* will have the following three properties:

- If $n > 2$, the histogram mean will be approximately 0.
- If $n > 3$, the histogram standard deviation will be approximately $\sqrt{(n-1)/(n-3)}$.
- Provided n is sufficiently large, the histogram density will approximate the pdf of a *Student*$(n-1)$ random variable.

Example 8.1.6

To illustrate Example 8.1.5 in more detail, the sequence of sample means and sample standard deviations from Example 8.1.2 were processed as *t*-statistics to form the two histograms illustrated in Figure 8.1.6.

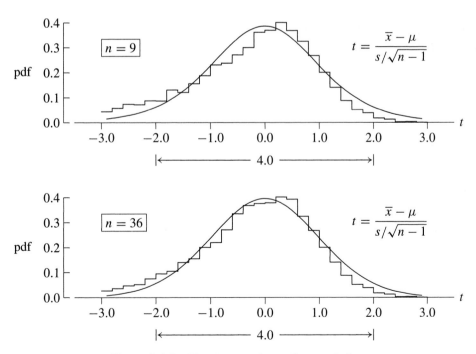

Figure 8.1.6   Histograms of sample *t*-statistics.

Consistent with Example 8.1.5, in both cases, the histogram mean is approximately 0.0 and the histogram standard deviation is approximately $\sqrt{(n-1)/(n-3)} \cong 1.0$. Moreover, the histogram density corresponding to the 36-point samples (the stairstep curve) matches the pdf of a *Student*(35) random variable (the continuous curve) relatively well. Similarly, the histogram density corresponding to the 9-point samples matches the pdf of a *Student*(8) random variable, but not quite as well.

8.1.2 Interval Estimation

The discussion up to this point has been focused on the central limit theorem and has led to Examples 8.1.5 and 8.1.6, which provide an experimental justification for the following theorem. In turn, this theorem is the theoretical basis for the important idea called *interval estimation*, which is the application focus of this chapter.

Theorem 8.1.2   If $x_1, x_2, \ldots, x_n$ is an (independent) random sample from a "source" of data with *unknown* finite mean μ, if $\overline{x}$ and s are the mean and standard deviation of this sample, and if n is large, then it is approximately true that the statistic

$$t = \frac{\overline{x} - \mu}{s/\sqrt{n-1}}$$

is a *Student*$(n-1)$ random variate.*

The significance of Theorem 8.1.2 is that it provides the justification for estimating an *interval* that is quite likely to contain the mean μ—that is, if T is a *Student*$(n-1)$ random variable, and if α is a "confidence parameter" with $0.0 < \alpha < 1.0$ (typically $\alpha = 0.05$), then there exists a corresponding positive real number t^* such that

$$\Pr(-t^* \le T \le t^*) = 1 - \alpha,$$

as illustrated in Figure 8.1.7.

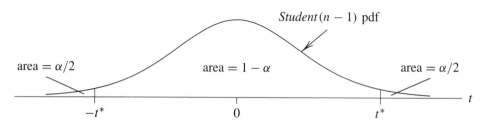

Figure 8.1.7   Student($n-1$) pdf.

---

*Recall, from Figure 7.4.9, that, as $n \to \infty$, a *Student*$(n-1)$ random variate becomes statistically indistinguishable from a *Normal*$(0, 1)$ random variate. Although Theorem 8.1.2 is an exact result if the random sample is drawn from a *Normal*(μ, σ) population, this theorem is stated here only as an approximate result.

Suppose that μ is *unknown*. From Theorem 8.1.2, since t is a *Student*$(n-1)$ random variate, the inequality

$$-t^* \leq \frac{\overline{x} - \mu}{s/\sqrt{n-1}} \leq t^*$$

will be (approximately) true with probability $1 - \alpha$. Multiplying by $s/\sqrt{n-1}$ yields

$$-\frac{t^* s}{\sqrt{n-1}} \leq \overline{x} - \mu \leq \frac{t^* s}{\sqrt{n-1}}.$$

Subtracting $\overline{x}$ and negating makes the inequality

$$\overline{x} - \frac{t^* s}{\sqrt{n-1}} \leq \mu \leq \overline{x} + \frac{t^* s}{\sqrt{n-1}},$$

which will be (approximately) true with probability $1 - \alpha$. This proves that there is an interval within which μ lies with approximate probability $1 - \alpha$, as is summarized by the following theorem.

Theorem 8.1.3   If $x_1, x_2, \ldots, x_n$ is an (independent) random sample from a "source" of data with finite *unknown* mean μ, if $\overline{x}$ and s are the mean and standard deviation of this sample, and if n is large, then, given a confidence parameter α with $0.0 < \alpha < 1.0$, there exists an associated positive real number t^* such that

$$\Pr\left(\overline{x} - \frac{t^* s}{\sqrt{n-1}} \leq \mu \leq \overline{x} + \frac{t^* s}{\sqrt{n-1}}\right) \cong 1 - \alpha.$$

Example 8.1.7
From Theorem 8.1.3, if $\alpha = 0.05$, then we are approximately 95% confident that μ lies *somewhere* in the interval defined by the endpoints

$$\overline{x} - \frac{t^* s}{\sqrt{n-1}} \qquad \text{and} \qquad \overline{x} + \frac{t^* s}{\sqrt{n-1}}.$$

For a fixed sample size n and level of confidence $1 - \alpha$, we can use the idf capability in the library `rvms` to compute the critical value to be $t^* = \texttt{idfStudent}(n-1, 1-\alpha/2)$. For example, if $n = 30$ and $\alpha = 0.05$, then the value of t^* is `idfStudent(29, 0.975)`, which is approximately 2.045.

Definition 8.1.2   The interval defined by the two endpoints

$$\overline{x} \pm \frac{t^* s}{\sqrt{n-1}}$$

is a $(1 - \alpha) \times 100\%$ confidence *interval estimate* for μ. The parameter $1 - \alpha$ is the *level of confidence* associated with this interval estimate, and t^* is the *critical value* of t.

> **Algorithm 8.1.1**   The following steps summarize the process of calculating an *interval estimate* for the unknown mean μ of that much larger population from which a random sample $x_1, x_2, \ldots, x_n$ of size $n > 1$ was drawn.
>
> - Pick a level of confidence $1 - \alpha$ (typically $\alpha = 0.05$; but $\alpha = 0.20, 0.10$, and 0.01 are also commonly used).
> - Calculate the sample mean $\overline{x}$ and the sample standard deviation s (Definition 4.1.1).
> - Calculate the critical value $t^* = \mathtt{idfStudent}(n - 1, 1 - \alpha/2)$.
> - Calculate the interval endpoints
>
> $$\overline{x} \pm \frac{t^* s}{\sqrt{n - 1}}.$$
>
> If n is sufficiently large, you are then $(1 - \alpha) \times 100\%$ confident that μ lies somewhere in the interval defined by these two endpoints. The associated point estimate of μ is the midpoint of this interval (which is $\overline{x}$).

Example 8.1.8

The data set

$$1.051 \quad 6.438 \quad 2.646 \quad 0.805 \quad 1.505 \quad 0.546 \quad 2.281 \quad 2.822 \quad 0.414 \quad 1.307$$

is a random sample of size $n = 10$ from a population with unknown mean μ. The sample mean and standard deviation are $\overline{x} = 1.982$ and $s = 1.690$.

- To calculate a 90% confidence interval estimate, we first evaluate

$$t^* = \mathtt{idfStudent}(9, \ 0.95) \cong 1.833$$

and then construct the interval estimate as

$$1.982 \pm (1.833) \frac{1.690}{\sqrt{9}} = 1.982 \pm 1.032.$$

Therefore, we are approximately 90% confident that the (unknown) value of μ is somewhere between 0.950 and 3.014. This is typically reported as the confidence interval $0.950 \le \mu \le 3.014$.

- Similarly, to calculate a 95% confidence interval estimate,

$$t^* = \mathtt{idfStudent}(9, \ 0.975) \cong 2.262,$$

and thus the interval is

$$1.982 \pm (2.262) \frac{1.690}{\sqrt{9}} = 1.982 \pm 1.274.$$

Therefore, we are approximately 95% confident that the (unknown) value of μ is somewhere in the interval $0.708 \le \mu \le 3.256$.

• And, to calculate a 99% confidence interval estimate,

$$t^* = \text{idfStudent}(9, 0.995) \cong 3.250$$

and thus the interval is

$$1.982 \pm (3.250)\frac{1.690}{\sqrt{9}} = 1.982 \pm 1.832.$$

Therefore, we are approximately 99% confident that the (unknown) value of μ is somewhere in the interval $0.150 \le \mu \le 3.814$.

In each case, the modifier "approximately" is used because n is not large. If we had knowledge that the ten data values were drawn from a *Normal*(μ, σ) population distribution, then the "approximately" modifier could be dropped and the confidence interval would be known as an "exact" confidence interval.

Trade-off—Confidence Versus Sample Size.   Example 8.1.8 illustrates the classic interval-estimation trade-off. For a given sample size, more confidence can be achieved *only* at the expense of a wider interval, and a narrower interval can be achieved *only* at the expense of less confidence. The only way to make the interval smaller without lessening the level of confidence is to increase the sample size—collect *more* data!

Collecting more data in simulation applications is usually easy to do. The question, then, is how much more data is enough? That is, how large should n be in Algorithm 8.1.1 to achieve an interval estimate $\overline{x} \pm w$ for a *user-specified* interval width w? The answer to this question is based on using Algorithm 4.1.1 in conjunction with Algorithm 8.1.1 to collect data iteratively until a specified interval width is achieved. This process is simplified by the observation that, if n is large, then the value of t^* is essentially independent of n, as illustrated in Figure 8.1.8.

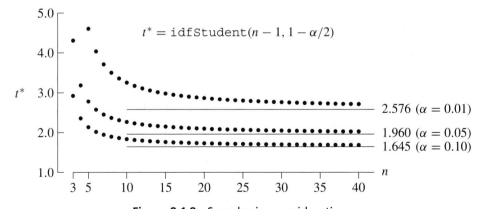

Figure 8.1.8 Sample-size considerations.

The asymptotic (large n) value of t^* indicated is

$$t^*_\infty = \lim_{n \to \infty} \text{idfStudent}(n - 1, 1 - \alpha/2) = \text{idfNormal}(0.0, 1.0, 1 - \alpha/2).$$

Unless α is very close to 0.0, if $n > 40$, then the asymptotic value t^*_∞ can be used. In particular, if $n > 40$ and we wish to construct a 95% confidence interval estimate for μ, then $t^*_\infty = 1.960$ can be used for t^* in Algorithm 8.1.1.

Example 8.1.9

Given a reasonable guess for s and a user-specified *half-width* parameter w, if t^*_∞ is used in place of t^*, then the value of n can be computed by solving the equation

$$w = \frac{t^*_\infty s}{\sqrt{n - 1}}$$

for n. The result is

$$n = \left\lfloor \left(\frac{t^*_\infty s}{w}\right)^2 \right\rfloor + 1,$$

provided that $n > 40$ (which will surely be the case if w/s is small). For example, if $s = 3.0$, $\alpha = 0.05$, and $w = 0.5$, then the sample size $n = 139$ should be used.

Example 8.1.10

Similarly, if a reasonable guess for s is not available, then w can be specified as a proportion of s, thereby eliminating s from the previous equation. For example, if w is 10% of s (so that $w/s = 0.1$) and if 95% confidence is desired, then the value $n = 385$ should be used to estimate μ to within $\pm w$.

Program estimate. By implementing Algorithm 8.1.1, program estimate automates the interval-estimation process. A typical application of this program is to estimate the value of an unknown population mean μ by using n replications to generate an *independent* random-variate sample $x_1, x_2, \ldots, x_n$. Let Generate() represent a discrete-event or Monte Carlo simulation program configured as a function that returns a random-variate output x. If n is specified, then the algorithm

```
for (i = 1; i <= n; i++)
    xi = Generate();
return x1, x2, ..., xn;
```

generates a random-variate sample. Given a user-defined level of confidence $1 - \alpha$, program estimate can then be used with $x_1, x_2, \ldots, x_n$, to compute an interval estimate for μ.

As an alternative to using program estimate directly, if the desired interval half-width w is specified instead of the sample size n, then repeated calls to Generate() can be used in conjunction with Algorithm 4.1.1 to collect data iteratively until the $2w$ interval width is achieved. This process is summarized by the following algorithm, which uses the t^* asymptotic approximation discussed previously to eliminate needless calls to idfStudent$(n - 1, 1 - \alpha/2)$ inside the while loop.

Algorithm 8.1.2   Given an interval half-width w and a level of confidence $1 - \alpha$, this algorithm computes the (large-sample-size) interval estimate for μ having the form $\bar{x} \pm w$.

```
t = idfNormal(0.0, 1.0, 1 - α/2);                    // this is t*∞
x = Generate();
n = 1;
v = 0.0;
x̄ = x;
while ((n < 40) or (t * sqrt(v / n) > w * sqrt(n - 1)) {
    x = Generate();
    n++
    d = x - x̄;
    v = v + d * d * (n - 1) / n;
    x̄ = x̄ + d / n;
}
return n, x̄;
```

Welford's algorithm (Algorithm 4.1.1) is contained in Algorithm 8.1.2, so that only one pass will be made through the data and the likelihood of overflow is minimized.

It is important to appreciate the need for sample *independence* in Algorithms 8.1.1 and 8.1.2. The problem here is that both of these algorithms can be used naively even if the sample is *not* independent. However, the resulting interval estimate is valid only if the sample is independent. If the sample is not independent, then the interval "coverage" will be incorrect, perhaps dramatically so. For example, an interval alleged to be 95% confident may be actually only, say, 37% confident.

The Meaning of Confidence.   One common mistake that occurs when novices begin working with confidence intervals is reflected in the following statements:

> "I made n runs of the simulation and created a 95% confidence interval for μ. The probability that the true mean μ is contained in my confidence interval is 0.95."

The problem with this statement is that the (unknown) true value of μ either *is* or *is not* in the confidence interval. A probability statement can be made, although it is painfully tedious:

> "I made n runs of the simulation and created a 95% confidence interval for μ. This particular interval may or may not contain μ. However, if I were to create many confidence intervals like this one many times over, then approximately 95% of these intervals would cover the true mean μ."

This discussion indicates why the phrase "confidence interval" is used rather than the phrase "probability interval."

Example 8.1.11

One hundred 95% confidence intervals associated with data drawn from a *Normal* (6, 3) population with sample size $n = 9$ are shown in Figure 8.1.9, each displayed as a vertical line. A horizontal line is drawn at the population mean $\mu = 6$. Three of the confidence intervals "miss low" (e.g., the upper limit is less than $\mu = 6$) and two of the confidence intervals "miss high" (e.g., the lower limit is greater than $\mu = 6$). The lower and upper *limits* of these confidence intervals are random—whereas μ is fixed. Since sampling is from a normal distribution, these confidence intervals are *exact*.

Figure 8.1.9   One hundred confidence intervals.

8.1.3 Exercises

8.1.1   This exercise illustrates the central limit theorem. (*a*) Generate 100 000 *Uniform* (2, 6) random-variate samples of size $n = 4$. For each sample, calculate the sample mean and then generate a 20-bin continuous-data histogram of the 100 000 sample means. (Use 10 bins on each side of 4.0, and choose the other histogram parameters appropriately.) Plot the histogram, and overlay the theoretical *Normal* pdf that should approximate it. (*b*) Repeat with $n = 9$. (*c*) Repeat with $n = 25$. (*d*) Comment.

8.1.2   To illustrate that the central limit theorem is an approximation, prove that, for an *iid* sequence of *Exponential* (μ) random variables $X_1, X_2, \ldots, X_n$, the pdf of the $\overline{X}$ is

$$f(\overline{x}) = \frac{n}{\mu(n-1)!} \left(\frac{n\overline{x}}{\mu}\right)^{n-1} \exp(-n\overline{x}/\mu) \qquad \overline{x} > 0.$$

8.1.3   (*a*) Generate 100 000 *Poisson* (10) random-variate samples of size $n = 4$. For each sample, calculate the variance, and then calculate the mean of these 100 000 sample variances. (*b*) Compare this mean with the theoretical variance of a *Poisson* (10) random variable. (*c*) Explain why there is a difference between these two variances.

8.1.4   This exercise is a continuation of Exercise 8.1.1. (*a*) Generate 100 000 *Uniform* (2, 6) random-variate samples of size $n = 4$. For each sample, calculate the sample mean $\overline{x}$, the standard deviation s, and

$$t = \frac{\overline{x} - 4}{s/\sqrt{n-1}}.$$

Then generate a 20-bin histogram of the $100\,000$ t's. (Use 10 bins on each side of 0.0, and choose the other histogram parameters appropriately.) Plot the histogram and overlay the theoretical *Student* pdf that should approximate it. (*b*) Repeat with $n = 9$. (*c*) Repeat with $n = 25$. (*d*) Comment.

8.1.5   This exercise illustrates interval estimation for the mean. (*a*) Generate $100\,000$ *Exponential* (4) random-variate samples of size $n = 4$, and, for each sample, calculate a 95% confidence-interval estimate of the mean. Compute the relative frequency of times (out of $100\,000$) that the interval estimate includes 4.0. (*b*) Repeat with $n = 9, 25, 49, 100$, and 1000. (*c*) Comment on how the "coverage" depends on the sample size.

8.1.6   Repeat Exercise 8.1.5, but using *Uniform* (2, 6) random variates instead.

8.1.7   Repeat Exercise 8.1.5, but using *Geometric* (0.8) random variates instead.

8.1.8   How does Exercise 8.1.6 relate to Exercises 8.1.1 and 8.1.4? Be specific.

8.1.9   Do Exercise 7.5.7, but calculate 95% confidence-interval estimates for the instantaneous (snapshot) number in the node at each of the $\lambda(t)$ time knots.

8.2 MONTE CARLO ESTIMATION

One of the primary applications of parameter estimation is Monte Carlo simulation, which is based upon the use of many independent replications of the same static, stochastic system to generate an estimate of some parameter of interest. If, as in Section 2.3, this estimate is just a *point* estimate, then there is no associated measure of precision. How close is the point estimate likely to be to the true (unknown) value of the parameter being estimated? If the Monte Carlo estimate is an *interval* estimate, however, then the precision of the estimate has been quantified by establishing an interval in which the true value of the parameter lies with high probability.

8.2.1 Probability Estimation

Example 8.2.1
The program craps was run once with $n = 100$ replications and an rngs initial seed of 12345. There were 56 wins and 44 losses; the corresponding *point* estimate of the probability of winning is 0.56. Recall, from Section 2.4.2, that the true (theoretical) probability of winning at craps is $244/495 \cong 0.493$; we see that this *point* estimate is not particularly accurate. However, by *interval* estimation, it can be shown that 56 wins in 100 tries corresponds to the 95% confidence-*interval* estimate 0.56 ± 0.10 (see Example 8.2.3, to follow). Therefore we are 95% confident that the true probability of winning is somewhere between 0.46 and 0.66. There are two important points in this case:

- the 0.56 ± 0.10 interval estimate is "correct," in the sense that it contains the true probability of winning;
- that the interval is quite wide indicates that (many) more than 100 replications should have been used.

The general interval-estimation technique developed in the previous section applies to the specific case of Monte Carlo simulation used to estimate a probability. Recall, from Section 4.1, that, if $x_1, x_2, \ldots, x_n$ is a random sample of size n with the property that each x_i is either 0 (a "loss") or 1 (a "win"), then the sample mean is

$$\overline{x} = \frac{1}{n} \sum_{i=1}^{n} x_i = \frac{\text{the number of 1's}}{n};$$

and that, since $x_i^2 = x_i$ for binary data, the sample standard deviation is

$$s = \sqrt{\left(\frac{1}{n} \sum_{i=1}^{n} x_i^2\right) - \overline{x}^2} = \sqrt{\overline{x} - \overline{x}^2} = \sqrt{\overline{x}(1 - \overline{x})}.$$

From Algorithm 8.1.1, if n is sufficiently large, the corresponding approximate interval estimate for the (unknown) population mean is

$$\overline{x} \pm t^* \sqrt{\frac{\overline{x}(1 - \overline{x})}{n - 1}},$$

where the critical value $t^* = \texttt{idfStudent}(n - 1, 1 - \alpha/2)$ is determined by the sample size and the level of confidence $1 - \alpha$.

Example 8.2.1 not withstanding, when Monte Carlo simulation is used to estimate a probability, the sample size (the number of replications) is almost *always* large. Moreover, the conventional level of confidence is 95%. Per the discussion in Section 8.1, if n is sufficiently large, and if $\alpha = 0.05$, then

$$t^* = t_\infty^* = \texttt{idfNormal}(0.0, 1.0, 0.975) \cong 1.960 \cong 2.0.$$

In addition, if n is large, then the distinction between n and $n - 1$ in the interval estimate equation is insignificant. Therefore, the approximate 95% confidence interval estimate can be written as $\overline{x} \pm 1.96\sqrt{\overline{x}(1 - \overline{x})/n}$, or

$$\overline{x} - 1.96\sqrt{\overline{x}(1 - \overline{x})/n} < p < \overline{x} + 1.96\sqrt{\overline{x}(1 - \overline{x})/n}.$$

Algorithm.   If p denotes the (unknown) value of the probability to be estimated, then it is conventional to use the notation $\hat{p}$, rather than $\overline{x}$, to denote the point estimate of p. Using this notation, Monte Carlo probability-interval estimation can be summarized by the following algorithm. [If p is very close to either 0.0 or 1.0, then more sophisticated approaches to interval estimation, such as those given in Hogg, McKean, and Craig (2005), are required.]

Algorithm 8.2.1   To estimate the unknown probability p of an event $\mathcal{A}$ by Monte Carlo simulation, replicate n times and let

$$\hat{p} = \frac{\text{the number of occurrences of event } \mathcal{A}}{n}$$

be the resulting point estimate of $p = \Pr(\mathcal{A})$. We are then approximately 95% confident that p lies in the interval having the endpoints

$$\hat{p} \pm 1.96 \sqrt{\frac{\hat{p}(1 - \hat{p})}{n}}.$$

Example 8.2.2

Program `craps` was run 200 times with $n = 100$. For each run, a 95% confidence interval estimate was computed via Algorithm 8.2.1. Each of these 200 interval estimates is illustrated as a vertical line in Figure 8.2.1; the horizontal line corresponds to the true probability $p \cong 0.493$. (The leftmost vertical line corresponds to the 0.56 ± 0.10 interval estimate in Example 8.2.1.)

Figure 8.2.1   Confidence intervals for craps.

As is consistent with the probability interpretation of a 95% confidence interval, we see that, in this case, 95.5% (191 of 200) of the intervals include p.

Approximate Interval Estimation.   If the point estimate $\hat{p}$ is close to 0.5, then $\sqrt{\hat{p}(1 - \hat{p})}$ is also close to 0.5. In this case, the 95% confidence interval estimate reduces to (approximately)

$$\hat{p} \pm \frac{1}{\sqrt{n}}.$$

The significance of this approximate interval estimate is that it is easily remembered, it is conservative, and it clearly highlights the "curse" of $\sqrt{n}$. In particular, unless p is close to 0.0 or 1.0, approximately

- 100 replications are necessary to estimate p to within ± 0.1;
- 10 000 replications are necessary to estimate p to within ± 0.01;
- 1 000 000 replications are necessary to estimate p to within ± 0.001.

Therefore, the general result is seen clearly—to achieve one more decimal digit of precision in the interval estimate, be prepared to do 100 times more work! (This is a worst-case scenario—see Example 8.2.4.)

Example 8.2.3

For a random sample of size $n = 100$, $\sqrt{n} = 10$. Therefore, if $\hat{p} = 0.56$, then the corresponding 95% confidence-interval estimate is approximately 0.56 ± 0.10.

Specified Precision

Example 8.2.4

When we are using Monte Carlo simulation to estimate a probability p, it is frequently the case that we want to know *a priori* how many replications n will be required to achieve a specified precision $\pm w$. Specifically, if it is desired to estimate p as $\hat{p} \pm w$ with 95% confidence, then by replacing 1.96 with 2, the equation

$$w = 2\sqrt{\frac{\hat{p}(1 - \hat{p})}{n}}$$

can be solved for n to yield

$$n = \left\lfloor \frac{4\hat{p}(1 - \hat{p})}{w^2} \right\rfloor.$$

A small run with, say, $n = 100$ can be used to get a *preliminary* value for $\hat{p}$. This preliminary value can then be used in the previous equation to calculate the approximate number of replications required to estimate p to within $\pm w$. In particular, suppose a small trial run produced the preliminary estimate $\hat{p} = 0.2$, and suppose we wish to estimate p to within ± 0.01. Then $w = 0.01$ and, in this case, the approximate number of replications required to achieve a 95% confidence interval estimate is

$$n = \left\lfloor \frac{4(0.2)(0.8)}{(0.01)^2} \right\rfloor = 6400$$

instead of the more conservative 10 000 replications suggested previously.

The sample-size calculations considered here apply also to political polls used to estimate the support for a particular issue or candidate. Monte Carlo estimation has two advantages over this more classical statistical estimation problem: (*i*) observations tend to be very cheap and can be collected quickly, and (*ii*) there is no concern over *sampling bias*, which might be generated by nonrepresentative sampling.* Sampling bias can be introduced in classical statistics by factors such as dependent sampling.

Algorithm 8.2.2, properly modified, can be used to automate the process illustrated in Example 8.2.4. In this case, the function `Generate()` represents a Monte Carlo program

---

*One of the more famous cases of biased sampling occurred after the pre-election polls of 1948, which predicted that Thomas Dewey would defeat incumbent Harry Truman in the U.S. presidential election. Three separate polls incorrectly predicted that Dewey would prevail. What went wrong? All three polls used a technique known as *quota sampling*, which resulted in somewhat more likely access to Republicans, who were more likely to have permanent addresses, own telephones, and so on (Trosset, 2005). A photograph of the victorious and jubilant Truman holding a copy of the *Chicago Tribune* with the headline "Dewey defeats Truman" was the result of this biased sampling.

configured to estimate the probability $p = \Pr(\mathcal{A})$ by returning the value $x = 1$ or $x = 0$, depending on whether event $\mathcal{A}$ occurred or not. The result is the following algorithm.

Algorithm 8.2.2   Given a specified interval half-width w and level of confidence $1 - \alpha$, the following algorithm computes a confidence-interval estimate for p to at least $\hat{p} \pm w$ precision.

```
t = idfNormal(0.0, 1.0, 1 - α / 2);
n = 1;
x = Generate();                              // returns 0 or 1
p̂ = x;
while ((n < 40) or (t * sqrt(p̂ * (1 - p̂)) > w * sqrt(n)) {
   n++;
   x = Generate();                           // returns 0 or 1
   p̂ = p̂ + (x − p̂) / n;
}
return n, p̂;
```

8.2.2 Monte Carlo Integration

As a second application of interval estimation, we will now consider the use of Monte Carlo simulation to estimate the value of the definite integral

$$I = \int_a^b g(x)\,dx.$$

This is a classic application having considerable intuitive appeal but limited practical value.* Let X be a *Uniform*(a, b) random variable, and use the function $g(\cdot)$ to define the random variable

$$Y = (b - a)g(X).$$

Because the pdf of X is $f(x) = 1/(b - a)$ for $a < x < b$, from Definition 7.1.5 the expected value of Y is

$$E[Y] = \int_a^b (b - a)g(x)f(x)\,dx = \int_a^b g(x)\,dx = I.$$

Therefore $I = E[Y]$, and so an interval estimate $\hat{I} \pm w$ for I can be calculated by using a slightly modified version of Algorithm 8.1.2 to compute an interval estimate for $E[Y]$, as summarized by Algorithm 8.2.3.

---

*One-dimensional Monte Carlo integration has limited practical value because there are much more efficient deterministic algorithms for doing the same thing (e.g., the trapezoid rule and Simpson's rule)—see Example 8.2.5.

Algorithm 8.2.3   Given a specified interval half-width w and level of confidence $1 - \alpha$, and given a need to evaluate the definite integral

$$I = \int_a^b g(x)\, dx,$$

the following algorithm computes an interval estimate for I to at least $\hat{I} \pm w$ precision.

```
t = idfNormal(0.0, 1.0, 1 - α / 2);
n = 1;
x = Uniform(a, b);                          // modify as appropriate
y = (b - a) * g(x);                         // modify as appropriate
ȳ = y;
v = 0.0;
while ((n < 40) or (t * sqrt(v / n) > w * sqrt(n - 1))) {
   n++;
   x = Uniform(a, b);                       // modify as appropriate
   y = (b - a) * g(x);                      // modify as appropriate
   d = y - ȳ;
   v = v + d * d * (n - 1) / n;
   ȳ = ȳ + d / n;
}
return n, ȳ;                                             // ȳ is Î
```

Example 8.2.5

Algorithm 8.2.3 was used to evaluate the integral

$$I = \frac{1}{\sqrt{2\pi}} \int_{-3}^{4} \exp(-x^2/2)\, dx$$

with $w = 0.01$ and $\alpha = 0.05$. With the rngs initial seed 12345 and

$$g(x) = \frac{\exp(-x^2/2)}{\sqrt{2\pi}} \qquad -3 < x < 4,$$

the number of replications $n = 37\,550$ produced the interval estimate

$$\hat{I} \pm w = 0.997 \pm 0.01.$$

That is, we are 95% confident that $0.987 < I < 1.007$. This interval estimate is consistent with the correct value of I, which is $I = \Phi(4) - \Phi(-3) \cong 0.9986$.* Recognize,

---

*See Section 7.4.4 for the definition of $\Phi(\cdot)$.

however, that 37 550 function evaluations were required to achieve this modest precision. In contrast, the trapezoid rule with just 30 function evaluations evaluates I to ± 0.0001 precision. To achieve this precision with Monte Carlo integration, approximately 3.8×10^8 function evaluations would be required. (Start early—your mileage could vary.)

Algorithm 8.2.3 uses

$$\hat{I} = \frac{(b-a)}{n} \sum_{i=1}^{n} g(x_i) \qquad \text{as a point estimate of} \qquad I = \int_a^b g(x)\,dx$$

where $x_1, x_2, \ldots, x_n$ is a *Uniform*(a, b) random-variate sample. As illustrated in Example 8.2.5, a huge sample size may be required to achieve reasonable precision. What can be done to improve things?

Importance Sampling.   Relative to Example 8.2.5, it should be clear that the slow convergence is due, in part, to the use of *Uniform*(a, b) random variates: It is wasteful to repeatedly sample and sum $g(x)$ for values of x where $g(x) \cong 0$. Instead, the integration process should concentrate on sampling and summing $g(x)$ for values of x where $g(x)$ is large, because those values contribute most to the integral. What is needed to accomplish this is a random variable X whose pdf is similar to $g(x)$. This particular *variance-reduction technique* is known as importance sampling.

In general, to derive a generalization of Algorithm 8.2.3, we can proceed as illustrated previously, *but* with X a general random variable with pdf $f(x) > 0$ defined for all possible values $x \in \mathcal{X} = (a, b)$. Then we define the new random variable

$$Y = g(X)/f(X)$$

and recognize that the expected value of Y is

$$E[Y] = \int_a^b \frac{g(x)}{f(x)} f(x)\,dx = \int_a^b g(x)\,dx = I.$$

As before, an interval estimate $\hat{I} \pm w$ for $I = E[Y]$ can be calculated by using a slightly modified version of Algorithm 8.2.3 to compute an interval estimate for $E[Y]$. The key feature here is a clever choice of X or, equivalently, the pdf $f(\cdot)$.

Example 8.2.6

In the spirit of Example 8.2.5, suppose we wish to evaluate the integral

$$I = \int_0^4 \exp(-x^2/2)\,dx.$$

If X is a *Uniform*(a, b) random variate, then this is, essentially, Example 8.2.5 all over again, complete with slow convergence. If, instead, X is an *Exponential*(1) random variate, truncated to the interval $\mathcal{X} = (0, 4)$, then better convergence should be achieved. In this case, the pdf of X is

$$f(x) = \frac{\exp(-x)}{F(4)} \qquad 0 < x < 4$$

where $F(4) = 1 - \exp(-4) \cong 0.9817$ is the cdf of an untruncated *Exponential*(1) random variable. Algorithm 8.2.3, with the two indicated x, y assignments modified accordingly, can be used to compute an interval estimate for I. The details are left as an exercise.

8.2.3 Time-Averaged Statistics

Recall from Section 1.2 that, by definition, the time-averaged number in a single-server service node is

$$\bar{l} = \frac{1}{\tau} \int_0^\tau l(t)\, dt,$$

where $l(t)$ is the number in the service node at time t and the average is over the interval $0 < t < \tau$. The following example, an application of Monte Carlo integration, clarifies the statement made in Section 1.2 that "if we were to observe (sample) the number in the service node at many different times chosen *at random* between 0 and τ and then calculate the arithmetic average of all these observations, the result should be close to $\bar{l}$."

Example 8.2.7
Let the random variable T be *Uniform*$(0, \tau)$, and define the new (discrete) random variable

$$L = l(T).$$

By definition, the pdf of T is $f(t) = 1/\tau$ for $0 < t < \tau$; therefore, the expected value of L is

$$E[L] = \int_0^\tau l(t) f(t)\, dt = \frac{1}{\tau} \int_0^\tau l(t)\, dt = \bar{l}.$$

Now, let the observation times $t_1, t_2, \ldots, t_n$ be generated as an n-point random-variate sample of T. Then $l(t_1), l(t_2), \ldots, l(t_n)$ is a random sample of L, and the mean of this random sample is an (unbiased) *point* estimate of the expected value $E[L] = \bar{l}$. If the sample size is large, the sample mean ("arithmetic average") should be close to $\bar{l}$.*

---

*There is a notational conflict in this example. Do not confuse the n (sample size) in Example 8.2.7 with the n (number of jobs) in the single-server service-node model presented in Section 1.2. The two numbers can be the same, but that is not necessarily the case.

Random Sampling.   The previous example applies equally well to any other function defined for $0 < t < \tau$. For example, the utilization can be estimated by sampling $x(t)$. Moreover, the random times t_i at which the $l(t_i)$ [or $x(t_i)$] samples are taken can be generated by the same algorithm used to generate a stationary Poisson arrival process (Algorithm 7.3.1)—that is, to generate n random sample times (in sorted order) with an average inter-sample time of δ, we can use the algorithm

```
t0 = 0;
for (i = 0; i < n; i++)
    t_{i+1} = t_i + Exponential(δ);
return t1, t2, ..., tn;
```

8.2.4 Exercises

8.2.1   (*a*) Modify program `buffon` so that it will use Algorithm 8.2.1 to produce a 95% confidence-interval estimate for the probability of a line-crossing. Then run this modified program 10 000 times (with a different `rngs` initial seed for each run), use $n = 100$ replications per run, and count the proportion of times out of 10 000 that the true (theoretical) probability of a line-crossing falls within the 95% confidence interval predicted. (*b*) Comment.

8.2.2   Work through the details of Example 8.2.6, estimating the value of I as $\hat{I} \pm w$, with 95% confidence and $w = 0.01$.

8.2.3   How many replications are required to estimate a probability whose value is expected to be 0.1 to within ± 0.001 with 95% confidence?

8.2.4   (*a*) Use Algorithm 8.2.3 to estimate the value of the integral

$$I = \int_0^2 x^7 (2 - x)^{13} \, dx$$

to within 1% of its true (unknown) value with 95% confidence. (*b*) Calculate the true value of I. (*c*) Comment.

8.2.5   (*a*) Evaluate the integral in Exercise 8.2.4, using the trapezoid rule. (*b*) Comment.

8.2.6   Modify program `ssq4` to use random sampling to generate 95% confidence-interval estimates for $\bar{l}$, $\bar{q}$, and $\bar{x}$. Sample at the rate 2λ (where λ is the arrival rate).

8.3 FINITE-HORIZON AND INFINITE-HORIZON STATISTICS

The discussion in this section is motivated by the following type of question and answer. Consider a single-server service node that processes a *large* number of jobs arriving as a stationary Poisson process with arrival rate $\lambda = 1/2$. If the service time is an *Erlang* $(5, 0.3)$ random variable, and so the service rate is $\nu = 2/3$, then what is the average wait in the node, the time-averaged number in the node, and the utilization? This question can be answered by using program `ssq4`. To do so, however, we must make some judgment about what constitutes a "large" number of jobs.

Moreover, we should be concerned about the possible effects on the simulation output of forcing the service node to begin and end in an idle state, as is done in program `ssq4`.*

8.3.1 Experimental Investigation

> **Definition 8.3.1** *Steady-state* system statistics are those statistics, if they exist, that are produced by simulating the operation of a *stationary* discrete-event system for an effectively infinite length of time.

Example 8.3.1

If program `sis4` is used to simulate a large number of time intervals, the resulting average inventory level per time interval is an estimate of a steady-state statistic. In particular, for an `rngs` initial seed of 12345 and for an increasing number of time intervals n, program `sis4` yields the following sequence of average-inventory-level estimates $\bar{l} = \bar{l}^{+} - \bar{l}^{-}$.

n	20	40	80	160	320	640	1280	2560
$\bar{l}$	25.98	26.09	25.49	27.24	26.96	26.36	27.19	26.75

As $n \to \infty$, it appears that $\bar{l}$ will converge to a steady-state value of approximately 26.75.

Since the $n \to \infty$ convergence in Example 8.3.1 can only be approximated, a better approach to estimating the steady-state value of $\bar{l}$ would be to develop an *interval*-estimation technique applicable when n is large, but finite. One way to do this is to use the method of *batch means*—which will be introduced in Section 8.4.

The "if they exist" phrase in Definition 8.3.1 is important. For some systems, steady-state statistics do not exist.

Example 8.3.2

If we run the simulation program `ssq4` with an arrival rate of $1/2$ and the *Erlang* (n, b) parameters specified (and so the service rate is $1/2$ or less), then we will find that the average wait and the average number in the service node will tend to increase without limit as the number of jobs is increased. The reason for this should be intuitive—unless the service rate is greater than the arrival rate then, on average, the server will not be able to keep up with the demand, resulting in an average queue length that continues to grow as more jobs arrive. In this case, the steady-state average wait and the average number in the node are both infinite.

---

*As an alternative, if we interpret "large" as *infinite*, then this question can also be answered by applying the mathematical tools of *steady-state* queuing theory. Some of these theoretical tools are summarized in Section 8.5.

Infinite-Horizon Versus Finite-Horizon Statistics.   Steady-state statistics are also known as *infinite-horizon* statistics. The idea here is that an infinite-horizon discrete-event simulation is one for which the simulated operational time is effectively infinite. In contrast, a *finite-horizon* discrete-event simulation is one for which the simulated operational time is finite. One of the characteristics of an infinite-horizon simulation is that the initial conditions are not important—as simulated time becomes infinite, the choice of the initial state of the simulated system on system statistics becomes irrelevant and the simulated system loses all memory of its initial state. For finite-horizon simulations, this is not true—the initial state of the system is important.

Definition 8.3.2   *Transient* system statistics are those statistics that are produced by a finite-horizon discrete-event simulation.

Example 8.3.3
With minor modifications, program `ssq4` can be used to simulate an initially idle $M/M/1$ service node processing a small number of jobs (say 100) and having a relatively high traffic intensity (say 0.8).* If this program is executed multiple times, varying *only* the `rngs` initial seed from replication to replication, then the average wait in the node will vary significantly from replication to replication. Moreover, for most (perhaps all) replications, the average wait will *not* be close to the steady-state (infinite-horizon) average wait in an $M/M/1$ service node. On the other hand, if a relatively large number of jobs (say 10 000) are used, then the replication-to-replication variability of the average wait in the node will be much less significant, and, for most replications, the average wait will be close to the steady-state average number in an $M/M/1$ service node.

Consistent with Example 8.3.3, if the number of jobs processed (or equivalently, the simulated operational time) is small, then the average wait in the service node will be strongly biased by the initial idle state of the system. However, as the number of jobs becomes infinite, this initial-condition bias disappears.

In addition to the importance of the initial-condition bias, there is another important distinction between steady-state and transient statistics. In an infinite-horizon simulation, it is virtually always true that the system "environment" is assumed to remain *static* (stationary). Thus, for example, if the system is a single-server service node, then both the arrival rate and the service rate are assumed to remain constant in time. Most real-world systems do not operate in a static environment for an extended period of time; for such systems, steady-state statistics might have little meaning. In a finite-horizon simulation, there is no need to assume a static environment—natural temporal (dynamic) changes in the environment can be incorporated into the discrete-event simulation, and the transient statistics so produced will then, properly, reflect the influence of this dynamic environment.

---

*Recall that an $M/M/1$ service node is a single-server service node with a Poisson arrival process and *Exponential* service times.

The choice between using steady-state or transient statistics is typically dictated by the system being simulated. An example of each type of system is given below.

- Consider a bank that opens at 9 AM and closes at 5 PM. A finite-horizon simulation over the 8-hour period produces *transient* statistics, which might be valuable in determining the optimal staffing of tellers throughout the day.
- Consider a fast-food restaurant with a drive-up window that experiences a lunch rush period between 11:45 AM and 1:15 PM with an arrival rate that remains constant over the rush period. This 90-minute period could be simulated for a much longer time period, producing *steady-state* statistics that might be valuable for estimating the average queue length at 1:15 PM under a particular set of operating procedures.

Steady-state statistics are better understood, because they are easier to analyze *mathematically*—for example, by using analytic methods, as in Section 8.5. Transient statistics are important, because steady state is often a convenient fiction—most real systems do not operate long enough in a stationary environment to produce steady-state statistics. Depending on the application, both transient and steady-state statistics may be important. For that reason, one of the most important skills that must be developed by a discrete-event-simulation specialist is the ability to decide, on a system-by-system basis, which kind of statistics best characterizes the system's performance.

Initial and Terminal Conditions.   Finite-horizon discrete-event simulations are also known as *terminating* simulations. This terminology emphasizes the point that there must be some terminal conditions that are used to define the simulated period of system operation. Terminal conditions are typically stated in terms either of a fixed number of events (e.g., processed jobs) or of a fixed amount of (simulated) elapsed time. In any case, for a finite-horizon discrete-event-simulation model, it is important to define the desired terminal system state clearly.

Example 8.3.4
In program ssq4, the system (service node) state is idle at the beginning and at the end of the simulation. The terminal condition is specified by the "close the door" time. Similarly, in program sis4, the state of the inventory system is the current and on-order inventory levels; these states are the same at the beginning and at the end of the simulation. The terminal condition is specified by the number of time intervals.

Infinite-horizon (steady-state) discrete-event simulations must be terminated. However, in an infinite-horizon simulation, the state of the system at termination should not be important, and so it is common to use whatever stopping conditions are most convenient. The basic idea here is that steady-state statistics are based on such a huge amount of data that a few "non-steady-state" data points accumulated at the beginning and the end of the simulation should have no significant impact (bias) on output statistics. Recognize, however, that there is a fair amount of wishful thinking involved here; a significant amount of research (e.g., Schruben, 1982; Gallagher, Bauer, and Maybeck, 1996) has been devoted to the problem of eliminating initial-condition (and terminal-condition) bias from simulation-generated steady-state statistics.

8.3.2 Formal Representation

To view the previous discussion in a more formal analytic setting, consider a discrete-event system whose state is characterized by a time-varying random variable $X(t)$ with associated realizations denoted by $x(t)$. The state variable $X(t)$ can be either discrete or continuous, and the initial state $X(0)$ can either be specified as a fixed value or be modeled as a random variable with a specified pdf. The state variable $X(\cdot)$ is known formally as a *stochastic process*.*

*   The typical objective of a finite-horizon simulation of this system would be to estimate the time-averaged *transient* statistic

$$\overline{X}(\tau) = \frac{1}{\tau} \int_0^\tau X(t)\, dt,$$

    where $\tau > 0$ is the terminal time.
*   In contrast, the typical objective of an infinite-horizon simulation of this system would be to estimate the time-averaged *steady-state* statistic

$$\overline{x} = \lim_{\tau \to \infty} \overline{X}(\tau) = \lim_{\tau \to \infty} \frac{1}{\tau} \int_0^\tau X(t)\, dt.$$

As the notation suggests, $\overline{X}(\tau)$ is a random variable but $\overline{x}$ is not: To estimate $\overline{X}(\tau)$ it is necessary to replicate a finite-horizon simulation many times, each time generating one estimate of $\overline{X}(\tau)$. In contrast, to estimate $\overline{x}$, it is, in principle, necessary to run the infinite-horizon simulation only once, but for a very long time.

> **Definition 8.3.3**   If a discrete-event simulation is repeated, varying *only* the rngs initial states from run to run, then each run of the simulation program is a *replication*, and the totality of replications is said to be an *ensemble*.

Replications are used to generate *independent* estimates of the same transient statistic. Therefore, the initial seed for each replication should be chosen so that there is no replication-to-replication overlap in the sequence of random numbers used. The standard way to prevent this overlap is to use the final state of each rngs stream from *one* replication as the initial state for the *next* replication. This is automatically accomplished by calling PlantSeeds once *outside* the main replication loop. The set of estimated time-averaged statistics is then an independent sample that can be used to produce an interval estimate, as in Section 8.2.

Independent Replications and Interval Estimation.   Suppose the finite-horizon simulation is replicated n times, each time generating a state–time history $x_i(t)$ and the corresponding time-averaged transient statistic

$$\overline{x}_i(\tau) = \frac{1}{\tau} \int_0^\tau x_i(t)\, dt,$$

---

*For a single-server service node with stochastic arrival and service times, $X(t)$ might be the number in the node at time t. For a simple inventory system with stochastic demand instances and delivery lags, $X(t)$ might be the inventory level at time t. For a stochastic arrival process, $X(t)$ might be the number of arrivals in the interval $(0, t)$.

where $i = 1, 2, \ldots, n$ is the replication index within the ensemble. Each data point $\bar{x}_i(\tau)$ is then an independent observation of the random variable $\overline{X}(\tau)$. If n is large enough, the pdf of $\overline{X}(\tau)$ can be estimated from a histogram of the $\bar{x}_i(\tau)$ data. In practice, however, it is usually only the expected value $E[\overline{X}(\tau)]$ that is desired, and a *point* estimate of this transient statistic is available as an *ensemble average*, even if n is not large—that is,

$$\frac{1}{n} \sum_{i=1}^{n} \bar{x}_i(\tau)$$

is a point estimate of $E[\overline{X}(\tau)]$. Moreover, an *interval* estimate for $E[\overline{X}(\tau)]$ can be calculated from the ensemble (sample) mean and standard deviation of the $\bar{x}_i(\tau)$ data, by using Algorithm 8.1.1. Independence is assured by the careful selection of seeds. Normality follows from the central limit theorem, since $\bar{x}_i(\tau)$ is an average.

Example 8.3.5

As an illustration of the previous discussion, a modified version of program `sis4` was used to produce 21 replications. For each replication, 100 time intervals of operation were simulated, with $(s, S) = (20, 80)$. The 21 transient time-averaged inventory levels that were produced are illustrated.

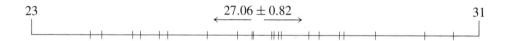

If $\overline{L}(\tau)$ denotes the time-averaged inventory level (with $\tau = 100$), then each of these 21 numbers is a *realization* of $\overline{L}(\tau)$. The mean and standard deviation of this sample are 27.06 and 1.76, respectively. If a 95%-confidence-interval estimate is desired, then the corresponding critical value of t is

$$t^* = \texttt{idfStudent}(20, 0.975) = 2.086.$$

Therefore, from program `estimate`, the 95% confidence-interval estimate is

$$27.06 \pm \frac{(2.086)(1.76)}{\sqrt{20}} = 27.06 \pm 0.82,$$

as illustrated above. We are 95% confident that $E[\overline{L}(\tau)]$ is somewhere between the interval endpoints $26.24 = 27.06 - 0.82$ and $27.88 = 27.06 + 0.82$—that is, we are 95% confident that, if program `sis4` were used to generate a *large* ensemble, then the ensemble average of the resulting transient time-averaged inventory levels would be somewhere between 26.24 and 27.88.

In some discrete-event simulations, the terminal time is not fixed. Instead, the terminal time is determined by other terminal conditions. For example, in program `ssq2`, the terminal condition is specified by the number of jobs processed, not the stopping time; because of this, the terminal time will vary randomly from replication to replication. The idea of multiple replications, with nothing varied except the initial state of the rngs stream, is still valid; each replication will produce an independent observation of the transient statistic of interest, and these observations can be combined to produce a point and an interval estimate.

Example 8.3.6

A modified version of program ssq2 was used to produce 20 replications of a simulation for which 100 jobs are processed through an initially idle $M/M/1$ service node, with arrival rate $\lambda = 1.0$ and service rate $\nu = 1.25$. The resulting 20 observations of the average wait in the node are illustrated.

From program estimate, the resulting 95%-confidence-interval estimate is 3.33 ± 1.04. Thus, we are 95% confident that, if we were to do millions of replications, the ensemble average of the resulting transient time-averaged number in the node would be somewhere between $2.29 = 3.33 - 1.04$ and $4.37 = 3.33 + 1.04$.

In Section 8.5, it is shown that, if the arrival rate is $\nu = 1.00$ and the service rate is $\nu = 1.25$ for an $M/M/1$ service node, then the steady-state average wait in the node is

$$\frac{1}{\nu - \lambda} = \frac{1}{1.25 - 1.00} = 4.0.$$

This exact steady-state value, found by analytic methods, is indicated in Example 8.3.6 with a '•'. Therefore, we see that it is *possible* that 100 jobs are enough to produce steady-state statistics. To explore this conjecture, the modified version of program ssq2 was used to produce 60 more replications. Consistent with the $\sqrt{n}$ rule, this four-fold increase in the number of replications (from 20 to 80) should result in an approximate two-fold decrease in the width of the interval estimate and if the ensemble mean remains close to 3.33, then 4.0 will no longer be interior to the interval estimate.

Example 8.3.7

The resulting 95%-confidence-interval estimate based upon 80 replications was 3.25 ± 0.39, as illustrated.

Now we see that, in this case, 100 jobs are clearly not enough to produce steady-state statistics—the bias of the initial idle state is still evident in the transient statistic.

Example 8.3.8

As a continuation of Example 8.3.6, the number of jobs per replication was increased from 100 to 1000. As before, 20 replications were used to produce 20 observations of the average wait in the node, as illustrated.

With this increase in the number of processed jobs, the 95%-confidence-interval estimate shifted to the right toward the steady-state average wait and shrank to 3.82 ± 0.37. The steady-state average wait '•' indicator is now interior to the interval estimate, suggesting that, in this case, 1000 jobs are enough to achieve statistics that are close to their steady-state values.

Relative to the 100-jobs-per-replication results in Example 8.3.6, reproduced here for convenience, note that the 1000-jobs-per-replication results in Example 8.3.8 exhibit a sample distribution that is much more symmetric about the sample mean.

That is, the 1000-jobs-per-replication results are more consistent with the underlying theory of interval estimation, dependent as it is on a sample mean distribution that is approximately $Normal(\mu, \sigma/\sqrt{n})$ and centered on the (unknown) population mean.

8.3.3 Exercises

8.3.1   (*a*) Repeat the experiment presented in Example 8.3.5, using a version of program sis4 modified so that estimates of $\bar{L}(10)$, $\bar{L}(20)$, $\bar{L}(40)$, and $\bar{L}(80)$ can also be produced. (*b*) Comment.

8.3.2   (*a*) Repeat the 100-jobs-per-replication experiment in Examples 8.3.6 and 8.3.7, using program ssq4 (but modified so that estimates of the average number in the node can be produced for 10, 20, 40, 80, 160, and 320 replications). (*b*) For the 320-replication case, construct a continuous-data histogram of the average waits. (*c*) Comment.

8.3.3   (*a*) Repeat the 1000-jobs-per-replication experiment presented in Example 8.3.8, using program ssq4 (but modified so that estimates of the average number in the node can be produced for 10, 20, 40, 80, 160, and 320 replications). (*b*) For the 320-replication case, construct a continuous-data histogram of the average waits. (*c*) Comment.

8.3.4   Illustrate graphically how well the two histograms in Exercise 8.3.2 and 8.3.3 can be approximated by a $Normal(\mu, \sigma)$ pdf, for appropriately chosen values of (μ, σ).

8.4 BATCH MEANS

As discussed in the previous section, there are two types of discrete-event simulation models: transient (finite horizon) and steady-state (infinite horizon). In a transient simulation, *interval* estimates can be constructed for the statistics of interest by replication. In a steady-state simulation, the general advice is to run the simulation for a *long* time in order to (hopefully) achieve a good *point* estimate for each statistic of interest. The last observation leads to an obvious question—is there a way to obtain *interval* estimates for steady-state statistics?

The answer is, yes, there are several ways to obtain interval estimates for steady-state statistics. The most common way to do so is by using the so-called method of *batch means*.

8.4.1 Experimental Investigation

Example 8.4.1

To investigate the issue of interval estimates for steady-state statistics, program ssq2 was modified to simulate an *M/M/1* single-server service node with $\lambda = 1.0$, $\nu = 1.25$. In this case, the steady-state utilization is $\rho = \lambda/\nu = 0.8$, and, from the results in Section 8.5, the expected steady-state wait is $E[W] = 4$. We can attempt to eliminate the initial-state bias by initializing the program variable departure to a nonzero value—in this case, 4.2. In other words, in Exercise 8.4.1, you are asked to prove that this initialization causes the simulation to begin in its *expected* steady-state condition, in the sense that the first job's expected wait is approximately 4. The multiple-stream capability of the library rngs was used (with initial seed 12345) to create 16 independent replications, with the first using stream 0 for both arrival times and service times, the second using stream 1 for both, and so on. The *finite-horizon* interval-estimation technique from the previous section was used to compute average-wait point and interval estimates for 8, 16, 32, . . . , 1024 processed jobs. The results are illustrated in Figure 8.4.1.

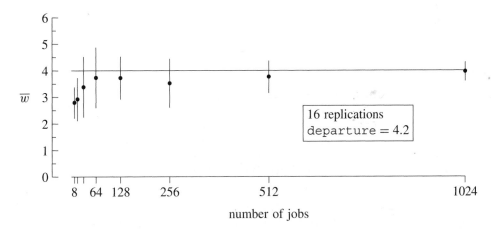

Figure 8.4.1   Confidence intervals for mean wait.

The horizontal line is the steady-state wait $E[W] = 4$. The vertical lines are the 95%-confidence-*interval* estimates for the sample size 16, and the • denotes the *point* estimate $\overline{w}$. As hoped, we see that, unless the number of jobs is very small, the *finite-horizon* interval estimates are accurate *steady-state* estimates, becoming increasingly more accurate estimates of the expected steady-state wait as the number of jobs increases.

Example 8.4.2
As a continuation of Example 8.4.1, in an attempt to further clarify the use of replicated finite-horizon statistics to estimate a steady-state statistic, two other initial values of departure were tried, 0.0 and 6.0. The initial value 0.0 can be justified on the basis that the *most likely* steady-state condition is an *idle* node. Indeed, in this case, an arriving job finds the service node idle with probability 0.2. The resulting interval estimates are presented in the plot on the left-hand side of Figure 8.4.2. The initial value 6.0 was chosen by experimentation and justified on the basis that it produced interval estimates consistent with the expected steady-state wait even when the number of jobs was small. The resulting interval estimates are on the right-hand side of Figure 8.4.2.

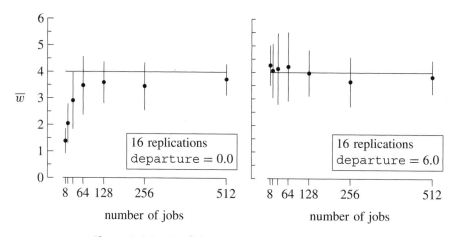

Figure 8.4.2   Confidence intervals for mean wait.

The graph on the left illustrates the potentially significant impact of the initial-state bias on the estimates when the number of jobs is small. The graph on the right illustrates that, at least in principle, it is possible to choose an initial state for the service node that minimizes the initial-state bias. As in Example 8.4.1, when the number of jobs is large, the estimates are very good and essentially are *independent* of the initial state of the service node. Of course, this is as it should be if they are to be steady-state estimates.

Replications.   Examples 8.4.1 and 8.4.2 illustrate that one way to obtain interval estimates for steady-state statistics is to use the replication-based finite-horizon interval-estimation technique from the previous section, with each replication corresponding to a long simulated period of operation. To do so, three interrelated issues must be resolved.

* What is the initial state of the system?
* How many replications should be used?
* What is the length of the simulated period of operation (per replication)?

Figures 8.4.1 and 8.4.2 provide insight into the first and third of these issues. The initial-state values 4.2, 0.0, and 6.0 for `departure` illustrated the initial states. The length of each simulation varied from 8 to 1024 jobs. The number of replications, however, was always fixed at 16. The following example provides insight into the second issue.

Example 8.4.3

To better understand the importance of the number of replications of the simulation model, the simulations in Examples 8.4.1 and 8.4.2 were repeated, but using 64 replications. Except for this four-fold increase in replications (using the first 64 `rngs` streams), all other parameters in the model were held at the same levels as before. By increasing the number of replications, the width of the interval estimates should be reduced and the estimates thereby improved. Specifically, because of the square-root rule (Section 8.1), we expect that a four-fold increase in the number of replications should reduce the interval width by approximately $1/2$.

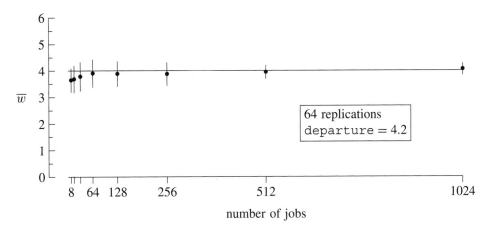

Figure 8.4.3 Confidence intervals for mean wait.

As illustrated in Figure 8.4.3, as compared with the results displayed in Figure 8.4.1, the interval estimates have been cut approximately in half. Moreover, even for a small number of jobs, the interval estimates now include the steady-state expected value, indicating that the initial-condition bias has been largely eliminated by initializing the program variable `departure` to 4.2—that is, the initial-state bias that *seemed* to be evident in Figure 8.4.1 (based on 16 replications) was actually just natural small-sample variation. Another run with a different seed might just as likely produce a bias that is above the theoretical steady-state value. A valid 95%-confidence-interval procedure will contain the true value 95% of the time. It is quite possible that the two confidence intervals in Figure 8.4.1 (based on 8 and 16 replications) that fell wholly below the steady-state value were among the 5% of the confidence intervals that did not contain the true value, 4. The endpoints of the confidence intervals are also positively correlated because they were calculated using common data values.

Example 8.4.4

As a continuation of the previous example, when compared to the corresponding results in Example 8.4.2 based on 16 replications, these figures based on 64 replications reveal that, again, the widths of the interval estimates have been halved approximately. Moreover, from the graph on the right-hand side of Figure 8.4.4, we see that choosing the initial value of departure to be 6.0, which *seemed* to work with just 16 replications, actually introduces some initial-state bias, causing the estimates (for a small number of jobs) to be larger than the theoretical steady-state value. That 6.0 seemed like a reasonable initial value of departure from Figure 8.4.2 was a false impression based on statistics from a misleadingly small sample.

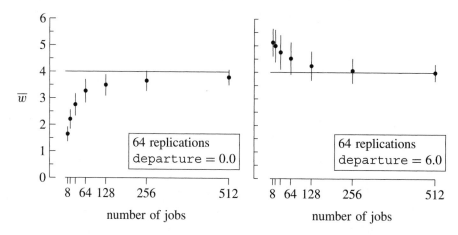

Figure 8.4.4   Confidence intervals for mean wait.

To summarize, replicated finite-horizon statistics can be used to generate accurate infinite-horizon (steady-state) interval estimates. However, there is an initial-bias problem with this technique. Therefore, we need another interval-estimation technique, one that avoids the initial-state-bias problem.*

8.4.2 Method of Batch Means

The reason the initial-state bias was an issue in the previous examples was that each replication was initialized *with the same system state*. Why not, instead, make *one* long run with, say, $64 \cdot 512$ jobs, but partition the wait times into 64 consecutive *batches*, each of length 512? An average wait can be computed for each batch of 512 jobs and an interval estimate can be computed from the 64 batch means. This is the method of *batch means*. With this method, the initial-state bias is eliminated because the statistics for each batch (other than the first) are naturally initialized to the state of the system at the time the statistical counters for the batch are reset.

---

*We were able to avoid the initial-state-bias problem in Example 8.4.3 (with departure initialized to 4.2) only because we already knew the steady-state result! In practice, the steady-state result is seldom known, for, if it were, there would be no point in simulating the system to compute the steady-state estimates.

Algorithm 8.4.1 The method of batch means, as applied to the sequence $x_1, x_2, \ldots, x_n$, consists of six steps. (Only the first two steps are unique to the method of batch means. Except for notation, the last four steps are identical to Algorithm 8.1.1.)

- Select a *batch size* $b > 1$. The number of batches is then $k = \lfloor n/b \rfloor$. (If b is not a divisor of n, discard the last $n - kb$ data points.)
- Group the sequence into the k batches

$$\underbrace{x_1, x_2, \ldots, x_b}_{\text{batch 1}}, \underbrace{x_{b+1}, x_{b+2}, \ldots, x_{2b}}_{\text{batch 2}}, \underbrace{x_{2b+1}, x_{2b+2}, \ldots, x_{3b}}_{\text{batch 3}}, \ldots.$$

These batches play a role analogous to that which *replications* played in the previous section. For each batch, calculate the *batch mean*

$$\overline{x}_j = \frac{1}{b} \sum_{i=1}^{b} x_{(j-1)b+i} \qquad j = 1, 2, \ldots, k.$$

- Compute the mean $\overline{x}$ and standard deviation s of the batch means $\overline{x}_1, \overline{x}_2, \ldots, \overline{x}_k$.
- Pick a *level of confidence* $1 - \alpha$ (typically $\alpha = 0.05$).
- Calculate the critical value $t^* = \mathtt{idfStudent}(k - 1, 1 - \alpha/2)$.
- Calculate the interval endpoints $\overline{x} \pm t^* s / \sqrt{k - 1}$.

You then claim to be $(1 - \alpha) \times 100\%$ confident that the true (unknown) steady-state mean lies somewhere in this interval. If the batch size b is large, your claim is probably true *even if the sample is autocorrelated*. (See Section 4.4 and Appendix F.) The method of batch means reduces the impact of autocorrelated data, since only the trailing observations of one batch are significantly correlated with the initial observations of the subsequent batch.

Provided that no points are discarded, the "mean of the means" is the same as the "grand sample mean":

$$\overline{x} = \frac{1}{k} \sum_{j=1}^{k} \overline{x}_j = \frac{1}{n} \sum_{i=1}^{n} x_i.$$

This is an important observation, because it demonstrates that the choice of the (b, k) parameters has *no* impact on the *point* estimate of the mean. Only the width of the *interval* estimate is affected by the choice of (b, k).

Example 8.4.5

A modified version of program $\mathtt{ssq2}$ was used to generate $n = 32\,768$ consecutive waits for an initially idle $M/M/1$ service node with $(\lambda, \nu) = (1, 1.25)$. The batch-means estimate of the steady-state wait $E[W] = 4.0$ was computed for several different (b, k)

values, with the following results:

(b, k)	$(8, 4096)$	$(64, 512)$	$(512, 64)$	$(4096, 8)$
$\overline{w}$	3.94 ± 0.11	3.94 ± 0.25	3.94 ± 0.29	3.94 ± 0.48

As illustrated, the point estimate 3.94 is independent of (b, k) but the width of the interval estimate clearly is not.

The variable width result in Example 8.4.5 is typical of the method of batch means. The obvious question then is, for a fixed value of $n = bk$, what is the best choice of (b, k)? The answer to this question is conditioned on the following important observations relative to the theory upon which the method of batch means is based.

- The validity of the $\overline{x} \pm t^* s / \sqrt{k - 1}$ interval estimate is based on the assumption that the batch means $\overline{x}_1, \overline{x}_2, \ldots, \overline{x}_k$ are (at least approximately) an *independent*, identically distributed sequence of *Normal* random variates.
- The method of batch means becomes increasingly more valid as the batch size b is increased. This occurs for two related reasons: (i) the mean of b random variables, independent or not, tends to become more *Normal* as b is increased and (ii) the autocorrelation that typically exists in time-sequenced data is reduced in the associated time-sequenced batch means, becoming effectively zero if b is sufficiently large.

Guidelines.   If the value of the product $n = bk$ is fixed, then the batch size b must not be allowed to be too large, for, if it is, the number of batches k will be too small, resulting in a wide interval estimate. (See Figure 8.1.8.) The following guidelines are offered for choosing (b, k).

- Schmeiser (1982) recommends a number of batches between $k = 10$ and $k = 30$, although this advice should be tempered by the two other guidelines.
- Pegden, Shannon, and Sadowski (1995, page 184) recommend a batch size that is at least ten times as large as the largest lag for which the autocorrelation function remains significant. (See Section 4.4.)
- Banks, Carson, Nelson, and Nicol (2005) recommend that the batch size be increased until the lag-one autocorrelation between batch means is less than 0.2.

The following examples provide additional insight into the choice of (b, k).

Example 8.4.6

An extrapolation of the autocorrelation results in Example 4.4.3 suggests that, for an $M/M/1$ service node with steady-state utilization $\lambda / \nu = 0.8$, the effective cut-off lag is approximately 100. Figure 8.4.5 illustrates that autocorrelation is present in a sequence of batch means for batch sizes $b = 4, 8, 16, 32, 64, 128$.[*] Although the second guideline is violated with $b = 256$, Figure 8.4.5 shows that the third guideline will be satisfied with $b = 256$.

---

[*]For each value of b, the results in this figure were produced by piping 1000 batch means into program acs.

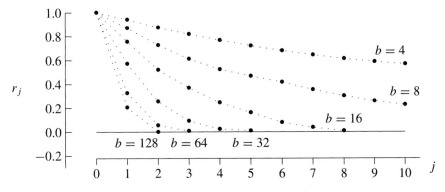

Figure 8.4.5   Sample autocorrelation function.

As b is increased, the autocorrelation in the batch means is reduced. For $b = 128$, the reduction is so nearly complete that there is no residual autocorrelation for lags greater than 1.

To summarize, we expect that, if $k = 64$ batches each of size $b = 256$ are used with Algorithm 8.4.1, then the batch-mean interval estimate of the steady-state wait for jobs in an $M/M/1$ service node with steady-state utilization $\lambda/\nu = 0.8$ will contain $E[W]$ with probability $1 - \alpha$. Therefore, in this case, $n = bk = 16\,384$ jobs must be processed to produce this estimate. If λ/ν were larger, a larger value of n would be required.

Iteration.   Since Algorithm 8.4.1 can be applied for *any* reasonable pair of (b, k) values, even if the underlying theory is not valid, it is common to find the method of batch means presented as a *black box* algorithm with an associated recommendation to experiment with increasing values of b until "convergence" is achieved. In this context, convergence means significant overlap in a sequence of increasingly more narrow interval estimates produced, typically, by using $k = 64$ batches and systematically doubling the batch size b. Although they were computed in a different way, the results produced by fixing $k = 64$ and systematically doubling b will be similar to those illustrated in Examples 8.4.3 and 8.4.4.

Example 8.4.7
The method of batch means, with $(b, k) = (256, 64)$, was applied to a simulation of an $M/M/1$ service node with $(\lambda, \nu) = (1.00, 1.25)$ to produce interval estimates of the steady-state wait $E[W] = 4$. Ten 95%-confidence-interval estimates were produced, corresponding to the first ten `rngs` streams.

As illustrated on the left-hand side of Figure 8.4.6, all but one of these interval estimates (the one corresponding to stream 2) brackets the true steady-state value, resulting in an *actual* coverage of 90%. In contrast, as illustrated on the right-hand side of Figure 8.4.6, if the interval estimates are computed naively and incorrectly by *not* batching but, instead, just piping all $n = 16\,384$ waits directly into program `esti-mate`, then the intervals are *far* too narrow, and only three out of ten estimates bracket

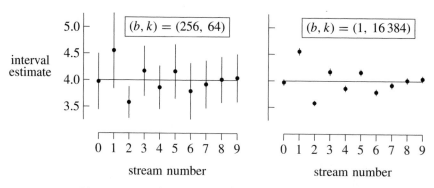

Figure 8.4.6   Confidence intervals for mean wait.

the true steady-state value, resulting in an actual coverage of 30%. The message is clear—failure to batch the data correctly can lead to unrealistically small variance estimates, resulting in a false sense of confidence about the accuracy of the interval estimate. For example, if $(b, k) = (1, 16\,384)$, then a confidence-interval estimate alleged to have 95% confidence appears to have a coverage of approximately 30%.

Some interval-estimation techniques other than batch means are the following: overlapping batch means, by Meketon and Schmeiser (1984); the autoregressive method, by Fishman (1971); standardized time series, by Schruben (1983); spectrum analysis, by Heidelberger and Welch (1981) and Lada and Wilson (2006); and the regenerative method, by Crane and Iglehart (1975) and Fishman (1973), outlined in Glynn (2006). These techniques are overviewed in Alexopoulos and Seila (1998), Goldsman and Nelson (1998), and Goldsman and Nelson (2006).

8.4.3 Exercises

8.4.1   In the notation of Section 1.2, let c_0 denote the departure time of the (virtual) 0th job (i.e., the initial value of `departure` in program `ssq2`). Assume a Poisson arrival process having rate λ, assume service times that are *Exponential*$(1/\nu)$, and let W_1 be the associated random wait of the first job. (*a*) Prove that

$$E[W_1] = c_0 + \frac{1}{\nu} - \frac{1}{\lambda}\big(1 - \exp(-\lambda c_0)\big).$$

(*b*) How does this relate to Examples 8.4.1 and 8.4.3?

8.4.2   (*a*) Repeat Example 8.4.5, except with observations obtained by $n = 32\,768$ calls to the function `Exponential(4.0)`. (*b*) Comment. (*c*) Why are the interval estimates so much more narrow? Do not just conjecture, explain.

8.4.3   (*a*) Produce a figure like Figure 8.4.6, but for the two cases $(b, k) = (64, 256)$ and $(b, k) = (16, 1024)$. (*b*) Comment on "coverage."

8.4.4   The stochastic process $X_1, X_2, X_3, \ldots$ is said to be *covariance stationary* provided that there is a common mean μ and variance σ^2

$$E[X_i] = \mu \quad \text{and} \quad E[(X_i - \mu)^2] = \sigma^2 \qquad i = 1, 2, 3 \ldots$$

and that the covariance is independent of i, so that

$$\frac{E[(X_i - \mu)(X_{i+j} - \mu)]}{\sigma^2} = \rho(j) \qquad\qquad j = 1, 2, 3, \dots.$$

For an integer batch size b, define the k batch means $\overline{X}_1, \overline{X}_2, \dots, \overline{X}_k$ as in Algorithm 8.4.1. The grand sample mean is then

$$\overline{X} = \frac{\overline{X}_1 + \overline{X}_2 + \cdots + \overline{X}_k}{k}.$$

(a) Prove that, for any value of b, the mean of $\overline{X}$ is $E\left[\overline{X}\right] = \mu$. (b) Prove that, if b is large enough to make $\overline{X}_1, \overline{X}_2, \dots, \overline{X}_k$ independent, then the variance of $\overline{X}$ is

$$E\left[(\overline{X} - \mu)^2\right] = \frac{\sigma^2}{n}\,\xi_b,$$

where $n = bk$ and

$$\xi_b = 1 + 2\sum_{j=1}^{b-1}(1 - j/b)\,\rho(j).$$

8.5 STEADY-STATE SINGLE-SERVER SERVICE-NODE STATISTICS

The purpose of this section is to provide insight into the mathematics of traditional queuing theory as it applies to characterizing the *steady-state* statistics for a single-server service node. In this section, we will develop mathematical characterizations of what happens to job-averaged statistics and time-averaged statistics in the limit as the number of processed jobs becomes infinite. This discussion is based on the single-server service-node model developed in Section 1.2.

8.5.1 General Case

In all the discussion in this section, we assume that the arrival process and service process are *stationary*—that is, that random-variate interarrival times can be generated by sampling from a fixed interarrival-time distribution with mean $1/\lambda > 0$. Therefore, as is consistent with Definition 7.3.1, the arrival process is stationary with rate λ. (Later in the section, as necessary, we will assume a *Poisson* arrival process.) Similarly, service times can be generated by sampling from a fixed service-time distribution with mean $1/\nu > 0$. By analogy to the arrival-process terminology, it is conventional to say that the service process is stationary with rate $\nu.$*

---

*For historical reasons, in queuing theory, it is conventional to use μ to denote the service rate. Unfortunately, this notational convention conflicts with the common use of μ to denote a mean or expected value. Because of this notational conflict, ν is used consistently throughout this book to denote a service rate.

Effective Service Rate.   Although it is conventional to refer to the reciprocal of the average service time as the "service rate" (see Definition 1.2.5), some explanation is necessary. Suppose we monitor jobs departing the service node.

- During a period of time when the server is always busy—as soon as one job departs, another enters the server—the average time between departures will be $1/\nu$ and the departure times will form a stationary process with rate ν. Therefore, during a continuously busy period, the service process *is* stationary with rate ν.
- During a period of time when the server is always idle, there are no departures, and so the service rate is 0. Therefore, during a continuously idle period, the service process is stationary with rate 0.

Recall that the server utilization $\bar{x}$ is the proportion of time (probability) that the server is busy and so $1 - \bar{x}$ is the proportion of time (probability) that the server is idle. Therefore, an estimate of the *expected* service rate is $\bar{x} \cdot \nu + (1 - \bar{x}) \cdot 0 = \bar{x}\nu$. In other words, because $\bar{x}$ is an estimate of the probability that the server is busy, the service rate is analogous to a discrete random variable with possible values ν (when busy) and 0 (when idle) and associated probabilities $\bar{x}$ and $1 - \bar{x}$. The expected value of this random variable is $\bar{x}\nu$. The point here is that ν is actually the *maximum* possible service rate. Because the server is occasionally idle, the *effective* service rate is $\bar{x}\nu$.

Flow Balance

Definition 8.5.1   The *traffic intensity* ρ is the ratio of the arrival rate to the service rate. That is, $\rho = \lambda/\nu$. (See Section 1.2.)

The traffic intensity is particularly relevant in steady-state queuing theory because of the following important theorem. Although a rigorous proof of this very general result is beyond the scope of this book, a specific instance of the theorem is proved later in this section.

Theorem 8.5.1   Steady-state (infinite-horizon) statistics exist for a stochastic, conservative single-server service node with infinite capacity and *any* queue discipline if and only if $\rho < 1$.

Assume that $\rho = \lambda/\nu < 1$, and let $\bar{x}$ be the steady-state utilization. The arrival rate is less than the service rate, so it is intuitive that, in this case, the service node is "flow balanced"—or, equivalently, that there is a "conservation of jobs." In other words, as illustrated in Figure 8.5.1,

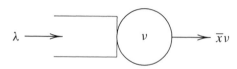

Figure 8.5.1   Single-server service node.

the average rate of flow of jobs *into* the service node is λ, the average rate of flow of jobs *out* of the service node is $\bar{x}\nu$, and "flow in = flow out"; so $\lambda = \bar{x}\nu$.

If the flow-balance equation $\lambda = \bar{x}\nu$ is compared with Definition 8.5.1, we see that it is equivalent to the equation $\bar{x} = \rho$: If $\rho = \lambda/\nu < 1$, then ρ is the *steady-state* server utilization. If, instead, $\rho \geq 1$, then $\bar{x}$ will approach 1.0 as the number of jobs increases, and the average queue length will grow without bound. If $\rho \geq 1$, then steady-state values for the statistics $\bar{d}$, $\bar{w}$, $\bar{q}$, and $\bar{l}$ are arbitrarily large.

Example 8.5.1

Use a modified version of program `ssq4` to simulate processing a large number of jobs, graphing the accumulated value of $\bar{x}$ every, say, 20 jobs (as in Example 3.1.3). If $\rho < 1$, then this sequence of $\bar{x}$'s will converge to ρ. If $\rho \geq 1$, then the convergence will be to 1.0.

Steady-state flow balance is a *long-term* property—if $\rho < 1$, then, in the long term, as many jobs will flow into the service node as flow out. Flow balance is generally *not* true in the short term. Even if the mean arrival rate and service rate are constant, there will be periods of time when the *effective* arrival rate and/or service rate are significantly different from their mean values, and, during these periods, "flow in $\neq$ flow out." During periods when the flow in is greater than the flow out, the queue length will build. When the opposite is true, the queue length will diminish. In either case, there will be a natural ebb and flow *that will persist forever*. "Steady state" does *not* mean that the state of the service node becomes constant—it means that the initial state of the system is not relevant.

Definition 8.5.2   Let $\bar{w}_n$ be the average wait experienced by the first n jobs to leave the service node:

$$\bar{w}_n = \frac{1}{n}\sum_{i=1}^{n} w_i;$$

and let $\bar{l}_n$ be the time-averaged number in the node over the time interval $(0, \tau_n)$:

$$\bar{l}_n = \frac{1}{\tau_n}\int_0^{\tau_n} l(t)\, dt,$$

where τ_n is the departure time of the nth job to leave the service node. Let the average delay $\bar{d}_n$, the average service time $\bar{s}_n$, the time-averaged number in the queue $\bar{q}_n$, and the server utilization $\bar{x}_n$ be defined similarly.

Note that there is a subtle but significant difference between Definition 8.5.2 and the corresponding definitions in Section 1.2. In Definition 8.5.2, the job index is referenced to jobs *leaving* the service node.

- If the queue discipline is FIFO, then $\tau_n = c_n$, and Definition 8.5.2 is identical to those used in Section 1.2.*

---

*Recall that c_n is the departure time of the nth job to *enter* the service node.

- If the queue discipline is not FIFO, then the order of arrival is different from the order of departure (for at least some jobs), so τ_n can differ from c_n.

By concentrating as it does on departing jobs, Definition 8.5.2 allows for *any* type of queue discipline. (See Exercise 1.2.7.)

By using arguments similar to those in Section 1.2, it can be shown that the six statistics in Definition 8.5.2 are related as follows:

$$\bar{w}_n = \bar{d}_n + \bar{s}_n,$$

$$\bar{l}_n = \bar{q}_n + \bar{x}_n,$$

$$\bar{l}_n \cong \left(\frac{n}{\tau_n}\right)\bar{w}_n,$$

$$\bar{q}_n \cong \left(\frac{n}{\tau_n}\right)\bar{d}_n,$$

$$\bar{x}_n \cong \left(\frac{n}{\tau_n}\right)\bar{s}_n,$$

where, for any finite value of n, the $\cong$ is an equality if and only if the service node is idle at $t = 0$ and at $t = \tau_n$. Note that, if the service node is idle at τ_n, then

$$\tau_n = \max\{c_1, c_2, \ldots, c_n\}.$$

What we will discuss next is what happens to $\bar{w}_n$, $\bar{l}_n$, $\bar{d}_n$, $\bar{q}_n$, $\bar{s}_n$, and $\bar{x}_n$ in the limit as $n \to \infty$, provided that $\rho < 1$.

Definition 8.5.3   Assume that $\rho < 1$ (so that Theorem 8.5.1 applies), and let

$$\bar{w} = \lim_{n \to \infty} \bar{w}_n = \lim_{n \to \infty} \frac{1}{n} \sum_{i=1}^{n} w_i$$

and

$$\bar{l} = \lim_{n \to \infty} \bar{l}_n = \lim_{n \to \infty} \frac{1}{\tau_n} \int_0^{\tau_n} l(t)\, dt.$$

Define $\bar{d}$, $\bar{s}$, $\bar{q}$, and $\bar{x}$ similarly. Each of these quantities represents a steady-state statistic for a single-server service node.

Steady-State Equations.   Given a knowledge of λ and ν with $\rho = \lambda/\nu < 1$, we want to derive equations that will relate the steady-state statistics in Definition 8.5.3 to λ, ν, and ρ. Since $\bar{s}_n$ is an unbiased point estimate of the expected service time,

$$\bar{s} = \lim_{n \to \infty} \bar{s}_n = 1/\nu.$$

In addition, we have already argued that, if $\rho < 1$, then

$$\bar{x} = \lim_{n \to \infty} \bar{x}_n = \rho.$$

Therefore, we know two of the six steady-state statistics in Definition 8.5.3. Moreover, if $\rho < 1$, then

$$\lim_{n \to \infty} \frac{n}{\tau_n} = \lambda.$$

This result can be reached intuitively from the observation that, in the time interval from 0 to τ_n, the number of jobs expected to *enter* the service node is $\lambda \tau_n$. If n is sufficiently large then, because $\rho < 1$, this product is also the expected number of jobs to *leave* the service node. Therefore, $n \cong \tau_n \lambda$, or, equivalently, $n/\tau_n \cong \lambda$, with equality in the limit as $n \to \infty$. From this result, we see that, in the limit as $n \to \infty$, the "equation"

$$\bar{x}_n \cong \left(\frac{n}{\tau_n} \right) \bar{s}_n$$

becomes

$$\bar{x} \cong \frac{\lambda}{\nu}.$$

However, $\rho = \lambda/\nu$, and so, as $n \to \infty$, we see that the $\cong$ will become an equality. It can be shown that the same comment also applies to the other two "equations"

$$\bar{l}_n \cong \left(\frac{n}{\tau_n} \right) \bar{w}_n \qquad \text{and} \qquad \bar{q}_n \cong \left(\frac{n}{\tau_n} \right) \bar{d}_n.$$

Therefore, we have the following theorem, which summarizes much of the previous discussion.

Theorem 8.5.2   For a conservative single-server service node with infinite capacity and *any* queue discipline, if $\rho = \lambda/\nu < 1$, then the steady-state server utilization and average service time are

$$\bar{x} = \rho \qquad \text{and} \qquad \bar{s} = 1/\nu.$$

Moreover, the four steady-state statistics $\bar{w}, \bar{l}, \bar{d}, \bar{q}$ are related by the four linear equations*

$$\bar{w} = \bar{d} + 1/\nu,$$
$$\bar{l} = \bar{q} + \rho,$$
$$\bar{l} = \lambda \bar{w},$$

and

$$\bar{q} = \lambda \bar{d}.$$

---

*The last two equations in Theorem 8.5.2 are known as *Little's equations*—they are an important part of the folklore of steady-state queuing theory.

At first glance, it might appear that the four linear equations in Theorem 8.5.2 relating $\overline{w}$, $\overline{l}$, $\overline{d}$, and $\overline{q}$ can be solved to determine unique values for these four steady-state statistics. That is *not* the case, however. These four equations can be written in matrix form as

$$
\begin{bmatrix}
\nu & 0 & -\nu & 0 \\
0 & 1 & 0 & -1 \\
\lambda & -1 & 0 & 0 \\
0 & 0 & \lambda & -1
\end{bmatrix}
\begin{bmatrix}
\overline{w} \\
\overline{l} \\
\overline{d} \\
\overline{q}
\end{bmatrix}
=
\begin{bmatrix}
1 \\
\rho \\
0 \\
0
\end{bmatrix},
$$

and, by applying elementary row operations, it can be shown that the coefficient matrix has only *three* linearly independent rows. That is, three of the four steady-state statistics can be computed by solving these equations, but the fourth statistic must be found some other way—either by discrete-event simulation or, in special cases, by analytic methods.

Example 8.5.2

Suppose $\lambda = 0.5$ (and so $1/\lambda = 2.0$) and $\nu = 0.625$ (and so $1/\nu = 1.6$). Then the steady-state utilization is $\overline{x} = \rho = 0.8$, and $\overline{s} = 1.6$. Suppose also that we have used discrete-event simulation to estimate that the steady-state average wait is $\overline{w} = 8.0$. Then the other three steady-state statistics are

$$
\overline{d} = \overline{w} - 1/\nu = 8.0 - 1.6 = 6.4
$$

$$
\overline{l} = \lambda\overline{w} = 8.0/2 = 4.0
$$

$$
\overline{q} = \overline{l} - \rho = 4.0 - 0.8 = 3.2.
$$

The discussion and results up to this point have been "distribution free": No distribution-specific stochastic assumptions were made about the interarrival times and service times. (In addition, the discussion and results to this point are valid for *any* queue discipline.) Given this generality, little more can be said without making additional distributional assumptions, as we will now do.

8.5.2 *M/M/1* Service Node

In the previous discussion, we saw that the four steady-state single-server service-node statistics $\overline{w}$, $\overline{l}$, $\overline{d}$, and $\overline{q}$ can be evaluated *provided* that any one of the four can be evaluated. We will now consider an important *analytic* method whereby $\overline{l}$ (and thus $\overline{w}$, $\overline{d}$, and $\overline{q}$) can be computed—provided that we are willing to make appropriate assumptions about the distribution of interarrival times and service times.

> **Definition 8.5.4**   An *M/M/1* service node has interarrival times and service times that are *iid* sequences of *Exponential*$(1/\lambda)$ and *Exponential*$(1/\nu)$ random variables, respectively.

Notation.   The first M in the $M/M/1$ notation refers to the distribution of the interarrival times, the second M refers to the distribution of the service times, and 1 denotes the number of parallel servers. M refers to the "memoryless" (Markov) property of the *Exponential* distribution. (See Section 7.4.) The standard notation used for probability distributions in queuing theory is as follows:

- M for *Exponential*;
- E for *Erlang*;
- U for *Uniform*;
- D for deterministic (not stochastic);
- G for general (a random variable with positive support not listed above).

Example 8.5.3
An $E/U/1$ service node has a single server with *Erlang* interarrival times and *Uniform* service times. Similarly, a $M/G/3$ service node has *Exponential* interarrival times and three statistically identical servers operating in parallel, each with the same general service-time distribution. Program ssq1 simulates a $G/G/1$ service node. Programs ssq2 and ssq3 simulate a $M/U/1$ service node. Program ssq4 simulates a $M/E/1$ service node. Program msq simulates a $M/U/3$ service node.

$M/M/1$ Steady-State Characteristics.
The *steady-state* characteristics of an $M/M/1$ service node have been extensively analyzed. One common approach to this analysis is based on the following definition.

Definition 8.5.5   Let the discrete random variable $L(t)$ denote the number of jobs in an $M/M/1$ service node at time $t > 0$. The possible values of $L(t)$ are $l = 0, 1, 2, \ldots$, and the associated pdf is

$$f(l, t) = \Pr\left(L(t) = l\right).$$

Note that the function $f(l, t)$ represents the pdf of the number of jobs at time t.

The pdf $f(l, t)$ can be *estimated* if we have a single-server service-node next-event-simulation program (like program ssq4) with the capability to report the number in the system at a specified time t. By using *many* replications and varying *only* the initial seed from replication to replication, we can generate many independent samples of $L(t)$. From these samples, we can form a discrete-data histogram that will estimate $f(l, t)$. In particular, if we simulate an $M/M/1$ service node in this way, if the steady-state utilization $\rho = \lambda/\nu$ is less than 1, and if t is very large then, from the following theorem (stated without proof), the histogram will approximate the pdf of a *Geometric*(ρ) random variable.

Theorem 8.5.3   For an $M/M/1$ service node with $\rho = \lambda/\nu < 1$

$$\lim_{t \to \infty} f(l, t) = (1 - \rho)\rho^l \qquad\qquad l = 0, 1, 2, \ldots.$$

This says that, if L is the steady-state number of jobs in an $M/M/1$ service node, then L is a $Geometric(\rho)$ random variable.[*]

$M/M/1$ Steady-State Equations.   Before we construct a proof of Theorem 8.5.3, observe that, if L is a $Geometric(\rho)$ random variable (see Section 6.4), then the steady-state average (expected) number in the service node is

$$\bar{l} = E[L] = \frac{\rho}{1 - \rho}.$$

Given this result and the steady-state equations from Theorem 8.5.2, it follows that

$$\bar{w} = \frac{\bar{l}}{\lambda} = \frac{1}{\nu - \lambda}, \qquad \bar{q} = \bar{l} - \rho = \frac{\rho^2}{1 - \rho}, \qquad \text{and} \qquad \bar{d} = \frac{\bar{q}}{\lambda} = \frac{\rho}{\nu - \lambda}.$$

In other words, these four equations, along with the three additional equations

$$\bar{x} = \rho = \frac{\lambda}{\nu}, \qquad \bar{r} = \frac{1}{\lambda}, \qquad \text{and} \qquad \bar{s} = \frac{1}{\nu},$$

represent the *steady-state* statistics for an $M/M/1$ service node.

Example 8.5.4

If we simulate an $M/M/1$ service node for thousands of jobs, printing intermediate statistics every, say, 20 jobs, and if $\rho < 1$ then as $n \to \infty$ we will see the intermediate statistics $\bar{l}_n$, $\bar{w}_n$, $\bar{q}_n$, $\bar{d}_n$, $\bar{x}_n$, $\bar{r}_n$, and $\bar{s}_n$ all eventually converge to the values given by the $M/M/1$ steady-state equations.

Example 8.5.5

For selected values of ρ, the steady-state[†] average number in the node, in the queue, and in service for an $M/M/1$ service node are as follows:

ρ	0.1	0.2	0.3	0.4	0.5	0.6	0.7	0.8	0.9	1.0
$\bar{l}$	0.11	0.25	0.43	0.67	1.00	1.50	2.33	4.00	9.00	∞
$\bar{q}$	0.01	0.05	0.13	0.27	0.50	0.90	1.63	3.20	8.10	∞
$\bar{x}$	0.10	0.20	0.30	0.40	0.50	0.60	0.70	0.80	0.90	1.0

---

[*]The capacity must be infinite, and the queue discipline cannot make use of any service-time information—that is, in terms of Definition 1.2.2, Theorem 8.5.3 is valid for FIFO, LIFO, and SIRO (service in random order) queue disciplines, but is *not* valid for a priority-queue discipline that is based on a knowledge of the service times.

[†]Although the emphasis in this section is on steady-state behavior of the statistics associated with a single-server queue, Kelton and Law (1985), for example, consider the transient behavior of the statistics.

Thus, for example, an *M/M/1* service node operating with a steady-state utilization of 0.8 will contain a steady-state average of 3.2 jobs in the queue and 0.8 jobs in service for a total of 4.0 jobs in the service node. As illustrated in Figure 8.5.2, $\bar{l}$ (as well as $\bar{w}$, $\bar{q}$, and $\bar{d}$) becomes infinite as ρ approaches 1. This is consistent with the observation that steady state is possible only if $\rho < 1$.

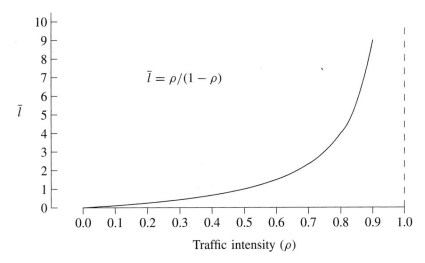

Figure 8.5.2   Average number of jobs in the queue.

Example 8.5.6

If L is a *Geometric*(ρ) random variable, then

$$\Pr(L = l) = (1 - \rho)\rho^l \qquad l = 0, 1, 2, \ldots .$$

Thus, even though the *expected* number of jobs in an *M/M/1* service node becomes infinite as $\rho \to 1$, if $\rho < 1$ then it is always true that the steady-state pdf is a monotone *decreasing* function of l. That is, for an *M/M/1* service node with $\rho < 1$, the *most likely* steady-state number in the node is 0, the next most likely number is 1, etc. This is a counter-intuitive result, very much dependent on the memoryless property of *Exponential* random variables.

Proof of Theorem 8.5.3.   To prove that the steady-state number of jobs in an *M/M/1* service node is a *Geometric*(ρ) random variable, we begin by observing that the *arrivals* to an *M/M/1* service node are a stationary Poisson process with rate λ, where $1/\lambda > 0$ is the mean interarrival time. Let $\delta > 0$ be the width of a small interval of time, and recall, from Section 7.3, that the number of arrivals in this time interval is a *Poisson*($\delta\lambda$) random variable. Therefore,

$$\Pr(x \text{ arrivals in an interval of width } \delta) = \frac{\exp(-\delta\lambda)(\delta\lambda)^x}{x!} = \left(1 - \delta\lambda + \frac{(\delta\lambda)^2}{2!} - \cdots\right)\frac{(\delta\lambda)^x}{x!}$$

for $x = 0, 1, 2, \ldots$. It turns out that, in the expansion of these $Poisson(\delta\lambda)$ probabilities, terms of order $(\delta\lambda)^2$ or higher can be ignored, because we will ultimately derive the pdf of $L(t)$ by using a limit as $\delta \to 0$. Therefore, if $\delta\lambda$ is small,

$$\Pr(0 \text{ arrivals in } \delta) \cong 1 - \delta\lambda$$

$$\Pr(1 \text{ arrival in } \delta) \cong \delta\lambda$$

$$\Pr(2 \text{ or more arrivals in } \delta) \cong 0.$$

Similarly, during a period when the server is continuously busy, then, because the service times are $Exponential(1/\nu)$, the *departures* from the service node are a stationary Poisson process with rate ν, where $1/\nu > 0$ is the mean service time. It follows from the previous discussion that, if $L(t) > 0$ and if $\delta\nu$ is small relative to 1, then

$$\Pr(0 \text{ departures in } \delta) \cong 1 - \delta\nu$$

$$\Pr(1 \text{ departure in } \delta) \cong \delta\nu$$

$$\Pr(2 \text{ or more departures in } \delta) \cong 0.$$

These probabilities are valid if there is at least one job in the node. If, instead, $L(t) = 0$, then the server is idle, no departure is possible, and so

$$\Pr(0 \text{ departures in } \delta) \cong 1$$

$$\Pr(1 \text{ or more departures in } \delta) \cong 0.$$

Let $f(l, t')$ denote the pdf of the random variable $L(t')$ with $t' = t + \delta$—that is, for $l = 0, 1, 2, \ldots$,

$$f(l, t) = \Pr\big(L(t) = l\big) \qquad \text{and} \qquad f(l, t') = \Pr\big(L(t') = l\big).$$

The small-$\delta\lambda$ and small-$\delta\nu$ probability equations derived previously are the key to proving Theorem 8.5.3, by determining how $f(l, t)$ and $f(l, t')$ are related.

Transient-Rate Equations.   We now use the arrival and departure probabilities established previously to derive the relationship between $f(l, t')$ and $f(l, t)$. We do so by assuming that δ is sufficiently small, implying that both $\delta\nu$ and $\delta\lambda$ are sufficiently small, and so terms of order δ^2 or smaller can be ignored.

First consider $f(l, t')$ for $l = 0$. This is the probability that the service node is idle at $t' = t + \delta$. If $L(t') = 0$, then it must be true that, at time t,

- either $L(t) = 0$ and there was no arrival in the next δ;
- or $L(t) = 1$ and there was one departure in the next δ.

Therefore, by conditioning on the two possible states of the service node at time t, the probability that $L(t') = 0$ is

$$f(0, t') \cong f(0, t)(1 - \delta\lambda) + f(1, t)\delta\nu.$$

Equivalently, this equation can be written in the form

$$\frac{f(0, t + \delta) - f(0, t)}{\delta} \cong \nu f(1, t) - \lambda f(0, t) \qquad l = 0.$$

In the limit as $\delta \to 0$, the $\cong$ becomes an equality, and the left-hand side of this equation represents the first derivative of $f(0, t)$ with respect to t.

Now, consider $f(l, t')$ for $l > 0$. If $L(t') = l$, then it must be true that, at time t,

- either $L(t) = l$ and there were no arrivals or departures in the next δ;
- or $L(t) = l + 1$ and there was one departure in the next δ;
- or $L(t) = l - 1$ and there was one arrival in the next δ.

Therefore, by conditioning on the three possible states of the service node at time t, the probability that $L(t') = l$ is

$$f(l, t') \cong f(l, t)(1 - \delta\lambda)(1 - \delta\nu) + f(l + 1, t)(1 - \delta\lambda)(\delta\nu) + f(l - 1, t)(\delta\lambda)(1 - \delta\nu).$$

If terms of order δ^2 or smaller are ignored, this equation can be rewritten as

$$f(l, t') \cong f(l, t)(1 - \delta\lambda - \delta\nu) + f(l + 1, t)\delta\nu + f(l - 1, t)\delta\lambda,$$

which is equivalent to

$$\frac{f(l, t + \delta) - f(l, t)}{\delta} \cong \nu f(l + 1, t) - (\lambda + \nu) f(l, t) + \lambda f(l - 1, t) \qquad l = 1, 2, \ldots.$$

As in the $l = 0$ case, in the limit as $\delta \to 0$, the $\cong$ becomes an equality, and the left-hand side of this equation represents the first derivative of $f(l, t)$ with respect to t.

Steady-State-Rate Equations. If $\rho < 1$, then, as $t \to \infty$, the pdf $f(l, t)$ will lose its "memory" of the *initial* state of the queue (at $t = 0$) and converge to a steady-state pdf: If t is large, then the instantaneous rate of change of the pdf will become zero, resulting in a pdf $f(l)$ that is *independent* of t and is characterized by the equations

$$0 = \nu f(1) - \lambda f(0) \qquad l = 0$$

$$0 = \nu f(l + 1) - (\lambda + \nu) f(l) + \lambda f(l - 1) \qquad l = 1, 2, \ldots.$$

These equations can be solved for $f(\cdot)$ as follows.

- First divide by ν, then write the equations as

$$f(1) = \rho f(0) \qquad l = 0$$

$$f(l + 1) = (1 + \rho) f(l) - \rho f(l - 1) \qquad l = 1, 2, \ldots.$$

- From these equations, we can argue inductively that

$$f(2) = (1 + \rho)f(1) - \rho f(0) = \rho^2 f(0)$$

$$f(3) = (1 + \rho)f(2) - \rho f(1) = \rho^3 f(0)$$

$$\vdots$$

$$f(l) = \cdots = \rho^l f(0) \qquad l = 0, 1, 2, \ldots.$$

- Finally, by applying the usual pdf-normalizing convention

$$1 = \sum_{l=0}^{\infty} f(l) = f(0) \sum_{l=0}^{\infty} \rho^l = f(0)(1 + \rho + \rho^2 + \cdots) = \frac{f(0)}{1 - \rho}$$

for $\rho < 1$, we find that the probability of an idle server is

$$f(0) = 1 - \rho,$$

which is consistent with the fact that $1 - \rho$ is the steady-state probability of an idle server. Therefore,

$$f(l) = (1 - \rho)\rho^l \qquad l = 0, 1, 2, \ldots,$$

which proves Theorem 8.5.3—the steady-state number in an *M/M/*1 service node is a *Geometric*(ρ) random variable.

Steady-State Waiting Time. Let the random variable W denote the steady-state waiting time in an *M/M/*1 service node. From results established earlier, we know that the expected value of W is

$$\overline{w} = \frac{1}{\nu - \lambda};$$

but what is the pdf of W?

If ρ is very small, then most jobs will experience little or no delay, and their only wait will be their service time. Therefore, if ρ is small, it seems intuitive that W will be *Exponential*$(1/\nu)$. On the other hand, if ρ is close to 1, then most jobs will experience large delays, and their wait will be determined primarily by their delay time. In this case, it is *not* intuitive that W should be *Exponential*. It is, however—provided that the queue discipline is FIFO. Indeed, we can prove the following theorem.

Theorem 8.5.4 If $\rho = \lambda/\nu < 1$, then the steady-state wait in a FIFO *M/M/*1 service node with infinite capacity is an *Exponential*$(\overline{w})$ random variable with $\overline{w} = 1/(\nu - \lambda)$.

Proof. Let l be the number in the service node at the instant a new job arrives. Since the steady-state number in the node is a *Geometric*(ρ) random variable, this occurs with probability $(1 - \rho)\rho^l$, for $l = 0, 1, 2, \ldots$. There are now $l + 1$ jobs in the node and, because the queue discipline is FIFO, the time this new job will spend in the node is the *sum* of its service time and the service time of all those l jobs ahead of it. Each of the l jobs in the queue (including the job that just arrived) has a service time that is *Exponential*$(1/\nu)$, and, because of the memoryless property of the *Exponential* distribution, the *remaining* service time of the job in service is also *Exponential*$(1/\nu)$. Thus the wait experienced by a job that arrives to find l jobs in a FIFO $M/M/1$ service node is the sum of $l+1$ *iid Exponential*$(1/\nu)$ random variables. This wait, then, is an *Erlang*$(l+1, 1/\nu)$ random variable W with possible values $w > 0$ and pdf

$$\frac{\nu}{l!}(\nu w)^l \exp(-\nu w) \qquad w > 0$$

(per Section 7.4). To construct the pdf of W, it is necessary to condition on all possible values of l. From the law of total probability, the pdf of W is

$$\sum_{l=0}^{\infty} \frac{\nu(1-\rho)\rho^l}{l!}(\nu w)^l \exp(-\nu w) = \nu(1-\rho)\exp(-\nu w)\sum_{l=0}^{\infty}\frac{1}{l!}(\rho \nu w)^l$$

$$= (\nu - \lambda)\exp(-\nu w)\exp(\lambda w)$$

$$= \frac{1}{\overline{w}}\exp(-w/\overline{w}) \qquad w > 0,$$

which proves that W is an *Exponential*$(\overline{w})$ random variable.   □

Caveat.   It is important to avoid becoming too taken with the elegance of $M/M/1$ steady-state queuing-theory results. These results provide valuable discrete-event simulation benchmarks and useful system-performance tools, but their applicability to real systems is limited for at least three reasons:

- The assumption of an *Exponential*$(1/\nu)$ server is not valid in many practical applications—not even approximately. This is particularly true of service-time distributions, which *rarely* have a mode of 0 in real-world systems.
- The assumption that both the arrival and service processes are *stationary* could be invalid, particularly if people are an integral part of the system.
- The number of jobs required to achieve statistics close to their steady-state value could be very large, particularly if the traffic intensity is close to 1.0.

Relative to the last point, see Exercise 8.5.4.

8.5.3 Exercises

8.5.1   Prove that the coefficient matrix

$$\begin{bmatrix} \nu & 0 & -\nu & 0 \\ 0 & 1 & 0 & -1 \\ \lambda & -1 & 0 & 0 \\ 0 & 0 & \lambda & -1 \end{bmatrix}$$

has rank 3.

8.5.2   If Q is the steady-state number in the queue of an $M/M/1$ service node (with $\rho < 1$), then what is the pdf of Q?

8.5.3   Simulate an initially idle $M/M/1$ service node with $\lambda = 1$ and $\nu = 1.25$. Generate the histograms corresponding to $f(l, t)$ for $t = 10$, $t = 100$, and $t = 1000$. Repeat if the service times are $Erlang(4, 0.2)$. Comment on the shape of the histograms.

8.5.4   Verify that, if an $M/M/1$ service node is simulated for thousands of jobs, printing intermediate statistics every 50 jobs, and if $\rho < 1$, then, as $n \to \infty$, the intermediate statistics $\bar{l}_n, \bar{w}_n, \bar{q}_n, \bar{d}_n, \bar{x}_n, \bar{r}_n$, and $\bar{s}_n$ will all eventually converge to the values given by the seven $M/M/1$ steady-state equations. Consider the cases $\rho = 0.3$, 0.6, and 0.9, and comment on the *rate* of convergence.

8.5.5   An $M/M/1$ service node produced a steady-state value of $\bar{l} = 5$. What was ρ?

8.5.6   If D is the steady-state delay in the queue of an $M/M/1$ service node (with $\rho < 1$), then what is the pdf of D?

9

Input Modeling

This chapter, which introduces procedures for developing credible input models, completes all of the building blocks for the modeling of a system via discrete-event simulation. We recommend reading Appendix G prior to reading this chapter for a discussion of the sources of error in discrete-event simulation modeling and for a development of a high-level framework for the modeling process.

In the various simulation models presented thus far in the text, the choices both of the distribution and of the associated parameters have been "pulled out of thin air." Program ssq2, for example, uses $Uniform(a, b)$ for the service-time distribution, with parameters $a = 1$ and $b = 2$. These parameters have been fictitious so as to concentrate our efforts on developing algorithms for the modeling of the system, rather than on identifying the origins of their values. In practice, representative "input models" are found by gathering data from the system under consideration and then analyzing that data to construct an appropriate probability distribution to mimic some aspect (e.g., service times) of the real-world system. This chapter introduces the input modeling process in an example-driven fashion.

Input modeling can be subdivided into that for stationary models, whose probabilistic mechanism does not vary with time, and that for nonstationary models, where time plays a substantive role in the model. Nonparametric methods for constructing stationary input models are introduced in Section 9.1, along with a discussion concerning the collecting of data. Parametric methods for constructing stationary models are taken up in Section 9.2. We introduce the *method of moments* and *maximum likelihood* as two techniques for estimating parameter values for a particular distribution. Section 9.3 considers methods for constructing nonstationary input models. The introduction to input modeling presented here is rather cursory, so the reader is encouraged to consult the more complete treatments in, for example, Banks, Carson, Nelson, and Nicol (2005, Chapter 9); Bratley, Fox, and Schrage (1987, Chapter 4); Fishman (2001, Chapter 10); Law and Kelton (2000, Chapter 6); Lewis and Orav (1989, Chapters 6 and 7); and Hoover and Perry (1989, Chapter 6); plus the more advanced input-modeling techniques in Nelson and Yamnitsky (1998).

9.1 MODELING STATIONARY PROCESSES: NONPARAMETRIC APPROACH

Most discrete-event simulation models have stochastic elements that mimic the probabilistic nature of the system under consideration. A close match between the input model and the true underlying probabilistic mechanism associated with the system is required for successful input modeling. The general question considered in this and in the next two sections is how to model an element (e.g., arrival process, service times) in a discrete-event-simulation model, given a data set collected on the element of interest. For brevity, we assume that there is an existing system from which data can be drawn. The example-driven approach used here examines only introductory techniques for input modeling.

There are five basic questions to be answered in the following order:

- Have the data been sampled in an appropriate fashion?
- Should a *nonparametric* (often called "trace-driven," although this text reserves that term for the data-driven models in Chapter 1) model or a *parametric* probability model be selected as an input model? If the latter is chosen, the following three questions arise.
  — What type of distribution seems to "fit" the sample? Equivalently, what type of random-variate generator seems to have produced the data? Does the *Exponential*(μ), *Gamma*(a, b), or *Lognormal*(a, b), for example, most adequately describe the data?
  — What are the value(s) of the parameter(s) that characterize the distribution? If the distribution is *Exponential*(μ), for example, then what is the value of μ?
  — How much confidence do we have in our answers to the two previous questions?

In statistical jargon, the processes involved in answering these five questions are known as

- assuring that appropriate *statistical sampling* procedures have been used for data collection,
- choosing between a *nonparametric* or *parametric* input model, and if a parametric model is chosen, then subsequently
  — *hypothesizing* a distribution,
  — *estimating* the distribution's parameter(s), and
  — *testing* for *goodness of fit*.

We will consider these five processes, in order.

9.1.1 Data Collection

There are two approaches that arise with respect to the collection of data. The first is the classical approach, where a designed experiment is conducted to collect the data. The second is the exploratory approach, where questions are addressed by means of existing data that the modeler had no hand in collecting. The first approach is generally better in terms of control; the second approach is generally better in terms of cost.

Collecting data on the appropriate elements of the system of interest is one of the initial and pivotal steps in successful input modeling. An inexperienced modeler, for example, improperly collects delay times on a single-server service node when the mean delay time is the performance measure of interest. Although these delay times are valuable for model validation, they do not contribute to the input model. The appropriate data elements to collect for an input model for a single-server service node are typically arrival and service times. The analysis of sample data collected on stationary processes is considered in this section and the next section, and the analysis of sample data collected on nonstationary processes is considered in the final section.

Even if the decision to sample the appropriate element is made correctly, Bratley, Fox, and Schrage (1987) warn that there are several things that can be "wrong" with a data set. Vending-machine sales will be used to illustrate the difficulties.

- Wrong amount of aggregation. We desire to model daily sales, but have only monthly sales.
- Wrong distribution in time. We have sales for this month and want to model next month's sales.
- Wrong distribution in space. We want to model sales at a vending machine in location A, but have sales figures only on a vending machine at location B.
- Insufficient distribution resolution. We want the distribution of the number of soda cans sold at a particular vending machine, but our data is given in cases, effectively rounding the data up to the next multiple of 24, the number of cans in a case.
- Censored data. We want to model *demand*, but we only have *sales* data. If the vending machine ever sells out, this constitutes a right-censored observation. The reliability and biostatistics literature contains techniques for accommodating censored data sets (Lawless, 2003).

All of these depictions of data problems exclude the issue of *data validity*. Henderson and Mason (2004), for example, developed a nonparametric simulation model for ambulance planning where the ambulance took just one second to load the patient into the ambulance. How did the modeler arrive at such unusually efficient emergency personnel? The source of the problem was in the data-collection procedures. The ambulance personnel had a "computer-aided dispatch" system, which included a button situated on the dashboard of the ambulance to collect the arrival time to and departure time from the scene. The ambulance driver often (understandably) got busy and forgot to record event times by pressing the button. Later, when the error was discovered, the driver would press the button several times in a row in order to "catch up" on missed events. This would result in data values collected at the scene as short as one second, which, of course, turned up in the simulation driven by the empirical distribution associated with the data. The tiny event times were sampled occasionally, resulting in the unusually efficient performance during the execution of the simulation. These errors not only corrupt the recorded time at the scene, but also corrupt adjacent travel times.

There is a lot more that could be written about the careful collection of data and the assortment of pitfalls that await the collector. We end our short and sketchy discussion here and refer the interested reader to Cochran (1977) for further reading.

9.1.2 Preliminary Data Analysis

We now begin a discussion of some preliminary steps that can be taken to analyze a data set. Following convention, the sample values are denoted by $x_1, x_2, \ldots, x_n$, and may be sampled from a discrete or continuous population. The following elementary example (which will be referenced throughout this section) is used to illustrate the types of decisions that often arise in input modeling for a stationary process.

Example 9.1.1
Consider a data set of $n = 23$ service times collected on a server in a single-server service node to construct an input model for a discrete-event-simulation model. The service times in seconds are

105.84	28.92	98.64	55.56	128.04	45.60	67.80	105.12	48.48
51.84	173.40	51.96	54.12	68.64	93.12	68.88	84.12	68.64
		41.52	127.92	42.12	17.88	33.00.		

The data values are collected in the order that they are displayed. [Although these service times actually come from the life-testing literature (Caroni, 2002; Lawless, 2003), the same principles apply to both input modeling and survival analysis.]

The first step is to assess whether the observations are independent and identically distributed. The data must be given in the order collected for independence to be assessed. Situations where the identically distributed assumption would *not* be valid include the following:

- A new teller has been hired at a bank, and the 23 service times represent a task that has a steep learning curve. The expected service time is likely to decrease as the new teller learns how to perform the task more efficiently.
- The service times represent 23 times to completion of a physically demanding task during an 8-hour shift. If fatigue is a significant factor, the expected time to complete the task is likely to increase with time.

If a simple linear regression of the service times versus the observation numbers shows a significant nonzero slope, then the identically distributed assumption is probably *not* appropriate, and a nonstationary model is appropriate. If the slope of the regression line does not differ statistically from zero, then a stationary model is appropriate. This simple linear regression differs slightly from those considered in Section 4.4 in that the observation number, which ranges from 1 to 23 and is known by statisticians as the *independent variable*, is fixed, rather than random.

Assume that there is a suspicion that a learning curve is present, which makes us suspect that the service times are decreasing. The scatterplot and least-squares regression line shown in Figure 9.1.1 indicate a slight downward trend in the service times. The regression line has slope -1.3 and y-intercept 87.9 seconds. But is this negative slope significantly different from zero? A hypothesis test[*] shows that the negative slope is

---

[*]The reader should consult any introductory statistics textbook if hypothesis-testing terminology and procedures are not familiar. Although beyond the prerequisites for this text, here are some of the details for the reader with a background in statistics. The null and alternative hypotheses $H_0 : \beta_1 = 0$ and $H_1 : \beta_1 < 0$ are associated with the linear model $Y = \beta_0 + \beta_1 X + \epsilon$, where X is the (fixed) observation number, Y is the associated (random) service time, β_0 is the intercept, β_1 is the slope, and ϵ is an error term. [See Kutner, Nachtsheim, Neter, and Wasserman (2003) for details.]

more likely to be due to sampling variability, rather than a systematic decrease in service times over the 23 values collected. For the remainder of this section, we will assume that a *stationary* model is appropriate and thus that the observations can be treated as 23 identically distributed observed service times.

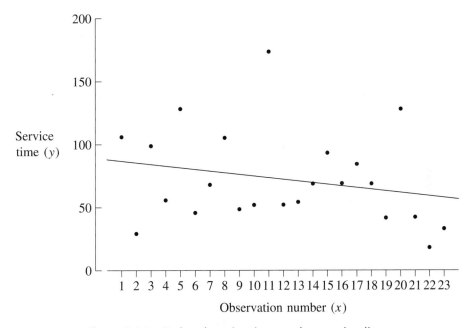

Figure 9.1.1   Ordered service times and regression line.

There are a number of graphical and statistical methods for assessing independence. These include analysis of the sample-autocorrelation function associated with the observations and a scatterplot of adjacent observations. The sample-autocorrelation function for the service times is plotted in Figure 9.1.2 for the first eight lags. The sample-autocorrelation function value at lag 1, for example, is the sample correlation for adjacent service times. The sample-autocorrelation function value at lag 4, for example, is the sample correlation for service times four customers apart. We plot these autocorrelation values as spikes, rather than just points as we did in Figure 4.4.6, although this is largely a matter of taste. The horizontal dashed lines at $\pm 2/\sqrt{n}$ are 95% bounds used to evaluate whether the spikes in the autocorrelation function are statistically significant. Since none of the spikes strayed significantly outside of these bounds for this particular small data set, we are satisfied that the observations are truly independent and identically distributed. Since only $n = 23$ data values were collected and there seems to be no pattern to the sample-autocorrelation function, we are likely to dismiss the spike at lag

The p-value associated with the hypothesis test is 0.14, which is insufficient evidence to conclude that there is a statistically significant learning curve present. The negative slope is probably due to sampling variability. The p-value may be small enough, however, to warrant further data collection.

6 (falling slightly above the limits) as being due to random sampling variability. Even if these were truly independent observations, we would expect 1 in 20 of the observations to fall outside of the 95% bounds. See Chatfield (2004) for further discussion of sample-autocorrelation functions.

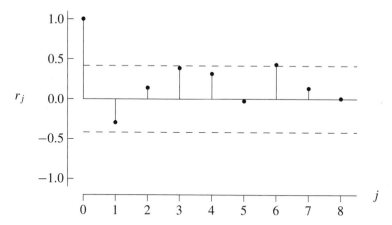

Figure 9.1.2   Sample-autocorrelation function of the first eight lags.

The next decision that needs to be made is whether a parametric or nonparametric input model should be used.

9.1.3 Nonparametric Modeling

The nonparametric approach allows a data set to "stand on its own," as opposed to being fit to a parametric model, such as the *Exponential* (μ) distribution. The decision about whether to use a nonparametric model or a parametric model can rest on the sample size n, on the number of unusual "gaps" that occur in the data that should be smoothed over, and the importance and use of the input model in the simulation. We will discuss discrete and continuous distributions separately.

Discrete Distributions.   When the simulation-modeling situation calls for a discrete model (e.g., number of cars at a dealership, number of operations in a service time, or number of puppies in a litter), the modeling and associated variate-generation algorithms that use the nonparametric approach are straightforward. Let $x_1, x_2, \ldots, x_n$ be the data values collected on a random variable that is discrete by its nature. One simple nonparametric input model for variate generation would be repeatedly to select one of the data values with probability $1/n$. This corresponds to an "empirical" pdf

$$\hat{f}(x) = \frac{1}{n} \qquad x = x_i; \ i = 1, 2, \ldots, n$$

if all of the data values are distinct. With a discrete random variable, of course, it is likely that there are tied data values. If d data values, for example, are tied at a particular x_i, then the pdf at that particular x_i is $\hat{f}(x) = d/n$. The empirical cdf can be defined by

$$\hat{F}(x) = \frac{N(x)}{n} \qquad x = x_i; \ i = 1, 2, \ldots, n,$$

where $N(x)$ is the number of data values that are less than or equal to x.

Variate generation in this case is straightforward. An algorithm for generating variates for a discrete-event simulation requires only two lines. Algorithm 9.1.1 is synchronized, efficient, clear, and, if the data values are sorted, monotone. The algorithm has reasonable storage requirements for moderate sample sizes n, which is typically the case for data collected by hand.

Algorithm 9.1.1   Given the (sorted or unsorted) data values $x_1, x_2, \ldots, x_n$ stored in x[1], x[2],..., x[n], the following algorithm returns one of the data values with probability $1/n$ for use in a nonparametric input model:

```
i = Equilikely(1, n);
return (x[i]);
```

Two disadvantages to the nonparametric approach are the interpolation and extrapolation problems, as illustrated in the following example.

Example 9.1.2
Consider the number of door panels that contain painting defects in a lot of ten panels. Such a random variable is naturally modeled by a discrete distribution, being a counting variable. Furthermore, it is clear that the support of this random variable is $\mathcal{X} = \{x \mid x = 0, 1, \ldots, 10\}$. The interpolation and extrapolation problems can be seen in the following scenario. Assume that $n = 50$ lots are sampled, yielding the data shown below.

x	0	1	2	3	4	5	6	7	8	9	10
counts	14	11	8	0	7	4	5	1	0	0	0

The empirical pdf is

$$\hat{f}(x) = \begin{cases} 14/50 & x = 0 \\ 11/50 & x = 1 \\ 8/50 & x = 2 \\ 7/50 & x = 4 \\ 4/50 & x = 5 \\ 5/50 & x = 6 \\ 1/50 & x = 7. \end{cases}$$

There is a "hole" in the data set at $x = 3$. Surprisingly, none of the 50 lots sampled contains exactly three door panels with defects. If we decide to take the nonparametric approach, the discrete-event simulation would never generate a lot that contains exactly three defects, which is almost certainly undesirable. This is known as the *interpolation* problem with the nonparametric approach. Likewise, it is not possible to generate a lot of 10 door panels with exactly 8, 9, or 10 panels with defects by using the nonparametric approach, which also could be undesirable. This is known as the *extrapolation* problem with the nonparametric approach. These two problems often lead modelers to consider the parametric approach for small data sets or those that exhibit significant random-sampling variability. The parametric approach will smooth over the rough spots and fill the gaps that might be apparent in the data. The parametric $Binomial(10, p)$ model, for example, might be a more appropriate model for this particular data set. Estimating an appropriate value of p is a topic taken up in the next section.

Continuous Distributions. When the simulation-modeling situation calls for a continuous model (e.g., interarrival times or service times at a service node), the nonparametric approach is analogous to the discrete case. Again let $x_1, x_2, \ldots, x_n$ be the data values collected on a random variable that is continuous by nature. One simple nonparametric input model for variate generation would be to repeatedly select one of the data values with probability $1/n$, just as in the discrete case. This again corresponds to an "empirical" pdf

$$\hat{f}(x) = \frac{1}{n} \qquad x = x_i; \; i = 1, 2, \ldots, n$$

if all of the data values are distinct. As before, if d data values are tied at a particular $x = x_i$ value, then the pdf at that particular x_i is $\hat{f}(x) = d/n$. The empirical cdf can be defined by

$$\hat{F}(x) = \frac{N(x)}{n} \qquad x = x_i; \; i = 1, 2, \ldots, n,$$

where $N(x)$ is the number of data values that are less than or equal to x. Variate generation is identical to the discrete case. The algorithm for generating variates for a discrete-event simulation is repeated here.

Algorithm 9.1.2 Given the (sorted or unsorted) data values $x_1, x_2, \ldots, x_n$ stored in x[1], x[2], …, x[n], the following algorithm returns one of the data values with probability $1/n$ for use in a nonparametric input model:

```
i = Equilikely(1, n);
return (x[i]);
```

Example 9.1.3
Returning to the $n = 23$ service times from Example 9.1.1, the empirical cdf is plotted in Figure 9.1.3. The geometric interpretation of Algorithm 9.1.2 is as follows: The random number generated in the call to $Equilikely(1, n)$ has a height on the $\hat{F}(x)$ axis that corresponds to a returned x-value associated with the stairstep empirical cdf.

    The nonparametric model that repeatedly selects one of the service times with probability $1/23$ has several drawbacks for this data set. First, the small size of the data set indicates that there will be significant sampling variability. Second, the tied value, 68.64 seconds, and the observation in the far right-hand tail of the distribution, 173.40 seconds, tend to indicate that the smoothing associated with a parametric analysis could be more appropriate. When these are combined with the interpolation and extrapolation problems inherent in nonparametric simulations, a modeler might be pushed toward a parametric analysis for this particular data set.

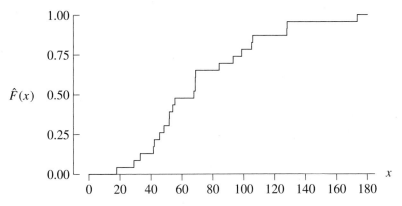

Figure 9.1.3 Empirical cdf for the service times.

There is a way to overcome the interpolation problem for a continuous data set. Using the $n - 1$ "gaps" created by the n data values, a piecewise-linear empirical cdf can be derived. These $n - 1$ gaps can each have a linear cdf that rises $1/(n - 1)$ between each of the sorted data values. Thus, from the "rise over run" definition of the slope of a line, each piecewise-linear segment will have the slope

$$\frac{1/(n - 1)}{x_{(i+1)} - x_{(i)}}$$

on $x_{(i)} \leq x < x_{(i+1)}$ for $i = 1, 2, \ldots, n - 1$, where $x_{(1)}, x_{(2)}, \ldots, x_{(n)}$ are the (distinct) sorted data values.*

Each of the piecewise segments must pass through the point $(x_{(i)}, (i - 1)/(n - 1))$, so, for $i = 1, 2, \ldots, n - 1$, some algebra yields the piecewise-linear empirical cdf

$$\hat{F}(x) = \frac{i - 1}{n - 1} + \frac{x - x_{(i)}}{(n - 1)(x_{(i+1)} - x_{(i)})} \qquad x_{(i)} \leq x < x_{(i+1)},$$

for $i = 1, 2, \ldots, n - 1$, and $\hat{F}(x) = 0$ for $x < x_{(1)}$, and $\hat{F}(x) = 1$ for $x \geq x_{(n)}$.[†] This estimator is still considered "nonparametric" because no parameters need to be estimated in order to arrive at the estimator.

Example 9.1.4

Returning to the $n = 23$ service times, the piecewise-linear empirical cdf is plotted in Figure 9.1.4. The empirical cdf has 22 linear segments, and, not surprisingly, has a shape that is similar to $\hat{F}(x)$ from Figure 9.1.3. This cdf allows x-values of any real value between $x_{(1)} = 17.88$ and $x_{(23)} = 173.40$, as opposed to just the data values, which was the case with the empirical cdf plotted in Figure 9.1.3. The cdf is not a function (but is plotted as a vertical line) at the tied data value 68.64. The piecewise-linear empirical cdf effectively solves the interpolation problem presented earlier.

---

[*]The values $x_{(1)}, x_{(2)}, \ldots, x_{(n)}$ are referred to as "order statistics" by statisticians. See David (2004) for an introduction and comprehensive survey of the literature.

[†]Notice that this estimator is undefined for tied data values. Fortunately, as will be seen later, this does not create problems with the associated variate-generation algorithm.

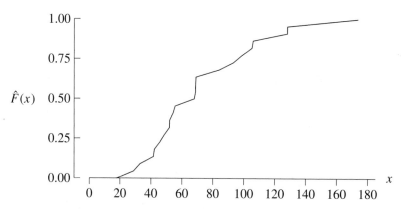

Figure 9.1.4   Piecewise-linear empirical cdf for the service times.

We now consider generating variates from the piecewise-linear empirical cdf. Since this cdf is linear, it can be inverted in closed form for distinct data values. When this equation is solved for x [replacing $\hat{F}(x)$ with u], the following equation is appropriate for generating a variate on segment i:

$$x = x_{(i)} + \big((n-1)u - (i-1)\big)(x_{(i+1)} - x_{(i)}) \qquad \frac{i-1}{n-1} \le u < \frac{i}{n-1},$$

for $i = 1, 2, \ldots, n-1$. A variate-generation algorithm must first find the appropriate piecewise segment, then invert the linear function, as shown above.

Algorithm 9.1.3   A one-time sort of the data values $x_1, x_2, \ldots, x_n$ into the order statistics $x_{(1)}, x_{(2)}, \ldots, x_{(n)}$ must be performed prior to executing this variate-generation algorithm. These sorted values are stored in x[1], x[2], ..., x[n]. This algorithm returns a variate associated with the piecewise-linear empirical cdf for use in a nonparametric input model.

```
u = Random();
i = ⌈(n - 1) * u⌉;
return (x[i] + ((n - 1) * u - (i - 1)) * (x[i + 1] - x[i]));
```

Algorithm 9.1.3 is monotone, synchronized, clear, and efficient. Further, the use of the piecewise-linear empirical cdf eliminates the interpolation problem associated with Algorithm 9.1.2. The geometry behind the three steps in Algorithm 9.1.3 is as follows: (i) a random number u is generated and plotted on the $\hat{F}(x)$ axis; (ii) the appropriate piecewise-linear segment corresponding to the random number is found; (iii) the appropriate piecewise-linear segment is inverted to yield a variate in the range $x_{(1)} < x < x_{(n)}$.[*]

---

[*]These are strict equalities on $x_{(1)} < x < x_{(n)}$ because we have assumed that the random-number generator will not generate exactly 0.0 or 1.0.

How does Algorithm 9.1.3 handle tied data values? These tied data values will be generated with probability d/n, where d is the number of values tied at a particular x-value. This is precisely what we would hope for in the case of tied values.

The piecewise-linear empirical cdf has solved the interpolation problem. Solving the extrapolation problem is difficult and requires more assumptions concerning the population.

This concludes our discussion of nonparametric modeling of stationary processes. The simple nonparametric techniques presented here let the data "speak for itself" rather than approximating the distribution with a parametric model such as the *Exponential* (μ) or *Normal* (μ, σ) distribution.

In terms of the modeling framework developed in Appendix G, nonparametric simulations suffer from three sources of error: (i) error due to a finite data set, embodied in $\mathcal{C}_r$; (ii) error due to a finite simulation-run length, embodied in $\mathcal{G}_r$; and (iii) error due to incorrect assumptions about the system, embodied in $\mathcal{A}$. The parametric approach, presented next, suffers from these three sources of error and more. Barton and Schruben (2001) list five additional disadvantages of the parametric approach:

- Parametric modeling requires more effort and expertise than nonparametric modeling;
- Error is introduced when an inaccurate parametric model is selected;
- Additional error is introduced in the estimation of parameters;
- Significant serial correlation and between-variable dependencies can oftentimes be lost;
- The parametric approach is typically more difficult to "sell" to management.

In spite of these problems, parametric modeling remains, for reasons hinted at in the introduction to Section 9.2, significantly more popular than nonparametric modeling in the simulation community. A novel mix of the two approaches, one using Bézier distributions, is given by Wagner and Wilson (1996). With their approach, the modeler chooses whether a "notch" in the empirical cdf, such as the one between $x = 60$ and $x = 70$ in Figure 9.1.3, is part of the target cdf of interest (and should be included in the model) or just random sampling variability (which should be smoothed over in the model).

9.1.4 Exercises

9.1.1   Consider a (tiny) fictitious data set with just $n = 3$ values collected: 1, 2, and 6. (a) Plot the empirical cdf. (b) Plot the piecewise-linear empirical cdf. (c) Generate 1000 samples of size $n = 3$ from the empirical cdf, and generate 1000 samples of size $n = 3$ from the piecewise-linear empirical cdf; then calculate the sample means and variances of each data set of 1000 values. (d) Comment.

9.1.2   Derive the equation,

$$\hat{F}(x) = \frac{i - 1}{n - 1} + \frac{x - x_{(i)}}{(n - 1)(x_{(i+1)} - x_{(i)})} \qquad x_{(i)} \le x < x_{(i+1)},$$

for $i = 1, 2, \ldots, n - 1$, associated with the piecewise-linear empirical cdf.

9.1.3   Prove that

$$x = x_{(i)} + \big((n - 1)u - (i - 1)\big)(x_{(i+1)} - x_{(i)}) \qquad \frac{i - 1}{n - 1} \le u < \frac{i}{n - 1},$$

for $i = 1, 2, \ldots, n - 1$, is the value of the idf associated with the piecewise-linear empirical cdf.

9.1.4   Consider program `ssq2`. (*a*) Modify the program to compute and print the average delay times for 50 replications having *Exponential* (2.0) interarrival times and *Uniform* (1.0, 2.0) service times. (*b*) Modify the program to compute and print the average delay times for 50 replications having *Uniform* (1.0, 2.0) service times and interarrival times that are drawn from an empirical cdf of $n = 4$ *Exponential* (2.0) random variates (using 50 such empirical cdfs). This is analogous to a nonparametric simulation in which the service-time distribution is known but only 4 interarrival times have been collected. (*c*) Compute the means and standard deviation of the two sets of numbers generated in parts (*a*) and (*b*). Plot an empirical cdf of the two sets of numbers. (*d*) Comment on the two empirical cdfs. (*e*) Would it have been a good or bad idea to have used a single random-number stream to generate the service times in parts (*a*) and (*b*) of this problem? Why?

9.2 MODELING STATIONARY PROCESSES: PARAMETRIC APPROACH

We now describe the parametric approach to the analysis of a stationary data set. The good news associated with the parametric approach is that the smoothing, interpolation, and extrapolation problems all vanish. The bad news is that more sophistication in probability and statistics is required (e.g., there is a wide array of parametric models, far beyond those described in this text, to choose from) and that additional error is introduced when a parametric model is fitted to a data set. As with making a copy of a copy on a copying machine, we are one more layer removed from the data values when we fit a model.

9.2.1 Hypothesizing a Distribution

In some respects, the most difficult aspect of fitting a distribution to data is the first step—hypothesizing a parametric model. This process is less "mechanical" than the others; consequently, it is more dependent on insight, theory, and experience. At best, there are just guidelines and questions. The answers to the questions posed below can be used to rule in or rule out certain probability models.

- What is the *source* of the data? Is the data inherently discrete or continuous? Is the data bounded? Is it nonnegative? Is the data produced by a stationary arrival process? Is the data the sum or product of random variables?
- Generate a histogram and look at its shape—frequently, this will eliminate many distributions from further consideration. Is the histogram "flat"? Is it symmetric about the mean? The shape of the histogram can change drastically for different histogram parameters, and you may need to experiment with different values for the histogram parameters. Obviously, this is best done interactively in a computer-graphics environment, where the histogram parameters can be varied easily.
- Calculate some simple sample statistics in order to help thin the field of potential input models. For example, (*i*) a coefficient of variation $s/\overline{x}$ close to 1, along with the appropriate histogram shape, indicates that the *Exponential* (μ) distribution is a potential input model

for an inherently continuous data set; (*ii*) a sample skewness close to 0 indicates that a symmetric distribution [e.g., a *Normal*(μ, σ) or *Uniform*(a, b) distribution] is a potential input model.

At this point, if you are not a statistician, consider finding one to guide you through this process.

If the collection of data values is either expensive, difficult, or time-consuming, you might have to address the question of what to do if you do not have enough data to generate a meaningful histogram. The answer is, not much—unless you can confidently hypothesize the distribution, primarily on the basis of a knowledge of the *source* of the data. For example, if the data are interarrival times and you have information that the process is stationary and likely to be a Poisson process, then the interarrival distribution is likely to be *Exponential*(μ), and you need only enough data to estimate the mean interarrival time μ accurately.

Example 9.2.1
A histogram with $k = 9$ cells for the $n = 23$ service times from Example 9.1.1 is given in Figure 9.2.1.

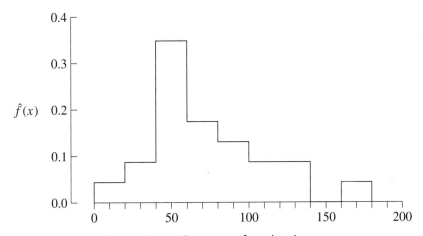

Figure 9.2.1   Histogram of service times.

It is very difficult to draw any meaningful conclusions from a data set that is this small. This histogram indicates that a skewed, bell-shaped pattern is appropriate for a pdf. The largest observation lies in the far right-hand tail of the distribution, so care must be taken to assure that it is representative of the population. The sample mean, standard deviation, coefficient of variation, and skewness are

$$\bar{x} = 72.22 \qquad s = 37.49 \qquad \frac{s}{\bar{x}} = 0.52 \qquad \frac{1}{n} \sum_{i=1}^{n} \left(\frac{x_i - \bar{x}}{s} \right)^3 = 0.88.$$

A data set of service times is inherently nonnegative, so a list of the potential models from the text would include the *Lognormal*(a, b) and the *Gamma*(a, b) distributions.

Another graphical device for distinguishing one distribution from another is a plot of the skewness on the vertical axis versus the coefficient of variation on the horizontal axis. This approach has the advantage of viewing several competing parametric models simultaneously. The *population* coefficient of variation $\gamma_2 = \sigma/\mu$ and the *population* skewness $\gamma_3 = E[(X - \mu)^3/\sigma^3]$ can be plotted for several parametric distributions. The *Gamma*(a, b) distribution, for example, has mean $\mu = ab$, standard deviation $\sigma = b\sqrt{a}$, coefficient of variation $\gamma_2 = 1/\sqrt{a}$, and skewness $\gamma_3 = 2/\sqrt{a}$. Thus a plot of γ_2 vs. γ_3 is linear. The special cases of the *Gamma*(a, b) distribution, namely the *Chisquare*(n), *Erlang*(n, b), and *Exponential*(μ) distributions, all lie on this line. Some distributions are represented as *points*, others lie in *lines*, still others fall in *regions*. A reasonable criterion for a hypothesized model is that the *sample* values for γ_2 and γ_3, calculated in the previous example as $\hat{\gamma}_2 = 0.52$ and $\hat{\gamma}_3 = 0.88$, should lie reasonably close to the point, curve, or region associated with the *population* values.

Example 9.2.2

A plot of γ_2 vs. γ_3 is given in Figure 9.2.2 for the *Exponential*(μ), *Gamma*(a, b), *Lognormal*(a, b), and *Weibull*(a, b) distributions. The *Weibull*(a, b) distribution will be introduced after this example. The point plotted for the $n = 23$ service-time data values is at $(\hat{\gamma}_2, \hat{\gamma}_3) = (0.52, 0.88)$ and lies reasonably close (based on the small sample size) to the *Gamma*, *Lognormal*, and *Weibull* models. All three are potential parametric input models.

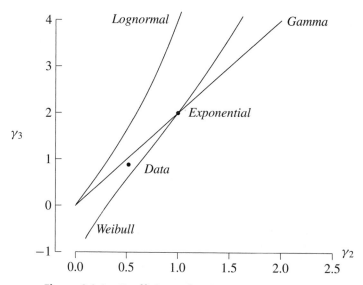

Figure 9.2.2   Coefficient of variation vs. skewness.

Weibull Random Variates.   Since the *Lognormal*, *Gamma*, and *Weibull* models are candidates for modeling the service times, we briefly diverge from our discussion of input modeling to introduce the *Weibull* distribution.

Definition 9.2.1   The continuous random variable X is *Weibull*(a, b) if and only if

- the real-valued shape parameter a satisfies $a > 0$,
- the real-valued scale parameter b satisfies $b > 0$,
- the possible values of X are $\mathcal{X} = \{x \mid x > 0\}$,
- the pdf of X is

$$f(x) = b^a a x^{a-1} \exp\left(-(bx)^a\right) \qquad x > 0,$$

- the cdf of X can be expressed in closed form as*

$$F(x) = 1 - \exp\left(-(bx)^a\right) \qquad x > 0,$$

- the mean of X is

$$\mu = \frac{1}{b} \Gamma\left(1 + \frac{1}{a}\right),$$

- the standard deviation of X is

$$\sigma = \frac{1}{b} \sqrt{\Gamma\left(1 + \frac{2}{a}\right) - \left[\Gamma\left(1 + \frac{1}{a}\right)\right]^2}.$$

The cdf of the *Weibull*(a, b) random variable can be inverted in closed form, resulting in efficient variate generation [unlike the *Gamma*(a, b) random variable, which requires the complicated acceptance–rejection algorithm outlined in Section 7.6]. The *Weibull*(a, b) distribution reduces to the *Exponential*(μ) distribution when $a = 1$. The pdf is monotone decreasing for $a < 1$ and somewhat bell-shaped for $a > 1$.

The next step in the process of input modeling is to calculate appropriate parameters for the hypothesized model—via a process known as parameter estimation.

9.2.2 Parameter Estimation

We will consider two basic approaches to estimating distribution parameters from a data set—the *method of moments* and *maximum-likelihood estimation* techniques. For many of the distributions considered in this book, the two approaches yield similar or identical results. The general setting is as follows. We have collected data values denoted by $x_1, x_2, \ldots, x_n$ and used the techniques from the previous section to decide on one or more potential input models. Let q denote the number of unknown parameters [e.g., $q = 1$ for the *Exponential*(μ) model and $q = 2$ for the *Lognormal*(a, b) model].

---

*Beware! There is no standard parameterization for the *Weibull* distribution. Both $F(x) = 1 - \exp\left(-(x/b)^a\right)$ and $F(x) = 1 - \exp(-bx^a)$ are also legitimate *Weibull* cdfs.

Definition 9.2.2   The *method of moments* is an intuitive algebraic method for estimating the parameters of a distribution by equating the first q *population* moments defined by

$$E\left[X^k\right]$$

for $k = 1, 2, \ldots, q$ with the corresponding first q *sample* moments defined by

$$\frac{1}{n}\sum_{i=1}^{n} x_i^k$$

for $k = 1, 2, \ldots, q$, and solving the $q \times q$ set of equations for the q unknown parameters.

Although it might initially seem awkward, to set a fixed quantity (a *population* moment) equal to a random quantity (a *sample* moment), the intuition associated with the method-of-moments technique is compelling—a fitted model will match the first q sample moments. We use hats to denote estimators. The method is easily learned by example.

Example 9.2.3
To estimate the single ($q = 1$) parameter μ for a *Poisson*(μ) model, equate the population mean to the sample mean,

$$\mu = \bar{x},$$

and solve for μ to obtain the method-of-moments estimator $\hat{\mu} = \bar{x}$.

Example 9.2.4
To estimate the single ($q = 1$) parameter μ for a *Exponential*(μ) model, equate the population mean to the sample mean,

$$\mu = \bar{x},$$

and solve for μ to obtain the method-of-moments estimator $\hat{\mu} = \bar{x}$. Even though the *Exponential*(μ) model was not on our short list of potential input models for the $n = 23$ service times from our analysis in the previous section, we will go through the process of fitting the distribution to make the point that the fitting process can be entered into blindly, with corresponding results. Since the sample mean is $\bar{x} = 72.22$, the method of moments parameter estimate is $\hat{\mu} = 72.22$. The fitted cdf is $\hat{F}(x) = 1 - \exp(-x/72.22)$ for $x > 0$. This fitted cdf is plotted along with the empirical cdf in Figure 9.2.3.

We can see that this is a very poor fit. The fitted *Exponential* cdf is much higher than the empirical cdf for small service times. This means that a discrete-event simulation using the *Exponential*(72.22) model will generate many more short service times than it should. Likewise, the fitted *Exponential* cdf is much lower than the empirical cdf for large service times. In statistical jargon, this is said to indicate that the fitted pdf has a heavier right-hand "tail" than the empirical. A discrete-event simulation using the *Exponential*(72.22) model will generate more long service times than it should. Fitting

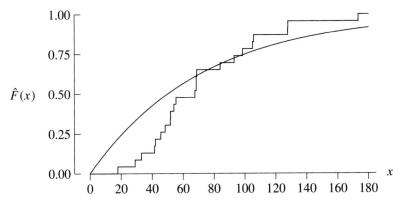

Figure 9.2.3   Fitted exponential cdf and empirical cdf.

the *Exponential* model in this case has resulted in a fitted model with significantly higher variability than the data would suggest (the fitted model has standard deviation 72.22, whereas the sample standard deviation is 37.49). Using this particular input model will introduce significant error into the discrete-event simulation. The accurate modeling of the right-hand tail, in particular, is important for queuing simulations, since these values constitute the unusually long service times that cause congestion.

Example 9.2.5
To estimate the two ($q = 2$) parameters μ and σ for a *Normal*(μ, σ) model, equate the first two population moments to the first two sample moments:

$$E[X] = \frac{1}{n} \sum_{i=1}^{n} x_i$$

$$E[X^2] = \frac{1}{n} \sum_{i=1}^{n} x_i^2.$$

For compactness, denote the right-hand sides of these two equations by m_1 and m_2. By using the relationship $\sigma^2 = E[X^2] - \mu^2$, the equations can be rewritten as

$$\mu = m_1$$

$$\sigma^2 + \mu^2 = m_2.$$

Solving for μ and σ yields the method-of-moments estimators

$$\hat{\mu} = m_1 = \overline{x}$$

$$\hat{\sigma} = \sqrt{m_2 - m_1^2} = s.$$

Thus the method-of-moments technique estimates the population mean by the sample mean and the population standard deviation by the sample standard deviation.

In the previous three examples, the population moments and the sample moments match up in a way that is intuitive. The next three examples illustrate less intuitive cases, where oftentimes more algebra is required to arrive at the method-of-moments estimators.

Example 9.2.6

To estimate the single ($q = 1$) parameter p for the *Geometric*(p) model, start by equating the population mean and the sample mean

$$\frac{p}{1-p} = \bar{x}.$$

Then solve for p to yield the estimate

$$\hat{p} = \frac{\bar{x}}{1+\bar{x}}.$$

The parameter p for the *Geometric*(p) model must satisfy $0 < p < 1$. Does the method of moments estimator satisfy this constraint on the parameter? If the data are drawn from a *Geometric*(p) population, then the data must assume the values $\mathcal{X} = \{0, 1, 2, \ldots\}$. This means that $\bar{x} \geq 0$. The only case where the method of moments experiences difficulty is a data set with $\bar{x} = 0$, which corresponds to a data set of all zeros.

Example 9.2.7

To estimate the two ($q = 2$) parameters a and b for the *Gamma*(a, b) model, introduced in Section 7.6, equate the first two population moments to the first two sample moments,

$$ab = m_1$$

$$ab^2 + (ab)^2 = m_2$$

using the relationship $\sigma^2 = E[X^2] - \mu^2$. Solving for a and b yields

$$\hat{a} = \frac{m_1^2}{m_2 - m_1^2} = \frac{\bar{x}^2}{s^2} \qquad \text{and} \qquad \hat{b} = \frac{m_2 - m_1^2}{m_1} = \frac{s^2}{\bar{x}}.$$

Since the *Gamma*(a, b) distribution was on our short list of potential input models for the service-time data, we estimate the parameters as:

$$\hat{a} = \frac{\bar{x}^2}{s^2} = \frac{72.22^2}{37.49^2} = 3.71 \qquad \text{and} \qquad \hat{b} = \frac{s^2}{\bar{x}} = \frac{37.49^2}{72.22} = 19.46,$$

yielding the fitted cdf displayed in Figure 9.2.4.

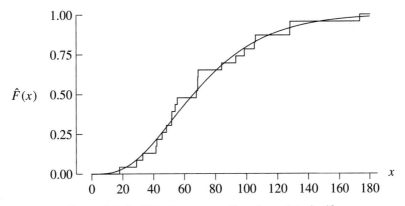

Figure 9.2.4 Fitted gamma cdf and empirical cdf.

The fitted *Gamma*(a, b) model is *far* superior to the fitted *Exponential*(μ) model illustrated in Figure 9.2.3. Both tails and the body display better fits and, since there are now *two* parameters in the fitted models, both the mean and the variance are now appropriate. It is up to the modeler to decide whether the "notch" in the empirical cdf between $x = 60$ and $x = 70$ is (*i*) an inherent and important part of the distribution that could require switching to a nonparametric approach or (*ii*) just an effect of sampling variability, in which case the *Gamma*(a, b) model appropriately smooths over the notch.

Since the *Exponential*(μ) distribution is a special case of the *Gamma*(a, b) distribution when $a = 1$, the *Gamma* distribution will typically "appear" to fit a data set better, even when the data are drawn from an *Exponential* population. The extra parameter gives the *Gamma* distribution more "flexibility" than the more rigid one-parameter *Exponential*. The extra parameter allows for an "S-shape" for the cdf, as illustrated in Figure 9.2.4, which is not possible for the cdf of an *Exponential* random variable.

Example 9.2.8
To estimate the two ($q = 2$) parameters a and b for the *Uniform*(a, b) distribution, start by equating the first two population and sample moments

$$\frac{a + b}{2} = m_1$$

$$\frac{(b - a)^2}{12} + \left(\frac{a + b}{2}\right)^2 = m_2.$$

Then solve for a and b to yield the estimates

$$\hat{a} = \bar{x} - \sqrt{3}\, s$$

$$\hat{b} = \bar{x} + \sqrt{3}\, s.$$

The a, b estimates for the *Uniform*(a, b) distribution provide a good illustration of the difference between method-of-moments estimates and maximum-likelihood estimates. It is intuitive that *if* the data are truly *Uniform*(a, b), then all the data must fall between a and b. The $\hat{a}$ and $\hat{b}$ values given above do not necessarily satisfy this criterion. The maximum-likelihood estimators for a and b are the smallest and largest data values. When maximum-likelihood estimates differ from method-of-moments estimates, it is generally considered desirable to use the former. Computational considerations are also important, however, and, when there is a difference, method-of-moments estimates are often easier to calculate.

Definition 9.2.3   The *maximum-likelihood estimators* are the parameter values associated with a hypothesized distribution that correspond to the distribution that is the most likely to have produced the data set $x_1, x_2, \ldots, x_n$. If $\theta = (\theta_1, \theta_2, \ldots, \theta_q)'$ is a vector of unknown parameters, then the maximum-likelihood estimators $\hat{\theta}$, often abbreviated MLE, maximize the likelihood function

$$L(\theta) = \prod_{i=1}^{n} f(x_i, \theta).$$

The vector of unknown parameters θ has been added to the pdf in order to emphasize the dependence of the likelihood function on θ.

The observations are independent, so the likelihood function, $L(\theta)$, is the product of the pdf evaluated at each data value, known to statisticians as the *joint* pdf. The maximum-likelihood estimator $\hat{\theta}$ is found by maximizing $L(\theta)$ with respect to θ, which typically involves some calculus. Thus the maximum-likelihood estimator corresponds to the particular value(s) of the parameters that are most likely to have produced the data values $x_1, x_2, \ldots, x_n$.

In practice, it is often easier to maximize the log-likelihood function, $\ln L(\theta)$, to find the vector of maximum-likelihood estimators; this approach is valid because the logarithm function is monotonic. The log-likelihood function is

$$\ln L(\theta) = \sum_{i=1}^{n} \ln f(x_i, \theta);$$

it is asymptotically normally distributed, by the central limit theorem, since it consists of the sum of n random, independent terms.

Example 9.2.9

Let $x_1, x_2, \ldots, x_n$ be a random sample from an *Exponential*(μ) population. The likelihood function is

$$L(\mu) = \prod_{i=1}^{n} f(x_i, \mu) = \prod_{i=1}^{n} \frac{1}{\mu} \exp(-x_i/\mu) = \mu^{-n} \exp\left(-\sum_{i=1}^{n} x_i/\mu\right).$$

The log-likelihood function is

$$\ln L(\mu) = -n \ln \mu - \sum_{i=1}^{n} x_i / \mu.$$

In order to maximize the log-likelihood function, differentiate with respect to μ.

$$\frac{\partial \ln L(\mu)}{\partial \mu} = -\frac{n}{\mu} + \frac{\sum_{i=1}^{n} x_i}{\mu^2}.$$

Equating to zero and solving for μ yields the maximum-likelihood estimator

$$\hat{\mu} = \frac{1}{n} \sum_{i=1}^{n} x_i,$$

which is the sample mean $\bar{x}$. In this particular case, the method-of-moments and maximum-likelihood estimators are identical. This is the case for many, but not all, of the distributions introduced in this text.

Example 9.2.10
Figure 9.2.5 illustrates the maximum-likelihood estimator for μ for an *Exponential* population drawn from a (fictitious) sample size of $n = 3$,

$$1 \qquad 2 \qquad 6.$$

(One could argue that these data values look more discrete than continuous, but we use integers to make the arithmetic easy.) Consider all the allowable μ values for an *Exponential*(μ) distribution (i.e., $\mu > 0$). Of the infinite number of $\mu > 0$, the $\hat{\mu} = \bar{x} = 3$ value depicted in Figure 9.2.5 (shown in the associated pdf) is that which maximizes the product of the pdf values at the data points, i.e., the product of the lengths of the vertical dashed lines.

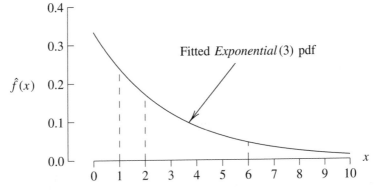

Figure 9.2.5 Maximum-likelihood-estimation geometry.

Any other choice of μ results in a smaller value of $L(\mu)$. The pdf plotted in Figure 9.2.5 is the one most likely to produce the observed data values. Although the next example violates one of the prerequisites for the text (single-variable calculus), we include it to illustrate how two-parameter distributions are handled.

Example 9.2.11

The *Weibull* (a, b) distribution was also listed as a potential model for the $n = 23$ service times. We derive the $q = 2$ maximum-likelihood estimates in the general case, then compute their numerical values associated with the service times. The *Weibull* distribution has pdf

$$f(x) = b^a a x^{a-1} \exp\left(-(bx)^a\right) \qquad x > 0,$$

where b is a positive scale parameter and a is a positive shape parameter. Let $x_1, x_2, \ldots, x_n$ denote the data values. The likelihood function is a function of the unknown parameters a and b:

$$L(a, b) = \prod_{i=1}^{n} f(x_i) = b^{an} a^n \left[\prod_{i=1}^{n} x_i\right]^{a-1} \exp\left(-\sum_{i=1}^{n}(bx_i)^a\right).$$

The mathematics is typically more tractable for maximizing a log-likelihood function, which, for the *Weibull* distribution, is

$$\ln L(a, b) = n \ln a + an \ln b + (a-1) \sum_{i=1}^{n} \ln x_i - b^a \sum_{i=1}^{n} x_i^a.$$

Differentiating these equations with respect to the unknown parameters a and b and equating to zero yields the following 2×2 set of nonlinear equations:

$$\frac{\partial \ln L(a, b)}{\partial a} = \frac{n}{a} + n \ln b + \sum_{i=1}^{n} \ln x_i - \sum_{i=1}^{n}(bx_i)^a \ln bx_i = 0,$$

and

$$\frac{\partial \ln L(a, b)}{\partial b} = \frac{an}{b} - ab^{a-1} \sum_{i=1}^{n} x_i^a = 0.$$

These simultaneous equations have no closed-form solution for $\hat{a}$ and $\hat{b}$. To reduce the problem to a single unknown, the second equation can be solved for b in terms of a:

$$b = \left(\frac{n}{\sum_{i=1}^{n} x_i^a}\right)^{1/a}.$$

Law and Kelton (2000, page 305) give an initial estimate for a, and Qiao and Tsokos (1994) give a fixed-point algorithm for calculating $\hat{a}$ and $\hat{b}$. Their algorithm is guaranteed to converge for any positive initial estimate for a.

For the 23 service times, the fitted *Weibull* distribution has maximum-likelihood estimators $\hat{a} = 2.10$ and $\hat{b} = 0.0122$. The log-likelihood function evaluated at the maximum-likelihood estimators is $\ln L(\hat{a}, \hat{b}) = -113.691$. Figure 9.2.6 shows the empirical cdf, along with the cdf of the *Weibull* fitted to the data.

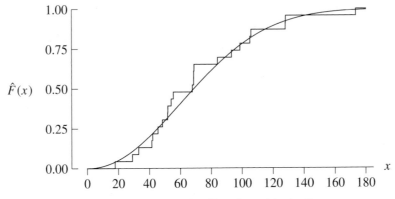

Figure 9.2.6   Weibull cdf and empirical cdf.

The difference between the fitted *Gamma*(a, b) cdf displayed in Figure 9.2.4 and the fitted *Weibull*(a, b) cdf displayed in Figure 9.2.6 might look so insignificant that the modeler might conclude that the two are interchangeable. This is an example of where a good knowledge of probability and statistics can help to differentiate between the two models. The *Gamma*(a, b) distribution has "exponential" right-hand tails, which means that, for large service times that have not been completed, the remaining service time is approximately exponentially distributed via the memoryless property. Is this an appropriate property for service times? That all depends on the situation. It is enough to say that the *Gamma*(a, b) model has a heavier right-hand tail and that the *Weibull*(a, b) model has a lighter right-hand tail for the particular fitted distribution given here. There may be specific information about the system that would help evaluate which model is superior. As mentioned earlier, precise modeling in the right-hand tail of a service-time distribution is crucial to the effective modeling of the service node.

The method-of-moments and maximum-likelihood estimators are special cases of what is known more generally in statistics as "point estimators" Hogg, McKean, and Craig (2005) and Casella and Berger (2002), for example, survey properties of point estimators known as unbiasedness, minimum variance, efficiency, and consistency.

Accuracy of Point Estimators.   Many argue that a point estimator is of little use by itself, since it does not contain any information about the accuracy of the estimator. Statisticians often use *interval estimation* to quantify the precision of estimates. When there is a single $(q = 1)$ parameter, a confidence interval is used to measure the precision of the point estimate. When there are several $(q > 1)$ parameters, a "confidence region" is used to measure the precision of the point estimates. Interval estimators incorporate the variability of the point estimator to indicate the accuracy of the point estimator. Not surprisingly, these intervals and regions tend to narrow as the sample size n increases.

Although the specifics of interval estimation are beyond the prerequisites for this text, we present an example of such an interval estimate and its associated interpretation for the service-time example.

Example 9.2.12

In the case of the $n = 23$ service times that were fitted by the *Weibull* model, the fact that the "likelihood ratio test statistic," defined by Casella and Berger (2002) as $2[\ln L(\hat{a}, \hat{b}) - \ln L(a, b)]$, is asymptotically *Chisquare*(2) means that a 95% confidence region for the parameters is the region containing all a and b satisfying

$$2[-113.691 - \ln L(a, b)] < 5.99,$$

where 5.99 is the 95th percentile of the *Chisquare*(2) distribution. The maximum-likelihood estimators and the associated 95%-confidence region are shown in Figure 9.2.7. The maximum-likelihood estimators are plotted at the point $(\hat{a}, \hat{b}) = (2.10, 0.0122)$. The line $a = 1$ (when the *Weibull* distribution collapses to the *Exponential* distribution) is not interior to the confidence region, which gives a third confirmation that the *Exponential* distribution is not an appropriate model for the service-time data set.

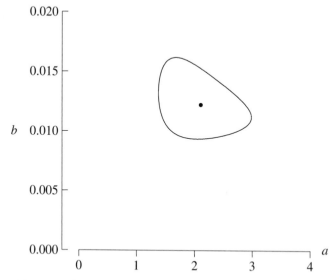

Figure 9.2.7   Confidence region for the estimated Weibull parameters.

A confidence region like the one plotted in Figure 9.2.7 can be quite helpful in evaluating the effect of the sampling variability associated with the data set drawn from the system on the simulation output. Assume that the *Weibull* (a, b) model is selected as an input model. It would be reasonable to make simulation runs both at the values of the maximum-likelihood estimates, $(\hat{a}, \hat{b}) = (2.10, 0.0122)$, and at a few selected points along the boundary of the confidence region. In this manner the effect of how far the parameters of the input model could stray from the maximum-likelihood estimators can be used to assess the impact on the output values from the simulation.

9.2.3 Goodness of Fit

There are two approaches to testing for goodness of fit—visual and analytic. The visual method has been emphasized thus far. A *visual* test for goodness of fit is an educated inspection of how well the histogram or empirical cdf is approximated (*fitted*) by a hypothesized distribution pdf or cdf whose parameters have been estimated. Visually comparing the fitted cdf of the *Exponential*(72.22) model with the empirical cdf in Figure 9.2.3, for example, allowed us to conclude that the *Exponential* model was not an appropriate service-time model. Other methods for visually comparing distributions include "P–P" (probability–probability) and "Q–Q" (quantile–quantile) plots—as described, for example, in Law and Kelton (2000). A P–P plot is the fitted cumulative distribution function at the ith order statistic $x_{(i)}$, $\hat{F}(x_{(i)})$, versus the adjusted empirical cumulative distribution function, $\tilde{F}(x_{(i)}) = (i - 0.5)/n$, for $i = 1, 2, \ldots, n$. A plot where the points fall close to the line passing through the origin and $(1, 1)$ indicates a good fit.

Example 9.2.13
For the $n = 23$ service times, a P–P plot for the *Weibull* fit is shown in Figure 9.2.8, along with a line connecting $(0, 0)$ and $(1, 1)$. P–P plots should be constructed for all competing models.

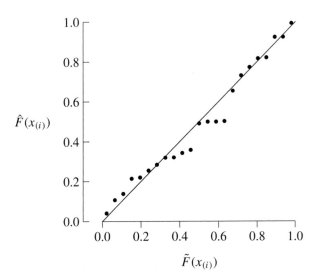

Figure 9.2.8   P–P plot for the fitted Weibull model.

We would like to have an objective technique for assessing fit, since visual methods are subjective. The analytic approach begins with two standard statistical tests: *chi-square* for discrete models, *Kolmogorov–Smirnov* for continuous models.

- If the data are discrete, the heights of the mass values on a discrete-data histogram are compared with the corresponding mass values associated with the fitted distribution. The chi-square goodness-of-fit test statistic is a function of the squared differences between these heights. (See Section 10.1 for more details.)
- If the data are continuous, the Kolmogorov–Smirnov goodness-of-fit test statistic is the largest vertical distance between the empirical and fitted cdfs. This largest difference is traditionally denoted by D_n. (See Section 10.1 for more details.)

Example 9.2.14

The Kolmogorov–Smirnov test statistics for four of the models considered in this chapter, all fitted by maximum likelihood, are given below. A large vertical difference in cdfs for the *Exponential* model, and smaller vertical differences for the three two-parameter models, confirm our visual goodness-of-fit results from Figures 9.2.3, 9.2.4, and 9.2.6.

Distribution	D_{23}
Exponential (μ)	0.307
Weibull (a, b)	0.151
Gamma (a, b)	0.123
Lognormal (a, b)	0.090

The Cramer–von Mises and Anderson–Darling goodness-of-fit tests (Lawless, 2003; D'Agostino and Stephens, 1986) are improvements upon the Kolmogorov–Smirnov test. These tests are often more sensitive to lack of fit in the tails of a distribution, which can often be critical in developing an input model.

Many of the discrete-event simulation packages, and several stand-alone software packages, automate all of the processes considered in this chapter with little or no expert intervention. These packages calculate parameter estimates for dozens of popular parametric distributions and perform goodness-of-fit tests so that the distribution that best fits the data set can quickly be identified.

In both simulation education and practice, it is typically the case that the input model for the discrete-event simulation is considered "exact" and that output analysis (considered in Chapter 9) estimates performance measures on the basis of this incorrect assumption. The correct (and more difficult) approach is to somehow, in the estimation of performance measures, combine the error associated with the input modeling process with the error involved in simulating the system. Chick (2001), for example, incorporates random-sampling variability from the data set in the analysis of the output from a discrete-event simulation.

9.2.4 Exercises

9.2.1   Consider program `ssq2`. (*a*) Modify the program to compute and print the average delay times for 50 replications, with *Exponential* (2.0) interarrival times and *Uniform* (1.0, 2.0) service times. (*b*) Modify the program to compute and print the average delay times for 50 replications but with *Exponential* ($\hat{\mu}$) interarrival times and *Uniform* (1.0, 2.0) service times, where $\hat{\mu}$ is the sample mean of $n = 4$ *Exponential* (2.0) random variates (use 50 such sample means). (*c*) Compute the means and standard deviations of the two sets of numbers generated in parts (*a*) and (*b*). Plot an empirical cdf of the two sets of numbers. (*d*) Comment.

9.2.2   Explain why the *Gamma* (*a*, *b*) distribution is generally preferred to the *Erlang* (*n*, *b*) distribution for modeling a continuous random variable.

9.2.3   Explain why you would be nervous about the validity of a discrete-event simulation model with a *Uniform* (*a*, *b*) service time.

9.2.4   Perform the algebra necessary to arrive at the method-of-moments estimators for *a* and *b* in Example 9.2.8.

9.2.5   (*a*) Use a modified version of program `ssq2` to estimate the expected steady-state wait for customers entering an *M*/*M*/1 service node with arrival rate $\lambda = 1$ and service rate $\nu = 1.25$. (*b*) Which of the continuous distributions we have studied seems to fit these data best? (*c*) Comment.

9.2.6   Derive the *Lognormal* (*a*, *b*) method-of-moments parameter-estimation equations.

9.2.7   Let $x_1, x_2, \ldots, x_n$ be a random sample drawn from an *inverse Gaussian* (Wald) population having positive parameters λ and μ and pdf (per Chhikara and Folks, 1989)

$$f(x) = \sqrt{\frac{\lambda}{2\pi}} x^{-3/2} \exp\left(-\frac{\lambda}{2\mu^2 x}(x - \mu)^2\right) \qquad x > 0.$$

Find the maximum-likelihood estimators for μ and λ.

9.2.8   Find the maximum-likelihood estimators for μ when *n* independent and identically distributed observations are drawn from a *Poisson* (μ) population.

9.2.9   Use a modified version of program `ssq3` to supply convincing numerical evidence that the steady-state number of customers in an *M*/*M*/1 service node is a *Geometric* (ρ) random variable.

9.2.10   (*a*) Use a modified version of program `msq` to estimate the pdf of the steady-state number of customers in an *M*/*M*/20 service node. (*b*) What discrete distribution fits this histogram best?

9.3 MODELING NONSTATIONARY PROCESSES

There are modeling situations that arise where one of the stationary input models from the previous section does not adequately describe a stochastic element of interest. A *nonstationary* model is needed to model the arrival times of patients to an emergency room, the failure times of an automobile, or the arrival times of customers to a fast-food restaurant. Any stochastic model that varies with time requires a nonstationary model. Therefore, accurate input modeling requires a careful evaluation of whether a stationary (no time dependence) or nonstationary model is appropriate.

9.3.1 An Example

Modeling customer arrival times to a lunch wagon will be used throughout this section to illustrate the decision-making process involved in selecting a nonstationary input model.

Example 9.3.1

Customer arrival times to a lunch wagon between 10:00 AM and 2:30 PM are collected on three days. These realizations were generated from a hypothetical arrival process given by Klein and Roberts (1984). A total of 150 arrival times were observed: 56 on the first day, 42 on the second day, and 52 on the third day. Defining $(0, 4.5]$ to be the time interval of interest (in hours), the three realizations are

$$0.2153 \quad 0.3494 \quad 0.3943 \quad 0.5701 \quad 0.6211 \quad \ldots \quad 4.0595 \quad 4.1750 \quad 4.2475,$$
$$0.3927 \quad 0.6211 \quad 0.7504 \quad 0.7867 \quad 1.2480 \quad \ldots \quad 3.9938 \quad 4.0440 \quad 4.3741,$$
and
$$0.4499 \quad 0.5495 \quad 0.6921 \quad 0.9218 \quad 1.2057 \quad \ldots \quad 3.5099 \quad 3.6430 \quad 4.3566.$$

One preliminary statistical issue concerning this data is whether the three days represent processes drawn from the same population. External factors such as the weather, day of the week, advertisement, and workload should be fixed. For this particular example, we assume that these factors have been fixed and that the three realizations are drawn from a representative target arrival process.

When arrival times are realizations of a continuous-time, discrete-state stochastic process, the remaining question concerns whether the process is stationary. If the process proves to be stationary, the techniques from the previous section, such as the drawing of a histogram and the choosing of a parametric or nonparametric model for the *interarrival* times, are appropriate. On the other hand, if the process is nonstationary, a nonstationary Poisson process, which was introduced in Section 7.5, might be an appropriate input model. Recall that a nonstationary Poisson process is governed by an event-rate function $\lambda(t)$ that gives an arrival rate [e.g., $\lambda(2) = 10$ means that the arrival rate is 10 customers per hour at time 2] that can vary with time. Although we place exclusive emphasis on the nonstationary-Poisson-process model in this section, there are dozens of other nonstationary models that are appropriate in other modeling situations. As in the previous two sections on the modeling of stationary processes, input models can be divided into nonparametric and parametric models. We begin with nonparametric models.

9.3.2 Nonparametric Modeling

This subsection describes three nonparametric procedures for estimating the event-rate function $\lambda(t)$: or, equivalently, the cumulative event-rate function

$$\Lambda(t) = \int_0^t \lambda(\tau)d\tau$$

from k realizations sampled from a "target" nonstationary Poisson process. The term *target* refers to the population process that we want to estimate and simulate. The first procedure can be used on "count" data. The second and third procedures are appropriate for "raw" data. In each of the procedures, we (i) find point estimates for the cumulative event-rate function, (ii) find interval estimates for the cumulative event-rate function, and (iii) develop algorithms for generating event times from the estimated process.

Count Data.   The event-rate function $\lambda(t)$ or cumulative event-rate function $\Lambda(t)$ is to be estimated on $(0, S]$, where S is a known constant. The interval $(0, S]$ might represent the time a system allows arrivals (e.g., 9:00 AM to 5:00 PM at a bank) or one period of a cycle (e.g., one day at an emergency room). There are k representative realizations collected on the target process on $(0, S]$.

For systems with high arrival rates (e.g., busy call centers or web sites), there is often so much data that *counts* of events that occur during *bins* (subintervals) are available, rather than the raw event times. Although this is less preferable than having the raw data, it is still possible to construct an estimate of the event-rate function and generate variates for a discrete-event simulation model. The time interval $(0, S]$ can be partitioned into m subintervals

$$(a_0, a_1], (a_1, a_2], \ldots, (a_{m-1}, a_m],$$

where $a_0 = 0$ and $a_m = S$. The subintervals do not necessarily have equal widths. Let $n_1, n_2, \ldots, n_m$ be the total number of observed events in the subintervals over the k realizations.

Example 9.3.2
Consider the following $m = 9$ equal-width subintervals associated with the lunch-wagon example:

$$(0.0, 0.5], (0.5, 1.0], \ldots, (4.0, 4.5].$$

The counts during each half-hour subinterval are

subinterval number, i	1	2	3	4	5	6	7	8	9
event count, n_i	5	13	12	24	32	27	21	10	6

The counts range from $n_1 = 5$ arrivals in the first subinterval to $n_5 = 32$ arrivals in the fifth subinterval, so there is a strong suspicion that a nonstationary model is appropriate, and we can proceed toward fitting a nonstationary Poisson process to the data set of counts.

To simplify the estimation process, assume that the target nonstationary Poisson process has an event-rate function $\lambda(t)$ that is piecewise constant on each subinterval of the partition $(a_0, a_1], (a_1, a_2], \ldots, (a_{m-1}, a_m]$. The average event-rate function on the interval $(a_{i-1}, a_i]$ is the rate per unit time of the events that occur on that interval, so the maximum-likelihood estimator is the average number of events that occurred on the interval, normalized for the length of the interval:

$$\hat{\lambda}(t) = \frac{n_i}{k(a_i - a_{i-1})} \qquad a_{i-1} < t \le a_i$$

for $i = 1, 2, \ldots, m$.

Example 9.3.3

For the lunch-wagon count data, there are five arrivals in the first subinterval. The estimated event rate during this subinterval is

$$\hat{\lambda}(t) = \frac{5}{3(0.5 - 0.0)} = 10/3 \qquad a_0 < t \leq a_1$$

customers per hour. Figure 9.3.1 shows the event-rate function for all nine of the subintervals, calculated in a similar manner.

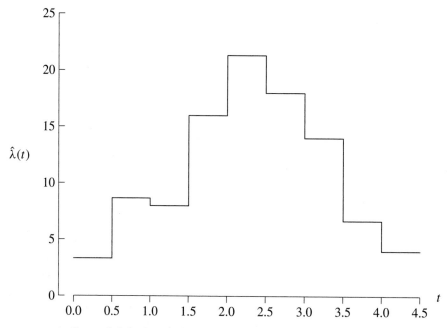

Figure 9.3.1   Lunch-wagon event-rate-function estimate.

We have connected the piecewise-constant segments with vertical lines, although this is largely a matter of taste. Even though the diagram in Figure 9.3.1 looks tantalizingly close to a histogram, it should not be called a "histogram" because of the time dependence.

The event-rate estimator is piecewise constant, so the associated cumulative event-rate-function estimator is a continuous, piecewise-linear function on $(0, S]$:

$$\hat{\Lambda}(t) = \int_0^t \hat{\lambda}(\tau)d\tau = \left(\sum_{j=1}^{i-1} \frac{n_j}{k} \right) + \frac{n_i(t - a_{i-1})}{k(a_i - a_{i-1})} \qquad a_{i-1} < t \leq a_i$$

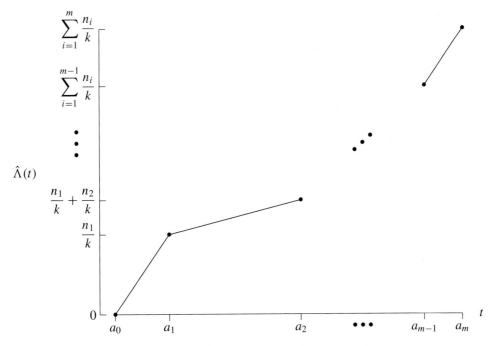

Figure 9.3.2   Cumulative event-rate-function estimate.

for $i = 1, 2, \ldots, m$.[*] This estimator passes through the points $\left(a_i, \sum_{j=1}^{i} n_j/k \right)$ for $i = 1, 2, \ldots, m$. A graph of a generic version of this estimator is shown in Figure 9.3.2. The tick marks on the horizontal axis correspond to the subinterval endpoints. The tick marks on the vertical axis are determined by the count data. In the unlikely case that the piecewise-linear segments are all parallel [i.e., the estimator is a line connecting $(0, 0)$ with $\left(S, \sum_{i=1}^{m} n_i/k \right)$], the model reduces from a nonstationary Poisson process to a stationary (homogeneous) Poisson process. Asymptotic properties of this estimator in the case of equal-width subintervals are considered by Henderson (2003).

Example 9.3.4

Figure 9.3.3 contains the piecewise-linear cumulative event-rate-function estimate associated with the lunch-wagon arrival counts. The estimator begins at $(0, 0)$ and ends at $(4.5, 150/3)$, as expected. Recalling that the interpretation of the cumulative event-rate function is the expected number of events by time t, we can say that the data suggest that there will be $n/k = 150/3 = 50$ arrivals to the lunch wagon per day between the hours of 10:00 AM and 2:30 PM. The S-shape for the cumulative event-rate-function

---

[*]If there are no events observed on interval i (i.e., $n_i = 0$), then the event-rate-function estimate is zero on interval i, and the cumulative event-rate-function estimate is constant on interval i. In the variate-generation algorithm to be described subsequently, no events will be generated for such an interval. This is useful for modeling an interval where no events should occur (e.g., lunch breaks).

estimator indicates that fewer customers tend to arrive at the beginning and end of the observation period.

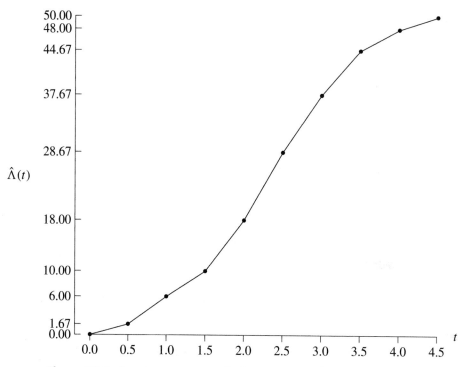

Figure 9.3.3   Lunch-wagon cumulative event-rate-function estimate.

As in the previous section, it is important to assess the accuracy of the estimators developed thus far; this assessment typically is done via confidence intervals. From the fact that the number of events by time t has a $Poisson\big(\Lambda(t)\big)$ distribution, an approximate, two-sided $(1 - \alpha)100\%$ confidence interval for $\Lambda(t)$ is

$$\hat{\Lambda}(t) - z_{\alpha/2}\sqrt{\frac{\hat{\Lambda}(t)}{k}} < \Lambda(t) < \hat{\Lambda}(t) + z_{\alpha/2}\sqrt{\frac{\hat{\Lambda}(t)}{k}},$$

for $0 < t \leq S$, where $z_{\alpha/2}$ is the $1 - \alpha/2$ fractile of the standard normal distribution. (The quantity $z_{\alpha/2}$ is equivalent to t^*_∞, which was introduced in Section 8.1.) The interval is always asymptotically exact at the endpoints, but asymptotically exact for all t in $(0, S]$ only when the target event-rate function $\lambda(t)$ is piecewise constant over each subinterval $(a_{i-1}, a_i]$ in the arbitrary partition of $(0, S]$. In most applications, this rather restrictive assumption is *not* satisfied.

Example 9.3.5
Figure 9.3.4 illustrates the cumulative event-rate-function estimator and the associated 95%-confidence limits. Not surprisingly, these limits expand with time, as we are less

certain of the number of arrivals by time 4.0, for example, than we are of the number of arrivals by time 1.0. Increasing the number of realizations k will narrow the width of the confidence limits.

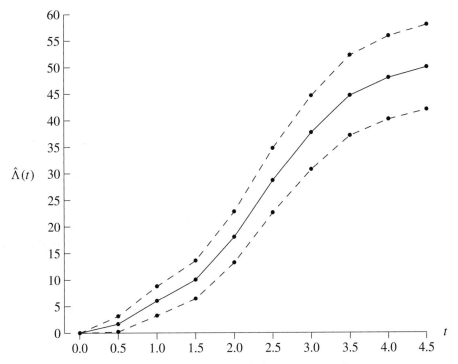

Figure 9.3.4   Lunch-wagon cumulative-event-rate-function
estimator and confidence limits.

At this point, the input model for the nonstationary Poisson process has been chosen. A realization of a nonstationary Poisson process can be generated from the cumulative event-rate-function estimate for modeling in a discrete-event simulation by inversion. Let $E_1, E_2, \ldots$ be the event times of a unit *stationary* Poisson process. Let $T_1, T_2, \ldots$ denote the associated event times for the nonstationary Poisson process with cumulative event-rate function $\hat{\Lambda}(t)$ generated on $(0, S]$. Algorithm 9.3.1 generates the nonstationary-Poisson-process event times from the inputs $a_0, a_1, a_2, \ldots, a_m$; $n_1, n_2, \ldots, n_m$; and k. A key element of this algorithm is selecting the appropriate subinterval (denoted in the algorithm by i) in which to perform the inversion.

Algorithm 9.3.1   Given: the number of realizations collected k; the subinterval boundaries $a_0, a_1, a_2, \ldots, a_m$; and the event counts $n_1, n_2, \ldots, n_m$; then the following algorithm returns, via inversion, a realization of a nonstationary Poisson process associated with the cumulative event-rate estimator $\hat{\Lambda}(t)$.

```
Λmax = Σⁱ₌₁ᵐ nᵢ/k;                           // upper bound for HPP
i = 1;                                       // initialize interval counter
j = 1;                                       // initialize variate counter
Λ = nᵢ/k;                              // initialize cumulative event rate
Eⱼ = Exponential(1.0);                  // generate first HPP event time
while (Eⱼ <= Λmax) {                    // while more events to generate
   while (Eⱼ > Λ) {                        // while in wrong interval
      i = i + 1;                          // increment interval counter
      Λ = Λ + nᵢ/k;                //increment cumulative event rate
   }
   Tⱼ = aᵢ - (Λ - Eⱼ) * k * (aᵢ - aᵢ₋₁) / nᵢ;
   j = j + 1;
   Eⱼ = Eⱼ₋₁ + Exponential(1.0);
}
return (T₁, T₂, ..., Tⱼ₋₁);                           // return event times
```

The geometry associated with Algorithm 9.3.1 is as follows. The *stationary* (homogeneous) unit-Poisson-process values $E_1, E_2, \ldots$ are generated along the vertical axis in Figure 9.3.3. Each E_j is associated with some subinterval i, $i = 1, 2, \ldots, m$. The appropriate piecewise-linear segment, when inverted, yields the associated *nonstationary*-Poisson-process value T_j. Algorithm 9.3.1 is monotone, synchronized, clear, and efficient.

Algorithm 9.3.1 is valid only when the target process has an event-rate function $\lambda(t)$ that is piecewise constant—a dubious assumption in a real-world setting. Any departure from this assumption results in an approximate estimator $\hat{\Lambda}(t)$ and an associated approximate variate-generation algorithm between the subinterval endpoints. The binning of the data into subintervals typically results in missed trends that occur between the subinterval endpoints. As will be seen subsequently, this problem is overcome by working with raw data.

One problem with Algorithm 9.3.1 is that there is no limit on the number of nonstationary-Poisson-process event times generated. The algorithm must be modified to accommodate a *next-event* simulation model, where the scheduling of the next event occurs when a current event is being processed. Algorithm 9.3.2 uses the next-event approach to schedule the next event when the current event is being processed. The algorithm has the same static inputs ($a_0, a_1, \ldots, a_m$; $n_1, n_2, \ldots, n_m$, and k) as Algorithm 9.3.1; but this algorithm returns the next event time, given that the current event occurs at the dynamic input time $T \in (0, S]$. The algorithm returns the time of the next nonstationary-Poisson-process event $\hat{\Lambda}^{-1}(\hat{\Lambda}(T) + E)$, where E is an *Exponential*(1.0) variate.

Algorithm 9.3.2   Given: the number of realizations collected k; the subinterval boundaries $a_0, a_1, a_2, \ldots, a_m$; the event counts $n_1, n_2, \ldots, n_m$; and the time of the current event T; then the following algorithm returns the time of the next event in a nonstationary Poisson process associated with the cumulative event-rate estimator $\hat{\Lambda}(t)$, via inversion.

```
Λmax = Σⱼ₌₁ᵐ nⱼ/k;                         // maximum cumulative event rate
j = 1;                                     // initialize interval index
while (T > aⱼ) {                           // while wrong interval
    j = j + 1;                             // find interval index
}
ΛT = Σᵢ₌₁ʲ⁻¹ nᵢ / k + nⱼ * (T - aⱼ₋₁) / (k * (aⱼ - aⱼ₋₁));
                                           // calculate Λ̂(T)
Λ = Σᵢ₌₁ʲ nᵢ / k;          // initialize cumulative event rate bound
ΛT = ΛT + Exponential(1.0);                // calculate Λ̂(t) + E
if (ΛT > Λmax)             // if there are more events to generate
    return (-1);          // -1 indicates no more events to generate
else {
    while (ΛT > Λ) {      // while Λ̂(T) is in the wrong interval
        j = j + 1;                         // increment interval counter
        Λ = Λ + nⱼ / k;    // increment cumulative event rate
    }
    return (aⱼ - (Λ - ΛT) * k * (aⱼ - aⱼ₋₁) / nⱼ);
                                 // next event time Λ̂⁻¹(Λ(T) + E)
}
```

The algorithm returns the next event time (or -1, if there are no further events to be generated). The variable Λ_T initially contains the cumulative event-rate function associated with the time of the current nonstationary-Poisson-process event T: $\hat{\Lambda}(T)$. This variable is updated to contain the cumulative event-rate function associated with the time of the next unit-stationary-Poisson-process event, $\hat{\Lambda}(T) + Exponential(1.0)$. At the end of the execution of this algorithm, the variable Λ contains the cumulative event-rate-function value at the right subinterval endpoint associated with the returned event time.

A more sophisticated implementation of this next-event algorithm would store $\Lambda_{\max}$, j, Λ, and Λ_T between the generation of events, effectively eliminating the first seven lines of the algorithm. The procedure would then begin with the updating of the Λ_T value, saving substantial execution time for large m.

Raw Data I.   We now proceed to the case where *raw event times* are available for the k realizations collected on $(0, S]$. The meaning of n_i now changes from the count-data case. We now let n_i, $i = 1, 2, \ldots, k$ be the number of observations in the ith realization, $n = \sum_{i=1}^{k} n_i$, and let $t_{(1)}, t_{(2)}, \ldots, t_{(n)}$ be the order statistics of the superposition of the event times in the k realizations, $t_{(0)} = 0$ and $t_{(n+1)} = S$.

Example 9.3.6

For the lunch-wagon example, the realizations were collected between time 0 (10:00 AM) and $S = 4.5$ (2:30 PM), yielding $n_1 = 56$, $n_2 = 42$, and $n_3 = 52$ observations on the $k = 3$ days. The superposition consists of $n = 150$ arrival times: $t_{(1)} = 0.2153$, $t_{(2)} = 0.3494$, $t_{(3)} = 0.3927$, $t_{(4)} = 0.3943, \ldots,$ and $t_{(150)} = 4.3741$. The superposition of the arrival times to the lunch wagon is plotted in Figure 9.3.5 [the endpoints $t_{(0)} = 0$ and $t_{(151)} = 4.5$ are not plotted]. A cursory visual inspection of the arrival pattern reveals a concentration of arrivals near the noon hour.

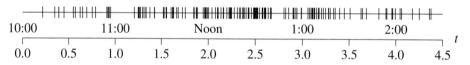

Figure 9.3.5   Lunch-wagon arrival times.

We now introduce the standard nonparametric estimate of $\Lambda(t)$, which is often known as the "step-function" estimator. For the jth independent replication of the target nonstationary Poisson process, $j = 1, 2, \ldots, n$, let $N_j(t)$ denote the number of events observed in the time interval $(0, t]$, and let

$$N_k^*(t) = \sum_{j=1}^{k} N_j(t) \qquad 0 < t \le S$$

denote the aggregated counting (or superposition) process so that $n = N_k^*(S)$. The step-function $N_j(t)$ is an appropriate estimator for the cumulative event-rate function for realization j only, $j = 1, 2, \ldots, k$. The step-function estimator of $\Lambda(t)$ that averages all of the k replications is

$$\hat{\Lambda}(t) = \frac{1}{k} \sum_{i=1}^{k} N_j(t) = \frac{N_k^*(t)}{k} \qquad 0 < t \le S.$$

By a result from Leemis (1991), the $N_1(t), N_2(t), \ldots, N_j(t)$ are independent and identically distributed $Poisson\big(\Lambda(t)\big)$ random variables. Also, for any arbitrary t in the interval $(0, S]$, we can conclude that

$$\lim_{k \to \infty} \hat{\Lambda}(t) = \Lambda(t)$$

with probability 1. In addition, we can construct the following asymptotically exact $100(1 - \alpha)\%$ confidence interval for $\Lambda(t)$, whose form is identical to the event-count-data case:*

$$\hat{\Lambda}(t) - z_{\alpha/2}\sqrt{\frac{\hat{\Lambda}(t)}{k}} < \Lambda(t) < \hat{\Lambda}(t) + z_{\alpha/2}\sqrt{\frac{\hat{\Lambda}(t)}{k}}.$$

The standard step-function estimator takes upward steps of height $1/k$ only at the event times in the superposition of the k realizations collected from the target process. This means that a variate-generation algorithm will produce realizations that contain only event times equal

---

*This interval is asymptotically exact for all t, unlike the interval for the count data.

to those collected in one of the k realizations. This is exactly the problem that we ran into in Section 9.1, which we called the "interpolation" problem. In a very similar fashion, we can overcome this problem by using a piecewise-linear cumulative event-rate-function estimator. Thus we skip the development of a variate-generation algorithm and move directly to the solution of the interpolation problem.

Raw Data II.   There are $n + 1$ "gaps" on $(0, S]$ created by the superposition $t_{(1)}, t_{(2)}, \ldots, t_{(n)}$. Setting $\hat{\Lambda}(S) = n/k$ yields a process where the expected number of events by time S is the average number of events in k realizations, since $\Lambda(S)$ is the expected number of events by time S. The piecewise-linear cumulative event-rate-function estimate rises by $n/[(n + 1)k]$ at each event time in the superposition. Thus the piecewise-linear estimator of the cumulative event-rate function between the time values in the superposition is

$$\hat{\Lambda}(t) = \frac{in}{(n+1)k} + \left[\frac{n(t - t_{(i)})}{(n+1)k(t_{(i+1)} - t_{(i)})} \right] \qquad t_{(i)} < t \leq t_{(i+1)}; i = 0, 1, 2, \ldots, n.$$

This estimator passes through the points $\left(t_{(i)}, in/(n+1)k \right)$, for $i = 1, 2, \ldots, n + 1$. This interval estimator was developed in Leemis (1991) and extended to overlapping intervals in Arkin and Leemis (2000).

In a fashion similar to the previous two estimators, a $(1 - \alpha)100\%$ asymptotically exact (as $k \to \infty$) confidence interval for $\Lambda(t)$ can be determined as

$$\hat{\Lambda}(t) - z_{\alpha/2}\sqrt{\frac{\hat{\Lambda}(t)}{k}} < \Lambda(t) < \hat{\Lambda}(t) + z_{\alpha/2}\sqrt{\frac{\hat{\Lambda}(t)}{k}},$$

where $z_{\alpha/2}$ is the $1 - \alpha/2$ fractile of the standard normal distribution.

Example 9.3.7
Figure 9.3.6 contains point and interval estimates for $\Lambda(t)$ on $(0, 4.5]$ for the lunch-wagon arrival times. The solid line denotes $\hat{\Lambda}(t)$, and the dotted lines denote the 95%-confidence-interval bounds. Lower confidence-interval bounds that fall below zero are equated to zero. The cumulative event-rate-function estimator at time 4.5 is $150/3 = 50$, which is the point estimator for the expected number of arriving customers per day. If $\hat{\Lambda}(t)$ is linear, a stationary model is appropriate. As before, customers are more likely to arrive to the lunch wagon between 12:00 ($t = 2.0$) and 12:30 ($t = 2.5$) than at other times, so the cumulative event-rate-function estimator has an S-shape and a nonstationary model is indicated. The point and interval estimates appear more jagged than in the case of the count data, because each of the three curves represents 151 (tiny) linear segments.

The fact that each of the piecewise segments rises by the same height can be exploited to create an efficient variate-generation algorithm.

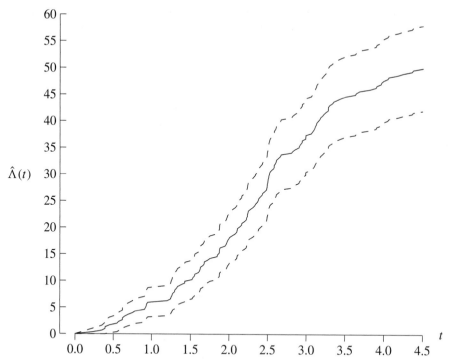

Figure 9.3.6   Lunch-wagon cumulative event-rate-function estimate
and confidence-interval limits.

Algorithm 9.3.3   Given: the number of realizations collected k; the number of observed events n; the interval length S; and the superposition times $t_{(1)}, t_{(2)}, \ldots, t_{(n)}$; then the following algorithm returns, via inversion, a realization of a nonstationary Poisson process associated with the piecewise-linear cumulative event-rate estimator $\hat{\Lambda}(t)$.

```
i = 1;
E_i = Exponential(1.0);
while (E_i <= n / k) {
    m = ⌊(n + 1) * k * E_i / n⌋;
    T_i = t_(m) + (t_(m+1) - t_(m)) * ((n + 1) * k * E_i / n - m);
    i = i + 1;
    E_i = E_(i-1) + Exponential(1.0);
}
return (T_1, T_2, ..., T_(i-1));
```

Algorithm 9.3.3 shows that it is a straightforward procedure to obtain a realization of $i - 1$ events on $(0, S]$ from i calls to *Exponential*(1.0). The values of $E_1, E_2, \ldots$ denote the values in the unit Poisson process, m denotes the index of the appropriate piecewise-linear segment to invert, and $T_1, T_2, \ldots, T_{i-1}$ denote the returned event times from the simulated nonstationary Poisson process. Inversion has been used to generate this nonstationary Poisson process, so certain variance-reduction techniques, such as antithetic variates or common random numbers, may be implemented. Tied values in the superposition pose no problem to this algorithm, although there could be tied values in the generated realization. As n increases, the amount of memory required increases, but the amount of execution time required to generate a realization depends only on the ratio n/k, the average number of events per realization. Thus, collecting more realizations (resulting in narrower confidence intervals) increases the amount of memory required, but does not affect the expected execution time for generating a realization. Converting Algorithm 9.3.3 to a next-event orientation is left as an exercise.

To summarize, this piecewise-linear cumulative event-rate-function estimator is in some sense ideal in that

- it uses raw data, to avoid the loss of accuracy imposed by breaking $(0, S]$ into arbitrary time subintervals;
- the point and interval estimates of $\Lambda(t)$ are in closed form and are easily computed;
- the point estimate of $\Lambda(t)$ is consistent [i.e., $\lim_{k \to \infty} \hat{\Lambda}(t) = \Lambda(t)$];
- the interval estimate of $\Lambda(t)$ is asymptotically exact as $k \to \infty$;
- the variate-generation algorithm that can be used to simulate a realization from $\hat{\Lambda}(t)$ is efficient, monotone, and synchronized.

This algorithm completes our discussion of nonparametric modeling of nonstationary processes. We will now consider parametric modeling of nonstationary processes.

9.3.3 Parametric Modeling

We will continue to focus on our lunch-wagon arrival times and the fitting of a nonstationary Poisson process as we switch the emphasis to parametric models. Maximum likelihood can again be used for estimating the parameters in a parametric nonstationary-Poisson-process model. The likelihood function for estimating the vector of unknown parameters $\boldsymbol{\theta} = (\theta_1, \theta_2, \ldots, \theta_q)'$ from a single realization of event times $t_1, t_2, \ldots, t_n$ drawn from a nonstationary Poisson process with event rate $\lambda(t)$ on $(0, S]$ is

$$L(\boldsymbol{\theta}) = \left[\prod_{i=1}^{n} \lambda(t_i)\right] \exp\left[-\int_0^S \lambda(t)dt\right].$$

Maximum-likelihood estimators can be found by maximizing $L(\boldsymbol{\theta})$ or its logarithm with respect to the q unknown parameters. Confidence regions for the unknown parameters can be found in a manner similar to the service-time example in Section 9.2.

Because of the additive property of the event-rate function for multiple realizations, the likelihood function for the case of k realizations is

$$L(\boldsymbol{\theta}) = \left[\prod_{i=1}^{n} k\lambda(t_i)\right] \exp\left[-\int_0^S k\lambda(t)dt\right].$$

There are many potential parametric models for nonstationary Poisson processes. We limit our discussion to procedures for fitting a *power-law process*, where the event-rate function is

$$\lambda(t) = b^a a t^{a-1} \qquad t > 0,$$

for shape parameter $a > 0$ and scale parameter $b > 0$. This event-rate function can assume monotone increasing ($a > 1$) and monotone decreasing ($a < 1$) shapes. Such event-rate functions can be used to model sequences of events that occur with always-increasing or always-decreasing frequency. The event-rate function is constant when $a = 1$, which corresponds to an ordinary (stationary) Poisson process. The likelihood function for k realizations is

$$L(a, b) = k^n b^{na} a^n \exp\left(- k(bS)^a\right) \prod_{i=1}^{n} t_i^{a-1}.$$

The log-likelihood function is

$$\ln L(a, b) = n \ln(ka) + na \ln b - k(bS)^a + (a - 1) \sum_{i=1}^{n} \ln t_i.$$

Differentiating with respect to a and b and equating to zero yields

$$\frac{\partial \ln L(a, b)}{\partial a} = n \ln b + \frac{n}{a} + \sum_{i=1}^{n} \ln t_i - k(bS)^a \ln(bS) = 0$$

and

$$\frac{\partial \ln L(a, b)}{\partial b} = \frac{an}{b} - kS^a a b^{a-1} = 0.$$

Unlike the Weibull distribution from the previous section, these equations can be solved in closed form. The analytic expressions for b and a are

$$\hat{a} = \frac{n}{n \ln S - \sum_{i=1}^{n} \ln t_i},$$

$$\hat{b} = \frac{1}{S} \left(\frac{n}{k}\right)^{1/a}.$$

We now apply this model to the lunch-wagon arrival-time data.

Example 9.3.8
Substituting the $n = 150$ lunch-wagon arrival times into these formulas yields the maximum-likelihood estimators $\hat{a} = 1.27$ and $\hat{b} = 4.86$. The cumulative event-rate function for the power-law process

$$\Lambda(t) = (bt)^a \qquad t > 0$$

is plotted along with the nonparametric estimator, in Figure 9.3.7. Note that the peak in customer arrivals around the noon hour makes the power-law process an inappropriate model: It is unable to approximate the event-rate function adequately. It is unable to model a nonmonotonic $\lambda(t)$ and hence is a poor choice for a model. In fact, the fitted cumulative event-rate function is so close to linear that the process would look almost like a stationary Poisson process—completely missing the noon-hour rush period! Other models can be fitted in a similar fashion.

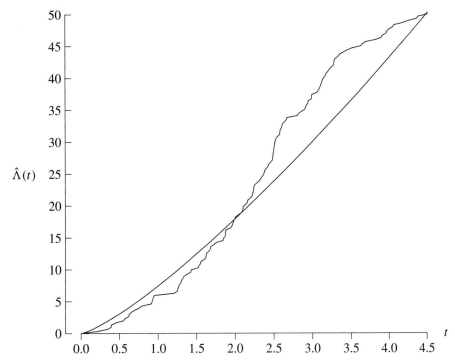

Figure 9.3.7   Lunch-wagon cumulative event-rate-function estimate and fitted power-law process.

A more appropriate two-parameter distribution to consider would be one with an event-rate function that increases initially, then decreases. The log-logistic process, for example, with event-rate function (Meeker and Escobar, 1998)

$$\lambda(t) = \frac{ba(bt)^{a-1}}{1 + (bt)^a} \qquad t > 0$$

for shape parameter $a > 0$ and scale parameter $b > 0$, would certainly be more appropriate than the power-law process.

The EPTMP (exponential-polynomial-trigonometric function with multiple periodicities) model, originally given by Lee, Wilson, and Crawford (1991) and generalized by

Kuhl, Damerdji, and Wilson (1998) and having event-rate function

$$\lambda(t) = \exp\left[\sum_{i=0}^{m} \alpha_i t^i + \sum_{j=1}^{p} \gamma_j \sin(\omega_j t + \phi_j)\right] \qquad t > 0$$

can also model a nonmonotonic event-rate function. The exp forces the event-rate function to be positive, the first summation models polynomial trend, and the second summation models sinusoidal periodicities in the event rate. The cyclic portion of the model has been used in discrete-event simulation applications to model the times that storms occur in the Arctic Sea and arrivals of donated livers for transplantation. Kuhl, Sumant, and Wilson (2006) generalize this approach to model periodic effects that are not necessarily modeled by a trigonometric function. Goodness-of-fit tests associated with the fitted models are given in Rigdon and Basu (2000) and in Ross (2002). Taaffe and Clark (1988), for example, consider nonstationary arrival processes via the axiomatic approach to probability.

9.3.4 Exercises

9.3.1   The lunch-wagon example was generated from the following cumulative event-rate function from Klein and Roberts (1984):

$$\Lambda(t) = \begin{cases} 5t^2 + t & 0 < t \le 1.5 \\ 16t - 11.25 & 1.5 < t \le 2.5 \\ -3t^2 + 31t - 30 & 2.5 < t \le 4.5. \end{cases}$$

(a) Find the rate function $\lambda(t)$. (b) Generate 10 realizations via the thinning algorithm and 10 realizations via inversion (both algorithms were introduced in Section 7.5) and provide convincing numerical evidence that the two algorithms were implemented correctly.

9.3.2   Derive the cumulative event-rate-function estimator from the event-rate-function estimator in the case of count data.

9.3.3   Perform the algebra necessary to confirm the appropriate assignment of T_j in Algorithm 9.3.1.

9.3.4   Draw a diagram that illustrates the geometry associated with Algorithm 9.3.2.

9.3.5   Use Algorithm 9.3.2 to determine the time of the next event in a nonstationary Poisson process when $m = 3$; $a_0 = 0$, $a_1 = 2$, $a_2 = 6$, $a_3 = 7$; $n_1 = 10$, $n_2 = 3$, $n_3 = 11$; $k = 2$; $T = 1.624$; $E = 5.091$.

9.3.6   Convert Algorithm 9.3.3 to a next-event orientation: Given a current event that occurs at time T, when should the next event be scheduled?

10

Projects

Projects related to the discrete-event simulation topics considered in this text compose this final chapter. These projects may be used as launching pads for a term project or can be used as miscellaneous topics at the end of the course.

Section 10.1 introduces six empirical tests of randomness for a random-number generator. These six tests, along with the Kolmogorov–Smirnov goodness-of-fit test, are applied to the Lehmer random-number generator $(a, m) = (48271, 2^{31} - 1)$. Section 10.2 introduces a *birth–death* process, which is a mathematical model for systems that can be characterized by a state that changes only by either increasing by one or decreasing by one. Population models are a popular example of birth–death models used by biologists. The number of jobs in a single-server service node is another example of a birth–death process. Section 10.3 introduces a finite-state *Markov chain*, which uses a state-transition matrix to define the probability mechanism for transitions from state to state. These states can represent a very broad range of real-world system states (e.g., the state of the weather on a particular day, the position of a job in a job shop, the social class of an individual, or some genetic characteristic of an individual). Finally, Section 10.4 extends the single-server service node, which has been used throughout the book, to a network of single-server service nodes.

10.1 EMPIRICAL TESTS OF RANDOMNESS

Although the $(a, m) = (16807, 2^{31} - 1)$ minimal standard Lehmer random-number generator introduced in Chapter 2 is less than ideal, there is no need to subject it to any of the standard tests of randomness—that has been done many times before. In contrast, the $(a, m) = (48271, 2^{31} - 1)$ generator used in this book is less well established in the literature, and so there is merit in testing it for randomness. Moreover, testing for randomness is such an important topic that it must be included in any comprehensive study of discrete-event simulation. With this in mind, empirical tests of randomness will be illustrated in this section by applying six of the standard tests to the output produced by the library `rngs`.

10.1.1 Empirical Testing

> **Definition 10.1.1**   An *empirical test* of randomness is a statistical test of the hypothesis that repeated calls to a random-number generator will produce an *iid* sample from a *Uniform* $(0, 1)$ distribution. An empirical test of randomness consists of three steps.
>
> - Generate a *sample* by repeated call to the generator.
> - Compute a *test statistic* whose statistical distribution (pdf) is known when the random numbers are truly *iid* *Uniform* $(0, 1)$ random variates.
> - Assess the likelihood of the observed (computed) value of the test statistic relative to the theoretical distribution from which it is assumed to have been drawn.

The test in Definition 10.1.1 is *empirical*, because it makes use of actual generated data and produces a conclusion that is valid only in a local, statistical sense. In contrast, *theoretical* tests are not statistical. Instead, as illustrated in Section 2.2, theoretical tests use the numerical value of the generator's modulus and multiplier to assess the global randomness of the generator.*

Chi-Square Statistic

> **Definition 10.1.2**   Let X' be a random variable (discrete or continuous), and suppose the possible values of X' are partitioned into a finite set of possible values (states) $\mathcal{X}$ to form a new discrete random variable X. Typically, these states correspond to the bins of a histogram. Given a large random sample, then, for each possible state $x \in \mathcal{X}$, define
>
> $e[x] =$ the *expected* number of times state x will occur, and
>
> $o[x] =$ the *observed* number of times state x occurred.
>
> The resulting nonnegative quantity
>
> $$v = \sum_x \frac{(o[x] - e[x])^2}{e[x]}$$
>
> is known as a *chi-square statistic*. The sum is over all $x \in \mathcal{X}$.

All the empirical tests of randomness in this section make use of some form of the chi-square statistic. The following test is typical.

---

*See Knuth (1998) for a comprehensive discussion of theoretical and empirical tests of randomness.

Test of Uniformity

Algorithm 10.1.1   The uniformity of a random-number generator's output can be tested by making n calls to Random and tallying this random-variate sample into an equally spaced k-bin histogram. In this case, x is the histogram bin index, and the bin counts $o[0], o[1], \ldots, o[k-1]$ can be tallied as follows:

```
for (x = 0; x < k; x++)
   o[x] = 0;
for (i = 0; i < n; i++) {
   u = Random();
   x = ⌊ u * k ⌋;
   o[x]++;
}
```

Because the expected number of points per bin is, in this case, $e[x] = n/k$ for all x, the chi-square statistic is

$$v = \sum_{x=0}^{k-1} \frac{\left(o[x] - n/k\right)^2}{n/k}.$$

The *Test of Uniformity* (frequency test), based on Algorithm 10.1.1, provides a statistical answer to the question "Is the histogram flat?" More precisely, is the variability among the histogram heights sufficiently small so as to conclude that the n random numbers were drawn from a *Uniform*(0, 1) population? The underlying theory behind this test is that, *if* the random sample is a truly *iid Uniform*(0, 1) random-variate sample and *if* n/k is sufficiently large—say, 10 or greater—then the test statistic v is (approximately) a *Chisquare*$(k-1)$ random variate. Therefore, if

$$v_1^* = \texttt{idfChisquare}(k-1, \alpha/2) \quad \text{and} \quad v_2^* = \texttt{idfChisquare}(k-1, 1-\alpha/2),$$

then $\Pr(v_1^* \le v \le v_2^*) \cong 1 - \alpha$. To apply the test of uniformity, pick $k \ge 1000$ and $n \ge 10k$ and then do the following:

- Use Algorithm 10.1.1 to calculate a k-bin histogram.
- Calculate the chi-square statistic v.
- Pick a $(1 - \alpha) \times 100\%$ level of confidence (typically, $\alpha = 0.05$).
- Compute the critical values v_1^* and v_2^*.

If $v < v_1^*$ or $v > v_2^*$, the test *failed*. (Failure occurs with approximate probability $1 - \alpha$ for a legitimate random-number generator.)

Just as in a jury trial, where people are assumed innocent until proven guilty, a random-number generator is assumed to be "good" until proven "bad." Therefore, if $v_1^* \le v \le v_2^*$, we haven't really proven that the generator is good—that was already assumed. Instead, we have only failed to prove it bad. Equivalently, an empirical test is never passed; instead it is either "failed" or "not failed." This is an important statistical point. Equivalent statistical jargon is "reject" the null hypothesis and "fail to reject" the null hypothesis.

Example 10.1.1

As an illustration of the test of uniformity, the library `rngs` was used (with initial seed 12345) to create the data for 256 tests, one for each possible stream.* For each test, the number of histogram bins was $k = 1000$, and the number of observations was $n = 10\,000$. Therefore, the expected number of observations per bin was $n/k = 10$. The resulting 256 chi-square statistics (the v's) are plotted in Figure 10.1.1, along with the corresponding critical values $v_1^* = 913.3$ and $v_2^* = 1088.5$.

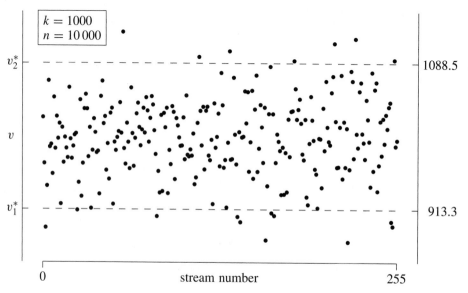

Figure 10.1.1   Test of uniformity.

As would be consistent with the (hoped-for) goodness of fit of the random-number generator in `rngs`, most of the 256 tests were not failed. Indeed, there were just 7 "the histogram is not uniform" ($v > v_2^*$) failures and 11 "the histogram is too uniform" ($v < v_1^*$) failures.

Two-Tailed Versus One-Tailed Tests.   We have formulated the test of uniformity as a *two-tailed* test: As is illustrated in Example 10.1.1, the test could be failed in two ways. If v is too small ($v < v_1^*$), then the observed and expected data agree *too* well—the histogram is too flat. At the other extreme, if v is too large ($v > v_2^*$), then the observed and expected data don't agree well enough. Intuitively, $v > v_2^*$ failures seem to be worse; consequently, some authors argue that v can't be too small. Instead, these authors advocate a *one-tailed* test based only on a check to see whether v is too large. Knuth (1998) advocates a two-tailed test; we agree.

---

*Test 0 corresponds to stream 0, test 1 to stream 1, and so on.

As the name suggests, the test of uniformity is nothing more than a test for histogram flatness—does the generator tend to fill each histogram bin equally? The *order* (or lack thereof) in which the bins are filled is not tested. Because of this, the test of uniformity is a statistically weak test of randomness. Moreover, it is perhaps the most obvious test to apply to a *Uniform*(0, 1) random-number generator. If you are tempted to use a generator whose quality is unknown (to you), then it would be prudent to apply the test of uniformity first. However, don't be surprised to find that the generator will produce an acceptably flat histogram (at least for selected values of k), and, in that case, do not jump to the conclusion that the generator is necessarily "good."

But what about the 18 failures in Example 10.1.1; do these failures indicate problems? The answer is "No." In fact, if the generator is good, then approximately $\alpha \times 100\%$ of the tests *should* be failed. In particular, in 256 tests with a 95% level of confidence, we would expect to have approximately $256 \cdot 0.05 \cong 13$ failures; significantly more *or less* than this number would be an indication of problems.

Test of Extremes.   The next empirical test is similar, in spirit, to the test of uniformity. It is a more powerful test, however, with theoretical support based on the following theorem.

Theorem 10.1.1   If $U_0, U_1, \ldots, U_{d-1}$ is an *iid* sequence of *Uniform*(0, 1) random variables, and if

$$R = \max\{U_0, U_1, \ldots, U_{d-1}\},$$

then the random variable $U = R^d$ is *Uniform*(0, 1).*

Proof.   Pick $0 < r < 1$. Since $R \leq r$ if and only if $U_j \leq r$ for $j = 0, 1, \ldots, d-1$,

$$\Pr(R \leq r) = \Pr\big((U_0 \leq r) \text{ and } (U_1 \leq r) \text{ and } \ldots \text{ and } (U_{d-1} \leq r)\big)$$

$$= \Pr(U_0 \leq r)\Pr(U_1 \leq r)\ldots\Pr(U_{d-1} \leq r)$$

$$= r \cdot r \cdot \ldots \cdot r$$

$$= r^d,$$

and so the cdf of the random variable R is $\Pr(R \leq r) = r^d$. However, $R \leq r$ if and only if $R^d \leq r^d$; therefore,

$$\Pr(R^d \leq r^d) = \Pr(R \leq r) = r^d.$$

Finally, let $U = R^d$ and $u = r^d$, so that the cdf of the random variable U is

$$\Pr(U \leq u) = \Pr(R^d \leq r^d) = r^d = u \qquad 0 < u < 1.$$

Since $\Pr(U \leq u) = u$ is the cdf of a *Uniform*(0, 1) random variable, the theorem is proven. $\square$

---

*Don't misinterpret Theorem 10.1.1. The random variable $U = R^d$ is *Uniform*(0, 1), but the random variable R is *not Uniform*(0, 1).

Algorithm 10.1.2 The "extreme" behavior of a random-number generator can be tested by grouping (batching) the generator's output d terms at a time, finding the maximum of each batch, raising this maximum to the dth power, and then tallying all the maxima so generated, as follows:

```
for (x = 0; x < k; x++)
    o[x] = 0;
for (i = 0; i < n; i++) {
    r = Random();
    for (j = 1; j < d; j++) {
        u = Random();
        if (u > r)
            r = u;
    }
    u = exp(d * log(r));          // u is r^d
    x = ⌊ u * k ⌋;
    o[x]++;
}
```

Because the expected number of points per bin is $e[x] = n/k$ for all x, in this case the chi-square statistic is

$$v = \sum_{x=0}^{k-1} \frac{\left(o[x] - n/k\right)^2}{n/k},$$

with critical values $v_1^* = \mathtt{idfChisquare}(k - 1, \alpha/2)$ and $v_2^* = \mathtt{idfChisquare}(k - 1, 1 - \alpha/2)$.

The *Test of Extremes* is based on Algorithm 10.1.2. To apply this test, pick $k \geq 1000$, $n \geq 10k$, and $d \geq 2$, and then do the following:

- Use Algorithm 10.1.2 to calculate a k-bin histogram.
- Calculate the chi-square statistic v.
- Pick a $(1 - \alpha) \times 100\%$ level of confidence (typically, $\alpha = 0.05$).
- Compute the critical values v_1^* and v_2^*.

If $v < v_1^*$ or $v > v_2^*$, the test *failed*. (Failure occurs with approximate probability $1 - \alpha$ for a legitimate random-number generator.) The test typically is applied for a range of values—say, $d = 2, 3, \ldots, 6$. For $d = 1$, the test of extremes reduces to the test of uniformity.

As formulated, the test of extremes is actually a test of *maxima*. To make the test of extremes into a test of *minima* involves only a minor extension to the theory—namely, that

$$S = \min\{U_0, U_1, \ldots, U_{d-1}\}$$

and $(1 - S)^d$ are *Uniform*$(0, 1)$ random variates. The details are left as an exercise.

Example 10.1.2

As in Example 10.1.1, the library `rngs` was used (with initial seed 12345) to create the data for 256 tests, one for each `rngs` stream. For each test, the number of histogram bins was

$k = 1000$, and the number of observations was $n = 10\,000$, where each observation was determined by the maximum of $d = 5$ random numbers. As in Example 10.1.1, the critical values denoted by horizontal lines in Figure 10.1.2 are $v_1^* = 913.3$ and $v_2^* = 1088.5$.

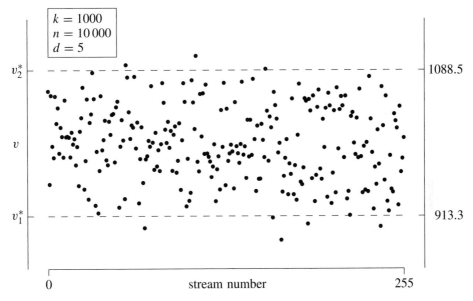

Figure 10.1.2   Test of extremes.

Consistent with expectations, most of the tests were not failed. The number of $v > v_2^*$ failures was three, and there were five $v < v_1^*$ failures. These results are further evidence of the goodness of the generator. Indeed, because of its multivariate character, the test of extremes (with $d > 1$) is statistically stronger than the test of uniformity, and so these results are more confidence-inspiring than are those in Example 10.1.1.

Runs-Up Test of Independence.   The next empirical test we will consider is the runs-up test of independence. Although the underlying theory is simple, the test is considered to be relatively powerful, particularly when applied to Lehmer generators with poorly chosen multipliers.

Definition 10.1.3   Let u_0, u_1, u_2, ... be a sequence of numbers produced by repeated calls to a random-number generator. A *run-up of length x* is a subsequence of length x with the property that each term in the subsequence is greater than its predecessor. By convention, each run-up is extended to its largest possible length and then, to make each run length independent, the next element is "thrown away" before beginning to compute the length of the next run-up. For example, the sequence of numbers (to just $d.dd$ precision)

$$0.21, \ 0.39, \ 0.47, \ \underline{0.12}, \ 0.87, \ \underline{0.21}, \ 0.31, \ 0.38, \ 0.92, \ 0.93, \ \underline{0.57}, \ \dots$$

consists of a run-up of length 3, followed by a run-up of length 1, followed by a run-up of length 4, etc. The "thrown away" points are underlined.

The following theorem provides theoretical support for the runs-up test of independence. The proof of this theorem is left as an exercise.

Theorem 10.1.2 If U_0, U_1, U_2, ... is an *iid* sequence of *Uniform*$(0, 1)$ random variables, then, beginning with U_0, the probability of a run-up of length x is

$$p[x] = \frac{x}{(x+1)!} \qquad x = 1, 2, \ldots.$$

The corresponding probability of a run-up of length k or greater is

$$\sum_{x=k}^{\infty} p[x] = \frac{1}{k!}.$$

Algorithm 10.1.3 This algorithm uses repeated calls to Random to tally n runs-up into a k-bin histogram of runs of length $x = 1, 2, \ldots, k$.*

```
for (x = 1; x <= k; x++)
   o[x] = 0;
for (i = 0; i < n; i++) {
   x = 1;
   u = Random();
   t = Random();
   while (t > u) {
      x++;
      u = t;
      t = Random();
   }
   if (x > k)
      x = k;
   o[x]++;
}
```

From the probabilities in Theorem 10.1.2, the appropriate chi-square statistic in this case is

$$v = \sum_{x=1}^{k} \frac{\left(o[x] - np[x]\right)^2}{np[x]},$$

---

*Note that $o[0]$ is not used and that $o[k]$ counts runs of length k *or greater*.

with $e[x] = np[x]$ for all x. The test statistic v is (approximately) a $Chisquare(k-1)$ random variate, and the critical values of v are

$$v_1^* = \text{idfChisquare}(k-1, \alpha/2) \qquad \text{and} \qquad v_2^* = \text{idfChisquare}(k-1, 1-\alpha/2).$$

A run-up of length $x > 6$ is so unlikely that it is conventional to choose $k = 6$ and consolidate all runs-up of length 6 or greater, as in Algorithm 10.1.3. The resulting probabilities are shown below.

x	1	2	3	4	5	6 or greater
$p[x]$	1/2	1/3	1/8	1/30	1/144	1/720

For n runs-up, the expected number of runs-up of length x is $np[x]$; so, if $k = 6$, then $n = 7200$ runs-up are required to raise $np[6]$ to 10.

The *Runs-Up Test of Independence* is based on Algorithm 10.1.3, with $n \geq 7200$ and $k = 6$. To apply this test, do the following.

- Use Algorithm 10.1.3 to calculate a k-bin histogram.
- Calculate the chi-square statistic v.
- Pick a $(1 - \alpha) \times 100\%$ level of confidence (typically, $\alpha = 0.05$).
- Compute the critical values v_1^* and v_2^*.

If $v < v_1^*$ or $v > v_2^*$, the test *failed*. (Failure occurs with approximate probability $1 - \alpha$ for a legitimate random-number generator.)

Example 10.1.3
As in Examples 10.1.1 and 10.1.2 (with initial `rngs` seed 12345 and $k = 6$), the results of 256 runs-up tests are plotted in Figure 10.1.3. For each test, the number of runs-up was $n = 7200$. The critical values are $v_1^* = 0.83$ and $v_2^* = 12.83$.

There were ten $v < v_1^*$ failures and thirteen $v > v_2^*$ failures, including the one failure indicated with a $\uparrow$ that was a spectacular off-scale $v = 26.3$ for stream number 58. This total of 23 failures is uncomfortably larger than expected and indicates possible problems with these particular portions of the generator.

Use of the chi-square statistic

$$v = \sum_{x=1}^{k} \frac{\left(o[x] - np[x]\right)^2}{np[x]}$$

in the runs-up test of independence is questionable for two (related) reasons.

- The expected number of observations per histogram bin is dramatically *non*uniform. Because the expected number of long runs is so small, the division by $np[x]$ causes the natural sample variation associated with long run lengths to have a much larger impact (bias) on v than does the natural sample variation associated with short run lengths. For the runs-up test, this bias is undesirable but inevitable. The effect of the bias can be minimized only at the expense of using a large value of n.

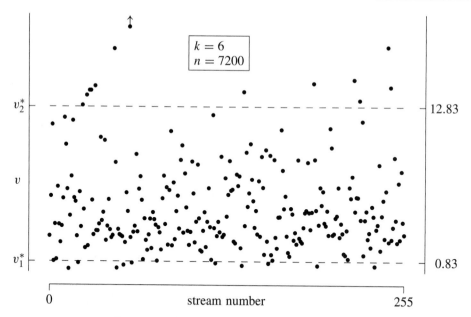

Figure 10.1.3 Runs-up test of independence.

- For this test, the chi-square statistic is only *approximately* a *Chisquare* $(k-1)$ random variate. The approximation becomes better as the expected number of observations per histogram bin is increased. This is particularly true because the number of bins $(k = 6)$ is so small.

Example 10.1.4
The tests illustrated in Figure 10.1.3 were repeated with n doubled to $n = 14\,400$. This doubles the expected number of observations in each histogram bin. The same initial rngs seed (12345) was used as before, and so the net effect was to add an additional 7200 observations to those generated previously. The results of the 256 runs-up tests are plotted in Figure 10.1.4.

In this case, as is consistent with expectations, the number of $v > v_2^*$ failures was seven, and there were also seven $v < v_1^*$ failures. Our confidence in the goodness of the rngs random-number generator is restored.

Three additional empirical tests of randomness will now be presented, the first of which is quite similar to the runs-up test. We will then consider the statistical question of how to use each of these six tests to answer the question, "Is the random-number generator good?"

Gap Test of Independence. The gap test is defined by two real-valued parameters a, b chosen so that $0 \le a < b \le 1$. These two parameters define a subinterval $(a, b) \subset (0, 1)$ with length $\delta = b - a$.

Given an *iid* sequence $U_0, U_1, U_2, \ldots$ of *Uniform* $(0, 1)$ random variables, the term U_i will be in the interval (a, b) with probability δ. If $U_i \in (a, b)$ then a *gap* of length 0 occurs if $U_{i+1} \in (a, b)$. This occurs with probability $f(0) = \delta$. If, instead, $U_{i+j} \notin (a, b)$ for $j = 1, 2, \ldots, x$ and $U_{i+j} \in (a, b)$ for $j = x + 1$, then a gap of length x has occurred. This occurs with probability

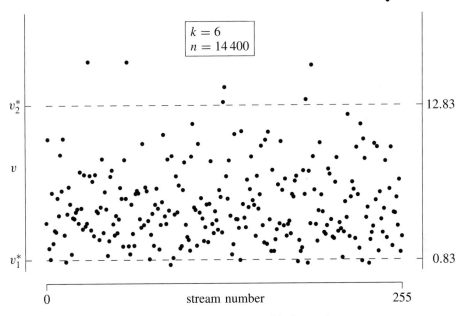

Figure 10.1.4   Runs-up test of independence.

$f(x) = \delta(1 - \delta)^x$. If X is the discrete random variable that counts gap lengths, it follows that X is *Geometric*$(1 - \delta)$. (See Section 6.4.3.) The length of a gap X is a count of the random numbers generated prior to a return to the interval (a, b).

Because X is *Geometric*$(1 - \delta)$, the expected gap length is $E[X] = (1 - \delta)/\delta$. As with the runs-up test, it is conventional to pick a positive integer k (which depends on δ) and consolidate all gaps of length $k - 1$ or greater. This consolidation occurs with probability

$$\Pr(X \geq k - 1) = \delta(1 - \delta)^{k-1}\big(1 + (1 - \delta) + (1 - \delta)^2 + \cdots\big) = \frac{\delta(1 - \delta)^{k-1}}{1 - (1 - \delta)} = (1 - \delta)^{k-1}.$$

Algorithm 10.1.4   Given two real-valued parameters a, b with $0.0 \leq a < b < 1.0$, this algorithm uses repeated calls to `Random` to tally n gaps into a k-bin histogram corresponding to gaps of length $x = 0, 1, 2, \ldots, k - 1$, as follows:

```
for (x = 0; x < k; x++)
  o[x] = 0;
for (i = 0; i < n; i++) {
  x = 0;
  u = Random();
  while (u ∉ (a,b)) {
    x++;
    u = Random();
```

```
    }
    if (x > k - 1)
        x = k - 1;
    o[x]++;
}
```

Much as in the runs-up algorithm, $o[k-1]$ counts gaps of length $k-1$ or greater.

Because the gap length is a *Geometric*$(1-\delta)$ random variate, if a total of n gaps are generated, then the expected number of gaps of length x is

$$e[x] = \begin{cases} n\delta(1-\delta)^x & x = 0, 1, 2, \ldots, k-2 \\ n(1-\delta)^{k-1} & x = k-1. \end{cases}$$

The smallest value of $e[x]$ occurs when $x = k-2$. Therefore, to make this expected number of gaps (and all others) be at least 10, it is necessary to choose k so that

$$e[k-2] = n\delta(1-\delta)^{k-2} \geq 10.$$

Solving for k yields the inequality

$$k \leq 2 + \left\lfloor \frac{\ln(10/n\delta)}{\ln(1-\delta)} \right\rfloor.$$

Typically $\delta = b - a$ is small, resulting in a relatively large value for k. For example, if $n = 10\,000$ and $\delta = 0.05$, then $k \leq 78$.*

The chi-square statistic corresponding to the gap test of independence is

$$v = \sum_{x=0}^{k-1} \frac{\left(o[x] - e[x]\right)^2}{e[x]},$$

and v is (approximately) a *Chisquare*$(k-1)$ random variate. The critical values of v are

$$v_1^* = \texttt{idfChisquare}(k-1, \alpha/2) \quad \text{and} \quad v_2^* = \texttt{idfChisquare}(k-1, 1-\alpha/2).$$

The *Gap Test of Independence* is based on Algorithm 10.1.4. To apply this test, pick $n \geq 10\,000$, $0 \leq a < b \leq 1$, and $k \leq 2 + \lfloor \ln(10/n\delta)/\ln(1-\delta) \rfloor$, with $\delta = b - a$. Then do the following:

- Use Algorithm 10.1.4 to calculate the k-bin histogram.

---

*The a, b parameters are usually chosen so that the subinterval (a, b) lies at one end of the $(0, 1)$ interval or the other. The reason for this is that, when the inverse-transformation method is used to generate random variates, points near the ends of the $(0, 1)$ interval are mapped to the tails of the variates' distribution. Because extreme points tend to have low probability but high "impact" in a discrete-event simulation, it is considered particularly important to test whether points in the tail of the distribution are generated correctly.

- Calculate the chi-square statistic v.
- Pick a $(1 - \alpha) \times 100\%$ level of confidence (typically, $\alpha = 0.05$).
- Compute the critical values v_1^* and v_2^*.

If $v < v_1^*$ or $v > v_2^*$, the test *failed*. (Failure occurs with approximate probability $1 - \alpha$ for a legitimate random-number generator.)

As it was for the three empirical tests considered previously, the library rngs can be used to generate the data for 256 gap tests, and a figure similar to Figures 10.1.1, 10.1.2, 10.1.3, and 10.1.4 can be created. Doing so is left as an exercise. Similarly, the construction of a corresponding figure for the following empirical test of randomness is left as an exercise.

Test of Bivariate Uniformity. The test of uniformity can be extended to pairs of points—that is, pairs of randomly generated points can be tallied into a two-dimensional $k \times k$-bin histogram via the following algorithm.

Algorithm 10.1.5 This algorithm uses the results of n pairs of calls to Random to tally a two-dimensional $k \times k$-bin histogram.

```
for (x₁ = 0; x₁ < k; x₁++)
   for (x₂ = 0; x₂ < k; x₂++)
      o[x₁, x₂] = 0;
for (i = 0; i < n; i++) {
   u₁ = Random();
   x₁ = ⌊ u₁ * k ⌋;
   u₂ = Random();
   x₂ = ⌊ u₂ * k ⌋;
   o[x₁, x₂]++;
}
```

The expected number of counts per bin is $e[x_1, x_2] = n/k^2$, and so the appropriate chi-square statistic is

$$v = \sum_{x_1=0}^{k-1} \sum_{x_2=0}^{k-1} \frac{\left(o[x_1, x_2] - n/k^2\right)^2}{n/k^2}.$$

The underlying theory is that the test statistic v is a *Chisquare*$(k^2 - 1)$ random variate. Therefore, the critical values of v are

$$v_1^* = \text{idfChisquare}(k^2 - 1, \alpha/2) \quad \text{and} \quad v_2^* = \text{idfChisquare}(k^2 - 1, 1 - \alpha/2).$$

The *Test of Bivariate Uniformity* is based on Algorithm 10.1.5. To apply this test, pick $k \geq 100$ and $n \geq 10k^2$, and then do the following:

- Use Algorithm 10.1.5 to calculate a $k \times k$-bin histogram.
- Calculate the chi-square statistic v.
- Pick a $(1 - \alpha) \times 100\%$ level of confidence (typically, $\alpha = 0.05$).
- Compute the critical values v_1^* and v_2^*.

If $v < v_1^*$ or $v > v_2^*$, the test *failed*. (Failure occurs with approximate probability $1 - \alpha$ for a legitimate random-number generator.)

The test of bivariate uniformity can be extended in an obvious way to triples, quadruples, and so on. Doing so with meaningful values of k and n, however, results in a requirement for a large d-dimensional array (with k^d elements) and an enormous number of u's. For example, even if d is only 2 and if $k = 100$, then the bivariate histogram is a 100×100 array, and at least $n = 2 \cdot 10 \cdot 100^2 = 200\,000$ calls to Random are required *each* time the test is applied. Of course, smaller values of k can be used but the resulting histogram-bin grid will be too coarse, resulting in a test that has little power to find bad random-number generators.

Permutation Test of Independence. The permutation test of independence makes use of the following result, which Knuth (1998) calls the "factorial number system." Let t be a positive integer, and consider the set of $t!$ integers between 0 and $t! - 1$, inclusive. For any such integer x from the set, there is an associated unique set of $t - 1$ integer coefficients $c_1, c_2, \ldots, c_{t-1}$, with $0 \le c_j \le j$, such that

$$x = c_{t-1}(t - 1)! + \cdots + c_2 2! + c_1 1!$$

For example, if $t = 5$, then $t! = 120$, and, for the following (arbitrarily selected) integers between 0 and 119, inclusive, we have

$$33 = 1 \cdot 4! + 1 \cdot 3! + 1 \cdot 2! + 1 \cdot 1!$$

$$47 = 1 \cdot 4! + 3 \cdot 3! + 2 \cdot 2! + 1 \cdot 1!$$

$$48 = 2 \cdot 4! + 0 \cdot 3! + 0 \cdot 2! + 0 \cdot 1!$$

$$119 = 4 \cdot 4! + 3 \cdot 3! + 2 \cdot 2! + 1 \cdot 1!$$

Given an *iid* sequence $U_0, U_1, \ldots, U_{t-1}$, Knuth cleverly uses the factorial number system to associate an integer x between 0 and $t! - 1$ uniquely with the sequence, as follows.

- Let $0 \le c_{t-1} \le t - 1$ be the index of the largest element of the sequence. Remove this largest element to form a new sequence with $t - 1$ terms.
- Let $0 \le c_{t-2} \le t - 2$ be the index of the largest element of the new sequence. Remove this largest element to form a new sequence with $t - 2$ terms.
- Repeat until all the coefficients $c_{t-1}, c_{t-2}, \ldots, c_1$ are determined (and the resulting sequence has just one term), then "encode" the original sequence with the integer

$$x = c_{t-1}(t - 1)! + \cdots + c_2 2! + c_1 1!$$

For the purposes of describing an algorithm for implementing the permutation test of independence, we refer to x as a "permutation index."

The corresponding theory behind the permutation test of independence is that the original sequence can have $t!$ equally likely possible relative orderings (permutations). Accordingly, the permutation index x is equally likely to have any value between 0 and $t! - 1$, that is, each possible value of x has probability $1/t!$.

Algorithm 10.1.6   To implement the permutation test of independence, group (batch) Random's output t terms at a time, calculate a permutation index x for each batch, and then tally all the indices so generated into a histogram with $k = t!$ bins, as follows:

```
for (x = 0; x < k; x++)
  o[x] = 0;
for (i = 0; i < n; i++) {
    for (j = 0; j < t; j++)          // build the permutation array
      u[j] = Random();
    r = t - 1;
    x = 0;
    while (r > 0) {
      s = 0;
        for (j = 1; j <= r; j++)     // max of u[0], u[1], ..., u[r]
          if (u[j] > u[s])
            s = j;
      x = (r + 1) * x + s;
      temp = u[s];                    // swap u[s] and u[r]
      u[s] = u[r];
      u[r] = temp;
      r--;
    }
    o[x]++;
}
```

The expected number of counts per bin is $e[x] = n/k$, and so the appropriate chi-square statistic is

$$v = \sum_{x=0}^{k-1} \frac{\left(o[x] - n/k\right)^2}{n/k}.$$

The test statistic v is a $Chisquare(k-1)$ random variate. Therefore, the critical values of v are $v_1^* = \texttt{idfChisquare}(k-1, \alpha/2)$ and $v_2^* = \texttt{idfChisquare}(k-1, 1-\alpha/2)$.

The *Permutation Test of Independence* is based on Algorithm 10.1.6. To apply this test, pick $t > 3$, $k = t!$, and $n \geq 10k$, and then do the following:

- Use Algorithm 10.1.6 to calculate a k-bin histogram.
- Calculate the chi-square statistic v.

- Pick a $(1 - \alpha) \times 100\%$ level of confidence (typically, $\alpha = 0.05$).
- Compute the critical values v_1^* and v_2^*.

If $v < v_1^*$ or $v > v_2^*$, the test *failed*. (Failure occurs with approximate probability $1 - \alpha$ for a legitimate random-number generator.)

Summary.   We have now introduced six empirical tests of randomness—how do we use them?

- If the generator to be tested is truly awful (some are), then it might be sufficient to use just one test—for example, the runs-up test of independence—a few times with $\alpha = 0.05$ and check to see whether failures are produced consistently. If so, repeat with one or two other tests, and, if failures again occur consistently, the generator should be discarded.
- If the generator to be tested is better than truly awful, but still not good, then *proving* (in a statistical sense) that it isn't good could require a lot of testing and some relatively careful analysis of the results. The recommended way to conduct this analysis is with the Kolmogorov–Smirnov statistic, introduced in the next subsection. Unfortunately, if the generator to be tested isn't good, there is no *guarantee* that any of the six empirical tests presented will detect the generator's deficiency, no matter how sophisticated the analysis of the test results. If the generator isn't good, then (by definition) there is *some* empirical test that will reveal the deficiency. There are, however, no simple rules for finding this "silver bullet" test.
- If the generator to be tested is good, then it *should* fail any of these tests approximately $\alpha \times 100\%$ of the time. In this case, it is relatively meaningless to select just one test and apply it a few times. If in 10 tests you observe 0, 1, or even 2 failures, what have you really learned? Extensive testing is required to conclude that a generator is good.

10.1.2 Kolmogorov–Smirnov Analysis

The Kolmogorov–Smirnov test statistic is the largest *vertical* distance between an empirical cdf calculated from a data set and a hypothesized cdf. Although the focus in this section has been on testing random numbers for uniformity, the KS statistic can be applied to distributions other than the *Uniform*$(0, 1)$ distribution, as indicated by the general definition given below.

Definition 10.1.4   Let $X_1, X_2, \ldots, X_n$ be an *iid* sequence of continuous random variables with a common set of possible values $\mathcal{X}$ drawn from a population with *theoretical* (or hypothesized) cdf $F(x)$. For x, define the associated *empirical* cdf as

$$F_n(x) = \frac{\text{the number of } X_i\text{'s that are} \leq x}{n}.$$

The *Kolmogorov–Smirnov* (KS) statistic is

$$D_n = \sup_x |F(x) - F_n(x)|$$

where the sup* is taken over all x.

---

*sup is an abbreviation for supremum, which plays a role analogous to that of maximum.

Example 10.1.5

We compute the KS test statistic D_6 for the first six random numbers produced by library `rngs` with an initial seed of 12345 and stream 0. The first $n = 6$ random numbers, shown to *d.ddd* precision, are:

$$0.277 \quad 0.726 \quad 0.698 \quad 0.941 \quad 0.413 \quad 0.720.$$

The first step in constructing the empirical cdf is to sort the data values:

$$0.277 \quad 0.413 \quad 0.698 \quad 0.720 \quad 0.726 \quad 0.941.$$

A plot of the empirical cdf (which takes an upward step of height $1/6$ at each data value) and the theoretical cdf [the cdf of a *Uniform*$(0, 1)$ random variable] are plotted in Figure 10.1.5.

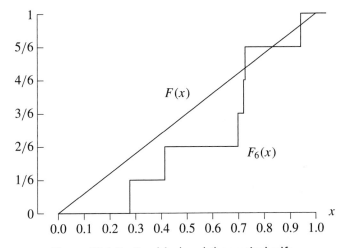

Figure 10.1.5 Empirical and theoretical cdfs.

Where does the largest vertical distance between $F(x)$ and $F_6(x)$ occur? It occurs just to the left (this is why sup, rather than max, is used in the definition of D_n) of 0.698. For this particular data set, $D_6 \cong 0.698 - 2/6 \cong 0.365$. Large values of D_6 indicate poor agreement between the theoretical cdf $F(x)$ and the empirical cdf $F_6(x)$.

As illustrated in Figure 10.1.5, because $F(\cdot)$ is a continuous monotone increasing function and $F_n(\cdot)$ is monotone increasing in a piecewise-constant fashion, the largest difference between $F(x)$ and $F_n(x)$ must occur at one of the data values. This means that a *computational* formula for computing the value of D_n is

$$D_n = \max_{i=1,2,\dots,n} \left\{ \left| F(x_i) - \frac{i}{n} \right|, \left| F(x_i) - \frac{i-1}{n} \right| \right\}.$$

Therefore, D_n can be computed by a simple linear search over the $2n$ values given in the computational formula.

The next question that must be addressed is how large D_n must be for the KS test to fail. This will require our knowing the distribution of D_n.

The Distribution of D_n.   Remarkably, it is known that the random variable (statistic) D_n depends *only* on n (Ross, 2002, page 203)—that is, for a fixed value of n, the shape of D_n's pdf (and associated cdf) is *independent* of $F(\cdot)$. For this reason, the KS statistic is said to be a *distribution-free* statistic. Recall that we saw an analogous result back in Section 8.1; the central limit theorem guarantees that, for large sample sizes, the sample mean is *Normal*, independent of the shape of the "parent" distribution. Unlike the central limit theorem, which is an asymptotic (large-sample-size) result, the distribution-free characteristic of the KS statistic is an *exact* result for *all* values of $n \geq 1$.[*]

For small-to-moderate values of n, Drew, Glen, and Leemis (2000) devised an algorithm that returns the cdf of D_n, which consists of nth order polynomials defined in a piecewise fashion. A "perfect fit" occurs when $F(x)$ intersects each riser of the stairstep function $F_n(x)$ at its midpoint, resulting in the D_n value $1/(2n)$. We illustrate the results of their algorithm for $n = 6$ in the next example.

Example 10.1.6

The cdf associated with D_6 is given by sixth-order polynomials defined in a piecewise fashion:

$$F_{D_6}(y) = \begin{cases} 0 & y < \frac{1}{12} \\ \frac{5}{324}(12y - 1)^6 & \frac{1}{12} \leq y < \frac{1}{6} \\ 2880y^6 - 4800y^5 + 2360y^4 - \frac{1280}{3}y^3 + \frac{235}{9}y^2 + \frac{10}{27}y - \frac{5}{81} & \frac{1}{6} \leq y < \frac{1}{4} \\ 320y^6 + 320y^5 - \frac{2600}{3}y^4 + \frac{4240}{9}y^3 - \frac{785}{9}y^2 + \frac{145}{27}y - \frac{35}{1296} & \frac{1}{4} \leq y < \frac{1}{3} \\ -280y^6 + 560y^5 - \frac{1115}{3}y^4 + \frac{515}{9}y^3 + \frac{1525}{54}y^2 - \frac{565}{81}y + \frac{5}{16} & \frac{1}{3} \leq y < \frac{5}{12} \\ 104y^6 - 240y^5 + 295y^4 - \frac{1985}{9}y^3 + \frac{775}{9}y^2 - \frac{7645}{648}y + \frac{5}{16} & \frac{5}{12} \leq y < \frac{1}{2} \\ -20y^6 + 32y^5 - \frac{185}{9}y^3 + \frac{175}{36}y^2 + \frac{3371}{648}y - 1 & \frac{1}{2} \leq y < \frac{2}{3} \\ 10y^6 - 38y^5 + \frac{160}{3}y^4 - \frac{265}{9}y^3 - \frac{115}{108}y^2 + \frac{4651}{648}y - 1 & \frac{2}{3} \leq y < \frac{5}{6} \\ -2y^6 + 12y^5 - 30y^4 + 40y^3 - 30y^2 + 12y - 1 & \frac{5}{6} \leq y < 1. \end{cases}$$

This cdf is plotted in Figure 10.1.6, and two important points are highlighted. The first value corresponds to the KS test statistic from the last example: $(0.365, 0.680)$, which indicates that, although the test statistic from the previous example seemed rather large, it was only at the 68th percentile of the distribution of D_6. The second value highlighted is $(0.519, 0.950)$, which corresponds to the D_6 value associated with the

---

[*]One other advantage that the KS test statistic possesses over the chi-square statistics is that it requires no *binning* from the modeler. All of the chi-square tests discussed thus far calculated a k-bin histogram (or a $k \times k$ bin in the test of bivariate uniformity), where k was determined arbitrarily. There is no such arbitrary parameter in the case of the KS statistic.

95th percentile of the distribution. Thus, a KS test on a sample size $n = 6$ at significance level $\alpha = 0.05$ fails when the test statistic D_6 exceeds the critical value $d_6^* = 0.519$. (Failure will occur 5% of the time for a valid random-number generator.)

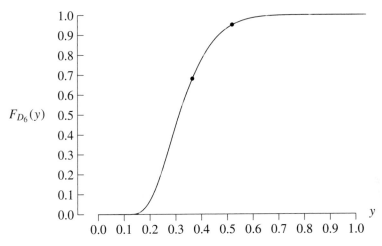

Figure 10.1.6   Cumulative distribution function of D_6.

For larger values of n, approximations have been developed in order to approximate particular critical values of the distribution of D_n. The following theorem describes an approximation developed by Stephens (1974).

Theorem 10.1.3   If D_n is a KS random variable (statistic), and if $\Pr(D_n \leq d_n^*) = 1 - \alpha$, then the (one-tailed) critical value d_n^* is given by

$$d_n^* = \frac{c_{1-\alpha}}{\sqrt{n} + 0.12 + 0.11/\sqrt{n}},$$

where, for selected values of α, the constant $c_{1-\alpha}$ is

α	0.100	0.050	0.025	0.010
$c_{1-\alpha}$	1.224	1.358	1.480	1.628.

Example 10.1.7

If $n = 6$ and $\alpha = 0.05$, then

$$d_6^* = \frac{1.358}{\sqrt{6} + 0.12 + 0.11/\sqrt{6}} \cong 0.519,$$

which is identical to the exact value from the previous example. Therefore,

$$\Pr(D_6 \leq 0.519) = 0.95.$$

Example 10.1.8

If $n = 256$ and $\alpha = 0.05$, then

$$d_{256}^* = \frac{1.358}{16 + 0.12 + 0.007} \cong 0.084.$$

Therefore, according to Stephens's approximation,

$$\Pr(D_{256} \leq 0.084) = 0.95,$$

and so the probability that a *single* realization of the KS random variable D_{256} will be less than 0.084 is 0.95. We fail the KS test when our computed test statistic D_{256} exceeds the critical value $d_{256}^* = 0.084$.

Example 10.1.8 suggests how to combine $n = 256$ empirical chi-square test statistics (the x's from the first part of this section) into one KS statistic d_{256} and, depending on whether the KS statistic is less than 0.084 or greater, with 95% confidence conclude whether the source of the empirical test statistics—the random-number generator in question—is random. This process can be done independently for each empirical test. The details follow.

Let $x_1, x_2, \ldots, x_n$ be a random-variate sample that is assumed to be *Chisquare*$(k-1)$ for an appropriate value of k.* Let D_n be the associated sample statistic defined by

$$D_n = \sup_x |F(x) - F_n(x)|$$

where $F(x)$ is the *Chisquare*$(k-1)$ cdf, the max is over all $x \in \mathcal{X}$, and

$$F_n(x) = \frac{\text{the number of } x_i\text{'s that are} \leq x}{n}.$$

Example 10.1.9

The 256 runs-up test chi-square statistics illustrated in Figure 10.1.4 were sorted and then used to generate $F_{256}(x)$. A portion of this empirical cdf is illustrated (as the piecewise-constant curve) along with the associated theoretical cdf $F(x)$ in Figure 10.1.7.

As indicated, in this case, the hypothesis is that the chi-square statistics are samples of a *Chisquare*(5) random variable. The empirical cdf and hypothesized theoretical cdf fit well. For reference, the •'s indicate the value of $F(x_i)$ for $i = 1, 2, \ldots, 13$.

---

*For example, for the test of uniformity, $k-1$ is one less than the number of histogram bins.

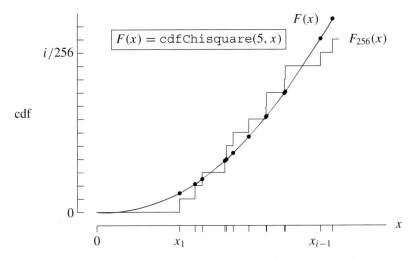

Figure 10.1.7 Empirical and theoretical chi-square cdf

Example 10.1.10

For the $n = 256$ chi-square statistics in Example 10.1.4, the computed KS statistic was $d_{256} = 0.044$, which is a value comfortably less than $d_{256}^* = 0.084$. This is strong evidence that the runs-up test of independence has detected *no* deficiencies in the rngs generator. Similarly, for the chi-square statistics in Examples 10.1.1 and 10.1.2, the computed KS statistics were 0.035 and 0.047, respectively; this is strong evidence that the test of uniformity and test of extremes also detected no deficiencies in the rngs generator.[*]

Most specialists consider the best random-number-generator empirical-testing strategy to consist of multiple applications ($n = 100$ or more) of multiple chi-square tests (at least five), using the KS statistic with a 95% (or 99%) level of confidence as the ultimate empirical test for randomness. Most of these specialists also have unrestricted access to computers, many hard-working graduate students, and much patience.

This section has contained a brief overview of random-number testing. We recommend Law and Kelton (2000, pages 417–427); Banks, Carson, Nelson, and Nicol (2005, pages 260–267); and Fishman (2001, Chapter 9) for alternative introductory treatments of the topic.

10.1.3 Exercises

10.1.1   (a) Implement the test of uniformity (with $\alpha = 0.05$) and apply it to each of the 256 streams of random numbers generated by rngs. Use any nine-digit number as initial seed for the generator. (b) For how many streams did the test fail? (c) Comment.

10.1.2   Same procedure as in the previous exercise, except using the test of extremes with $d = 5$.

---

[*]For the chi-square statistics in Example 10.1.3, the KS statistic was 0.056. Therefore, even though the number of test failures in this example was unexpectedly high, when all 256 tests results are viewed in the aggregate, there is no significant evidence of nonrandomness.

10.1.3   (*a*) Prove that, if $U_0, U_1, \ldots, U_{d-1}$ is an *iid* sequence of d *Uniform*$(0, 1)$ random variables and if

$$S = \min\{U_0, U_1, \ldots, U_{d-1}\},$$

then $(1 - S)^d$ is a *Uniform*$(0, 1)$ random variable. (*b*) Use this result to modify the test of extremes accordingly, and then repeat Exercise 10.1.2.

10.1.4   (*a*) Prove Theorem 10.1.2. *Hint*: First prove that the probability that a runs-up will be longer than x is $1/(x + 1)!$. (*b*) What is the expected number of calls to Random to produce n runs?

10.1.5   (*a*) Repeat the procedure in Exercise 10.1.1, using the runs-up test of independence. (*b*) How would you turn this test into a runs-down test?

10.1.6   Relative to the equation in Definition 10.1.2, if n is the number of observations, in the sample prove that an equivalent equation for v is

$$v = \sum_x \left(\frac{o^2[x]}{e[x]} \right) - n.$$

10.1.7   (*a*) Implement the gap test of independence, generate 256 chi-square statistics, and compute the corresponding KS statistic. (*b*) Comment.

10.1.8   Repeat the previous exercise for the bivariate test of uniformity.

10.1.9   Relative to the gap test of independence, what is the expected number of calls to Random to produce n gaps?

10.1.10   (*a*) Is the runs-up test of independence valid if the *iid* samples are drawn from another continuous distribution? (*b*) Experiment with draws from an *Exponential*(1) distribution. (*c*) Comment.

10.1.11   Repeat Exercise 10.1.7 for the permutation test of independence.

10.1.12   Repeat Exercise 10.1.10 for the permutation test of independence.

10.1.13   (*a*) Explain why one random number is skipped between runs in the runs-up test of independence. (*b*) If the random number between runs is *not* skipped, do you expect the value of the test statistic v to increase or decrease? Explain your reasoning.

10.1.14   What is the set of possible values $\mathcal{X}$ for the KS random variable D_n corresponding to sample size n?

10.1.15   What is the distribution of D_2, the KS statistic in the case of a sample size of $n = 2$?

10.1.16   If 256 goodness-of-fit tests are conducted with a 95% level of confidence, then one expects $256 \cdot 0.05 \cong 13$ failures. How far away from 13 failures (high and low) would arouse suspicions concerning an empirical test of randomness?

10.2   BIRTH–DEATH PROCESSES

The goal of this section is to introduce a particular type of stochastic process known as a *birth–death process*. Because the time evolution of such a process is usually defined in terms of an *Exponential* random variable, we begin with a brief review of the characteristics of this random variable from Section 7.4.2, with the cosmetic change that X is replaced by T in order to emphasize the time dependence.

10.2.1 Exponential Random Variables

Definition 10.2.1   If the random variable T is *Exponential*(μ), then

- the real-valued scale parameter μ satisfies $\mu > 0$,
- the possible values of T are $\mathcal{T} = \{t \mid t > 0\}$,
- the pdf of T is

$$f(t) = \frac{1}{\mu} \exp(-t/\mu) \qquad t > 0,$$

- the cdf of T is

$$F(t) = \Pr(T \le t) = 1 - \exp(-t/\mu) \qquad t > 0,$$

- the idf of T is

$$F^{-1}(u) = -\mu \ln(1 - u) \qquad 0 < u < 1,$$

- the mean and standard deviation of T are both μ.

In reliability analysis, if the random variable T represents the *lifetime* or *failure time* of a process—the elapsed time between the initialization of a process and its time of (first) failure—then it is common to assume that T is *Exponential*(μ). In this application, the parameter μ represents the expected lifetime, and the probability that the lifetime will exceed some threshold value t is $\Pr(T > t) = 1 - \Pr(T \le t) = \exp(-t/\mu)$.*

Example 10.2.1

If the lifetime T of a process (in seconds) is *Exponential*(100), then its expected lifetime is 100 seconds, and, for example, the probability that this process will last more that 120 seconds is $\Pr(T > 120) = \exp(-120/100) \cong 0.30$, as illustrated in Figure 10.2.1.

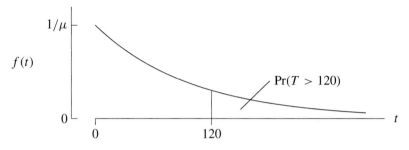

Figure 10.2.1   *Exponential*(100) pdf.

---

*$S(t) = \Pr(T > t)$ is known as the "survivor function" in reliability analysis.

Exponential Random Variables Are Memoryless

Theorem 10.2.1 If T is $Exponential(\mu)$, then T has the *memoryless* property—that is, for any $\tau > 0$ and $t > 0$,

$$\Pr(T > t + \tau \mid T > \tau) = \Pr(T > t) = \exp(-t/\mu).$$

Proof. As a consequence of the definition of conditional probability,

$$\Pr(T > t + \tau \mid T > \tau) = \frac{\Pr\big((T > t + \tau) \text{ and } (T > \tau)\big)}{\Pr(T > \tau)}$$

$$= \frac{\Pr(T > t + \tau)}{\Pr(T > \tau)}$$

$$= \frac{\exp\big(-(t + \tau)/\mu\big)}{\exp(-\tau/\mu)}$$

$$= \exp(-t/\mu)$$

$$= \Pr(T > t).$$

To paraphrase Theorem 10.2.1, $\Pr(T > t + \tau \mid T > \tau)$ is *independent* of τ for all $\tau > 0$; T has *no* memory of its past. Equivalently, T is said to be *Markovian*—the *future* stochastic behavior of T is determined uniquely by its *current* state. In most respects, people are decidedly non-Markovian; we remember our past, and that memory dictates, in part, our future behavior. To a surprising extent, however, many common natural stochastic processes are Markovian. $\square$

Example 10.2.2

As a continuation of Example 10.2.1, if the lifetime T of a process (in seconds) is $Exponential(100)$ and if, for example, $t = 120$, $\tau = 200$ then

$$\Pr(T > 120 + 200 \mid T > 200) = \Pr(T > 120) \cong 0.30.$$

Therefore, this process is just as likely to survive for at least 120 more seconds when it has survived 200 seconds as it was to survive for at least 120 seconds when it was *new*.

It can be shown that the *Exponential* random variable is the *only* continuous random variable with the memoryless property. Indeed, the memoryless property can be viewed as the defining characteristic of this random variable: It can be proven that, if T is a positive continuous random variable with mean $\mu = E[T]$ and if

$$\Pr(T > t + \tau \mid T > \tau) = \Pr(T > t),$$

for all $t > 0$, $\tau > 0$, then T is $Exponential(\mu)$. The proof is left as an exercise.[*]

---

[*]In the discrete case, a similar result applies to a *Geometric(p)* random variable.

The theory we have covered so far is a review of previous material. The following important theorem is new. The proof is deferred until after the examples.

Theorem 10.2.2 If $T_1, T_2, \ldots, T_n$ are independent *Exponential* (μ_i) random variables (the μ_i may be different), and if T is the smallest of these—

$$T = \min\{T_1, T_2, \ldots, T_n\}$$

—then T is *Exponential* (μ), where

$$\frac{1}{\mu} = \frac{1}{\mu_1} + \frac{1}{\mu_2} + \cdots + \frac{1}{\mu_n}.$$

In addition, the probability that T_i is the smallest of $T_1, T_2, \ldots, T_n$ is

$$\Pr(T = T_i) = \frac{\mu}{\mu_i} \qquad i = 1, 2, \ldots, n.$$

Example 10.2.3
Suppose that T_1 and T_2 are independent *Exponential* random variables with means $\mu_1 = 2$ and $\mu_2 = 4$ respectively. Then $\mu = 4/3$, and the random variable $T = \min\{T_1, T_2\}$ is *Exponential* $(4/3)$. Moreover,

$$\Pr(T = T_1) = \frac{2}{3} \qquad \text{and} \qquad \Pr(T = T_2) = \frac{1}{3}.$$

If you were to generate many (t_1, t_2) samples of (T_1, T_2) then t_1 would be smaller than t_2 approximately $2/3$ of the time. Moreover, a histogram of $t = \min\{t_1, t_2\}$ would converge to the pdf of an *Exponential* $(4/3)$ random variable for a large number of replications.

Example 10.2.4
A computer uses 16 RAM chips, indexed $i = 1, 2, \ldots, 16$. If these chips fail independently (by natural aging rather than, say, because of a power surge), what is the expected time of the first failure?

- If each chip has a lifetime T_i (in hours) that is *Exponential* $(10\,000)$, then the first failure occurs at $T = \min\{T_1, T_2, \ldots, T_{16}\}$, which is an *Exponential* (μ) random variable with $\mu = 10\,000/16$. Therefore, the expected time of the *first* failure is $E[T] = 10\,000/16 = 625$ hours.
- If, instead, one of the 16 chips has a lifetime that is *Exponential* (1000), then $E[T]$ is

$$\frac{1}{\mu} = \frac{15}{10\,000} + \frac{1}{1000} = \frac{25}{10\,000} = \frac{1}{400}$$

and so the expected time of the first failure is 400 hours. Moreover, when the first failure occurs, the probability that it will be caused by the *Exponential* (1000) chip is $400/1000 = 0.4$.

Example 10.2.5

As a continuation of the previous example, suppose there are 17 identical *Exponential* (10 000) RAM chips, including one spare. If the spare is installed as a replacement at the time of the first failure, then, because of the memoryless property, the 15 old surviving chips are statistically identical to the one new chip. Therefore, the expected time to the *next* failure is also 625 hours.

Proof of Theorem 10.2.2.   To prove the first result in Theorem 10.2.2, let $T = \min\{T_1, T_2, \ldots, T_n\}$, where T_i is an *Exponential* (μ_i) random variable, $i = 1, 2, \ldots, n$, and $T_1, T_2, \ldots, T_n$ are independent. For any $t > 0$, the cdf of T is

$$F(t) = \Pr(T \le t)$$

$$= 1 - \Pr(T > t)$$

$$= 1 - \Pr\big((T_1 > t) \text{ and } (T_2 > t) \text{ and } \ldots \text{ and } (T_n > t)\big)$$

$$= 1 - \Pr(T_1 > t)\Pr(T_2 > t) \ldots \Pr(T_n > t)$$

$$= 1 - \exp(-t/\mu_1)\exp(-t/\mu_2) \ldots \exp(-t/\mu_n)$$

$$= 1 - \exp(-t/\mu),$$

where

$$\frac{1}{\mu} = \frac{1}{\mu_1} + \frac{1}{\mu_2} + \cdots + \frac{1}{\mu_n}.$$

This proves that T is *Exponential* (μ).

To prove the second part of Theorem 10.2.2, let $T' = \min\{T_2, T_3, \ldots, T_n\}$. From the previous result, T' is *Exponential* (μ'), where

$$\frac{1}{\mu'} = \frac{1}{\mu_2} + \frac{1}{\mu_3} + \cdots + \frac{1}{\mu_n}.$$

Moreover, T_1 and T' are independent. Now recognize that $T_1 = T$ if and only if $T_1 \le T'$ and that, by conditioning on the possible values of T',

$$\Pr(T_1 = T) = \Pr(T_1 \le T') = \int_0^\infty \Pr(T_1 \le t') f(t') \, dt',$$

where

$$f(t') = \frac{1}{\mu'} \exp(-t'/\mu') \qquad\qquad t' > 0$$

is the pdf of T' and $\Pr(T_1 \le t') = 1 - \exp(-t'/\mu_1)$ is the cdf of T_1. Therefore,

$$\Pr(T_1 = T) = \frac{1}{\mu'} \int_0^\infty \big(1 - \exp(-t'/\mu_1)\big) \exp(-t'/\mu') \, dt'$$

$$\vdots$$

$$= 1 - \frac{\mu}{\mu'}$$

$$= \mu \left(\frac{1}{\mu} - \frac{1}{\mu'} \right)$$

$$= \frac{\mu}{\mu_1}.$$

This proves that $\Pr(T_i = T) = \mu/\mu_i$ for $i = 1$. Because the argument is equally valid for any i, the second part of Theorem 10.2.2 is proven. $\square$

Theorem 10.2.2 has an equivalent formulation expressed in terms of *rates* $\lambda_i = 1/\mu_i$ rather than expected times μ_i. The rate formulation is more intuitive and, for that reason, more easily remembered.

Theorem 10.2.3   If $T_1, T_2, \ldots, T_n$ is an independent set of *Exponential* $(1/\lambda_i)$ random variables (the λ_i may be different), and if T is the smallest of these—

$$T = \min\{T_1, T_2, \ldots, T_n\}$$

—then T is *Exponential* $(1/\lambda)$, where

$$\lambda = \lambda_1 + \lambda_2 + \cdots + \lambda_n.$$

In addition, the probability that T_i is the smallest of $T_1, T_2, \ldots, T_n$ is

$$\Pr(T_i = T) = \frac{\lambda_i}{\lambda} \qquad i = 1, 2, \ldots, n.$$

Example 10.2.6
Consider n independent streams of Poisson arrivals, with arrival rates $\lambda_1, \lambda_2, \ldots, \lambda_n$ respectively, that merge to form a single stream, as illustrated in Figure 10.2.2.

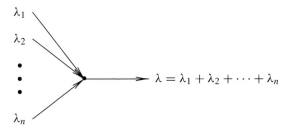

Figure 10.2.2 Merging n independent arrival streams.

From Theorem 10.2.3, the single stream is a Poisson arrival process with rate λ. Moreover, the next arrival at the merge point will come from stream i with probability λ_i/λ.*

---

*Similarly, if a single stream of Poisson arrivals with rate λ splits into n streams independently with probability $p_1, p_2, \ldots, p_n$, then each of the streams so formed is a Poisson process having rate $p_i \lambda$, for $i = 1, 2, \ldots, n$.

Example 10.2.7

Consider a multiple-server service node with c independent servers, all of which are busy at time t. Assume that, for $s = 1, 2, \ldots, c$, the service time for server s is *Exponential*$(1/v_s)$ with service rate v_s. From Theorem 10.2.3, the next departure from the service node will occur at $t + T$, where T is an *Exponential*$(1/v)$ random variable and

$$v = v_1 + v_2 + \cdots + v_c$$

is the sum of the individual service rates. Moreover, this next departure will come from server s with probability v_s/v.

10.2.2 Continuous-Time Birth–Death Processes

A birth–death processes is defined in terms of a discrete state-space $\mathcal{X}$ that may be either finite or infinite. Without loss of generality, we assume throughout that, if the state-space is *finite*, then

$$\mathcal{X} = \{0, 1, 2, \ldots, k\}$$

for some positive integer k; but, if the state-space is *infinite*, then

$$\mathcal{X} = \{0, 1, 2, \ldots\}.$$

Definition 10.2.2   Let X be a discrete random variable, indexed by time t as $X(t)$, that evolves in time as follows.

- At any time $t > 0$, $X(t) \in \mathcal{X}$.
- If $x \in \mathcal{X}$ is the state of X at time t, then X will remain in this state for a random time T that is *Exponential* with mean $\mu(x)$.
- When a state transition occurs, then, independent of the time spent in the current state, the random variable X will either shift from its current state x to state $x + 1 \in \mathcal{X}$ with probability $p(x)$ or shift from state x to state $x - 1 \in \mathcal{X}$ with probability $1 - p(x)$.

The shift from state x to $x + 1$ is called a *birth*; the shift from x to $x - 1$ is called a *death*. The stochastic process defined by $X(t)$ is called a *continuous-time birth–death process.*[*]

A continuous-time birth–death process is completely characterized by the initial state $X(0) \in \mathcal{X}$ at $t = 0$ and by the two functions $\mu(x)$ and $p(x)$ defined for all $x \in \mathcal{X}$; hence,

$$\mu(x) > 0 \qquad \text{and} \qquad 0 \le p(x) \le 1,$$

with the understanding that $p(0) = 1$, because a death cannot occur when X is in the state $x = 0$. Moreover, if the state-space is finite, then $p(k) = 0$, because a birth cannot occur when X is in the state $x = k$.

---

[*]Discrete-time birth–death processes are considered later in this section.

Continuous-Time Birth–Death Processes Are Memoryless. It is important to recognize that, because of the very special memoryless property of the *Exponential* random variable, a birth–death process has no memory of its past—this is the Markovian assumption. At any time, the future-time evolution of X depends *only* on the current state. The length of (past) time that X has been in this state, or in any other state, is completely irrelevant in terms of predicting when the next (future) state transition will occur and whether it will be a birth or a death. As we will see, for any $c \geq 1$, an $M/M/c$ service node represents an important special case of a continuous-time birth–death process, with $X(t)$ corresponding to the number in the node at time t and with births and deaths corresponding to arrivals and departures, respectively.

We can use the fact that the function `Bernoulli(p)` in the library `rvgs` will return the value 1 or 0 with probability p and $1 - p$ respectively to simulate the time evolution of a birth–death process with Algorithm 10.2.1. Each replication of this simulation will generate one *realization* $x(t)$ of the birth–death process $X(t)$; repeated replication generates an *ensemble* of realizations.

Algorithm 10.2.1 Given the state-space $\mathcal{X}$, the initial state $X(0) \in \mathcal{X}$, and the functions $\mu(x)$, $p(x)$ defined for all $x \in \mathcal{X}$ with $\mu(x) > 0$ and $0 \leq p(x) \leq 1$, this algorithm simulates the time evolution of the associated continuous-time birth–death process for $0 < t < \tau$.

```
t = 0;
x = X(0);
while (t < τ) {
    t += Exponential(μ(x));
    x += 2 * Bernoulli(p(x)) - 1;
}
```

Birth Rates and Death Rates. Continuous-time birth–death processes frequently are characterized in terms of *rates* —

$$\lambda(x) = \frac{1}{\mu(x)} \qquad x \in \mathcal{X}$$

—rather than in terms of the corresponding expected *times* $\mu(x)$. The reason for this preference is that

$$b(x) = \lambda(x)p(x) \qquad \text{and} \qquad d(x) = \lambda(x)\big(1 - p(x)\big)$$

can then be thought of as the *birth* and *death* rates (respectively) when X is in the state x. Note that $\lambda(x) = b(x) + d(x)$ is the rate at which the process will "flow" *out* of state x to one of the adjoining states ($x + 1$ or $x - 1$).[*]

Because a birth–death process is stochastic, it is a mistake to think of $b(x)$ and $d(x)$ as *instantaneous* birth and death rates. Instead they represent *infinite-horizon* rates. Over a long period of time, $b(x)$ and $d(x)$ represent the average rate of flow from state x into the states $x + 1$ and $x - 1$, respectively. This flow analogy is particularly compelling when the processes are represented by a directed graph ("digraph"), as illustrated in Figure 10.2.3.

---
[*]If $\lambda(x) = 0$, then state x is said to be an *absorbing* state, a state from which the birth–death process can't escape.

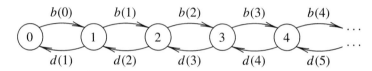

Figure 10.2.3   Rate diagram for a birth–death process.

Example 10.2.8

An $M/M/1$ service node with infinite capacity, arrival rate λ, and service rate ν defines an infinite-state continuous-time birth–death process. In this case, $X(t)$ represents the number in the node at time $t > 0$, and the births and deaths correspond to arrivals and departures respectively.

- If $X(t) = 0$, then the (idle) node will remain in this state until the next arrival. Therefore, in this state, only births are possible, and they occur at the rate λ; so

$$p(0) = 1 \qquad \text{and} \qquad \lambda(0) = \lambda.$$

- If $X(t) = x > 0$, then the transition to the next state will occur at the rate

$$\lambda(x) = \lambda + \nu \qquad x = 1, 2, \ldots,$$

and it will be an arrival (birth) with probability

$$p(x) = \frac{\lambda}{\lambda + \nu} \qquad x = 1, 2, \ldots,$$

or a departure (death) with probability

$$1 - p(x) = \frac{\nu}{\lambda + \nu} \qquad x = 1, 2, \ldots.$$

Therefore, the birth rate is

$$b(x) = \lambda(x)p(x) = \lambda \qquad x = 0, 1, 2, \ldots,$$

and the death rate is

$$d(x) = \lambda(x) - b(x) = \begin{cases} 0 & x = 0 \\ \nu & x = 1, 2, \ldots. \end{cases}$$

The corresponding (infinite) directed-graph representation is shown in Figure 10.2.4.

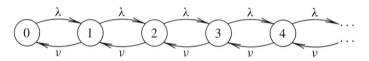

Figure 10.2.4   Rate diagram for an $M/M/1$ service node.

Example 10.2.9

As a continuation of the previous example, the notation $M/M/1/k$ denotes an $M/M/1$ service node with *finite* capacity k. Note that $k \geq 1$ is the maximum possible number of jobs in the service node, *not* in the queue. In this case, the equations in Example 10.2.8 must be modified slightly to reflect the fact that the state-space is $\mathcal{X} = \{0, 1, 2, \ldots, k\}$. The associated (finite) directed-graph representation is shown in Figure 10.2.5.

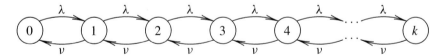

Figure 10.2.5   Rate diagram for an $M/M/1/k$ service node.

10.2.3 Steady-State Characterization

Definition 10.2.3   Let $f(x, t) = \Pr(X(t) = x)$ denote the probability (pdf) that a continuous-time birth–death process $X(t)$ is in the state $x \in \mathcal{X}$ at time $t > 0$. The *steady-state* pdf (if it exists) is defined by

$$f(x) = \lim_{t \to \infty} f(x, t) \qquad x \in \mathcal{X}.$$

Steady-State Flow Balance.   As always when dealing with probabilities, it must be true that $f(x) \geq 0$ for all $x \in \mathcal{X}$ and that

$$\sum_x f(x) = 1,$$

where the sum is over all $x \in \mathcal{X}$. Recall that, when the birth–death process is in state x, the rate of flow into the states $x + 1$ and $x - 1$ is $b(x)$ and $d(x)$ respectively. Because $f(x)$ represents the steady-state probability that the process is in state x, it follows that, for each $x \in \mathcal{X}$,

* $b(x) f(x)$ is the steady-state rate of flow from state x to state $x + 1$;
* $d(x) f(x)$ is the steady-state rate of flow from state x to state $x - 1$.

In steady state, the net rate of flow *into* state x must be balanced by the net rate of flow *out* of state x, as illustrated in Figure 10.2.6.

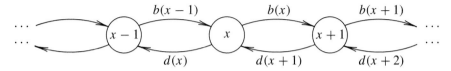

Figure 10.2.6   Generic flow diagram about node x.

That is, for state $x \in \mathcal{X}$, steady state is characterized by the *flow-balance equation*

(flow in) $d(x + 1) f(x + 1) + b(x - 1) f(x - 1) = b(x) f(x) + d(x) f(x)$ (flow out).

This equation is *linear* in the unknown steady-state probabilities. Moreover, there is one equation for each $x \in \mathcal{X}$, and so the steady-state probabilities can be computed by solving a (possibly infinite) system of linear equations.

Example 10.2.10

From Example 10.2.8, for an $M/M/1$ service node with infinite capacity, the balance equations are

$$\nu f(1) = \lambda f(0)$$

$$\nu f(x+1) + \lambda f(x-1) = (\lambda + \nu) f(x) \qquad x = 1, 2, \ldots.$$

Recognize that, except for cosmetic differences, these equations are identical to the steady-state equations developed in Section 8.5.

By looking at the steady-state rate of flow into and out of each state, we can write the flow-balance equations as

$$\text{flow in} = \text{flow out}$$

$$d(1)f(1) = b(0)f(0) \qquad\qquad \text{state 0}$$

$$d(2)f(2) + b(0)f(0) = b(1)f(1) + d(1)f(1) \qquad\qquad \text{state 1}$$

$$d(3)f(3) + b(1)f(1) = b(2)f(2) + d(2)f(2) \qquad\qquad \text{state 2}$$

$$\vdots$$

$$d(x+1)f(x+1) + b(x-1)f(x-1) = b(x)f(x) + d(x)f(x) \qquad\qquad \text{state } x$$

$$\vdots$$

If the first $x + 1$ flow-balance equations are added and then like terms are canceled, it can be shown that the resulting equation is

$$d(x+1)f(x+1) = b(x)f(x) \qquad x = 0, 1, 2, \ldots.$$

Note that, as illustrated in Figure 10.2.7, this last equation could have been written "by inspection" if we had applied flow balance to the digraph *arc* connecting states x and $x + 1$ rather than to the digraph *node* x.

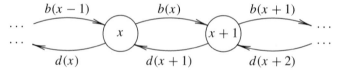

Figure 10.2.7 Generic flow diagram between nodes x and $x + 1$.

Steady-State Probabilities.   As summarized by the following theorem, the equations

$$d(x+1)f(x+1) = b(x)f(x) \qquad x = 0, 1, 2, \ldots$$

provide a recursive algorithm to solve for $f(1)$, $f(2)$, $f(3)$, ... in terms of $f(0)$.

Theorem 10.2.4   For a continuous-time birth–death process with $\mathcal{X} = \{0, 1, 2, \ldots, k\}$ (where k may be infinite), the steady-state probabilities (if they exist) can be calculated recursively in terms of $f(0)$ as

$$f(x+1) = \frac{b(x)b(x-1)\ldots b(1)b(0)}{d(x+1)d(x)\ldots d(2)d(1)} f(0) \qquad x = 0, 1, 2, \ldots, k-1.$$

The remaining steady-state probability $f(0)$ is determined by the equation

$$\sum_x f(x) = 1$$

where the sum is over all $x \in \mathcal{X}$.

Algorithm 10.2.2   The following three-step algorithm, based on Theorem 10.2.4 with $\mathcal{X} = \{0, 1, 2, \ldots, k\}$, where k may be infinite, computes the steady-state probabilities (if they exist) for a continuous-time birth–death process.

- With $f'(0) = 1$, define the *unnormalized* steady-state probabilities $f'(x)$ recursively as

$$f'(x+1) = \frac{b(x)}{d(x+1)} f'(x) \qquad x = 0, 1, 2, \ldots, k-1.$$

- Sum over all possible states $x \in \mathcal{X}$ to calculate the normalizing constant

$$\alpha = \sum_x f'(x) > 0.$$

- Compute the (normalized) steady-state probabilities as*

$$f(x) = \frac{f'(x)}{\alpha} \qquad x = 0, 1, 2, \ldots, k.$$

Finite State Space.   If the number of possible states is *finite*, then Algorithm 10.2.2 is always valid, provided that $b(k) = 0$ and $d(x) \neq 0$ for $x = 1, 2, \ldots, k$, where the finite state space is $\mathcal{X} = \{0, 1, 2, \ldots, k\}$—that is, steady state will always be achieved as $t \to \infty$.

---

*Note that $\alpha = 1/f(0)$.

Example 10.2.11

As a continuation of Example 10.2.10, for an $M/M/1/k$ service node with capacity $k = 1, 2, \ldots$, we have (using l in place of x)

$$f'(l + 1) = \rho f'(l) \qquad l = 0, 1, 2 \ldots, k - 1,$$

where $\rho = \lambda/\nu$. If follows that

$$f'(l) = \rho^l \qquad l = 0, 1, 2, \ldots, k$$

and that

$$\alpha = \sum_{l=0}^{k} f'(l) = 1 + \rho + \cdots + \rho^k = \frac{1 - \rho^{k+1}}{1 - \rho} \qquad (\rho \neq 1).$$

Therefore, the steady-state pdf for an $M/M/1/k$ service node is

$$f(l) = \left(\frac{1 - \rho}{1 - \rho^{k+1}} \right) \rho^l \qquad l = 0, 1, 2, \ldots, k,$$

provided that $\rho \neq 1$. If $\rho = 1$, then $f(l) = 1/(k + 1)$ for $l = 0, 1, 2, \ldots, k$.

The results in Example 10.2.11 are summarized by the following theorem. Note that, if $\rho < 1$, then $\rho^{k+1} \to 0$ as $k \to \infty$. Therefore, as expected, if $\rho < 1$, then, as the service-node capacity becomes infinite, Theorem 10.2.5 reduces to Theorem 8.5.3.

Theorem 10.2.5 For an $M/M/1/k$ service node with $\rho = \lambda/\nu \neq 1$, the steady-state probability of l jobs in the service node is

$$f(l) = \left(\frac{1 - \rho}{1 - \rho^{k+1}} \right) \rho^l \qquad l = 0, 1, 2, \ldots, k.$$

If $\rho = 1$, then $f(l) = 1/(k + 1)$ for $l = 0, 1, 2, \ldots, k$.

Example 10.2.12

As the equation in Theorem 10.2.5 indicates, steady-state can *always* be achieved for an $M/M/1/k$ service node, even if $\rho = \lambda/\nu > 1$. The reason for this is that, if the service-node capacity is finite, then jobs will be rejected if they arrive to find the service node full (holding k jobs).

As illustrated in Figure 10.2.8 for $k = 10$, if $\rho < 1$, then $f(l)$ is a monotone-decreasing function of l. When $\rho = 1$, all of the $k + 1$ possible service-node states are equally likely. For values of $\rho > 1$, $f(l)$ increases monotonically with increasing l. The probability of rejection is

$$f(k) = \left(\frac{1 - \rho}{1 - \rho^{k+1}} \right) \rho^k,$$

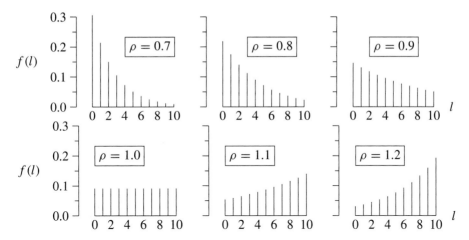

Figure 10.2.8   Steady-state probabilities for the $M/M/1/k$ queue.

and the probability of an idle service node is

$$f(0) = \left(\frac{1 - \rho}{1 - \rho^{k+1}} \right).$$

As ρ increases, the probability of rejection increases and the probability of an idle service node decreases.

Steady-State Statistics.   From the equation in Theorem 10.2.5, for an $M/M/1/k$ service node with $\rho \neq 1$, the steady-state expected number in the service node $E[L] = \lim_{t \to \infty} \bar{l}$ is

$$E[L] = \sum_{l=0}^{k} l\, f(l) = \sum_{l=1}^{k} l\, f(l) = \left(\frac{1 - \rho}{1 - \rho^{k+1}} \right) \left(\rho + 2\rho^2 + \cdots + k\rho^k \right).$$

From the identity

$$\rho + 2\rho^2 + \cdots + k\rho^k = \rho \frac{d}{d\rho} \left(1 + \rho + \rho^2 + \cdots + \rho^k \right) = \frac{\rho(1 - \rho^{k+1}) - (k+1)\rho^{k+1}(1 - \rho)}{(1 - \rho)^2},$$

which is valid for all $\rho \neq 1$, it follows that

$$E[L] = \frac{\rho}{1 - \rho} - \frac{(k+1)\rho^{k+1}}{1 - \rho^{k+1}}.$$

If $\rho = 1$, then $E[L] = k/2$.

Similarly, the steady-state expected number in the queue $E[Q] = \lim_{t \to \infty} \bar{q}$ is

$$E[Q] = \sum_{l=1}^{k} (l - 1) f(l) = \sum_{l=1}^{k} l\, f(l) - \sum_{l=1}^{k} f(l) = \cdots = E[L] - \left(1 - f(0) \right).$$

Little's equations can be used to compute the corresponding expected wait and delay. To use them, note that, because rejection occurs when the service node is full, the rate at which jobs flow into the service node is $(1 - f(k))\lambda$, not λ. Therefore, the steady-state expected wait $E[W] = \lim_{t\to\infty} \overline{w}$ is

$$E[W] = \frac{E[L]}{(1 - f(k))\lambda},$$

and the steady-state expected delay $E[D] = \lim_{t\to\infty} \overline{d}$ is

$$E[D] = \frac{E[Q]}{(1 - f(k))\lambda}.$$

In the three equations for $E[Q]$, $E[W]$, and $E[D]$, the probabilities $f(0)$ and $f(k)$ are given in Example 10.2.12.

Infinite State Space.   If the number of possible states for a continuous-time birth–death process is *infinite* then, even if $d(x) \neq 0$ for all x, it is possible that the sum of unnormalized steady-state probabilities will not converge—the parameter α could be infinite. In this case, Algorithm 10.2.2 will fail, because the continuous-time birth–death process can never achieve steady state. It turns out that the condition $\alpha < \infty$ is both a necessary and a sufficient condition for steady state. As we will see, the $\alpha < \infty$ condition is the continuous-time birth–death-process analog of the familiar $\rho < 1$ steady-state condition for infinite-capacity, single-server and multiple-server service nodes.

Example 10.2.13
As a continuation of Examples 10.2.11 and 10.2.12, consider the limiting case of an $M/M/1/k$ service node with $k = \infty$ (i.e., a conventional infinite-capacity $M/M/1$ service node). In this case,

$$\alpha = \sum_{l=0}^{\infty} f'(l) = 1 + \rho + \rho^2 + \cdots,$$

and (since $\rho > 0$) this harmonic series converges if and only if $\rho < 1$—that is, if $\rho < 1$, then this series converges to

$$\alpha = \frac{1}{1 - \rho},$$

and so the steady-state probability of l jobs in an $M/M/1$ service node is

$$f(l) = \frac{f'(l)}{\alpha} = (1 - \rho)\rho^l \qquad l = 0, 1, 2, \ldots$$

In other words, in steady state, the number of jobs in an $M/M/1$ service node is a *Geometric*(ρ) random variable. This is Theorem 8.5.3.

Multiple-Server Service Node. Let us now consider a (more challenging mathematically) case: an infinite-capacity $M/M/c$ service node with arrival rate λ and c identical servers, each having service rate v. In this case (using l in place of x), an $M/M/c$ service node defines a continuous-time birth–death process with

$$
p(l) = \begin{cases} \dfrac{\lambda}{\lambda + lv} & l = 0, 1, 2, \ldots, c-1 \\[2mm] \dfrac{\lambda}{\lambda + cv} & l = c, c+1, c+2, \ldots \end{cases}
$$

and

$$
\lambda(l) = \begin{cases} \lambda + lv & l = 0, 1, 2, \ldots, c-1 \\ \lambda + cv & l = c, c+1, c+2, \ldots . \end{cases}
$$

Therefore, the birth rate is

$$
b(l) = p(l)\lambda(l) = \lambda \qquad l = 0, 1, 2, \ldots,
$$

and the (state-dependent) death rate is

$$
d(l) = \lambda(l) - b(l) = \begin{cases} lv & l = 0, 1, 2, \ldots, c-1 \\ cv & l = c, c+1, c+2, \ldots . \end{cases}
$$

Algorithm 10.2.2 can then be applied, as is illustrated in the following example.

Example 10.2.14
For an $M/M/c$ service node, define $\rho = \lambda/cv$. Then,

$$
\frac{b(l-1)}{d(l)} = \begin{cases} c\rho/l & l = 1, 2, \ldots, c-1 \\ \rho & l = c, c+1, c+2, \ldots ; \end{cases}
$$

and, with $f'(0) = 1$, from Algorithm 10.2.2,

$$
f'(l) = \begin{cases} (c\rho)^l/l! & l = 0, 1, 2, \ldots, c-1 \\ (c\rho)^c \rho^{l-c}/c! & l = c, c+1, c+2, \ldots . \end{cases}
$$

If $\rho < 1$, then the results in the following theorem are based on evaluating the constant

$$
\alpha = \frac{1}{f(0)}
$$

$$
= \sum_{l=0}^{\infty} f'(l)
$$

$$
= \sum_{l=0}^{c-1} \frac{(c\rho)^l}{l!} + \sum_{l=c}^{\infty} \frac{(c\rho)^c \rho^{l-c}}{c!}
$$

$$
= \sum_{l=0}^{c-1} \frac{(c\rho)^l}{l!} + \frac{(c\rho)^c}{c!}(1 + \rho + \rho^2 + \cdots)
$$

$$= \sum_{l=0}^{c-1} \frac{(c\rho)^l}{l!} + \frac{(c\rho)^c}{(1-\rho)c!}$$

and calculating

$$f(l) = \frac{f'(l)}{\alpha} \qquad l = 0, 1, 2, \ldots.$$

Theorem 10.2.6   For an $M/M/c$ service node having $c > 1$ and $\rho = \lambda/cv < 1$, the steady-state probability of l jobs in the service node is $f(l)$, where, for $l = 0$,

$$f(0) = \left(\frac{(c\rho)^c}{(1-\rho)\,c!} + \sum_{l=0}^{c-1} \frac{(c\rho)^l}{l!} \right)^{-1},$$

and, for $l = 1, 2, \ldots$,

$$f(l) = \begin{cases} \dfrac{(c\rho)^l}{l!} f(0) & l = 1, 2, \ldots, c-1 \\[2ex] \dfrac{(c\rho)^c}{c!} \rho^{l-c} f(0) & l = c, c+1, c+2, \ldots. \end{cases}$$

If $c = 1$, this theorem reduces to Theorem 8.5.3.

Steady-State Statistics.   From the equations in Theorem 10.2.6, for an $M/M/c$ service node, the steady-state expected number in the queue is

$$E[Q] = \sum_{l=c}^{\infty} (l - c) f(l)$$

$$= \sum_{l=c}^{\infty} (l - c) \frac{(c\rho)^c}{c!} \rho^{l-c} f(0)$$

$$= \frac{(c\rho)^c}{c!} f(0) \sum_{l=c}^{\infty} (l - c)\rho^{l-c}$$

$$= \frac{(c\rho)^c}{c!} f(0) \left(\rho + 2\rho^2 + 3\rho^3 + \cdots \right)$$

$$\vdots$$

$$= \frac{(c\rho)^c}{c!} \cdot \frac{\rho}{(1-\rho)^2} f(0).$$

In particular, it can be shown that

$$\lim_{\rho \to 1} E[Q] = \frac{\rho}{1 - \rho}.$$

Therefore, as in the single-server case, we see that, as the steady-state utilization of each server approaches 1, the average number in the queue (and thus in the service node) approaches infinity. This result is consistent with the simulations in Figure 1.2.9.

As in Section 8.5, for an $M/M/c$ service node having $\rho = \lambda/c\nu$, the four steady-state statistics $E[W]$, $E[L]$, $E[D]$, and $E[Q]$ are related by the four equations

$$E[W] = E[D] + \frac{1}{\nu} \qquad E[L] = E[Q] + c\rho \qquad E[L] = \lambda E[W] \qquad E[Q] = \lambda E[D],$$

and any three of these can be calculated whenever the fourth is known. Therefore, from the fact that the steady-state average number in the queue is

$$E[Q] = \frac{(c\rho)^c}{c!} \frac{\rho}{(1-\rho)^2} \left(\frac{(c\rho)^c}{(1-\rho)\,c!} + \sum_{l=0}^{c-1} \frac{(c\rho)^l}{l!} \right)^{-1},$$

the other steady-state statistics for an $M/M/c$ service node can be calculated as

$$E[L] = E[Q] + c\rho \qquad \text{expected number in the node,}$$

$$E[D] = E[Q]/\lambda \qquad \text{expected delay in the queue, and}$$

$$E[W] = E[L]/\lambda \qquad \text{expected wait in the node.}$$

If $\rho \geq 1$, then steady-state statistics do not exist.

Example 10.2.15

For an $M/M/4$ service node having $\lambda = 1.0$ and $\rho = 0.75$ (so that the service rate for each server is $\nu = 1/3$), the steady-state probability of an idle node is

$$f(0) = \left(\frac{(c\rho)^c}{(1-\rho)\,c!} + \sum_{s=0}^{c-1} \frac{(c\rho)^s}{s!} \right)^{-1} = \left(\frac{3^4}{(1/4)\,4!} + \sum_{s=0}^{3} \frac{3^s}{s!} \right)^{-1} = \cdots = \frac{2}{53},$$

and the steady-state average number in the queue is

$$E[Q] = \frac{(c\rho)^c}{c!} \cdot \frac{\rho}{(1-\rho)^2} f(0) = \frac{3^4(3/4)}{4!\,(1/4)^2} \left(\frac{2}{53} \right) = \frac{81}{53} \cong 1.53.$$

Therefore, because $c\rho = 3.0$, the other three steady-state statistics are

$$E[L] = \frac{81}{53} + 3 = \frac{240}{53} \cong 4.53,$$

$$E[D] = \frac{81}{53} \cong 1.53, \text{ and}$$

$$E[W] = \frac{240}{53} \cong 4.53.$$

Algorithm 10.2.3   Given the arrival rate λ, the service rate ν (per server), and the number of servers c, then, if $\rho = \lambda/c\nu < 1$, this algorithm computes the first-order steady-state statistics for an $M/M/c$ service node. The symbols $\bar{q}$, $\bar{l}$, $\bar{d}$, and $\bar{w}$ denote $E[Q]$, $E[L]$, $E[D]$, and $E[W]$.

```
ρ = λ / (c * ν);
sum = 0.0;
s = 0;
t = 1.0;
while (s < c) {
    sum += t;
    s++;
    t *= (c * ρ / s);
}                                                    // at this point t = (cρ)^c/c!
sum += t / (1.0 - ρ);                                // and sum is 1/f(0)
q̄ = t * ρ / ((1.0 - ρ) * (1.0 - ρ) * sum);
l̄ = q̄ + (c * ρ);
d̄ = q̄ / λ;
w̄ = l̄ / λ;
return q̄, l̄, d̄, w̄, ρ;
```

Finite Capacity Multiple-Server Service Node.   An $M/M/c$ service node with *finite* capacity $c \leq k < \infty$ is denoted $M/M/c/k$. As in the special case $c = 1$, k is the capacity of the service node, not the queue. A proof of the following theorem is left as an exercise.

Theorem 10.2.7   For an $M/M/c/k$ service node with $c = 2, 3, \ldots, k$ and $\rho = \lambda/c\nu \neq 1$, the steady-state probability of l jobs in the service node is $f(l)$, where, for $l = 0$,

$$f(0) = \left(\frac{(c\rho)^c}{(1 - \rho)\, c!} \left(1 - \rho^{k-c+1} \right) + \sum_{l=0}^{c-1} \frac{(c\rho)^l}{l!} \right)^{-1},$$

and, for $l = 1, 2, \ldots, k$,

$$f(l) = \begin{cases} \dfrac{(c\rho)^l}{l!} f(0) & l = 1, 2, \ldots, c - 1 \\[2ex] \dfrac{(c\rho)^c}{c!} \rho^{l-c} f(0) & l = c, c + 1, c + 2, \ldots, k. \end{cases}$$

If $\rho = 1$, then

$$f(0) = \left(\frac{c^c}{c!}(k - c + 1) + \sum_{l=0}^{c-1} \frac{c^l}{l!} \right)^{-1}$$

and

$$f(l) = \begin{cases} \dfrac{c^l}{l!} f(0) & l = 1, 2, \ldots, c - 1 \\[2ex] \dfrac{c^c}{c!} f(0) & l = c, c + 1, c + 2, \ldots, k. \end{cases}$$

If $c = 1$, then this theorem reduces to Theorem 10.2.5.

Steady-State Statistics.   Much as in the case of an $M/M/c$ service node, for an $M/M/c/k$ service node, steady-state statistics can be computed from the equations in Theorem 10.2.7—that is, the steady-state number in the service node is

$$E[L] = \sum_{l=0}^{k} l \, f(l)$$

and the steady-state number in the queue is

$$E[Q] = \sum_{l=c}^{k} (l - c) \, f(l).$$

These two equations can be evaluated numerically for specific values of ρ, c, and k. The equations can also be simplified somewhat algebraically, but the resulting expressions are too complex to be useful for most applications.

   Given that $E[L]$ and $E[Q]$ have been evaluated, either numerically or algebraically, the corresponding steady-state wait in the service node and delay in the queue are

$$E[W] = \frac{E[L]}{\big(1 - f(k)\big)\lambda} \qquad \text{and} \qquad E[D] = \frac{E[Q]}{\big(1 - f(k)\big)\lambda},$$

respectively.

Infinitely Many Servers.   The $M/M/c$ steady-state results in Theorem 10.2.6 are valid for any $c \geq 1$. If the number of servers is (effectively) infinite ($c \to \infty$)—or, equivalently, if the jobs are *self-serving*, each with a expected service time of $1/\nu$—then

$$\frac{b(l-1)}{d(l)} = \frac{\lambda}{l\nu} \qquad l = 1, 2, 3, \ldots;$$

and, with $f'(0) = 1$, from Algorithm 10.2.2,

$$f'(l) = \frac{(\lambda/\nu)^l}{l!} \qquad l = 0, 1, 2, \ldots.$$

In this case, the normalizing constant is

$$\alpha = \frac{1}{f(0)} = \sum_{l=0}^{\infty} f'(l) = \sum_{l=0}^{\infty} \frac{(\lambda/\nu)^l}{l!} = \exp(\lambda/\nu).$$

Therefore, in steady state, the probability of l jobs in an $M/M/\infty$ service node is

$$f(l) = \frac{\exp(-\lambda/\nu)(\lambda/\nu)^l}{l!} \qquad l = 0, 1, 2, \ldots.$$

Thus, we have proven the following theorem valid for *any* value of λ/ν.

> **Theorem 10.2.8** In steady state, the number of jobs in an $M/M/\infty$ service node is a *Poisson* (λ/ν) random variable.

The $M/M/\infty$ terminology is potentially confusing. An $M/M/\infty$ "queue" will in fact *never* have a queue; the number of servers is infinite, and so *any* arrival rate, no matter how large, can be accommodated. The steady-state expected number of jobs in an $M/M/\infty$ service node is $E[L] = \lambda/\nu$, and this ratio can be arbitrarily large.

It is interesting to compare the two limiting cases associated with an $M/M/c$ service node: $c = 1$ and $c = \infty$.

- If $\lambda/\nu < 1$, then the steady-state number in an $M/M/1$ service node is a *Geometric* (λ/ν) random variable.
- For any value of λ/ν, the steady-state number in an $M/M/\infty$ service node is a *Poisson* (λ/ν) random variable.

Except for these two limiting cases, there is no simple discrete-random-variable characterization for the steady-state number of jobs in an $M/M/c$ service node.

10.2.4 Discrete-Time Birth–Death Processes

For some stochastic models, time is a *continuous* variable with, for example, possible values $t \geq 0$; for other stochastic models, time is a *discrete* variable with, for example, possible values $t = 0, 1, 2, \ldots$. To make the distinction between discrete-time and continuous-time models clearer, this section concludes with a discussion of the discrete-time analog of the continuous-time birth–death-process model discussed previously.

As in the continuous-time case, a discrete-time birth–death process is defined in terms of a discrete state-space $\mathcal{X}$ that may be either finite or infinite. Without loss of generality, we assume that, if the state-space is *finite*, then

$$\mathcal{X} = \{0, 1, 2, \ldots, k\}$$

for some positive integer k; but, if the state-space is *infinite*, then

$$\mathcal{X} = \{0, 1, 2, \ldots\}.$$

Formulation.   There are two equivalent formulations of a discrete-time birth–death process. We begin with a three-state "birth–death–remain" formulation that is intuitive, but not directly analogous to Definition 10.2.2. We then show that this formulation is equivalent to an alternative formulation that is analogous to the definition of a continuous-time birth–death process.

> **Definition 10.2.4** Let X be a discrete random variable, indexed by time t as $X(t)$, that evolves in time as follows.
>
> - $X(t) \in \mathcal{X}$ for all $t = 0, 1, 2, \ldots$.
> - State transitions can occur only at the discrete times $t = 0, 1, 2, \ldots$.
> - When a state transition *can* occur, then the random variable X either *will* shift from its current state x to state $x + 1 \in \mathcal{X}$ with probability $b(x)$ or will shift from state x to state $x - 1 \in \mathcal{X}$ with probability $d(x)$ or will remain in its current state x with probability $c(x) = 1 - b(x) - d(x)$.
>
> The shift from state x to $x + 1$ is called a *birth*; the shift from x to $x - 1$ is called a *death*. The stochastic process defined by $X(t)$ is called a *discrete-time birth–death process*.

A discrete-time birth–death process is completely characterized by the initial state $X(0) \in \mathcal{X}$ at $t = 0$ and by the functions $b(x)$, $c(x)$, $d(x)$ defined for all $x \in \mathcal{X}$ such that

$$b(x) \geq 0, \qquad c(x) \geq 0, \qquad d(x) \geq 0, \qquad b(x) + c(x) + d(x) = 1,$$

with the understanding that $d(0) = 0$, because a death can't occur when X is in the state $x = 0$. Moreover, if the state-space is finite, then $b(k) = 0$, because a birth can't occur when X is in the state $x = k$.

Because the $b(x)$, $c(x)$, and $d(x)$ probabilities sum to 1, it is sufficient to specify any two of them and then solve for the third. So, for example, one could specify the birth probability $b(x)$ and death probability $d(x)$ only. The "remain in state" probability $c(x)$ is then defined by

$$c(x) = 1 - b(x) - d(x)$$

—provided, of course, that $c(x) \geq 0$. The significance of $c(x)$ is discussed later.

As in the continuous-time case, a discrete-time birth–death process has no memory of its past—this is the Markovian assumption. The values of the future states $X(t + 1)$, $X(t + 2)$, $\ldots$ depend *only* on the value of the current state $X(t)$. Moreover, we can use a modified version of Algorithm 10.2.1 to simulate a discrete-time birth–death process, as follows.

> **Algorithm 10.2.4** Given the state-space $\mathcal{X}$, the initial state $X(0) \in \mathcal{X}$, and the probabilities $b(x) \geq 0$, $c(x) \geq 0$, and $d(x) \geq 0$ defined for all $x \in \mathcal{X}$ with $b(x) + c(x) + d(x) = 1$, this algorithm simulates the time evolution of the associated discrete-time birth–death process for $t = 0, 1, 2, \ldots, \tau$.
>
> ```
> t = 0;
> x = X(0);
> while (t < τ) {
>     t++;
>     u = Random();
> ```

```
    if (u < d(x))
        s = -1;
    else if (u < d(x) + c(x))
        s = 0;
    else
        s = 1;
    x += s;
}
```

Alternative Formation. Note that, if $c(x) = 0$, then, when the discrete-time birth–death process enters state x, it will remain in this state for exactly 1 time unit—that is, if $c(x) = 0$ and $X(t) = x$ then, with probability 1, $X(t + 1) \neq x$. If, however, $c(x) > 0$, then, once state x is entered, the discrete-time birth–death process will remain in this state for a (discrete) random amount of time. Indeed, because of independence, the probability of remaining in this state for exactly $n + 1$ time units is

$$c^n(x)\left(1 - c(x)\right) \qquad n = 0, 1, 2, \ldots.$$

Therefore, if $c(x) > 0$ and $X(t) = x$, then the amount of time the discrete-time birth–death process will remain in state x is $1 + T$, where T is a *Geometric*$(c(x))$ random variable (per Definition 6.4.3). It follows, in this case, that, when the process enters state x, the *expected* time the process will remain in state x is

$$\mu(x) = E[1 + T]$$
$$= 1 + E[T]$$
$$= 1 + \frac{c(x)}{1 - c(x)}$$
$$= \frac{1}{1 - c(x)}.$$

Equivalently, $1 - c(x) = b(x) + d(x)$, and so the expected time in state x is

$$\mu(x) = \frac{1}{b(x) + d(x)}.$$

Moreover, when a state change does occur, it will be a birth with probability

$$p(x) = \frac{b(x)}{b(x) + d(x)},$$

but a death with probability $1 - p(x)$.*

---

*It is important to appreciate the difference between $p(x)$ and $b(x)$. By definition,

$$b(x) = \Pr\left(X(t + 1) = x + 1 \mid X(t) = x\right)$$

The following theorem summarizes the previous discussion. Recognize that, as an alternative to Definition 10.2.4, this theorem provides an equivalent characterization of a discrete-time birth–death process. The conceptual advantage of this alternative characterization is that it parallels the *continuous-time* birth–death process definition (Definition 10.2.2) and illustrates nicely that the (memoryless) *Geometric* random variable is the discrete analog of the continuous *Exponential* random variable.

Theorem 10.2.9   Given $p(x) = b(x)/\big(b(x) + d(x)\big)$ and $\mu(x) = 1/\big(b(x) + d(x)\big)$ for all $x \in \mathcal{X}$, if X is a discrete random variable that evolves in time as follows, then the stochastic process defined by $X(t)$ is a discrete-time birth–death process.

- At any time, $t = 0, 1, 2, \ldots, X(t) \in \mathcal{X}$.
- State transitions can occur only at the discrete times $t = 0, 1, 2, \ldots$; and, if x is the state of X at time t, then X will remain in this state for a random time $1 + T$, where T is *Geometric* with mean $\mu(x) - 1$.
- When a state transition occurs, then, independent of the time spent in the current state, the random variable X either will shift from its current state x to state $x + 1 \in \mathcal{X}$ with probability $p(x)$ or will shift from state x to state $x - 1 \in \mathcal{X}$ with probability $1 - p(x)$.

If the process is specified in terms of $b(x)$ and $d(x)$, as in Definition 10.2.4, then $p(x)$ and $\mu(x)$ are defined by the equations in Theorem 10.2.9. On the other hand, if the process is specified in terms of $p(x)$ and $\mu(x)$ directly, then $b(x)$ and $d(x)$ are defined by

$$b(x) = \frac{p(x)}{\mu(x)} \qquad d(x) = \frac{1 - p(x)}{\mu(x)}.$$

The following algorithm is an implementation of Theorem 10.2.9. This algorithm, which is the discrete analog of Algorithm 10.2.1, is stochastically equivalent to Algorithm 10.2.4. Note that, because the parameter that defines a *Geometric* random variable is a probability, not a mean, Algorithm 10.2.5 is stated in terms of the "remain in state" probability $c(x)$, which can be shown to be related to $\mu(x)$ by the equation

$$c(x) = \frac{\mu(x) - 1}{\mu(x)} \qquad x \in \mathcal{X},$$

with the understanding that $\mu(x) \geq 1$ for all $x \in \mathcal{X}$.

---

is the probability of a state transition from state x at time t to state $x + 1$ at time $t + 1$. Instead,

$$p(x) = \Pr\big(X(t + 1) = x + 1 \mid (X(t) = x) \text{ and } (X(t + 1) \neq x)\big)$$

is $b(x)$ *conditioned* on the knowledge that there was a state change. This second probability can't be smaller than the first; indeed, if $c(x) > 0$, then $p(x) > b(x)$.

Algorithm 10.2.5   Given the state-space $\mathcal{X}$, the initial state $X(0) \in \mathcal{X}$, and the probabilities $p(x)$, $c(x)$ defined for all $x \in \mathcal{X}$ by the equations $p(x) = b(x)/\bigl(b(x) + d(x)\bigr)$ and $c(x) = 1 - b(x) - d(x)$, this algorithm simulates the time evolution of the associated discrete-time birth–death process for $t = 0, 1, 2, \ldots, \tau$.

```
t = 0;
x = X(0);
while (t < τ) {
    t += 1 + Geometric(c(x));
    x += 2 * Bernoulli(p(x)) - 1;
}
```

Example 10.2.16

As an example of a discrete-time birth–death process with an infinite state space, consider a data structure (e.g., a linked list) for which one item is either inserted or deleted each time the data structure is accessed. For each access, the probability of an insertion is p, independent of past accesses; the probability of a deletion is $q = 1 - p$, provided that the data structure is not empty. If accesses are indexed $t = 0, 1, 2, \ldots$, and $X(t)$ is the number of items in the data structure after the tth access, then $X(t)$ is a discrete-time birth–death process. We can use the characterization of this stochastic process in Definition 10.2.4 to write

$$b(x) = p \quad x = 0, 1, 2, \ldots \quad \text{and} \quad d(x) = \begin{cases} 0 & x = 0 \\ q & x = 1, 2, \ldots . \end{cases}$$

Equivalently, the characterization of this stochastic process in Theorem 10.2.9 can be used to write

$$p(x) = \begin{cases} 1 & x = 0 \\ p & x = 1, 2, \ldots \end{cases} \quad \text{and} \quad \mu(x) = \begin{cases} 1/p & x = 0 \\ 1 & x = 1, 2, \ldots . \end{cases}$$

In either case,

$$c(x) = \begin{cases} q & x = 0 \\ 0 & x = 1, 2, \ldots . \end{cases}$$

Flow-Balance Equations.   As in the continuous-time case, the expected times $\mu(x)$ can be converted to rates via the equation

$$\lambda(x) = \frac{1}{\mu(x)} \qquad x \in \mathcal{X}.$$

These rates can then be multiplied by $p(x)$ and $1 - p(x)$, respectively, to define the birth and death *rates*,

$$b(x) = \lambda(x)\, p(x) \qquad \text{and} \qquad d(x) = \lambda(x)\bigl(1 - p(x)\bigr),$$

which, for a discrete-time birth–death process, are numerically equivalent to the birth and death *probabilities*, as specified in Definition 10.2.4. It follows that the stochastic process can be represented by a labeled directed graph, as illustrated in Figure 10.2.9.

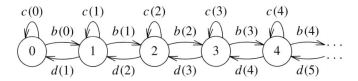

Figure 10.2.9 Transition probabilities.

The steady-state probability $f(x)$ of finding the process in state x can be found, as in the continuous-time case, by solving the flow-balance equations

$$d(x + 1) f(x + 1) = b(x) f(x) \qquad x \in \mathcal{X}.$$

Example 10.2.17
As a continuation of Example 10.2.16, consider the labeled directed-graph representation of the corresponding discrete-time birth–death stochastic process given in Figure 10.2.10.

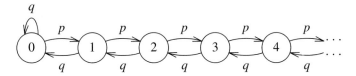

Figure 10.2.10 Transition probabilities.

The corresponding flow-balance equations are

$$q\, f(x + 1) = p\, f(x) \qquad x = 0, 1, 2, \ldots.$$

Except for cosmetic differences ($p \leftrightarrow \lambda$, $q \leftrightarrow \nu$) these are *exactly* the same equations as those characterizing the steady-state number in an $M/M/1$ service node. Therefore, it follows that, if $\rho < 1$,

$$\rho = \frac{p}{q} = \frac{p}{1 - p},$$

and that, if $\rho < 1$,

$$f(x) = (1 - \rho)\, \rho^x \qquad x = 0, 1, 2, \ldots.$$

Thus we see that, if the probability of insertion p is less than the probability of deletion q, the steady-state number of items in the data structure is a *Geometric*(ρ) random variable. In particular, in this case, the expected number of items in the data structure is

$$E[X] = \frac{\rho}{1 - \rho} = \frac{p}{q - p}.$$

As a specific example, if $(p, q) = (0.48, 0.52)$ then $E[X] = 12$ and the steady-state distribution of the size of the data structure is as illustrated in Figure 10.2.11.

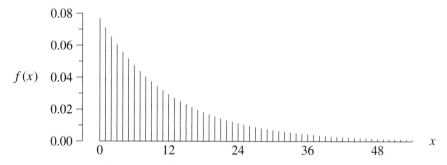

Figure 10.2.11   Steady-state size of the data structure.

If p is greater than or equal to q, then no steady state can be achieved, because the expected length of the data structure will grow without bound. If, however, the size of the data structure is physically restricted to be no larger than $k < \infty$, then the state space is finite and, as for an $M/M/1/k$ service-node model, a steady state will be possible.

10.2.5 Exercises

10.2.1   Suppose that T_1 and T_2 are independent *Exponential* random variables having means μ_1 and μ_2 respectively. Define $T = \max\{T_1, T_2\}$. (*a*) What is the pdf of T? (*b*) What is $E[T]$? (*c*) Use Monte Carlo simulation to verify the correctness of your math for the case $\mu_1 = 1$, $\mu_2 = 2$.

10.2.2   Complete the derivation of Theorem 10.2.2 by proving that

$$\frac{1}{\mu'} \int_0^\infty \left(1 - \exp(-t'/\mu_1)\right) \exp(-t'/\mu') \, dt' = \frac{\mu}{\mu_1}.$$

10.2.3   (*a*) If $T_1, T_2, \ldots, T_n$ is an independent sequence of *Exponential* (μ) random variables and $T = \max\{T_1, T_2, \ldots, T_n\}$, what is the cdf and pdf of T? (*b*) Use Monte Carlo simulation to support the correctness of this pdf for the case $n = 5$, $\mu = 1$. (*c*) Construct an algorithm that will generate the random variate T with just one call to `Random`.

10.2.4   (*a*) Use Algorithm 10.2.1 and Example 10.2.8 to construct (yet another) program to simulate an $M/M/1$ service node. (*b*) How would you verify that this program is correct? (*c*) Illustrate for the case $\lambda = 3$, $\nu = 4$.

10.2.5   (*a*) Use Algorithm 10.2.1 and Example 10.2.9 to construct a program to simulate an $M/M/1/k$ service node. This program should estimate steady-state values for the expected number in the service node and the steady-state probabilities $f(l)$ for $l = 0, 1, 2, \ldots, k$. (*b*) How would you verify that the program is correct? (*c*) Illustrate for the case $\lambda = 3$, $\nu = 4$, $k = 4$.

10.2.6   (*a*) Use Algorithm 10.2.3 to construct a program to calculate the first-order $M/M/c$ steady-state statistics, given values of λ and ν and a range of c values. (*b*) Calculate and print a table of these four steady-state statistics for the case $\lambda = 30$, $\nu = 2$ and $c = 20, 19, 18, 17, 16, 15$. (*c*) Comment.

10.2.7   If $\rho = 1$, derive the steady-state pdf for an $M/M/1/k$ service node.

10.2.8   Let $Q(t)$ be the number of jobs in the *queue* of an $M/M/c$ service node at time $t > 0$. Derive an equation for the steady-state ($t \to \infty$) pdf $\Pr(Q(t) = q)$ for $q = 0, 1, 2, \ldots$.

10.2.9   Use Algorithm 10.2.1 to verify by simulation that the results in Theorem 10.2.7 are correct in the case $c = 4$, $\lambda = 8$, $\nu = 2$, $k = 10$.

10.2.10   (*a*) Find an easily computed condition involving c, λ, and ν that will characterize when the steady-state distribution of the number in an $M/M/c$ service node and an $M/M/\infty$ service node are essentially identical. (The condition "c is real big" is not enough.) (*b*) Present simulation data in support of your condition for the case $\lambda = 10$, $\nu = 1$.

10.2.11   A classic example* of a discrete-time finite-state birth–death process is based on a game of chance played as follows. Two players, $\mathcal{P}$ and $\mathcal{Q}$, start with a fixed amount of money, k (dollars), with a belonging to $\mathcal{P}$ and $k - a$ to $\mathcal{Q}$. On each play of the game, either $\mathcal{P}$ will win one dollar (from $\mathcal{Q}$) with probability p, independent of the result of previous games, or $\mathcal{Q}$ will win one dollar with probability $q = 1 - p$. *Play continues until one player has no more money.* Let $f(x)$ represent the steady-state probability that player $\mathcal{P}$ has x dollars. (*a*) Set up the flow-balance equations for this process, and argue why $f(x) = 0$ for all $x = 1, 2, \ldots, k - 1$. (*b*) For the case $k = 1000$ with $p = 0.51$, estimate $f(0)$ and $f(k)$ by simulation to within ± 0.01 with 95% confidence for each of the cases $a = 10, 20, 30, 40, 50$. (*c*) Also, estimate the expected number of plays. (*d*) Comment.

10.2.12   As a continuation of Exercise 10.2.11, derive equations for $f(0)$ and $f(k)$. *Hint*: To derive the equations, define $\Pr(a) = f(k)$ as the probability that player $\mathcal{P}$ ultimately ends up with all the money, given that $\mathcal{P}$ has a dollars initially, and then argue that the 3-term recursion equation

$$\Pr(a) = q \Pr(a - 1) + p \Pr(a + 1) \qquad a = 1, 2, \ldots, k - 1$$

is valid, with the boundary conditions $\Pr(0) = 0$, $\Pr(k) = 1$.

10.2.13   Prove Theorem 10.2.7.

---

*This problem is well known in probability as the *gambler's ruin* problem. A special case of the gambler's ruin problem, known as the problem of *duration of play*, was proposed to Dutch mathematician Christian Huygens (1629–1695) by French mathematician Pierre de Fermat (1601–1665) in 1657. The general form of the gambler's ruin problem was solved by mathematician Jakob Bernoulli (1654–1705) and published eight years after his death. Ross (2006, pages 95–99) considers the problem, using the analytic approach to probability.

10.2.14   Consider whether an *Exponential*(μ) distribution is an appropriate probabilistic model for the lifetimes of the following entities: (*a*) fuse, (*b*) candle, (*c*) cat, (*d*) light bulb, (*e*) software company. Base your discussion of the appropriateness of an exponential model on the memoryless property (i.e., would a used entity have an identical lifetime distribution to a new entity?).

10.3 FINITE-STATE MARKOV CHAINS

This section introduces a particular type of stochastic process known as a *finite-state Markov chain*. As we will see, a finite-state birth–death process (see Section 10.2) is a special case of a finite-state Markov chain. The primary distinction is that, in a birth–death process, state transitions are restricted to neighboring states; but in a Markov chain, there is no such restriction. Because a much larger set of state transitions is possible, a Markov chain is a much richer stochastic-process model. There is, however, a price to pay for this increased modeling flexibility—an added level of mathematical sophistication is required to analyze steady-state behavior.

10.3.1 Discrete-Time Finite-State Markov Chains

In reverse order of presentation relative to Section 10.2, in this section we consider *discrete-time* Markov chains before *continuous-time* Markov chains.

Definition 10.3.1   Let X be a discrete random variable, indexed by time t as $X(t)$, that evolves in time as follows.

- $X(t) \in \mathcal{X}$ for all $t = 0, 1, 2, \ldots$.
- State transitions can occur only at the discrete times $t = 0, 1, 2, \ldots$; and, at these times, the random variable $X(t)$ will shift from its current state $x \in \mathcal{X}$ to another state, say $x' \in \mathcal{X}$, with fixed probability $p(x, x') = \Pr\big(X(t+1) = x' \mid X(t) = x\big) \geq 0$.

If $|\mathcal{X}| < \infty$, the stochastic process defined by $X(t)$ is called a *discrete-time finite-state Markov chain*. Without loss of generality, throughout this section we assume that the finite state space is $\mathcal{X} = \{0, 1, 2, \ldots, k\}$, where k is a finite, but perhaps quite large, integer.

A discrete-time finite-state Markov chain is completely characterized by the initial state at $t = 0$, $X(0)$ and by the function $p(x, x')$ defined for all $(x, x') \in \mathcal{X} \times \mathcal{X}$. When the stochastic process leaves the state x, the transition must be either to state $x' = 0$ with probability $p(x, 0)$, or to state $x' = 1$ with probability $p(x, 1), \ldots$, or to state $x' = k$ with probability $p(x, k)$, and the sum of these probabilities must be 1:

$$\sum_{x'=0}^{k} p(x, x') = 1 \qquad x = 0, 1, 2, \ldots, k.$$

Because $p(x, x')$ is independent of t for all (x, x'), the Markov chain is said to be *homogeneous* or *stationary*.

Definition 10.3.2   The *state-transition probability* $p(x, x')$ represents the probability of a transition *from* state x *to* state x'. The corresponding $(k + 1) \times (k + 1)$ matrix

$$
\mathbf{p} = \begin{bmatrix}
p(0, 0) & p(0, 1) & \cdots & p(0, k) \\
p(1, 0) & p(1, 1) & \cdots & p(1, k) \\
\vdots & \vdots & \ddots & \vdots \\
p(k, 0) & p(k, 1) & \cdots & p(k, k)
\end{bmatrix},
$$

with elements $p(x, x')$, is called the *state-transition matrix*.

The elements of the state-transition matrix $\mathbf{p}$ are nonnegative, and the elements of each row sum to 1.0. Hence, each row of the state-transition matrix defines a pdf for a discrete random variable. (A matrix with these properties is said to be a *stochastic* matrix.) The diagonal elements of the state-transition matrix need not be zero. If $p(x, x) > 0$, then, when $X(t) = x$, it will be possible for the next state "transition" to leave the state of the system unchanged [i.e., $X(t + 1) = x$]. This is sometimes called *immediate feedback*.

Example 10.3.1
Consider a Markov chain having four possible states ($k = 3$, $\mathcal{X} = \{0, 1, 2, 3\}$) and the state-transition matrix

$$
\mathbf{p} = \begin{bmatrix}
0.3 & 0.2 & 0.4 & 0.1 \\
0.0 & 0.5 & 0.3 & 0.2 \\
0.0 & 0.4 & 0.0 & 0.6 \\
0.4 & 0.0 & 0.5 & 0.1
\end{bmatrix}.
$$

For example, the probability of a transition from state 2 to 3 is $p(2, 3) = 0.6$, and the probability of immediate feedback when the system is in state 1 is $p(1, 1) = 0.5$.

As illustrated in the following example, a finite-state Markov chain can be represented as a directed graph.

- The nodes of this graph are the possible states (the state space $\mathcal{X}$).
- The labeled arcs, including the immediate-feedback circular arcs, are defined by the state-transition probabilities $p(x, x')$.

The sum of the probabilities associated with all of the arcs emanating from any node, including the circular (feedback) arc, must be 1.0. Why?

Example 10.3.2
A "probability" directed-graph representation of the discrete-time Markov chain in Example 10.3.1 is shown in Figure 10.3.1.

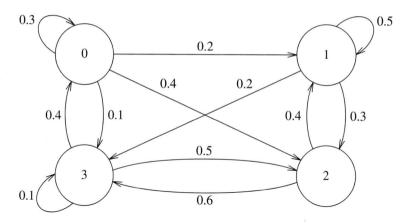

Figure 10.3.1   State-transition diagram for a discrete-time, finite-state Markov chain.

Definition 10.3.3   Define the *cumulative* state-transition function $P(x, x')$ to be

$$P(x, x') = p(x, 0) + p(x, 1) + \cdots + p(x, x') \qquad (x, x') \in \mathcal{X} \times \mathcal{X}.$$

Clearly, $P(x, x')$ is the probability of a transition *from* state x *to* a new state between 0 and x' inclusive. The associated $(k + 1) \times (k + 1)$ matrix

$$\mathbf{P} = \begin{bmatrix} P(0,0) & P(0,1) & \cdots & 1.0 \\ P(1,0) & P(1,1) & \cdots & 1.0 \\ \vdots & \vdots & \ddots & \vdots \\ P(k,0) & P(k,1) & \cdots & 1.0 \end{bmatrix}$$

with elements $P(x, x')$ is called the *cumulative* state-transition matrix.

Example 10.3.3
Relative to Example 10.3.2, the state-transition matrix and cumulative state-transition matrix are

$$\mathbf{p} = \begin{bmatrix} 0.3 & 0.2 & 0.4 & 0.1 \\ 0.0 & 0.5 & 0.3 & 0.2 \\ 0.0 & 0.4 & 0.0 & 0.6 \\ 0.4 & 0.0 & 0.5 & 0.1 \end{bmatrix} \qquad \text{and} \qquad \mathbf{P} = \begin{bmatrix} 0.3 & 0.5 & 0.9 & 1.0 \\ 0.0 & 0.5 & 0.8 & 1.0 \\ 0.0 & 0.4 & 0.4 & 1.0 \\ 0.4 & 0.4 & 0.9 & 1.0 \end{bmatrix}.$$

Given $\mathbf{p}$, the following algorithm can be used to compute the cumulative state-transition matrix.

Algorithm 10.3.1 This algorithm constructs the *cumulative* state-transition matrix $\mathbf{P}$ from the state-transition matrix $\mathbf{p}$.

```
for (x = 0; x < k + 1; x++) {              // loop over rows
   P[x, 0] = p[x, 0];            // first column elements identical
   for (x' = 1; x' < k; x'++)                // loop over columns
      P[x, x'] = P[x, x' - 1] + p[x, x'];       // pdf → cdf
   P[x, k] = 1.0;                      // last column elements
}
return P;
```

Note that the elements in each row of a cumulative state-transition matrix are monotone increasing. Also, $P[x, k] = 1.0$ for all $x \in \mathcal{X}$, a condition that is enforced in Algorithm 10.3.1 to guard against the possibility that accumulated floating-point round-off error would yield a computed value of $P[x, k]$ that is close to, but slightly less or more than 1.0. This is important, because a computed value of $P[x, k] < 1.0$ would create the possibility that Algorithm 10.3.2 (to follow) might fail.

If $\mathbf{p}$ is a stochastic matrix, then each row represents the pdf of a discrete random variable. Specifically, the x-row of $\mathbf{p}$ is the pdf of the "to" state when the "from" state is x. Correspondingly, the rows of the cumulative state-transition matrix $\mathbf{P}$ are cdfs. Because of this, the cumulative state-transition matrix can be used with the discrete idf method of random-variate generation (Algorithm 6.2.1) to construct the following algorithm for the function NextState.

Algorithm 10.3.2 Given the cumulative state-transition matrix $\mathbf{P}$ and $x \in \mathcal{X}$, this algorithm is an implementation of the function $x' = \text{NextState}(x)$ that uses inversion.

```
u =  Random();
x' = 0;
while (P[x, x'] <= u)
   x'++;
return x';
```

Because of the simple linear search in Algorithm 10.3.2, the $O(k)$ time complexity of the function NextState could be a problem if $|\mathcal{X}| = k + 1$ is large. For larger values of k, Algorithm 6.2.2 or, perhaps, a binary search should be used instead. Also, if k is large, it is quite likely that the state-transition matrix $\mathbf{p}$ will be *sparse*—each node in the associated digraph will have only a few adjacent nodes. If k is large and $\mathbf{p}$ is sparse, then a two-dimensional array is *not* the right data structure to use. Instead, a more appropriate data structure, for example a one-dimensional array of pointers to linked lists, and an associated idf search algorithm are required. In any case, given the function NextState, the following algorithm simulates the time evolution of the corresponding discrete-time finite-state Markov chain.

Algorithm 10.3.3   Given the initial state $X(0) \in \mathcal{X}$, this algorithm simulates the time evolution of the associated discrete-time finite-state Markov chain for $t = 0, 1, 2, \ldots, \tau$.

```
t = 0;
x = X(0);
while (t < τ) {
    t++;
    x = NextState(x);
}
```

Example 10.3.4

Using the state-transition matrix from Example 10.3.3, the initial state $X(0) = 1$, and $\tau = 100\,000$, Algorithm 10.3.3 produced the following output:

state	0	1	2	3
proportion	0.160	0.283	0.284	0.273

The second row of this table is the proportion of time spent in each state. Equivalently, because τ is large, this row is an *estimate* of the steady-state probability of finding the Markov chain in each state.

For all $x \in \mathcal{X}$, let

$$f(x, t) = \Pr\big(X(t) = x\big) \qquad t = 0, 1, 2, \ldots$$

be the probability that the Markov chain is in state x at time t. In other words, $f(x, t)$ is the pdf of $X(t)$. We assume that the *initial* distribution [the pdf of $X(0)$],

$$f(x, 0) = \Pr\big(X(0) = x\big) \qquad x \in \mathcal{X},$$

is known. The issue now is how to use this initial distribution and the state-transition matrix **p** to construct the pdf of $X(t)$ for $t = 1, 2, \ldots$. To do this, it is convenient (and conventional) to define these pdfs as a sequence of *row* vectors

$$\mathbf{f}_t = \begin{bmatrix} f(0, t) & f(1, t) & \ldots & f(k, t) \end{bmatrix} \qquad t = 0, 1, 2, \ldots.$$

In particular, $\mathbf{f}_0$ is the known initial distribution.

Chapman–Kolmogorov Equation.   By using matrix/vector notation, from the law of total probability we can write the pdf of $X(1)$ as

$$f(x', 1) = \Pr\big(X(1) = x'\big)$$

$$= \sum_{x=0}^{k} \Pr\big(X(0) = x\big) \Pr\big(X(1) = x' \mid X(0) = x\big)$$

$$= \sum_{x=0}^{k} f(x, 0)\, p(x, x') \qquad x' \in \mathcal{X},$$

which is equivalent to the matrix equation

$$\mathbf{f}_1 = \mathbf{f}_0\,\mathbf{p}.$$

Similarly, it follows that the pdf of $X(2)$ is

$$\mathbf{f}_2 = \mathbf{f}_1\,\mathbf{p} = (\mathbf{f}_0\,\mathbf{p})\,\mathbf{p} = \mathbf{f}_0\,\mathbf{p}^2$$

and (by induction) that the pdf of $X(t)$ is

$$\mathbf{f}_t = \mathbf{f}_0\,\mathbf{p}^t \qquad\qquad t = 1, 2, 3, \ldots.$$

This last matrix/vector equation, know as the *Chapman–Kolmogorov equation*, is of fundamental importance—the pdf of $X(t)$ is constructed by raising the state-transition matrix to the tth power, then multiplying the result, on the left, by the row vector corresponding to the pdf of $X(0)$.

Example 10.3.5

As a continuation of Examples 10.3.1 and 10.3.2, it can be verified that

$$\mathbf{p} = \begin{bmatrix} 0.300 & 0.200 & 0.400 & 0.100 \\ 0.000 & 0.500 & 0.300 & 0.200 \\ 0.000 & 0.400 & 0.000 & 0.600 \\ 0.400 & 0.000 & 0.500 & 0.100 \end{bmatrix}$$

$$\mathbf{p}^2 = \begin{bmatrix} 0.130 & 0.320 & 0.230 & 0.320 \\ 0.080 & 0.370 & 0.250 & 0.300 \\ 0.240 & 0.200 & 0.420 & 0.140 \\ 0.160 & 0.280 & 0.210 & 0.350 \end{bmatrix}$$

$$\mathbf{p}^4 = \begin{bmatrix} 0.149 & 0.296 & 0.274 & 0.282 \\ 0.148 & 0.297 & 0.279 & 0.277 \\ 0.170 & 0.274 & 0.311 & 0.245 \\ 0.150 & 0.295 & 0.269 & 0.287 \end{bmatrix}$$

$$\mathbf{p}^8 = \begin{bmatrix} 0.155 & 0.290 & 0.284 & 0.272 \\ 0.155 & 0.290 & 0.284 & 0.271 \\ 0.156 & 0.289 & 0.285 & 0.270 \\ 0.155 & 0.290 & 0.284 & 0.272 \end{bmatrix}.$$

If, as in Example 10.3.4, the initial state is $X(0) = 1$ with probability 1, then, from the Chapman–Kolmogorov equation, it follows that

$$\mathbf{f}_0 = \begin{bmatrix} 0.000 & 1.000 & 0.000 & 0.000 \end{bmatrix}$$

$$\mathbf{f}_1 = \mathbf{f}_0\,\mathbf{p}^1 = \begin{bmatrix} 0.000 & 0.500 & 0.300 & 0.200 \end{bmatrix}$$

$$\mathbf{f}_2 = \mathbf{f}_0\,\mathbf{p}^2 = \begin{bmatrix} 0.080 & 0.370 & 0.250 & 0.300 \end{bmatrix}$$

$$\mathbf{f}_4 = \mathbf{f}_0\,\mathbf{p}^4 = \begin{bmatrix} 0.148 & 0.297 & 0.279 & 0.277 \end{bmatrix}$$

$$\mathbf{f}_8 = \mathbf{f}_0\,\mathbf{p}^8 = \begin{bmatrix} 0.155 & 0.290 & 0.284 & 0.271 \end{bmatrix}.$$

If, instead, the initial state could be any of the four possible values $X(0) = 0, 1, 2, 3$ with equal probability, then

$$\mathbf{f}_0 = \begin{bmatrix} 0.250 & 0.250 & 0.250 & 0.250 \end{bmatrix}$$

$$\mathbf{f}_1 = \mathbf{f}_0\,\mathbf{p}^1 = \begin{bmatrix} 0.175 & 0.275 & 0.300 & 0.250 \end{bmatrix}$$

$$\mathbf{f}_2 = \mathbf{f}_0\,\mathbf{p}^2 = \begin{bmatrix} 0.153 & 0.293 & 0.278 & 0.278 \end{bmatrix}$$

$$\mathbf{f}_4 = \mathbf{f}_0\,\mathbf{p}^4 = \begin{bmatrix} 0.154 & 0.290 & 0.283 & 0.273 \end{bmatrix}$$

$$\mathbf{f}_8 = \mathbf{f}_0\,\mathbf{p}^8 = \begin{bmatrix} 0.155 & 0.290 & 0.284 & 0.271 \end{bmatrix}.$$

In either case, we see that, for increasing t, the pdf vector $\mathbf{f}_t$ converges rapidly to a pdf that is consistent with the estimated steady-state pdf in Example 10.3.4.

Matrix Power Method. It follows immediately from the Chapman–Kolmogorov equation that, if the steady-state pdf vector $\lim_{t \to \infty} \mathbf{f}_t = \mathbf{f}_\infty$ exists, then the matrix

$$\mathbf{p}^\infty = \lim_{t \to \infty} \mathbf{p}^t$$

must also exist. Example 10.3.5 points to the fact that the matrix $\mathbf{p}^\infty$ has a very special structure—all rows are equal to the steady-state pdf vector. The reason for this is that the equation

$$\mathbf{f}_\infty = \mathbf{f}_0\,\mathbf{p}^\infty$$

must be valid for *any* initial pdf vector $\mathbf{f}_0$. In particular, if the initial state of the Markov chain is $X(0) = 0$ with probability 1, then the initial pdf vector is

$$\mathbf{f}_0 = \begin{bmatrix} 1 & 0 & 0 & \dots & 0 \end{bmatrix},$$

and $\mathbf{f}_0\,\mathbf{p}^\infty = \mathbf{f}_\infty$ will be the *first* (0th) row of $\mathbf{p}^\infty$. Similarly, if $X(0) = 1$ with probability 1, then

$$\mathbf{f}_0 = \begin{bmatrix} 0 & 1 & 0 & \dots & 0 \end{bmatrix},$$

and we see that the *second* row of $\mathbf{p}^\infty$ must also be $\mathbf{f}_\infty$. This argument applies to all rows of $\mathbf{p}^\infty$ and so proves the following theorem.

Theorem 10.3.1 For a discrete-time finite-state Markov chain having state-transition matrix $\mathbf{p}$, if a unique steady-state pdf vector $\mathbf{f}_\infty = \begin{bmatrix} f(0) & f(1) & \dots & f(k) \end{bmatrix}$ exists, then

$$\mathbf{p}^\infty = \lim_{t \to \infty} \mathbf{p}^t = \begin{bmatrix} f(0) & f(1) & \cdots & f(k) \\ f(0) & f(1) & \cdots & f(k) \\ \vdots & \vdots & & \vdots \\ f(0) & f(1) & \cdots & f(k) \end{bmatrix}.$$

Algorithm 10.3.4 If a unique steady-state pdf vector exists, then Theorem 10.3.1 provides a numerical algorithm for computing it by computing $\mathbf{p}^t$ for increasing values of t until all rows of $\mathbf{p}^t$ converge to a common vector of nonnegative values that sum to 1.

Example 10.3.6
If the successive powers $\mathbf{p}$, $\mathbf{p}^2$, $\mathbf{p}^3$, $\mathbf{p}^4$, ... are computed for the state-transition matrix in Example 10.3.1 and displayed with $d.dddd$ precision, we find that, for $t > 16$, there is *no* change in the displayed values of $\mathbf{p}^t$. Thus we conclude that, in this case,

$$\mathbf{p}^\infty = \begin{bmatrix} 0.1549 & 0.2895 & 0.2844 & 0.2712 \\ 0.1549 & 0.2895 & 0.2844 & 0.2712 \\ 0.1549 & 0.2895 & 0.2844 & 0.2712 \\ 0.1549 & 0.2895 & 0.2844 & 0.2712 \end{bmatrix}$$

and that the steady-state pdf vector for this Markov chain is

$$\mathbf{f}_\infty = \begin{bmatrix} 0.1549 & 0.2895 & 0.2844 & 0.2712 \end{bmatrix}.$$

Eigenvector Method.   Algorithm 10.3.4 is easily understood and easily implemented. The primary drawback to this algorithm is that the multiplication of one $(k+1) \times (k+1)$ matrix by another requires $(k+1)^3$ multiplications, and, to compute $\mathbf{p}^t$ sequentially, this process must be repeated t times. Thus the time complexity of this algorithm is $O(t\,k^3)$. Some efficiency can be gained by using "successive squaring" to compute, instead, the sequence

$$\mathbf{p}, \mathbf{p}^2, \mathbf{p}^4, \mathbf{p}^8, \dots.$$

But this algorithm is still $O\big(\log(t)\,k^3\big)$, and so efficiency is an issue if the state space is large. Accordingly, we consider an alternative approach based on the fact that

$$\mathbf{f}_{t+1} = \mathbf{f}_t\,\mathbf{p} \qquad t = 0, 1, 2, \dots.$$

From this equation, it follows that, in the limit as $t \to \infty$, a steady-state pdf vector must satisfy the eigenvector equation

$$\mathbf{f}_\infty = \mathbf{f}_\infty\,\mathbf{p}.$$

By using this fact, we obtain the following theorem.

Theorem 10.3.2   For a discrete-time finite-state Markov chain having state-transition matrix $\mathbf{p}$, if a steady-state pdf vector $\mathbf{f}_\infty = \begin{bmatrix} f(0) & f(1) & \dots & f(k) \end{bmatrix}$ exists, then $\mathbf{f}_\infty$ is a (left) eigenvector of $\mathbf{p}$ with eigenvalue 1:

$$\mathbf{f}_\infty\,\mathbf{p} = \mathbf{f}_\infty.$$

Equivalently, the steady-state pdf vector is characterized by the following set of $(k+1)$ linear *balance* equations

$$\sum_{x=0}^{k} f(x)\,p(x, x') = f(x') \qquad x' = 0, 1, 2, \dots, k.$$

Algorithm 10.3.5 If a unique steady-state pdf vector exists, then Theorem 10.3.2 provides a second obvious algorithm for computing it.

- Solve the balance equations or, equivalently, compute a left eigenvector of **p** having eigenvalue 1.0.
- Check that all elements of the eigenvector are nonnegative.
- Normalize the eigenvector so that the elements sum to 1.0.

Not only is Algorithm 10.3.5 inherently more efficient that Algorithm 10.3.4, it is also the preferred way to produce *symbolic* (algebraic) steady-state solutions to finite-state Markov chain models whenever such solutions are possible. The following example is an illustration.

Example 10.3.7

The balance equations for the 4-state Markov chain in Example 10.3.1 (redisplayed in Figure 10.3.2 for convenience) are

$$0.3f(0) \qquad\qquad\qquad +0.4f(3) = f(0)$$

$$0.2f(0) + 0.5f(1) + 0.4f(2) \qquad\quad = f(1)$$

$$0.4f(0) + 0.3f(1) \qquad\quad +0.5f(3) = f(2)$$

$$0.1f(0) + 0.2f(1) + 0.6f(2) + 0.1f(3) = f(3).$$

By inspection $f(0) = f(1) = f(2) = f(3) = 0$ is a solution to the balance equations. Therefore, if steady state is possible, this linear system of equations must be singular; otherwise, $f(0) = f(1) = f(2) = f(3) = 0$ would be the *only* possible solution. Being singular, the *four* balance equations row-reduce to the equivalent *three* equations

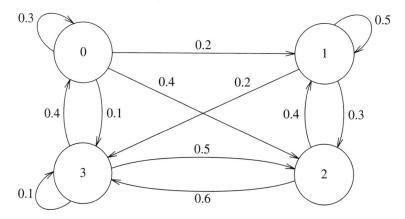

Figure 10.3.2 State-transition diagram for a discrete-time finite-state Markov chain.

$$f(0) \qquad\qquad -\frac{152}{266} f(3) = 0$$

$$f(1) \qquad\qquad -\frac{284}{266} f(3) = 0$$

$$f(2) - \frac{279}{266} f(3) = 0.$$

By back-substitution, $f(2)$, $f(1)$, and $f(0)$ can be expressed in terms of $f(3)$ and then combined with the normalization equation $f(0) + f(1) + f(2) + f(3) = 1$ to yield the steady-state probabilities:

$$f(0) = \frac{152}{981} \qquad f(1) = \frac{284}{981} \qquad f(2) = \frac{279}{981} \qquad f(3) = \frac{266}{981}.$$

To the accuracy displayed in Example 10.3.6, these probabilities agree exactly with the steady-state pdf vector.

Flow-Balance Equations.   Discrete-time Markov chains can be characterized in terms of expected times:*

$$\mu(x) = \frac{1}{1 - p(x, x)} \qquad\qquad x \in \mathcal{X};$$

or, equivalently, in terms of *rates*:

$$\lambda(x) = \frac{1}{\mu(x)} = 1 - p(x, x) \qquad\qquad x \in \mathcal{X}.$$

When $X(t) = x$, the term $\mu(x)$ represents the expected time the Markov chain will remain in state x; equivalently, $\lambda(x)$ represents the rate of "flow" out of this state. Accounting for immediate feedback in this way makes it important to recognize that the corresponding state-transition matrix must be modified accordingly. The state-transition matrix is now

$$\mathbf{q} = \begin{bmatrix} 0 & q(0, 1) & \cdots & q(0, k) \\ q(1, 0) & 0 & \cdots & q(1, k) \\ \vdots & \vdots & \ddots & \vdots \\ q(k, 0) & q(k, 1) & \cdots & 0 \end{bmatrix},$$

where the modified state-transition probabilities (which exclude immediate feedback) are

$$q(x, x') = \begin{cases} \dfrac{p(x, x')}{1 - p(x, x)} & x' \neq x \\ 0 & x' = x. \end{cases}$$

---

*This assumes that no states are *absorbing*—that is, $p(x, x) < 1$ for all $x \in \mathcal{X}$.

Then, when $X(t) = x$, the product

$$\lambda(x)\, q(x, x') = \begin{cases} p(x, x') & x' \neq x \\ 0 & x' = x \end{cases}$$

represents the rate of flow out of state x into state x'. Moreover, the balance equations

$$f(x')\, p(x', x') + \sum_{x \neq x'} f(x)\, p(x, x') = f(x') \qquad x' = 0, 1, \dots, k$$

can then be expressed as

$$\sum_{x \neq x'} f(x)\, \lambda(x)\, q(x, x') = f(x')\bigl(1 - p(x', x')\bigr) \qquad x' = 0, 1, \dots, k.$$

In other words, the balance equations in Theorem 10.3.2 are equivalent to the balance equations

$$\underbrace{\sum_{x \neq x'} f(x)\, \lambda(x)\, q(x, x')}_{\text{flow in}} = \underbrace{f(x')\, \lambda(x')}_{\text{flow out}} \qquad x' = 0, 1, \dots, k.$$

When $X(t) = x$, the product $\lambda(x)\, q(x, x') = p(x, x')$ represents the *rate* at which the flow out of state x goes into state $x' \neq x$. It is consistent with this observation that, as an alternative to a directed-graph representation like that in Example 10.3.7, a usually superior directed-graph representation of a discrete-time finite-state Markov chain is as follows:

- Each node $x \in \mathcal{X}$ is labeled by $\lambda(x)$.
- The arcs are labeled by the state-transition *rates* $p(x, x') = \lambda(x)\, q(x, x')$.

In this representation, at each node x, the sum of all the labeled out-arcs is $\lambda(x)$. There are no feedback arcs.

Example 10.3.8
As an alternative to the approach in Example 10.3.7, for the "flow rate" directed-graph representation shown in Figure 10.3.2, the balance equations are

$$\begin{aligned}
&& + 0.4 f(3) &= 0.7 f(0) \\
0.2 f(0) && + 0.4 f(2) && &= 0.5 f(1) \\
0.4 f(0) + 0.3 f(1) && && + 0.5 f(3) &= 1.0 f(2) \\
0.1 f(0) + 0.2 f(1) + 0.6 f(2) && && &= 0.9 f(3).
\end{aligned}$$

By inspection, these equations are equivalent to those in Example 10.3.7.

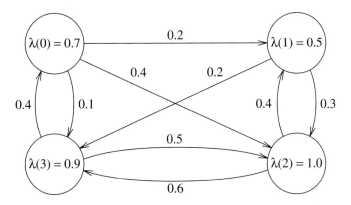

Figure 10.3.3   State-transition diagram without self-loops for a discrete-time finite-state Markov chain.

Existence and Uniqueness.   One important theoretical question remains—what conditions are necessary and sufficient to guarantee that a finite-state Markov chain has a unique steady-state pdf vector? Although it is beyond the scope of this book to answer this question in complete generality, the following general 2-state-Markov-chain example illustrates the kinds of situations that can arise.

Example 10.3.9

The most general 2-state $(k = 1)$ Markov chain has the state-transition matrix

$$\mathbf{p} = \begin{bmatrix} 1 - \alpha & \alpha \\ \beta & 1 - \beta \end{bmatrix},$$

where α, β are nonnegative parameters no larger than 1. Consider the following cases.

- If $\alpha = \beta = 0$, then the Markov chain will remain forever in either state 0 or state 1, the one that was its initial state. In this case, the Markov chain is said to be *reducible*, and there is more than one steady-state pdf vector.
- If either $0 < \alpha < 1$ or $0 < \beta < 1$ or both, then there is exactly one steady-state pdf vector

$$\mathbf{f}_\infty = \begin{bmatrix} \dfrac{\beta}{\alpha + \beta} & \dfrac{\alpha}{\alpha + \beta} \end{bmatrix}.$$

- If $\alpha = \beta = 1$, then the Markov chain will cycle forever between states 0 and 1 with period 2. In this case, the Markov chain is said to be *periodic* and there is no steady-state pdf vector.

The three cases illustrated in Example 10.3.9 generalize to *any* finite-state discrete-time Markov chain, as follows. Let $p^t[x, x']$ denote the $[x, x']$ element of the matrix $\mathbf{p}^t$.

- If $p^t[x, x'] > 0$ for some $t > 0$, the state x' is said to be *accessible* from state x (in t steps).
- If x' is accessible from x and, in addition, x is accessible from x', then states x and x' are said to *communicate*.

- State communication defines an *equivalence relation* on the state space $\mathcal{X}$: The state space naturally partitions into disjoint sets of communicating states.
- If there is only one set in the state-space partition defined by state communication, then the Markov chain is said to be *irreducible*—that is, a Markov chain is irreducible if and only if all states communicate. A *sufficient* condition for a Markov chain to be irreducible is the existence of some $t > 0$ such that $p^t[x, x'] > 0$ for all $[x, x'] \in \mathcal{X} \times \mathcal{X}$.
- If there is a partition of the state space $\mathcal{X} = \mathcal{X}_1 \cup \mathcal{X}_2 \cup \ldots \cup \mathcal{X}_r$ with the property that the Markov chain will transition forever as indicated

$$\cdots \to \mathcal{X}_1 \to \mathcal{X}_2 \to \cdots \to \mathcal{X}_r \to \mathcal{X}_1 \to \cdots$$

then the Markov chain is said to be *periodic* with period $r > 1$. If no such partition exists, the Markov chain is said to *aperiodic*. A *sufficient* condition for a Markov chain to be aperiodic is $p[x, x] > 0$ for at least one $x \in \mathcal{X}$.

The following theorem provides an important condition sufficient to guarantee the existence of a unique steady-state vector.

Theorem 10.3.3   An irreducible, aperiodic, finite-state Markov chain has one and only one steady-state pdf vector.

10.3.2 Continuous-Time Finite-State Markov Chains

In the remainder of this section, we consider *continuous-time finite-state* Markov chains. The emphasis is on demonstrating that "within" any continuous-time Markov chain is an associated "embedded" discrete-time Markov chain. As in the discrete-time case, we assume a finite state space $\mathcal{X}$, which is, without loss of generality,

$$\mathcal{X} = \{0, 1, 2, \ldots, k\}.$$

The state space $\mathcal{X}$ remains the same as before, but now the dwell time in a state is *continuous*, rather than discrete.

Definition 10.3.4   Let X be a discrete random variable, indexed by time t as $X(t)$, that evolves in time as follows.

- At any time $t > 0$, $X(t) \in \mathcal{X}$.
- If $x \in \mathcal{X}$ is the state of X at time t, then X will remain in this state for a random time T that is *Exponential* with mean $\mu(x)$.
- When a state transition occurs then, independently of the time spent in the current state, the random variable X will shift from its current state x to a different state, say $x' \neq x$, with fixed probability $p(x, x')$.

The stochastic process defined by $X(t)$ is called a *continuous-time finite-state Markov chain.**

---

*Compare with Definitions 10.2.2 and 10.3.1.

A continuous-time finite-state Markov chain is completely characterized by the initial state $X(0) \in \mathcal{X}$, the expected time-in-state function $\mu(x)$ defined for all $x \in \mathcal{X}$ so that $\mu(x) > 0$, and the state-transition probability function defined so that

$$p(x, x') \geq 0 \qquad (x, x') \in \mathcal{X} \times \mathcal{X}$$

with $p(x, x) = 0$ for all $x \in \mathcal{X}$ and

$$\sum_{x'=0}^{k} p(x, x') = 1 \qquad x \in \mathcal{X}.$$

Equivalently, because $p(x, x) = 0$, the last equation can be written as

$$\sum_{x' \neq x} p(x, x') = 1 \qquad x \in \mathcal{X},$$

where the summation is over all $x' \in \mathcal{X}$ except state x.

As in the discrete-time case, the analysis of the steady-state behavior of a continuous-time finite-state Markov chain is based on the state-transition matrix **p**, defined as follows. (Compare with Definition 10.3.2.)

Definition 10.3.5   The *state-transition probability* $p(x, x')$ represents the probability of a transition *from* state x *to* state x'. The corresponding $(k + 1) \times (k + 1)$ matrix

$$\mathbf{p} = \begin{bmatrix} 0.0 & p(0, 1) & \cdots & p(0, k) \\ p(1, 0) & 0.0 & \cdots & p(1, k) \\ \vdots & \vdots & \ddots & \vdots \\ p(k, 0) & p(k, 1) & \cdots & 0.0 \end{bmatrix}$$

with elements $p(x, x')$ is called the *state-transition matrix*.

The elements of the state-transition matrix **p** are nonnegative, and the elements of each row sum to 1.0. Hence, each row of the state-transition matrix defines the pdf for a discrete random variable. By convention, immediate feedback at each state is accounted for by the *Exponential* time-in-state model, so the diagonal elements of the state transition matrix are zero.

Continuous-time, finite-state Markov chains are usually characterized in terms of *rates*,

$$\lambda(x) = \frac{1}{\mu(x)} \qquad x \in \mathcal{X},$$

rather than in terms of the corresponding expected *times*, $\mu(x)$. The reason for this is that, when $X(t) = x$, $\lambda(x)$ represents the infinite-horizon rate at which X will "flow" out of state x. Therefore, when $X(t) = x$, the product $\lambda(x) p(x, x')$ represents the rate of flow out of state x into state $x' \neq x$. The sum of these rates of flow is $\lambda(x)$.

Example 10.3.10

Consider a Markov chain having four possible states ($k = 3$), the state-transition matrix

$$\mathbf{p} = \begin{bmatrix} 0.0 & 0.3 & 0.6 & 0.1 \\ 0.0 & 0.0 & 0.6 & 0.4 \\ 0.0 & 0.3 & 0.0 & 0.7 \\ 0.4 & 0.0 & 0.6 & 0.0 \end{bmatrix},$$

and the rates

$$\lambda(0) = 1, \qquad \lambda(1) = 2/3, \qquad \lambda(2) = 1/2, \qquad \text{and} \quad \lambda(3) = 2.$$

In this example, the probability of a transition from state 2 to 3 is $p(2, 3) = 0.7$; and, when $X(t) = 2$, the expected rate of flow into state 3 is $\lambda(2) \, p(2, 3) = (0.5)(0.7) = 0.35$.

As is illustrated in the following example, a continuous-time, finite-state Markov chain can be represented as a directed graph.

- The nodes of this graph are the possible states (the state space $\mathcal{X}$). Each node $x \in \mathcal{X}$ is labeled by $\lambda(x)$.
- The labeled arcs are defined by the state-transition probabilities.

This kind of labeled digraph provides a convenient graphical representation of a (small) continuous-time Markov chain. For each node (state), the sum of all the labeled *out*-arcs is 1.0.

Example 10.3.11

The Markov chain in Example 10.3.10 can be represented by the labeled digraph shown in Figure 10.3.4.

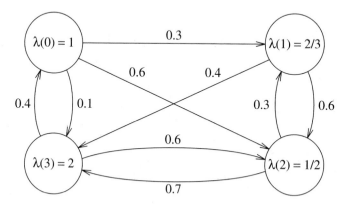

Figure 10.3.4   State-transition diagram for a continuous-time finite-state Markov chain.

The row elements of the state-transition matrix $\mathbf{p}$ can be summed to form the *cumulative* state-transition matrix $\mathbf{P}$. (See Algorithm 10.3.1.) This matrix can then be used, as in Algorithm 10.3.2, to define the function NextState(x), which returns a random state $x' \in \mathcal{X}$ (with $x' \neq x$) that is consistent with the pdf defined by the x-row of the state-transition matrix. This function can then be used to simulate the time evolution of a continuous-time Markov chain (for τ time units) with the following algorithm. (Compare with Algorithm 10.3.3.)

Algorithm 10.3.6   Given the initial state $X(0) \in \mathcal{X}$ and the functions $\mu(x)$ defined for all $x \in \mathcal{X}$, this algorithm simulates the time evolution of the associated continuous-time, finite-state Markov chain for $0 < t < \tau$. [See Nicol and Heidelberger (1995) for speed-ups that use algorithms for parallel architectures.]

```
t = 0;
x = X(0);
while (t < τ) {
    t += Exponential(μ(x));
    x = NextState(x);
}
```

As is illustrated by the following example, Algorithm 10.3.6 will produce a *piecewise-constant* state–time history $x(t)$ with step discontinuities when the state transitions occur.

Example 10.3.12
Algorithm 10.3.6, when applied to the Markov chain in Example 10.3.11 [with $X(0) = 0$ and $\tau = 25$], yield the state–time history shown in Figure 10.3.5.*

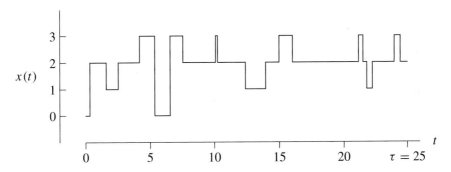

Figure 10.3.5   Realization of a continuous-time finite-state Markov chain.

Example 10.3.13
An implementation of Algorithm 10.3.6 was run with $\tau = 10\,000$, initial state $X(0) = 0$, and an rngs initial seed of 12345. The resulting proportion of time spent in each state is shown below.

state	0	1	2	3
proportion	0.100	0.186	0.586	0.128

---

*This time history was generated by using the rngs initial seed 12345. Of course, with a different initial seed, the time history would have been different.

As we will see, the proportions in Example 10.3.13 form a (good) estimate of the steady-state pdf for the continuous-time Markov chain in Examples 10.3.11. For future reference, note that (with *d.ddd* precision)

$$\mathbf{p}^\infty = \begin{bmatrix} 0.135 & 0.153 & 0.375 & 0.337 \\ 0.135 & 0.153 & 0.375 & 0.337 \\ 0.135 & 0.153 & 0.375 & 0.337 \\ 0.135 & 0.153 & 0.375 & 0.337 \end{bmatrix},$$

where **p** is the state-transition matrix for this Markov chain (from Example 10.3.10). Clearly, the common rows of this matrix, which represent the steady-state distribution of *something*, are not in agreement with the result in Example 10.3.13. The question then is how, if at all, are the two pdf vectors

$$\begin{bmatrix} 0.100 & 0.186 & 0.586 & 0.128 \end{bmatrix}$$

and

$$\begin{bmatrix} 0.135 & 0.153 & 0.375 & 0.337 \end{bmatrix}$$

related?

Embedded Chain. The answer to the question just posed is based on the observation that "within" any *continuous*-time Markov chain is an associated *embedded discrete*-time Markov chain. This is made clear by comparing Algorithm 10.3.6 (reproduced here for convenience), which simulates a continuous-time Markov chain—

```
t = 0;
x = X(0);
while (t < τ) {
    t += Exponential(μ(x));        // continuous-time increment
    x = NextState(x);
}
```

—with Algorithm 10.3.3, which simulates the associated embedded discrete-time Markov chain:

```
t = 0;
x = X(0);
while (t < τ) {
    t += 1;                         // discrete-time increment
    x = NextState(x);
}
```

The numerical value of τ in these two algorithms could differ. The only other difference is the time-increment assignment. Therefore, with a proper choice of τ's and the use of multiple rngs streams, the state-to-state transitions for both the continuous-time Markov chain and its associated embedded discrete-time Markov chain would be *exactly* the same. The only difference would be the *Exponential*$(\mu(x))$ amount of time the continuous-time Markov chain spends in each state, before the next state transition occurs.

This embedded-chain observation suggests that the answer to the question posed previously is that the steady-state pdf vector $\mathbf{f}_\infty$ for the *embedded* discrete-time Markov chain is related to the steady-state pdf vector $\mathbf{f}$ for the corresponding continuous-time Markov chain by the equation

$$f(x) = \frac{f_\infty(x)}{\alpha\,\lambda(x)} \qquad x \in \mathcal{X},$$

where

$$\alpha = \sum_x \frac{f_\infty(x)}{\lambda(x)}$$

is the real-valued parameter required to produce the normalization

$$\sum_x f(x) = 1$$

and the sums are over all $x \in \mathcal{X}$.

Example 10.3.14
As discussed previously, the steady-state pdf vector for the embedded discrete-time Markov chain corresponding to the continuous-time Markov chain in Example 10.3.10 is (with *d.ddd* precision)

$$\mathbf{f}_\infty = \begin{bmatrix} 0.135 & 0.153 & 0.375 & 0.337 \end{bmatrix}.$$

Starting from the definition of the unnormalized pdf,

$$f'(x) = \frac{f_\infty(x)}{\lambda(x)} \qquad x = 0, 1, 2, 3,$$

it follows that

$$f'(0) \cong \frac{0.135}{1.0000}, \qquad f'(1) \cong \frac{0.153}{0.6667}, \qquad f'(2) \cong \frac{0.375}{0.5000}, \qquad \text{and} \quad f'(3) \cong \frac{0.337}{2.0000}.$$

In this case, the numerical value of the parameter α is

$$\alpha = \sum_{x=0}^{3} f'(x) \cong 1.283,$$

so the (normalized) steady-state pdf vector for the continuous-time Markov chain is

$$\mathbf{f} = \frac{1}{\alpha}\mathbf{f}' \cong \begin{bmatrix} 0.105 & 0.179 & 0.585 & 0.131 \end{bmatrix}$$

which compares well to the steady-state *estimate*

$$\begin{bmatrix} 0.100 & 0.186 & 0.586 & 0.128 \end{bmatrix}$$

in Example 10.3.13.

10.3.3 Exercises

10.3.1   (*a*) Use Algorithms 10.3.2 and 10.3.3 to construct a program, dtmc, that simulates one realization $x(t)$ of a discrete-time finite-state Markov chain. Program dtmc should estimate the proportion of time spent in each state and report the statistic

$$\bar{x} = \frac{1}{\tau} \int_0^\tau x(t) \, dt.$$

(*b*) How does this statistic relate to the computed proportions? (*c*) Comment. (Simulate the Markov chain in Example 10.3.1 with $X(0) = 1$ and $\tau = 10\,000$.)

10.3.2   (*a*) As an extension of Exercise 10.3.1, modify program dtmc to produce multiple realizations (replications). (*b*) With $X(0) = 1$, $\tau = 1, 2, 4, 8, 16, 32, 64$, and using 100 replications, compute a 95%-confidence-interval estimate for the transient statistic

$$\overline{X}(\tau) = \frac{1}{\tau} \int_0^\tau X(t) \, dt.$$

(*c*) How does this statistic relate to the statistic $\bar{x}$ computed in Exercise 10.3.1? (*d*) Comment.

10.3.3   (*a*) As an alternative to always using $X(0) = 1$, repeat Exercise 10.3.2, except that, for each replication, draw the value of $X(0)$ at random from the initial distribution

$$\mathbf{f}_0 = \begin{bmatrix} 0.155 & 0.290 & 0.284 & 0.271 \end{bmatrix}.$$

(*b*) Comment.

10.3.4   (*a*) Use Algorithms 10.3.2 and 10.3.6 to construct a program, ctmc, that simulates one realization $x(t)$ of a continuous-time finite-state Markov chain. Program ctmc should estimate the proportion of time spent in each state and report the statistic

$$\bar{x} = \frac{1}{\tau} \int_0^\tau x(t) \, dt.$$

(*b*) How does this statistic relate to the computed proportions? (*c*) Comment. (Simulate the Markov chain in Example 10.3.10 with $X(0) = 1$ and $\tau = 10\,000$.)

10.3.5   (*a*) As an extension of Exercise 10.3.4, modify program ctmc to produce multiple realizations (replications). (*b*) With $X(0) = 1$, $\tau = 1, 2, 4, 8, 16, 32, 64$, and using 100 replications, compute a 95%-confidence-interval estimate for the transient statistic

$$\overline{X}(\tau) = \frac{1}{\tau} \int_0^\tau X(t) \, dt.$$

(*c*) How does this statistic relate to the statistic $\bar{x}$ computed in Exercise 10.3.4? (*d*) Comment.

10.3.6   (*a*) Use program ctmc (from Exercise 10.3.4) to estimate the steady-state pdf of the number of jobs in a $M/M/2/5$ service node if $\lambda = 2.5$, if the two servers are distinguishable because $\nu_1 = 1$ and $\nu_2 = 2$, and if the server discipline is to always choose the fastest available server. *Hint*: Represent the state of the service node as a 3-tuple of the form (x_1, x_2, q), where the binary variables x_1, x_2 represent the number of jobs with server 1, 2, respectively, and q represents the number in the queue. (*b*) What did you do to convince yourself that your results are correct?

10.3.7   Work through the details of Example 10.3.7. Keep the constants as fractions, rather than converting to decimal, so that your solutions will be expressed as

$$f(0) = \frac{152}{981} \qquad f(1) = \frac{284}{981} \qquad f(2) = \frac{279}{981} \qquad f(3) = \frac{266}{981}.$$

10.3.8   If $0 < \alpha < 1$ and $0 < \beta < 1$, then the discrete-time Markov $X(t)$ chain defined by

$$\mathbf{p} = \begin{bmatrix} 0 & 1-\alpha & \alpha & 0 \\ \beta & 0 & 0 & 1-\beta \\ 0 & 0 & 0 & 1 \\ 0 & 1 & 0 & 0 \end{bmatrix}$$

is irreducible and aperiodic. (*a*) Construct its steady-state pdf vector and find the steady-state expected value,

$$\lim_{t \to \infty} E[X(t)].$$

(*b*) If $X(0) = 2$ with probability 1, find $E[X(t)]$ for $t = 1, 2, 3, 4$.

10.3.9   If $0 < \alpha_x < 1$ for $x = 0, 1, \ldots, k$, the discrete-time Markov chain defined by

$$\mathbf{p} = \begin{bmatrix} 1-\alpha_0 & \alpha_0 & 0 & 0 & \cdots & 0 & 0 & 0 \\ 0 & 1-\alpha_1 & \alpha_1 & 0 & \cdots & 0 & 0 & 0 \\ \vdots & \vdots & \vdots & \vdots & \ddots & \vdots & \vdots & \vdots \\ 0 & 0 & 0 & \cdots & 0 & 1-\alpha_{k-1} & \alpha_{k-1} \\ \alpha_k & 0 & 0 & 0 & \cdots & 0 & 0 & 1-\alpha_k \end{bmatrix}$$

is irreducible and aperiodic. (*a*) Construct its steady-state pdf vector. *Hint:* Express your answer in terms of the parameters

$$\mu_x = \frac{1}{\alpha_x} \qquad x = 0, 1, \ldots, k.$$

(*b*) Comment.

10.3.10   The discrete-time Markov chain defined by

$$\mathbf{p} = \begin{bmatrix} 1/2 & 1/2 & 0 & 0 \\ 1/2 & 1/2 & 0 & 0 \\ 0 & 1/3 & 1/3 & 1/3 \\ 0 & 0 & 0 & 1 \end{bmatrix}$$

has *two* steady-state pdf vectors. What are they?

10.4 A NETWORK OF SINGLE-SERVER SERVICE NODES

In this section, we will discuss how to construct a discrete-event simulation model of a network of $k \geq 1$ single-server service nodes. This simulation will be based upon the following model.

- The single-server service nodes are indexed by $s = 1, 2, \ldots, k$. The index $s = 0$ is reserved for the "super node" that represents the *exterior* of the network—the *source* of jobs flowing into the network and the *sink* for jobs flowing out of the network. The set of service nodes is denoted $\mathcal{S} = \{1, 2, \ldots, k\}$ with $\mathcal{S}_0 = \{0\} \cup \mathcal{S} = \{0, 1, 2, \ldots, k\}$.
- There is a $(k+1) \times (k+1)$ *node-transition matrix* **p** defined in such a way that each job leaving node $s \in \mathcal{S}_0$ will transition to node $s' \in \mathcal{S}_0$ with probability

$$p[s, s'] = \Pr(\text{transition from node } s \text{ to node } s').$$

By convention $p[0, 0] = 0.0$.
- Each service node has its own queue and its own type of queueing discipline (FIFO, LIFO, etc.), its own service-time distribution (*Exponential*, *Erlang*, etc.), and infinite capacity. The service rate of node $s \in \mathcal{S}$ is ν_s.
- The net flow of jobs *into* the network—the *external* arrivals—is assumed to be a Poisson process with rate $\lambda_0(t)$. External arrivals occur at node $s' \in \mathcal{S}$ with probability $p[0, s']$, and so

$$p[0, s'] \lambda_0(t)$$

is the external arrival rate at node $s' \in \mathcal{S}$, provided that $p[0, s'] > 0$. Being consistent with this assumption, the external arrivals at each service node form a Poisson process.
- As the notation suggests, the external arrival rate $\lambda_0(t)$ could vary with time to form a nonstationary arrival process.[*] However, for each $s' \in \mathcal{S}$, the probability $p[0, s']$ is constant in time. Indeed, all the $p[s, s']$ probabilities are constant in time.

Definition 10.4.1   If $p[0, s'] = 0$ for all $s' \in \mathcal{S}$, and if $p[s, 0] = 0$ for all $s \in \mathcal{S}$, then the network is said to be *closed*. (A closed network is one for which the number of jobs in the network is fixed. A network that is not closed is said to be *open*.)

By allowing for a time-varying external arrival rate $\lambda_0(t)$, the discrete-event simulation model we are developing will apply equally well both to open and to closed networks. To simulate a closed network, it is sufficient to prevent departures from the network, by choosing the node-transition probabilities so that $p[s, 0] = 0$ for all $s \in \mathcal{S}$ and by choosing the external arrival rate so that $\lambda_0(t)$ becomes zero when the expected (or actual) number of jobs in the network reaches a desired level. Thus, from a simulation perspective, a closed network is just a special case of an open network.

There is an appealing conceptual simplicity to modeling external arrivals as transitions *from* the 0th node and departures from the network as transitions *to* the 0th node. In this way, by viewing the *exterior* of the network as one (super) node, all transitions, including external arrivals and network departures, can be neatly characterized by a single $(k+1) \times (k+1)$ node-transition matrix **p** having the structure indicated. This matrix defines the *topology* of the network, in the sense that $p[s, s'] > 0$ if and only if jobs can transition directly from node s to s'. The $p[0, s']$

---

[*]Techniques for simulating a *nonstationary* Poisson-arrival process are discussed in Section 7.5. Techniques for fitting a nonstationary Poisson-arrival process to a data set of arrival times are discussed in Section 9.3.

row of this matrix represents the external arrival probabilities; the $p[s, 0]$ column represents the network departure probabilities.*

$$\mathbf{p} = \begin{bmatrix} 0.0 & p[0,1] & p[0,2] & p[0,3] & \cdots & p[0,k] \\ p[1,0] & p[1,1] & p[1,2] & p[1,3] & \cdots & p[1,k] \\ p[2,0] & p[2,1] & p[2,2] & p[2,3] & \cdots & p[2,k] \\ \vdots & \vdots & \vdots & \vdots & \ddots & \vdots \\ p[k,0] & p[k,1] & p[k,2] & p[k,3] & \cdots & p[k,k] \end{bmatrix}.$$

Each row of $\mathbf{p}$ must sum to 1.0. Just as in a simulation of a Markov chain (Section 10.3), we can convert $\mathbf{p}$ to a *cumulative* node-transition matrix $\mathbf{P}$ (via Algorithm 10.3.1). This cumulative node-transition matrix can then be used to simulate the node-to-node transitions of jobs as they move into, through, and out of the network (via Algorithm 10.3.2).

Examples.   The following four examples illustrate the node-transition matrix associated with simple networks of single-server service nodes. The graph-like network representation illustrated in these examples is relatively standard. Given that, it is important to be able to translate this representation into the corresponding node-transition matrix.

Example 10.4.1
The $k = 1$ single-server model upon which the ssq series of programs is based corresponds to the node-transition matrix

$$\mathbf{p} = \begin{bmatrix} 0 & 1 \\ 1 & 0 \end{bmatrix}.$$

As an extension, this single-server model with feedback probability β from Section 3.3 is illustrated in Figure 10.4.1.

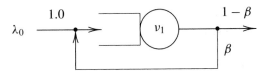

Figure 10.4.1   Single-server service node with feedback probability β.

This diagram corresponds to the node-transition matrix

$$\mathbf{p} = \begin{bmatrix} 0 & 1 \\ 1-\beta & \beta \end{bmatrix}.$$

---

*As an alternative approach, if the network is *closed*, then the first row and column of $\mathbf{p}$ are null. In this case, $\mathbf{p}$ is effectively a $k \times k$ matrix.

Small networks like those in the following three examples are commonly used to model *job shops*. The idea is that raw material enters the job shop at one (or more) nodes. There it is processed, then moved to other network nodes for further processing, and so on, until eventually a finished product leaves the network. Travel time from node to node is assumed to be negligible and hence is not modeled explicitly.

Example 10.4.2

The conventional *tandem-server* network model consists of k servers in series (sequence), as illustrated in Figure 10.4.2.

Figure 10.4.2 Tandem-server network.

This diagram corresponds to the node-transition matrix

$$\mathbf{p} = \begin{bmatrix} 0 & 1 & 0 & \cdots & 0 \\ 0 & 0 & 1 & \cdots & 0 \\ \vdots & \vdots & \vdots & \ddots & \vdots \\ 0 & 0 & 0 & \cdots & 1 \\ 1 & 0 & 0 & \cdots & 0 \end{bmatrix}.$$

Example 10.4.3

A network with $k = 4$ servers is shown in Figure 10.4.3.

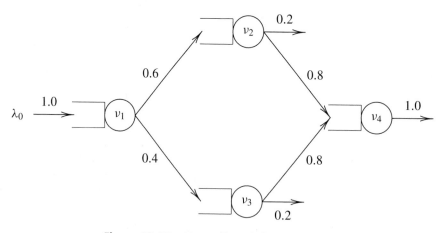

Figure 10.4.3 Network with $k = 4$ servers.

This network corresponds to the node-transition matrix

$$\mathbf{p} = \begin{bmatrix} 0.0 & 1.0 & 0.0 & 0.0 & 0.0 \\ 0.0 & 0.0 & 0.6 & 0.4 & 0.0 \\ 0.2 & 0.0 & 0.0 & 0.0 & 0.8 \\ 0.2 & 0.0 & 0.0 & 0.0 & 0.8 \\ 1.0 & 0.0 & 0.0 & 0.0 & 0.0 \end{bmatrix}.$$

Example 10.4.4

A network with $k = 5$ servers is shown in Figure 10.4.4.

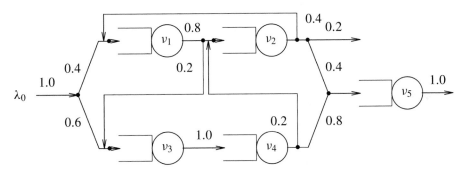

Figure 10.4.4   Network with $k = 5$ servers.

This network corresponds to the node-transition matrix

$$\mathbf{p} = \begin{bmatrix} 0.0 & 0.4 & 0.0 & 0.6 & 0.0 & 0.0 \\ 0.0 & 0.0 & 0.8 & 0.2 & 0.0 & 0.0 \\ 0.2 & 0.4 & 0.0 & 0.0 & 0.0 & 0.4 \\ 0.0 & 0.0 & 0.0 & 0.0 & 1.0 & 0.0 \\ 0.0 & 0.0 & 0.2 & 0.0 & 0.0 & 0.8 \\ 1.0 & 0.0 & 0.0 & 0.0 & 0.0 & 0.0 \end{bmatrix}.$$

As is indicated by the relatively large number of nonzero elements in $\mathbf{p}$, the topology of this network is significantly more complex than that of the previous three networks.

- Although there is no *immediate* feedback (i.e., the diagonal elements of $\mathbf{p}$ are all 0.0), feedback is certainly possible. For example, a job leaving node 1 will go directly to node 2 with probability 0.8; from there will be fed back to node 1 with probability 0.4. The probability of this two-step $1 \rightarrow 2 \rightarrow 1$ feedback is 0.32.
- More complex feedback paths are possible. For example, a job could circulate along the path $1 \rightarrow 3 \rightarrow 4 \rightarrow 2 \rightarrow 1$ and, because of independence, having done so once, traverse the same path again and again. The probability of this path is so low (0.016), however, that multiple cycles on this path by any job will virtually never occur.

States and Events.   To characterize the state of a network of single-server service nodes at any time, we need to know the number of jobs (if any) in each of the k service nodes. As is consistent with this concept, the events—those things that cause the state of the network to change—are the following:

- the external arrivals;
- the service-node departures.

Event List.   As is consistent with the previous discussion, the associated list of next-event times,

next external arrival

next departure from node 1

⋮

next departure from node k,

consists of (at most) $k + 1$ elements, one for each event type. The detailed structure of this event list (array, linked list, binary tree, heap, etc.) is an (important) implementation issue that we need not resolve at this modeling level. Note that, at any event time t, the numbers of jobs (if any) in each of the k service nodes, together with the contents of the event list, provide a comprehensive snapshot of the system. In other words, if all this information were recorded, a next-event simulation of a network of single-server service nodes could be stopped at any event time and then restarted later on by using *only* this recorded information to reinitialize the simulation. This stop–restart process would produce dynamic system behavior that would be statistically indistinguishable from that produced by the system if it were not stopped. This kind of conceptual check is an important part of discrete-event system modeling.

In addition to the state variables and the event list, to allow for various queueing disciplines, it will be necessary to maintain k queue data structures, one for each service node. Again, the detailed structure of these queues (circular arrays, linked lists, etc.) is an (important) implementation detail that need not be resolved at this point.

Next-Event Simulation.   Consistent with the model just constructed at the conceptual level, we can now construct a next-event simulation of a network of single-server service nodes at the specification level. Recognize that, at least in principle,

- the number of network service nodes can be arbitrarily large, and
- the network topology can be arbitrarily complex.

In other words, the next-event approach to this discrete-event simulation places no inherent bound on either the size or the topological complexity of the network. Of course, there will be practical bounds, but these will be imposed by hardware limitations and by the choice of data structures and of supporting algorithms, not by the next-event approach to the simulation.

The following algorithm makes use of an auxiliary function `ProcessArrival(s')` that handles the arrival of a job at node $s' \in \mathcal{S}$:

- if the server at service node s' is idle, this function will place the arriving job in service and schedule the corresponding departure from the service node;

- otherwise, the server is busy, and this function will place the arriving job in the queue.

Algorithm 10.4.1   The main loop of a next-event simulation of a network of single-server service nodes is

```
while ( some stopping criteria is not met ) {
    s = DetermineNextEvent();
    if (s == 0)
        CreateJob();
    else
        ProcessDeparture(s);
}
```

where the action of each function can be characterized as follows.

The function `DetermineNextEvent()` will

- check the event list to locate the next (most imminent) event type $s \in \mathcal{S}_0$ and associated time t, then
- advance the clock to t.

The function `CreateJob()` will

- create an *external* arrival by using the 0th row of the cumulative transition matrix $\mathbf{P}$ to identify at which service node $s' \in \mathcal{S}$ this arrival will occur, and then call the function `ProcessArrival(s')`, and then
- schedule the next external arrival.

The function `ProcessDeparture(s)`* will

- remove a job from service node $s \in \mathcal{S}$, use the sth row of the cumulative transition matrix $\mathbf{P}$ to identify where this job goes next (the new node $s' \in \mathcal{S}_0$), and then (unless $s' = 0$ in which case the job exits the network) call the function `ProcessArrival(s')` to process the arrival of this job at service node $s' \in \mathcal{S}$; then
- update the status of service node s by selecting a job from the queue at node s (if one exists) by setting the server's status to idle (otherwise); and, finally,
- if the server at node s is not idle, schedule the next departure from this node.

10.4.1 Steady State

Flow balance can be used to characterize steady state for a network of single-server service nodes with constant arrival rates. This characterization is based upon the following definition and on Theorem 10.4.1.

Definition 10.4.2   For each service node $s \in \mathcal{S}$, define λ_s to be the *total* arrival rate. The rate λ_s is computed by summing all the *internal* arrival rates—jobs from network service nodes (including, possibly, s itself)—with the *external* arrival rate.

---

*The processing of a job that experiences immediate feedback must be implemented carefully in the first two steps of `ProcessDeparture(s)`.

If the external arrival rate λ_0 is constant, from Definition 10.4.2, it follows that steady-state flow balance at all k service nodes is characterized by the k "flow in equals flow out" equations:

$$\underbrace{\lambda_0 p[0, s']}_{\text{external}} + \underbrace{\lambda_1 p[1, s'] + \cdots + \lambda_{s'} p[s', s'] + \cdots + \lambda_k p[k, s']}_{\text{internal}} = \lambda_{s'} \qquad s' = 1, 2, \ldots, k.$$

Moreover, it is intuitive that flow balance can be achieved only if $\lambda_s < \nu_s$ for $s = 1, 2, \ldots, k$. This is summarized by the following theorem.

Theorem 10.4.1   If steady state is possible, then the following observations hold:

- The total flow rate $\lambda_1, \lambda_2, \ldots, \lambda_k$ into and out of each service node is characterized by the k linear *balance equations**

$$\sum_{s=0}^{k} \lambda_s \, p[s, s'] = \lambda_{s'} \qquad s' = 1, 2, \ldots, k.$$

- The steady-state *utilization* of each server is

$$\rho_s = \frac{\lambda_s}{\nu_s} < 1 \qquad s = 1, 2, \ldots, k.$$

Example 10.4.5

As a continuation of Example 10.4.1, for the single-server service node with external arrival rate λ_0, service rate ν_1, and feedback probability β shown in Figure 10.4.5, the single balance equation is

$$\lambda_0 + \lambda_1 \beta = \lambda_1.$$

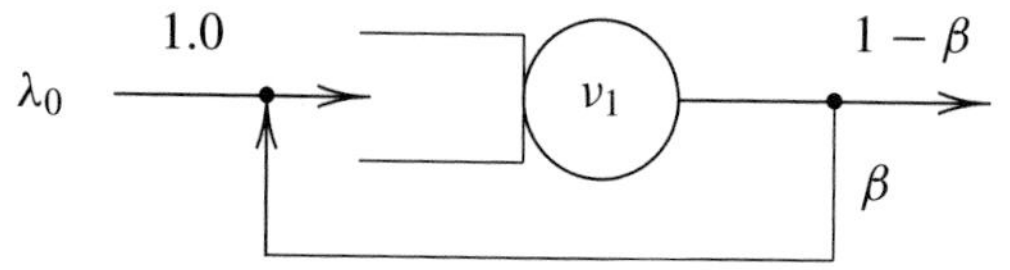

Figure 10.4.5   Single-server service node with feedback probability β.

Therefore $\lambda_1 = \lambda_0/(1 - \beta)$, and we see that the steady-state utilization of a single-server service node with feedback is $\rho_1 = \lambda_1/\nu_1 = \lambda_0/(1 - \beta)\nu_1$, provided that $\rho_1 < 1$.

Example 10.4.6

As a continuation of Example 10.4.2, for the tandem-server network illustrated in Figure 10.4.6, the balance equations are

$$\lambda_{s-1} = \lambda_s \qquad s = 1, 2, \ldots, k.$$

---

*There are k equations and k unknowns; the constant external arrival rate λ_0 is *not* unknown.

Figure 10.4.6   Tandem-server network.

Hence, $\lambda_s = \lambda_0$ for all $s = 1, 2, \ldots, k$. Therefore, the steady-state utilization of each service node in a tandem-server network is $\rho_s = \lambda_0/v_s$, provided that $\rho_s < 1$ for $s = 1, 2, \ldots, k$.

Example 10.4.7
As a continuation of Example 10.4.3, for the network in Figure 10.4.7, the balance equations are

$$\lambda_0 = \lambda_1 \qquad \text{node 1,}$$

$$0.6\lambda_1 = \lambda_2 \qquad \text{node 2,}$$

$$0.4\lambda_1 = \lambda_3 \qquad \text{node 3, and}$$

$$0.8\lambda_2 + 0.8\lambda_3 = \lambda_4 \qquad \text{node 4.}$$

It is easily verified that the (unique) solution to these four equations is

$$\lambda_1 = \lambda_0, \qquad \lambda_2 = 0.6\lambda_0, \qquad \lambda_3 = 0.4\lambda_0, \qquad \text{and} \qquad \lambda_4 = 0.8\lambda_0,$$

which defines the steady-state utilization of each service node as

$$\rho_1 = \frac{\lambda_0}{v_1}, \qquad \rho_2 = 0.6\frac{\lambda_0}{v_2}, \qquad \rho_3 = 0.4\frac{\lambda_0}{v_3}, \qquad \text{and} \qquad \rho_4 = 0.8\frac{\lambda_0}{v_4},$$

provided that $\rho_s < 1$ for $s = 1, 2, 3, 4$.

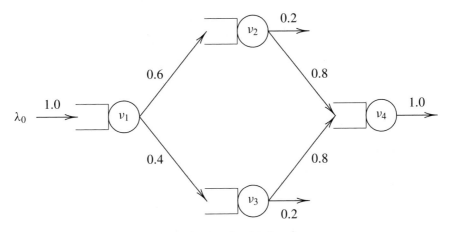

Figure 10.4.7   Network with $k = 4$ servers.

Example 10.4.8

As a continuation of Example 10.4.4, the balance equations are

$$0.4\lambda_0 + 0.4\lambda_2 = \lambda_1 \qquad \text{node 1,}$$

$$0.8\lambda_1 + 0.2\lambda_4 = \lambda_2 \qquad \text{node 2,}$$

$$0.6\lambda_0 + 0.2\lambda_1 = \lambda_3 \qquad \text{node 3,}$$

$$\lambda_3 = \lambda_4 \qquad \text{node 4, and}$$

$$0.4\lambda_2 + 0.8\lambda_4 = \lambda_5 \qquad \text{node 5.}$$

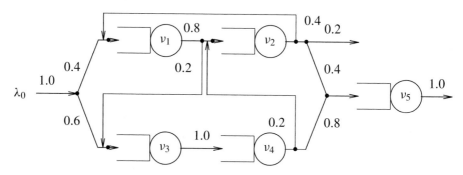

Figure 10.4.8   Network with $k = 5$ servers.

It can be verified that the (unique) solution to these five equations is

$$\lambda_1 = \frac{560}{830}\lambda_0, \quad \lambda_2 = \frac{570}{830}\lambda_0, \quad \lambda_3 = \frac{610}{830}\lambda_0, \quad \lambda_4 = \frac{610}{830}\lambda_0, \quad \text{and} \quad \lambda_5 = \frac{716}{830}\lambda_0,$$

which defines the steady-state utilization of each service node as

$$\rho_1 = \frac{560}{830}\cdot\frac{\lambda_0}{\nu_1}, \quad \rho_2 = \frac{570}{830}\cdot\frac{\lambda_0}{\nu_2}, \quad \rho_3 = \frac{610}{830}\cdot\frac{\lambda_0}{\nu_3}, \quad \rho_4 = \frac{610}{830}\cdot\frac{\lambda_0}{\nu_4}, \quad \text{and} \quad \rho_5 = \frac{716}{830}\cdot\frac{\lambda_0}{\nu_5},$$

provided that $\rho_s < 1$ for $s = 1, 2, 3, 4, 5$.

Consistency Check.   In addition to the k flow-balance equations in Theorem 10.4.1, there is one additional equation,

$$\sum_{s=1}^{k} \lambda_s p[s, 0] = \lambda_0;$$

it characterizes flow balance for the 0th node. This might seem to be a new equation, independent of the others, but it is not; this equation can be derived from the k equations in Theorem 10.4.1. The proof of this result is left as an exercise. The point is that the 0th node-balance equation provides no *new* information. However, it does provide a good consistency check on the correctness of the steady-state λ's.

Example 10.4.9

To illustrate the previous discussion:

- for Example 10.4.1, the 0th node-balance equation is $\lambda_1(1 - \beta) = \lambda_0$;
- for Example 10.4.2, the 0th node-balance equation is $\lambda_k = \lambda_0$;
- for Example 10.4.3, the 0th node-balance equation is $0.2\lambda_2 + 0.2\lambda_3 + \lambda_4 = \lambda_0$;
- for Example 10.4.4, the 0th node-balance equation is $0.2\lambda_2 + \lambda_5 = \lambda_0$.

It can verified that, in each case, this equation is satisfied.

Jackson Networks

Definition 10.4.3   A network of single-server service nodes is a *Jackson network* if and only if all the following hold:

- the external arrival process is a stationary Poisson process (λ_0 is constant in time);
- the service-time distribution at each service node is *Exponential*$(1/\nu_s)$ (the service rate ν_s is constant in time);
- all service-time processes and the external arrival process are statistically independent;
- at each service node, the queue discipline assumes no knowledge of service times (e.g., FIFO, LIFO, and SIRO, which were defined in Section 1.2, are valid queue disciplines, but SJF is not).

If we are willing to make all the assumption associated with a Jackson network, then the following remarkable theorem, *Jackson's theorem*, is valid. The proof of this theorem is beyond the scope of this text.

Theorem 10.4.2   Each service node in a Jackson network operates *independently* as an $M/M/1$ service node in steady state.*

By combining Theorems 10.4.1 and 10.4.2, we have the following algorithm for analyzing the steady-state behavior of an *open* Jackson network of single-server service nodes.

---

*This theorem is valid for both *open* and *closed* Jackson networks.

> **Algorithm 10.4.2**   The steady-state behavior of an open Jackson network of k single-server service nodes can be analyzed as follows.
>
> - Set up and solve the k balance equations for $\lambda_1, \lambda_2, \ldots, \lambda_k$.
> - Check that the 0th node-balance equation is satisfied.
> - Check that steady-state can occur (i.e., $\rho_s = \lambda_s / \nu_s < 1$ for all $s = 1, 2, \ldots, k$).
> - Analyze each node as a statistically independent $M/M/1$ service node having arrival rate λ_s, service rate ν_s, and utilization ρ_s.

Example 10.4.10

As a continuation of Example 10.4.5, for an external Poisson arrival process with rate λ_0, an *Exponential* $(1/\nu_1)$ server, and feedback probability β as shown in Figure 10.4.9, the steady-state number in the service node, L_1, is a *Geometric* (ρ) random variable, provided that $\rho_1 = \lambda_0 / (1 - \beta)\nu_1 < 1$.* Therefore, in particular, if $\lambda_0 = 1$, $\nu_1 = 3$, and $\beta = 0.5$, then $\rho_1 = 2/3$, and it follows that the expected value of L_1 is

$$E[L_1] = \frac{\rho_1}{1 - \rho_1} = \frac{2/3}{1 - 2/3} = 2.$$

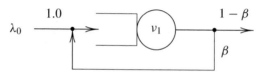

Figure 10.4.9   Single-server service node with feedback probability β.

Note the following:

- If there is no feedback ($\beta = 0$), then $\rho_1 = 1/3$ and $E[L_1] = 0.5$.
- If the feedback probability approaches $2/3$ ($\beta \to 2/3$), then $\rho_1 \to 1$ and $E[L_1] \to \infty$.

Example 10.4.11

As a continuation of Example 10.4.6, for an external Poisson arrival process with rate λ_0 and a tandem-server network of k independent *Exponential* $(1/\nu_s)$ servers shown in Figure 10.4.10, the steady-state number in each service node, L_s, is a *Geometric* (ρ_s) random variable if $\rho_s = \lambda_0 / \nu_s < 1$ for all $s = 1, 2, \ldots, k$. The steady-state number in the network is $L = L_1 + L_2 + \cdots + L_k$. Therefore, the expected value of L is

$$E[L] = \frac{\rho_1}{1 - \rho_1} + \frac{\rho_2}{1 - \rho_2} + \cdots + \frac{\rho_k}{1 - \rho_k}.$$

Moreover, if all the service rates are *equal* (i.e., $\nu_s = \nu$ for all s), and if $\rho = \lambda_0 / \nu < 1$, then

$$E[L] = \frac{k\rho}{1 - \rho}.$$

---

*The arrival, service, and feedback processes must be statistically independent.

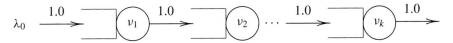

Figure 10.4.10   Tandem-server network.

Because L is a sum of k independent *Geometric*(ρ) random variables, we have the following theorem.

Theorem 10.4.3   Given a stationary Poisson arrival process with rate λ to a tandem-server network of k independent *Exponential*$(1/\nu)$ servers with $\rho = \lambda/\nu < 1$, the steady-state number in the network is a *Pascal*(k, ρ) random variable.

In general, for any open Jackson network with k service nodes, the steady-state expected number in the network is

$$E[L] = \frac{\rho_1}{1 - \rho_1} + \frac{\rho_2}{1 - \rho_2} + \cdots + \frac{\rho_k}{1 - \rho_k}.$$

The $E[L] = \lambda E[W]$ form of Little's equation is applicable to the entire network, so the expected *wait* (time spent in the network) is

$$E[W] = \frac{E[L]}{\lambda_0}.$$

Example 10.4.12
As a continuation of Example 10.4.7, if the network depicted in Figure 10.4.11 is a Jackson network with

$$\lambda_0 = \frac{1}{120}, \qquad \nu_1 = \frac{1}{84}, \qquad \nu_2 = \frac{1}{170}, \qquad \nu_3 = \frac{1}{225}, \qquad \text{and} \qquad \nu_4 = \frac{1}{120},$$

then it can be shown that

$$E[L] = \frac{7}{3} + \frac{17}{3} + 3 + 4 = 15.$$

It follows that the expected wait in the network is

$$E[W] = \frac{E[L]}{\lambda_0} = 120 \cdot 15 = 1800.$$

Exercise 10.4.4 illustrates another way to compute $E[W]$.

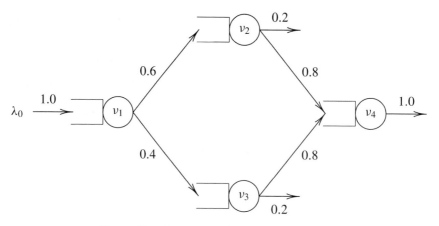

Figure 10.4.11 Network with $k = 4$ servers.

10.4.2 Exercises

10.4.1 Consider an open Jackson network having the node-transition matrix

$$\mathbf{p} = \begin{bmatrix} 0.00 & 1.00 & 0.00 & 0.00 \\ 0.20 & 0.00 & 0.72 & 0.08 \\ 0.00 & 1.00 & 0.00 & 0.00 \\ 0.00 & 1.00 & 0.00 & 0.00 \end{bmatrix}$$

and the parameters

$$\lambda_0 = 0.15, \qquad \nu_1 = 1.00, \qquad \nu_2 = 0.90, \qquad \text{and} \qquad \nu_3 = 0.12.$$

(*a*) Verify that this system can achieve steady state, and compute the steady-state utilization of each service node. (*b*) Verify that the 0th node-balance equation is satisfied. (*c*) Compute $E[L]$ and $E[W]$.

10.4.2 Do the same as in Exercise 10.4.1, except for the Jackson network defined by

$$\mathbf{p} = \begin{bmatrix} 0.00 & 0.50 & 0.00 & 0.50 \\ 0.00 & 0.00 & 1.00 & 0.00 \\ 0.00 & 0.25 & 0.25 & 0.50 \\ 1.00 & 0.00 & 0.00 & 0.00 \end{bmatrix}$$

and having the parameters

$$\lambda_0 = 2, \qquad \nu_1 = 2, \qquad \nu_2 = 3, \qquad \text{and} \qquad \nu_3 = 3.$$

10.4.3 (*a*) As an extension of Theorem 10.4.3, prove that, if the queue discipline at each service node is FIFO, then the steady-state wait in a k-node tandem-server network is an *Erlang*$(k, 1/(\nu - \lambda))$ random variable. (*b*) Is this result consistent with Example 10.4.11 and Little's equation $E[L] = \lambda_0 E[W]$?

10.4.4   (*a*) As a continuation of Example 10.4.12, compute the expected steady-state wait $E[W_s]$ for $s = 1, 2, 3, 4$, and then verify that

$$E[W] = E[W_1] + 0.6\, E[W_2] + 0.4\, E[W_3] + 0.8\, E[W_4] = 1800.$$

(*b*) What special property of the network in Example 10.4.12 facilitates this approach to computing $E[W]$?

10.4.5   Construct a next-event simulation of the network in Exercise 10.4.1. Assume all queues are FIFO.

10.4.6   Construct a next-event simulation of the network in Exercise 10.4.2. Assume all queues are FIFO.

Appendix A. Simulation Languages

This text emphasizes the use of a high-level algorithmic language for the implementation of discrete-event simulation models. Many discrete-event simulation modelers prefer to use a simulation-programming language, to save on model-development time. This appendix surveys the history of these simulation-programming languages and illustrates the use of one of these languages to model the single-server service-node from Section 1.2.

A.1 History

The use of a general-purpose simulation-programming language (SPL) expedites model development, input modeling, output analysis, and animation. In addition, SPLs have accelerated the use of simulation as an analysis tool by bringing down the cost of developing a simulation model. Nance (1993) gives a history of the development of SPLs from 1955 to 1986. He defines six elements that must be present in an SPL:

- random-number generation;
- variate generation;
- list-processing capabilities, so that objects can be created, altered, and deleted;
- statistical-analysis routines;
- summary-report generators;
- a timing executive or event calendar, to model the passage of time.

These SPLs may be (*i*) a set of subprograms in a high-level algorithmic language such as FORTRAN, Java, or C that can be called to meet the six requirements, (*ii*) a preprocessor that converts statements or symbols to lines of code in a high-level algorithmic language, or (*iii*) a conventional programming language.

The historical period is divided into five distinct eras. The names of several languages that came into existence in each era (subsequent versions of one particular language are not listed) are:

- 1955–1960. The era of search: GSP

- 1961–1965. The advent: CLP, CSL, DYNAMO, GASP, GPSS, MILITRAN, OPS, QUIKSCRIPT, SIMSCRIPT, SIMULA, SOL
- 1966–1970. The formative era: AS, BOSS, Q-GERT, SLANG, SPL
- 1971–1978. The expansion era: DRAFT, HOCUS, PBQ, SIMPL
- 1979–1986. Consolidation and regeneration: INS, SIMAN, SLAM

The General Purpose System Simulator (GPSS) was first developed on various IBM computers in the early 1960's (Karian and Dudewicz, 1991). Its block semantics were ideally suited for queuing simulations. Algol-based SIMULA was also developed in the 1960's and had features that were ahead of its time, including abstract data types, inheritance, the co-routine concept, and quasi-parallel execution.

SIMSCRIPT was developed by the RAND Corporation with the purpose of decreasing model- and program-development times. SIMSCRIPT models are described in terms of entities, attributes, and sets. The syntax and program organization were influenced by FORTRAN. The Control and Simulation Language (CSL) takes an "activity scanning" approach to language design, where the activity is the basic descriptive unit. The General Activity Simulation Program (GASP), as with several of the other languages, used flow-chart symbols to bridge the gap between personnel unfamiliar with programming and programmers unfamiliar with the application area. Although originally written in Algol, GASP provided FORTRAN subroutines for list-processing capabilities (e.g., queue insertion). GASP was a forerunner to the Simulation Language for Alternative Modeling (SLAM) and SIMulation ANalysis (SIMAN) languages. SLAM (Pritsker, 1995) was the first language to include three modeling perspectives in one language: network (process orientation), discrete-event, and continuous (state variables). SIMAN was the first major SPL executable on an IBM PC.

Languages that have been developed since Nance's timeline include AutoMod, Csim, Extend, Flexsim, Micro Saint, QUEST, SIMUL8, WITNESS, ProModel (Harrell, Ghosh, and Bowden, 2000), SIGMA (Schruben, 1992), @RISK (Seila, Ceric, and Tadikamalla, 2003), and Arena (a combination of SIMAN and the animator Cinema; see Kelton, Sadowski, and Sturrock, 2004). A survey of the current SPLs is given by Swain (2005). More detail on current popular SPLs is given in Chapter 4 of Banks, Carson, Nelson, and Nicol (2005).

A.2 Sample Model

Although the choice of languages is arbitrary, we illustrate the use of the SLAM (Pritsker, 1995) language, which came into prominence between the era of SPLs consisting entirely of subprograms and that of the current "point-and-click" GUIs now available with most SPLs. The language's market share peaked in the late 1980's and early 1990's. We model the single-server service node in SLAM. The single-server service-node model was presented in Section 1.2 via a "process interaction" world view and later presented in Section 5.1 (program `ssq3`) via a "next-event" world view. Like program `ssq3`, SLAM uses a next-event world view.

SLAM Nodes.   The simplest way to model a single-server service node in SLAM is to use a *network* orientation with *nodes* and *branches* used to represent the system. SLAM uses the term *entity* to refer generically to objects passing through the network. Other languages often use the term *transaction*. Depending on the model, an entity could be a job, a person, an object, a unit of information, and so on. Branches, or activities, connect the nodes in a network diagram and represent the paths along which entities move. Activities are used to model the passage of time. Nodes can be used for creating entities, delaying entities at queues, assigning variables, and so on. For the single-server service-node model, an entity is a job. There are 20 types of nodes in the SLAM language. We will use three of these node types to model a single-server service node.

A CREATE node, used to create entities, is illustrated in Figure A.1. The node parameters are:

- TF, the time of the first entity creation (default: 0);
- TBC, the time between entity creations (default: ∞);
- MA, the attribute number associated with the entity where the creation time is saved, known as the "mark attribute" (default: don't save the creation time);
- MC, maximum number of entities created at this node (default: ∞);
- M, maximum number of emanating branches to which a created entity can be routed (default: ∞).

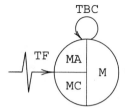

Figure A.1   CREATE node.

Entities are created by using a next-event approach with a deterministic or probabilistic time between creations, as specified by the TBC parameter. Each entity has a user-specified number of *attributes* that it maintains as it passes through the network. Typical attributes are the arrival time to the system, routing information, and service-time characteristics. The MC parameter determines when the CREATE node will stop creating entities. If the entity is to be "cloned" and duplicate copies sent out over several branches emanating from the CREATE node, the M parameter designates the maximum number of branches that can be taken.

A QUEUE node, used to store entities that are waiting for a server, is illustrated in Figure A.2. The node parameters are:

- IQ, initial number in the queue (default: 0);
- QC, queue capacity (default: ∞);
- IFL, a file number used by the SLAM filing system (no default).

Figure A.2   QUEUE node.

The QUEUE node is used for delaying entities prior to a "service" activity. If the initial number in the queue is positive, SLAM will place entities in the queue at the beginning of the simulation run and begin the simulation run with all servers busy. If the queue capacity is finite, three things can happen to an entity that arrives to a full queue: the arriving entity can be destroyed; the arriving entity can balk to another node; the entity can block another server from performing its activity, for lack of space in the queue. The file number IFL indicates where SLAM will store entities that are delayed because of busy server(s). SLAM maintains a filing system that is a FORTRAN-based doubly linked list to store entities waiting in queues and calendar events.

A TERMINATE node is illustrated in Figure A.3. The node parameter TC specifies a simulation termination condition (default: ∞). When the TCth entity arrives to the TERMINATE node, the simulation ends, and a SLAM output report is printed.

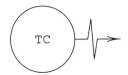

Figure A.3   TERMINATE node.

SLAM Network Discrete-Event Simulation Model.   Our attention now turns to using these three nodes to model a single-server service node. Recall that the times between arrivals to the node were independent *Exponential* (2.0) random variates and that service times were independent *Uniform* (1.0, 2.0) random variates. A SLAM network diagram for a single-server service node is illustrated in Figure A.4.

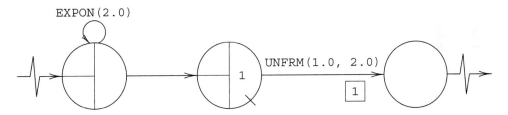

Figure A.4   Single-server service-node model.

The first job is created at time zero; subsequent jobs are created with an inter-creation time that is exponentially distributed with a mean of 2 minutes.* Jobs traverse the branch (activity) between the CREATE and QUEUE node in 0 time units. In the QUEUE node, jobs wait in file 1 (numbered for the sake of identification on the output report) for the server. The server is modeled by the activity (branch) number 1 (numbered for output-report identification) following the QUEUE node. The time to traverse this branch is *Uniform* (1, 2) and corresponds to a service time. The TERMINATE node is where the entity representing the job is "destroyed."

---

*SLAM uses a Lehmer random-number generator with ten streams and changeable initial seed values. SLAM also provides random-variate generators for several popular distributions (e.g., exponential, normal, Poisson, Weibull).

The SLAM philosophy is that models are constructed as network diagrams at the on-screen iconic level. For most implementations of SLAM, software is available for automatically converting an iconic model (network diagram) to executable FORTRAN-like statements. The executable SLAM statements corresponding to the single-server service-node network diagram are shown below.

```
GEN, LEEMIS, SSQ3, 9/30/05, 1, Y, Y, Y, Y, Y, 72;
LIMITS, 1, 2, 100;
NETWORK;
    CREATE, EXPON(2.0);                          CREATE JOBS
    QUEUE(1);                          WAIT FOR THE SERVER HERE
    ACTIVITY/1, UNFRM(1.0, 2.0);                SERVICE TIME
    TERMINATE;                                JOBS LEAVE NODE
    ENDNETWORK;
INIT, 0, 10000;                            RUN FOR 10 000 MINUTES
FIN;
```

The statements that begin in column 1 are called SLAM control statements. SLAM provides 20 control statements to create initial or terminal conditions, allocate memory, run traces, and so on. Five such statements are illustrated in this example. The statements that are indented correspond to the network diagram. All statements end in a semicolon, possibly followed by optional comments.

The GEN(erate) control statement has a parameter list that begins with the modeler's name, model name, and a date. The 1 following the date indicates the number of replications (called "runs" in SLAM) of the simulation model. The fields following the number of replications (with Y parameters indicating "YES") indicate that defaults are to be taken for printing an echo report, input statements, and so on. The last parameter tells SLAM to print a 72-column output report.

The LIMITS control statement has three parameters: the number of files used, the number of attributes per entity, and the maximum number of concurrent entities in the system. SLAM uses a file system based on static allocation of memory at compilation time. The NETWORK control statement indicates the beginning of the SLAM network statements. The INIT control statement indicates that the simulation should be run between times 0 and 10 000. The FIN control statement indicates the end of the SLAM control statements.

SLAM provides the user with an "echo" report, which gives information such as ranking criterion in files (the FIFO default is used here), random-number streams and associated seed values, model-initialization options, termination criteria, and memory allocation. The echo report has been omitted, since its contents simply reflect that the model has been properly specified.

The SLAM summary report, which is printed after execution of the model is complete, follows.

```
          S L A M    I I   S U M M A R Y   R E P O R T

 SIMULATION PROJECT SSQ3                    BY LEEMIS
 DATE   9/30/2005                           RUN NUMBER    1 OF    1

 CURRENT TIME    0.1000E+05
 STATISTICAL ARRAYS CLEARED AT TIME   0.0000E+00
```

```
            **FILE STATISTICS**

FILE                    AVERAGE   STANDARD   MAXIMUM   CURRENT  AVERAGE
NUMBER    LABEL/TYPE    LENGTH    DEVIATION  LENGTH    LENGTH   WAIT TIME

   1         QUEUE       1.181      1.839       16        0       2.349
   2       CALENDAR      1.756      0.429        3        1       1.273

          **SERVICE ACTIVITY STATISTICS**

ACT ACT LABEL OR  SER AVERAGE    STD   CUR AVERAGE MAX IDL MAX BSY  ENT
NUM START NODE    CAP   UTIL     DEV   UTIL BLOCK  TME/SER TME/SER  CNT

  1       QUEUE     1   0.756   0.43    0   0.00    16.75  138.05  5026
```

The report echos the model name, modeler's name, and date, then indicates the simulation clock time (current time) when this report was generated. The file-statistics section shows that there were two files used in the model: the first for queuing up jobs, the second to hold the event list (calendar). The statistics for the queue show that the average number in the queue (length of the queue) was 1.181, with a standard deviation of 1.839. The queue length reached a maximum of 16 jobs. The average delay (wait) time for a job was 2.349 minutes. The queue is currently (at time 10 000) empty. The statistics for the calendar show that there were an average of 1.756 events on the calendar during the simulation, with a standard deviation of 0.429. There were a maximum of three events on the calendar at one time: the next arrival, an end-of-service event, and the pseudo-event at time 10 000 to end the simulation. There is currently one event left on the calendar (which must be the next arrival since the pseudo-event has been processed and the server is idle at the end of the simulation). Finally, the average calendar event resides on the calendar for 1.273 time units.

The service activity statistics indicate that the server utilization was 0.756. The server is currently (at time 10 000) idle. The longest time that the server was idle during the simulation was 16.75 minutes; the longest time that the server was busy during the simulation was 138.05 minutes. A total of 5026 jobs were served during the simulation, a number consistent with the arrival rate 0.5 jobs per minute.

A.3 CONCLUSIONS

There are advantages to using SLAM (or any other SPL). First, many lines of code are saved over implementing a model in a high-level algorithmic language. This can be helpful in terms of debugging. Second, an SPL has routines for managing files and the event calendar which allow the modeler to focus on *modeling*, as opposed to sequencing events and establishing queue priorities. Third, standard summary reports, including histograms and other statistical displays, are generated automatically. Fourth, animation of a model, a powerful tool for communication with management, is typically included as a part of modern simulation languages. This would be a custom and rather intricate programming project for someone using a high-level algorithmic language. Templates are now available in many languages that customize an animation to a particular application area (e.g., banking, medicine, manufacturing). Fifth, most ordinary embellishments (e.g., multiple servers or a new queue discipline) are an easy matter in a simulation language. SLAM, for example, allows a modeler to write FORTRAN or C code for any embellishments that are not built into the language.

There are several disadvantages to using an SPL. First, simulation languages tend to have a slant toward one particular type of model. SLAM, for example, is very strong in modeling queuing-type models (e.g., inventory models). It has not been designed well for modeling most reliability systems, however, and a modeler might need to write code in a high-level algorithmic language for modeling series and parallel systems. Second, the modeler does not have complete control over the data structures that are used in a simulation language. In some models, this could be burdensome. Third, memory might be wasted in a simulation language. SLAM, for example, requires that all entities have the same number of attributes—something that a programmer might prefer to avoid. Finally, the modeler does not have access to the internal algorithms contained in an SPL, so the modeler can easily make an incorrect assumption about the behavior of the language under certain circumstances, which can result in incorrect conclusions. Schriber and Brunner (1998) survey the internal assumptions and associated algorithms present in most modern SPLs.

Recent advances in SPLs include object-oriented simulation languages (Joines and Roberts, 1998), web-based simulation languages (Kilgore, 2002), and parallel and distributed simulation (Fujimoto, 1998).

A.4 Exercises

A.1   (*a*) Program the $M/G/1$ queue described in this section in the SPL of your choice. (*b*) If possible, control the random-number-generator seed so that your output matches that of program `ssq3`.

A.2   (*a*) Use an Internet search engine to identify SPLs that support discrete-event simulation. (*b*) Classify these languages by modeling features of your choice.

Appendix B. Integer Arithmetic

The purpose of this appendix is to summarize the arithmetic and mathematical properties of integers that are most relevant to an understanding of random-number generation. The primary emphasis is on integer division and factoring.

Terminology. Because some properties summarized in this appendix apply to, for example, the positive integers only, we begin with the following standard clarifying terminology:

- the *positive integers* are $1, 2, 3, \ldots$;
- the *nonnegative integers* (positive and zero) are $0, 1, 2, 3, \ldots$;
- the *negative integers* are $\ldots, -3, -2, -1$;
- the *integers* (positive, negative, and zero) are $\ldots, -3, -2, -1, 0, 1, 2, 3, \ldots$.

So, for example, the positive integers have the property that they are closed under addition and multiplication, but not under subtraction. In contrast, the integers are closed under all three operations.

Theory. There are two (related) theoretical properties of the positive integers that are listed here for completeness:

- *well ordering*—any nonempty set of positive integers has a smallest element;
- *mathematical induction*—when S is a set of positive integers, if $1 \in S$ and if $n + 1 \in S$ for each $n \in S$, then *all* the positive integers are in S.

All of the standard integer-arithmetic existence theorems are based on one or both of these properties. The following important theorem is an example.

Integer Division

Theorem B.1 *Division Theorem*—if b is an integer and a is a positive integer, then there exists a unique pair of integers q, r with $0 \le r < a$ such that $b = aq + r$.

Example B.1

The division theorem is an *existence* theorem—that is, the integer pair (q, r) is guaranteed to exist. The theorem does not address at all the algorithmic question of *how* to determine (q, r). Of course, we were all taught to do division as children and, in that sense, the algorithm for finding (q, r) is second nature—divide a into b to get the quotient q with remainder r.

- If $(a, b) = (7, 17)$ then $(q, r) = (2, 3)$. That is, $17 = 7 \cdot 2 + 3$.
- If $(a, b) = (7, -17)$ then $(q, r) = (-3, 4)$. That is, $-17 = 7 \cdot (-3) + 4$.

In terms of implementation, the computation of (q, r) is directly related to the following two functions.

Floor and Ceiling Functions

Definition B.1   For any real-valued number x,

- the value of the *floor* function $\lfloor x \rfloor$ is the largest integer n such that $n \leq x$;
- the value of the *ceiling* function $\lceil x \rceil$ is the smallest integer n such that $x \leq n$.

Example B.2

For example, $\lfloor 7.2 \rfloor = \lfloor 7.7 \rfloor = 7$ and $\lfloor -7.2 \rfloor = \lfloor -7.7 \rfloor = -8$. Similarly, $\lceil 7.2 \rceil = \lceil 7.7 \rceil = 8$ and $\lceil -7.2 \rceil = \lceil -7.7 \rceil = -7$.

    The floor ("round to the left") and ceiling ("round to the right") functions are related. (This figure represents the real number line with tick marks at the integers.)

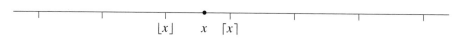

Specifically, the floor and ceiling functions satisfy the following properties.

- If x is an integer, then $\lceil x \rceil = \lfloor x \rfloor$.
- If x is not an integer, then $\lceil x \rceil = \lfloor x \rfloor + 1$.
- In general, $\lceil x \rceil = -\lfloor -x \rfloor$.

Because of these properties, it is sufficient for a programming language to provide just one of these functions, which is usually the floor function.

Example B.3

In ANSI C, both the floor and ceiling functions are provided, albeit in a potentially confusing way. A floating-point value is converted to an integer value by "truncation of any fractional part." So, for example, if x is a `double` and n is a `long` then the assignment n = x and the type conversion coerced by the cast (long) x will both produce $\lfloor x \rfloor$ if $x \geq 0.0$ or $\lceil x \rceil$ if $x < 0.0$. Therefore, one has to be careful about the sign of x if a float-to-int conversion is used to compute the floor or ceiling function. As an alternative, although the functions `floor(x)` and `ceil(x)` in the standard library

`<math.h>` return a floating-point result, they are otherwise a correct implementation of $\lfloor x \rfloor$ and $\lceil x \rceil$ respectively, even if x is negative.

An important characterization that is used at several places in the body of this text is that, for any number x, integer or not

$$\lfloor x \rfloor \leq x < \lfloor x \rfloor + 1.$$

So, for example, $3 = \lfloor 3.4 \rfloor \leq 3.4 < \lfloor 3.4 \rfloor + 1 = 4$. Moreover, for any integer n

$$\lfloor x + n \rfloor = \lfloor x \rfloor + n$$

but, in general,

$$\lfloor nx \rfloor \neq n \lfloor x \rfloor.$$

So, for example, $\lfloor 3.4 + 8 \rfloor = \lfloor 3.4 \rfloor + 8 = 11$ but $\lfloor 8 \cdot 3.4 \rfloor = \lfloor 27.2 \rfloor = 27 \neq 8 \cdot \lfloor 3.4 \rfloor = 24$.

Modulus Function

> **Definition B.2** Relative to Theorem B.1 and Definition B.1, if b is an integer and a is a positive integer, then the *remainder* is $r = b - \lfloor b/a \rfloor a$, where the *quotient* is $q = \lfloor b/a \rfloor$. Equivalently, the *modulus* (mod) function is effectively defined by the division theorem, with an implementation based on the floor function as
>
> $$b \bmod a = b - \lfloor b/a \rfloor a.$$

Example B.4
As in Example B.1, if $(a, b) = (7, 17)$ then the quotient is $q = \lfloor 17/7 \rfloor = 2$ and the remainder is $r = 17 \bmod 7 = 3$. Provided a is a positive integer and b is a nonnegative integer, in ANSI C the computation of q and r is direct as

```
q = b / a;                          // q is ⌊b/a⌋
r = b % a;                          // r is b mod a
```

If, however, b (or a) is a negative integer, then the ANSI C standard allows the computation of q and r to be implementation dependent. In this case, the value of q and r may be different from the mathematical definition of $\lfloor b/a \rfloor$ and $b \bmod a$.*

---

*It is the case, however, that the division and remainder operations are consistent, in that the value of $(b \ / \ a) * a + (b \ \% \ a)$ is required to be equal to b for all integer values of a and b, provided that a is not zero.

Although the following theorem is not as fundamental as either Theorem B.1 or B.3 (presented later), it is an important result that relates directly to the implementation of Lehmer random-number generators, introduced in Section 2.1. From an implementation perspective, the significance of this theorem is that the mod function can be applied term-by-term to help prevent integer overflow in intermediate calculations.

Theorem B.2   If a is a positive integer and $b_1, b_2, \ldots, b_n$ are integers, then
- $(b_1 + b_2 + \cdots + b_n) \bmod a = \big((b_1 \bmod a) + (b_2 \bmod a) + \cdots + (b_n \bmod a)\big) \bmod a;$
- $(b_1 b_2 \ldots b_n) \bmod a = \big((b_1 \bmod a)(b_2 \bmod a) \ldots (b_n \bmod a)\big) \bmod a.$

Example B.5

If a is a positive integer and b, c are integers, then, because $a \bmod a = 0$ and $0 \le b \bmod a < a$, it follows that

$$(b + ac) \bmod a = \big((b \bmod a) + (ac \bmod a)\big) \bmod a$$

$$= \big((b \bmod a) + ((a \bmod a)(c \bmod a) \bmod a)\big) \bmod a$$

$$= (b \bmod a) \bmod a$$

$$= b \bmod a.$$

Divisors and Primes

Definition B.3   If r is zero in Theorem B.1, so that $b = aq$, then a is said to be a *divisor* (or *factor*) of b. Equivalently, a is said to *divide* b. (Although not used in this book, the notation $a \mid b$ is commonly used to denote "a divides b.")

Example B.6

Given the positive integers a, b, and c:

- if a divides b and a divides c, then a divides $b + c$;
- if a divides b and a divides c with $b > c$, then a divides $b - c$;
- if a divides b or a divides c, then a divides bc;
- if a divides b and b divides c, then a divides c.

Definition B.4   A positive integer $p > 1$ is *prime* if and only if the only positive integers that divide p are p and 1.[*]

Example B.7

By convention, 1 is not a prime integer. The first few primes are 2, 3, 5, 7, 11, 13, 17, 19, 23, 29, ... The program `sieve`, described subsequently, implements the classic *prime*

---

[*]A positive integer that is not prime is said to be *composite*.

sieve of Eratosthenes, a simple algorithm for finding all the prime numbers between 2 and a user-specified upper limit.

Theorem B.3   *Fundamental Theorem of Arithmetic*—any positive integer $n > 1$ can be uniquely written as

$$n = p_1^{k_1} p_2^{k_2} \cdots p_r^{k_r}$$

where $p_1 < p_2 < \cdots < p_r$ are the r distinct prime integer divisors of n with corresponding positive integer exponents $k_1, k_2, \ldots, k_r$.

Example B.8
Relative to Theorem B.3, if n is prime, then $r = 1$ and $p_1 = n$ with $k_1 = 1$. If n is not prime, then $r \geq 2$. For example

$$38 = 2 \cdot 19$$

$$39 = 3 \cdot 13$$

$$40 = 2^3 \cdot 5$$

$$41 = 41 \qquad \text{(41 is prime)}$$

$$60 = 2^2 \cdot 3 \cdot 5$$

$$2100 = 2^2 \cdot 3 \cdot 5^2 \cdot 7$$

In general, given a table of primes between 2 and $\lfloor \sqrt{n} \rfloor$ (see program `sieve`), the prime factors of n can be found by systematically trying 2, 3, 5, 7, 11, $\ldots$, in order.

Definition B.5   If a, b are positive integers, then the positive integer d is a *common divisor* if and only if d divides both a and b. The largest such common divisor is called the *greatest common divisor*, denoted $\gcd(a, b)$. In general, $\gcd(a, b) = \gcd(b, a)$.

Example B.9
The two integers 12, 30 have 2, 3, and 6 as common divisors. The largest of these is 6, and so $\gcd(12, 30) = 6$. Note that $\gcd(12, 24) = 12$.

Definition B.6   The two positive integers a, b are *relatively prime* if and only if a, b have no common prime divisors, or, equivalently, $\gcd(a, b) = 1$.

Example B.10
The integers 10, 18 are not relatively prime because they have 2 as a common prime factor. Indeed, $\gcd(10, 18) = 2$. The integers 10, 21 are relatively prime because they have no common prime factors, i.e., $\gcd(10, 21) = 1$.

Example B.11

If both of the distinct positive integers a, b are prime, then a, b are relatively prime, i.e., $\gcd(a, b) = 1$.

The definition of the greatest common divisor of two positive integers a and b is clear. Figuring out the greatest common divisor is not. The *Euclidean Algorithm* can be used to compute the greatest common divisor of a and b.

Algorithm B.1 *Euclidean Algorithm*—Given two positive integers a, b, this algorithm computes the greatest common divisor $\gcd(a, b)$.

```
r = a % b;
while (r > 0) {
    a = b;
    b = r;
    r = a % b;
}
return b;
```

Theorem B.4 (*Fermat's "little" theorem*) If m is prime and a is an integer such that a/m is not an integer, then

$$a^{m-1} \bmod m = 1.$$

A small variation allows this theorem to be stated with a simpler hypothesis.

Theorem B.5 (*Fermat's "little" theorem, second statement*) If m is prime, then

$$a^m \bmod m = a \bmod m$$

for all integers a.

Sieve of Eratosthenes. The discussion thus far concerning prime numbers has focused on the definition of a prime number and associated results rather than on how to find prime numbers. Named after a Greek scientist who devised the algorithm, the sieve of Eratosthenes finds all prime numbers between 2 and some specified positive integer N. We begin with a conceptual description of the sieve.

Assume that we want a list of all the prime numbers between 2 and N. One way to proceed is as follows. For illustration, assume that $N = 100$. First, write all the integers between 2 and 100 in order:

$$2 \quad 3 \quad 4 \quad 5 \quad 6 \quad 7 \quad 8 \quad 9 \quad 10 \quad 11 \quad 12 \quad 13 \quad 14 \quad 15 \quad \ldots \quad 100$$

The sieve works from left to right on this string of integers in the following fashion. A pointer initially points to 2. We know that 2 is prime, but all other even numbers can't be prime, so we cross out $4, 6, 8, \ldots, 100$.

$$\downarrow$$
$$2 \quad 3 \quad \cancel{4} \quad 5 \quad \cancel{6} \quad 7 \quad \cancel{8} \quad 9 \quad \cancel{10} \quad 11 \quad \cancel{12} \quad 13 \quad \cancel{14} \quad 15 \quad \ldots \quad \cancel{100}$$

The pointer now advances to the next integer that has not been crossed out: 3, which is prime. The multiples of 3, namely $6, 9, 12, \ldots, 99$ can't be prime, so they are crossed out. (This will result in the even multiples of 3 being crossed out twice.)

$$\downarrow$$
$$2 \quad 3 \quad \cancel{4} \quad 5 \quad \cancel{6} \quad 7 \quad \cancel{8} \quad \cancel{9} \quad \cancel{10} \quad 11 \quad \cancel{12} \quad 13 \quad \cancel{14} \quad \cancel{15} \quad \ldots \quad \cancel{100}$$

The pointer now advances to 5, the next integer that has not been crossed out. The integers $10, 15, 20, \ldots, 100$ can't be prime, so they are crossed out.

$$\downarrow$$
$$2 \quad 3 \quad \cancel{4} \quad 5 \quad \cancel{6} \quad 7 \quad \cancel{8} \quad \cancel{9} \quad \cancel{10} \quad 11 \quad \cancel{12} \quad 13 \quad \cancel{14} \quad \cancel{15} \quad \ldots \quad \cancel{100}$$

Finally, the pointer is advanced to 7, and the integers $14, 21, 28, \ldots, 98$ are crossed out.

$$\downarrow$$
$$2 \quad 3 \quad \cancel{4} \quad 5 \quad \cancel{6} \quad 7 \quad \cancel{8} \quad \cancel{9} \quad \cancel{10} \quad 11 \quad \cancel{12} \quad 13 \quad \cancel{14} \quad \cancel{15} \quad \ldots \quad \cancel{100}$$

The sieve only needs to advance the pointer to $\sqrt{N} = \sqrt{100} = 10$ to cross out all of the composite integers between 2 and 100. The 25 prime numbers that remain are:

$$2, 3, 5, 7, 11, 13, 17, 19, 23, 29, 31, 37, 41, 43, 47, 53, 59, 61, 67, 71, 73, 79, 83, 89, 97.$$

At the specification level, we define an array `prime` that will eventually contain 0 for composite numbers, but 1 for prime numbers. To initialize, `prime[0]` and `prime[1]` are set to zero, and `prime[2]` through `prime[N]` are set to one. The pointer from the conceptual development is the outside `for` loop index n. Whenever a prime number n is encountered, `prime[2 * n]`, `prime[3 * n]`, etc. are set to zero.

Algorithm B.2   Given some positive integer N, this algorithm implements the sieve of Eratosthenes, assigning the array `prime` to binary values (0 for composite, 1 for prime) in order to indicate all of the prime numbers between 2 and N.

```
prime[0] = 0;                              // zero is composite
prime[1] = 0;                              // one is composite
for (n = 2;  n <= N;  n++)
   prime[n] = 1;
for (n = 2; n <= √N; n++)
   if (prime[n])
      for (s = 2; s <= (N/n); s++)
         prime[s * n] = 0;
```

Program `sieve` implements this algorithm and prints the primes between 2 and N in a tabular format.

The operation of the sieve of Eratosthenes suggests that primes could be rarer for larger integers. Experimentation indicates that this might indeed be the case. There are 25 primes between 1 and 100, 21 primes between 101 and 200, 16 primes between 1001 and 1100, 11 primes between 10001 and 10 100, and only 6 primes between 100001 and 100 100. Do we ever "run out" of primes (i.e., is there a "largest" prime)? The answer is no. There are an infinite number of primes, as proved by Euclid. The distribution of primes over the positive integers is a central topic in "analytic number theory." The *prime number theorem*, which was proved independently by Hadamard and de la Vallée–Poussin, states that

$$\lim_{x \to \infty} \frac{\pi(x)}{x / \ln(x)} = 1$$

where $\pi(x)$ is the number of primes less than or equal to x.

The presentation of integer arithmetic given in this appendix is intentionally elementary. For more information, we recommend the following texts, in order of increasing level of sophistication: Chapter 3 of Epp (1990), Vanden Eynden (2001), Chapters 3 and 4 of Graham, Knuth, and Patashnik (1989), and Niven, Zuckerman, and Montgomery (1991).

B.1 Exercises

B.1 Implement the Euclidean Algorithm (Algorithm B.1). Test the algorithm with (*a*) $a = 12$ and $b = 30$. (*b*) $a = 48271$ and $b = 2^{31} - 1$.

B.2 Modify program `sieve` to *count* the number of primes between 1 and x, defined earlier as $\pi(x)$, in order to check the validity of the prime number theorem for $x = 10, 100, 1000, 10\,000, 100\,000,$ and $1\,000\,000$.

Appendix C.
Parameter-Estimation
Summary

This appendix contains a summary of the parameter estimates for the discrete and continuous stationary parametric models presented in this text.

- See Section 9.2.1 for a discussion of how to hypothesize a parametric model [e.g., use the *Poisson* (μ) distribution only when $\bar{x}/s^2 \cong 1.0$];
- See Section 9.2.2 for a general discussion of parameter-estimation techniques (i.e., the method-of-moments and maximum-likelihood estimation techniques, whose results are listed in this appendix);
- See Sections 9.2.2 and 10.1.2 for a brief introduction to goodness-of-fit tests, which should be conducted after a parametric model has been selected and fitted (e.g., the Kolmogorov–Smirnov test).

Discrete Random Variables. Six discrete distributions were summarized in Section 6.4, two of which are special cases of the others. The following is a summary of parameter estimates (denoted with hats) for all six of these discrete distributions. In each case, $\bar{x}$ and s denote the sample mean and standard deviation, respectively. The integer parameter n in the *Binomial* (n, p) and the *Pascal* (n, p) distributions is not necessarily the same as the sample size.

- To estimate the two parameters in a *Binomial* (n, p) distribution, use

$$\hat{n} = \left\lfloor \frac{\bar{x}^2}{\bar{x} - s^2} + 0.5 \right\rfloor \qquad \text{and} \qquad \hat{p} = \frac{\bar{x}}{\hat{n}}.$$

These estimates are valid only if $n \geq 1$ and $0.0 < p < 1.0$.

- Since the *Bernoulli* (p) distribution corresponds to the special case $n = 1$ of a *Binomial* (n, p) distribution, use

$$\hat{p} = \bar{x}.$$

- To estimate the two parameters in a *Pascal* (n, p) distribution, use

$$\hat{n} = \left\lfloor \frac{\bar{x}^2}{s^2 - \bar{x}} + 0.5 \right\rfloor \qquad \text{and} \qquad \hat{p} = \frac{\bar{x}/\hat{n}}{1 + \bar{x}/\hat{n}}.$$

These estimates are valid only if $n \geq 1$ and $0.0 < p < 1.0$.
- Since the *Geometric* (p) distribution corresponds to the special case $n = 1$ of a *Pascal* (n, p) distribution, use

$$\hat{p} = \frac{\bar{x}}{1 + \bar{x}}.$$

- To estimate the one parameter in a *Poisson* (μ) distribution, use

$$\hat{\mu} = \bar{x}.$$

- To estimate the two parameters in an *Equilikely* (a, b) distribution, use

$$\hat{a} = \min\{x_1, x_2, \ldots, x_n\} \qquad \text{and} \qquad \hat{b} = \max\{x_1, x_2, \ldots, x_n\}.$$

Continuous Random Variables. Seven continuous distributions were summarized in Section 6.4, one of which is a special case of another. [The *Normal* $(0,1)$ distribution has no parameters to estimate.] The following is a summary of parameter estimates (denoted by hats) for all seven of these continuous distributions. In each case, $\bar{x}$ and s denote the sample mean and standard deviation, respectively. The integer parameter n in the *Erlang* (n, b), the *Chisquare* (n), and the *Student* (n) distributions is not necessarily the same as the sample size.

- To estimate the two parameters in an *Erlang* (n, b) distribution, use

$$\hat{n} = \left\lfloor \frac{\bar{x}^2}{s^2} + 0.5 \right\rfloor \qquad \text{and} \qquad \hat{b} = \frac{\bar{x}}{\hat{n}}.$$

These estimates are valid only if $n \geq 1$ and $b > 0.0$.
- Since the *Exponential* (μ) distribution corresponds to the special case $n = 1$ of an *Erlang* (n, b) distribution, use

$$\hat{\mu} = \bar{x}.$$

- To estimate the two parameters in a *Normal* (μ, σ) distribution, use

$$\hat{\mu} = \bar{x} \qquad \text{and} \qquad \hat{\sigma} = s.$$

- To estimate the two parameters in a *Lognormal* (a, b) distribution, use

$$\hat{a} = \frac{1}{2} \ln\left(\frac{\overline{x}^4}{\overline{x}^2 + s^2}\right) \qquad \text{and} \qquad \hat{b} = \sqrt{\ln\left(\frac{\overline{x}^2 + s^2}{\overline{x}^2}\right)}.$$

- To estimate the one parameter in a *Chisquare* (n) distribution, use

$$\hat{n} = \lfloor \overline{x} + 0.5 \rfloor.$$

This estimate is valid only if $n \geq 1$ and $s^2 / \overline{x} \cong 2.0$.

- To estimate the one parameter in a *Student* (n) distribution, use

$$\hat{n} = \left\lfloor \frac{2s^2}{s^2 - 1} + 0.5 \right\rfloor.$$

This estimate is valid only if $\overline{x} \cong 0.0$ and $s > 1.0$.

- To estimate the two parameters in a *Uniform* (a, b) distribution, use

$$\hat{a} = \min\{x_1, x_2, \ldots, x_n\} \qquad \text{and} \qquad \hat{b} = \max\{x_1, x_2, \ldots, x_n\}.$$

Appendix D. Random-Variable Models

This appendix defines the Random Variable ModelS library `rvms`. The library contains functions to compute the pdf, cdf, and idf for the six discrete parametric random-variable models summarized in Section 6.4 and for the seven continuous parametric random-variable models summarized in Section 7.4. This appendix also introduces several *special functions* that are called in library `rvms`.

Discrete Random-Variable Models.   The library contains 18 functions to compute the pdf, cdf, and idf for six discrete parametric random-variable models.

Bernoulli (p)

- `double pdfBernoulli(double` p`, long` x`)`
- `double cdfBernoulli(double` p`, long` x`)`
- `long idfBernoulli(double` p`, double` u`)`

Binomial (n, p)

- `double pdfBinomial(long` n`, double` p`, long` x`)`
- `double cdfBinomial(long` n`, double` p`, long` x`)`
- `long idfBinomial(long` n`, double` p`, double` u`)`

Equilikely (a, b)

- `double pdfEquilikely(long` a`, long` b`, long` x`)`
- `double cdfEquilikely(long` a`, long` b`, long` x`)`
- `long idfEquilikely(long` a`, long` b`, double` u`)`

Geometric (p)

- `double pdfGeometric(double` p`, long` x`)`

- `double cdfGeometric(double p, long x)`
- `long idfGeometric(double p, double u)`

Pascal (n, p)

- `double pdfPascal(long n, double p, long x)`
- `double cdfPascal(long n, double p, long x)`
- `long idfPascal(long n, double p, double u)`

Poisson (μ)

- `double pdfPoisson(double μ, long x)`
- `double cdfPoisson(double μ, long x)`
- `long idfPoisson(double μ, double u)`

Continuous Random-Variable Models.   The library contains 21 functions to compute the pdf, cdf, and idf for seven continuous parametric random-variable models.

Chisquare (n)

- `double pdfChisquare(long n, double x)`
- `double cdfChisquare(long n, double x)`
- `double idfChisquare(long n, double u)`

Erlang (n, b)

- `double pdfErlang(long n, double b, double x)`
- `double cdfErlang(long n, double b, double x)`
- `double idfErlang(long n, double b, double u)`

Exponential (μ)

- `double pdfExponential(double μ, double x)`
- `double cdfExponential(double μ, double x)`
- `double idfExponential(double μ, double u)`

Lognormal (a, b)

- `double pdfLognormal(double a, double b, double x)`
- `double cdfLognormal(double a, double b, double x)`
- `double idfLognormal(double a, double b, double u)`

Normal (μ, σ)

- `double pdfNormal(double μ, double σ, double x)`
- `double cdfNormal(double μ, double σ, double x)`
- `double idfNormal(double μ, double σ, double u)`

Student (n)

- `double pdfStudent(long n, double x)`

- double cdfStudent(long *n*, double *x*)
- double idfStudent(long *n*, double *u*)

Uniform(*a*, *b*)

- double pdfUniform(double *a*, double *b*, double *x*)
- double cdfUniform(double *a*, double *b*, double *x*)
- double idfUniform(double *a*, double *b*, double *u*)

In addition to 39 pdf, cdf, and idf functions, the library rvms also contains two functions for evaluating the natural logarithm of the factorial function $n! = n \cdot (n-1) \cdot \ldots \cdot 2 \cdot 1$ and the natural logarithm of the binomial coefficient

$$C(n, m) = \binom{n}{m} = \frac{n!}{m! \, (n-m)!}.$$

Specifically, the functions

- double LogFactorial(long *n*)
- double LogChoose(long *n*, long *m*)

evaluate $\ln(n!)$ and $\ln\big(C(n, m)\big)$ respectively.

D.1 Special Functions

Many of the 41 functions in the library rvms make use of one or more private (local to the library) functions belonging to an important class of mathematical functions that, for historical reasons, are known as *special functions*. These functions are typically defined in terms of summations and integrals that cannot be evaluated in "closed form." Very much like such familiar functions as $\sin(\cdot)$, $\ln(\cdot)$ and $\exp(\cdot)$, however, special functions can be evaluated numerically with essentially negligible error.

Gamma Function

Definition D.1   The *gamma function* is defined for all real values $a > 0$ as

$$\Gamma(a) = \int_0^\infty t^{a-1} \exp(-t) \, dt.$$

In Definition D.1, the phrase "is defined for all real values $a > 0$" means it can be proven mathematically that the integral converges to a finite value for all $a > 0$. For most values of a, however, this integral *cannot* be evaluated in closed form—that is part of what makes the gamma function special. The gamma function is a generalized factorial. In particular, it can be shown that:

- if $a > 0$, then $\Gamma(a + 1) = a \, \Gamma(a)$;
- if $a = 1, 2, 3, \ldots$, then $\Gamma(a) = (a - 1)!$

For most values of a, $\Gamma(a)$ is *extremely* large. For example,

a	1	2	4	8	16	32	64	128
$\Gamma(a)$	1	1	6	5040	1.31×10^{12}	8.22×10^{33}	1.98×10^{87}	3.01×10^{213}

Therefore, in practice (even if a is an integer) it is usually the natural logarithm, $\ln(\Gamma(a))$, that is calculated, *not* $\Gamma(a)$.

To evaluate the natural logarithm of $\Gamma(a)$ for $a > 0$, one can use the following excellent approximation, first published by Lanczos (1964):

$$\ln(\Gamma(a)) \cong (a - 0.5)\ln(a + 4.5) - (a + 4.5) + \ln(s(a)) \qquad a > 0,$$

where

$$s(a) = \sqrt{2\pi}\left(1 + \frac{c_0}{a} + \frac{c_1}{a + 1} + \cdots + \frac{c_5}{a + 5}\right)$$

and the constants $c_0, c_1, \ldots, c_5$ were (carefully) chosen by Lanczos to yield a relative error in $\ln(\Gamma(a))$ that is less than 2×10^{-10} for all $a > 0$. Although the gamma function is actually defined (i.e., the integral in Definition D.1 converges) for *all* values of a except for the integers $a = 0, -1, -2, -3, \ldots$, Lanczos's approximation is *not* valid when $a \leq 0$. Fortunately, that is not an issue for any of the applications considered in this book.

The function $\ln(\Gamma(\cdot))$ is in the library as the private (`static`) function `LogGamma`. Locally to the file `rvms.c`, this function can be evaluated as

```
x = LogGamma(a);
return x;
```

In the event that $\Gamma(a)$ must be evaluated, use

```
x = exp(LogGamma(a));
return x;
```

Beware of overflow, however, if a is very large.*

Factorial Function.   As a particular case of $\ln(\Gamma(\cdot))$, the function $\ln(n!)$ is available in the library as the function `LogFactorial`. To evaluate $\ln(n!)$ use

```
x = LogFactorial(n);
return x;
```

If you need to evaluate $n!$ (as a floating-point number) use

```
x = exp(LogFactorial(n));
return x;
```

Again, beware of overflow if n is very large.*

---

*Gamma functions and factorials frequently occur as terms in both the numerator and denominator of a fraction, in which case significant cancellation of like factors often occurs. If so, it is not necessary (or desirable) to evaluate the individual terms of the fraction.

Another special function, the beta function, is closely related to the gamma function. Unlike the gamma function, the integration in Definition D.2 is over the finite interval $(0, 1)$. For this particular integrand, convergence of the integral is not an issue.

Beta Function

> **Definition D.2**   The *beta function* is defined for all real values $a > 0$ and $b > 0$ as
>
> $$B(a, b) = \int_0^1 t^{a-1}(1-t)^{b-1}\, dt.$$

The beta function is symmetric in its arguments:

$$B(a, b) = B(b, a).$$

Indeed, it can be shown that

$$B(a, b) = \frac{\Gamma(a)\,\Gamma(b)}{\Gamma(a+b)},$$

or, equivalently, that

$$\ln\big(B(a, b)\big) = \ln\big(\Gamma(a)\big) + \ln\big(\Gamma(b)\big) - \ln\big(\Gamma(a+b)\big).$$

Thus, symmetry is obvious, as is the observation that the way to calculate the beta function is by using the function LogGamma.

The function $\ln\big(B(a, b)\big)$ is in the library as the private (static) function LogBeta. Locally to the file rvms.c, this function can be evaluated as

```
x =  LogBeta(a, b);
return x;
```

In the event that $B(a, b)$ must be evaluated, use

```
x = exp(LogBeta(a, b));
return x;
```

The beta function is used in the definition of the beta distribution (Casella and Berger, 2002).

Binomial Coefficient.   The beta function is particularly useful for evaluating the *binomial coefficient* $C(n, m)$ when the integer-valued arguments $0 \le m \le n$ are large. The relevant equations are listed in Exercise D.2. The function $\ln\big(C(n, m)\big)$ is available in the library as the function LogChoose. To evaluate $\ln\big(C(n, m)\big)$, use

```
x = LogChoose(n, m);
return x;
```

If you need to evaluate $C(n, m)$ (as a floating-point number), use

```
x = exp(LogChoose(n, m));
return x;
```

Incomplete Gamma Function

> **Definition D.3** The *incomplete gamma function* is defined for all real values $a > 0$ and $x > 0$ as
> $$P(a, x) = \frac{1}{\Gamma(a)} \int_0^x t^{a-1} \exp(-t) \, dt.$$

For a fixed value of $a > 0$, the incomplete gamma function $P(a, x)$ is a monotone-increasing function of x that is bounded between 0 and 1 and has the limiting values

$$P(a, 0) = 0 \qquad \text{and} \qquad P(a, \infty) = 1.$$

There is a *power series* representation,

$$P(a, x) = x^a \exp(-x) \sum_{n=0}^{\infty} \frac{x^n}{\Gamma(a + n + 1)},$$

that is valid (converges) for all $x > 0$. Because the convergence of this series is particularly rapid if $x < a + 1$, this representation provides a good way to evaluate $P(a, x)$ if x is small. There is also a *continued fraction* representation,

$$P(a, x) = 1 - \frac{x^a \exp(-x)}{\Gamma(a)} \left(\cfrac{1}{x + \cfrac{1-a}{1 + \cfrac{1}{x + \cdots}}} \right),$$

which we write as the more compact

$$P(a, x) = 1 - \frac{x^a \exp(-x)}{\Gamma(a)} \left(\frac{1}{x+} \frac{1-a}{1+} \frac{1}{x+} \frac{2-a}{1+} \frac{2}{x+} \frac{3-a}{1+} \cdots \right),$$

that is valid (converges) for all $x > 0$. The convergence of this continued fraction is rapid if $x > a + 1$; thus, this representation provides a good way to evaluate $P(a, x)$ if x is large. The function $P(a, x)$ is in the library as the private (`static`) function InGamma. Locally to the file rvms.c, this function can be evaluated as

```
x = InGamma(a, x);
return x;
```

The incomplete gamma function also happens to be the cdf for the gamma distribution and, hence, is called to construct the cdf of two special cases of the gamma distribution: the *Erlang* (n, b) distribution and the *Chisquare* (n) distribution.

Incomplete Beta Function

Definition D.4   The *incomplete beta function* is defined for all real $a > 0, b > 0$ and $0 < x < 1$ as

$$I(a, b, x) = \frac{1}{B(a, b)} \int_0^x t^{a-1}(1 - t)^{b-1} \, dt.$$

For fixed values of $a > 0$, $b > 0$, the incomplete beta function $I(a, b, x)$ is a monotone-increasing function of x that is bounded between 0 and 1 and has the limiting values

$$I(a, b, 0) = 0 \quad \text{and} \quad I(a, b, 1) = 1.$$

It can be shown that the incomplete beta function satisfies the following *symmetry equation*:

$$I(a, b, x) = 1 - I(b, a, 1 - x).$$

There is a *continued-fraction* representation

$$I(a, b, x) = \frac{x^a(1 - x)^b}{a B(a, b)} \left(\frac{1}{1+} \frac{c_1}{1+} \frac{c_2}{1+} \frac{c_3}{1+} \cdots \right)$$

having the coefficients

$$c_{2n+1} = -\frac{(a + n)(a + b + n)x}{(a + 2n)(a + 2n + 1)}$$

and

$$c_{2n} = \frac{n(b - n)x}{(a + 2n - 1)(a + 2n)}$$

that converges rapidly for $x < (a + 1)/(a + b + 1)$. Therefore this continued fraction provides an efficient way to evaluate $I(a, b, x)$ when x is less than $(a + 1)/(a + b + 1)$. If, instead, $x > (a + 1)/(a + b + 1)$, then $1 - x < (b + 1)/(a + b + 1)$. Therefore, if x is greater than $(a + 1)/(a + b + 1)$, the symmetry equation $I(a, b, x) = 1 - I(b, a, 1 - x)$ can be used to evaluate $I(a, b, x)$ by efficiently evaluating $I(b, a, 1 - x)$ instead.

The special function $I(a, b, x)$ is available in the library as the private (`static`) function `InBeta`. Locally to the file `rvms.c`, this function can be evaluated as

```
x = InBeta(a,b,x);
return x;
```

References. The standard encyclopedic reference on special functions is Abramowitz and Stegum (1964). Two good numerically oriented references on special functions are Chapter 6 of Press, Flannery, Teukolsky, and Vetterling (1992) and Chapter 5 of Kennedy and Gentle (1980). There is also a significant amount of relevant material in Knuth (1997).

D.2 Exercises

D.1 (*a*) Prove that $\Gamma(1) = 1$. (*b*) Use integration by parts to prove that $\Gamma(a + 1) = a\Gamma(a)$ for all $a > 0$. (*c*) Use an induction argument to conclude that $\Gamma(n) = (n-1)!$ for all positive integers n.

D.2 Use the fact that

$$B(a, b) = \frac{\Gamma(a)\,\Gamma(b)}{\Gamma(a + b)}$$

to prove that

$$C(n, m) = \binom{n}{m} = \frac{1}{m\,B(m,\ n - m + 1)} \qquad (m \neq 0)$$

and

$$C(n, m) = \binom{n}{m} = \frac{1}{(n - m)\,B(m + 1,\ n - m)} \qquad (m \neq n).$$

D.3 Use Definition D.2 to prove that $B(a, b) = B(b, a)$.

D.4 Use Definition D.2 to prove that

$$B(a, 1) = B(1, a) = \frac{1}{a}.$$

D.5 (*a*) Use Definition D.3 to prove that

$$P(1, x) = 1 - \exp(-x)$$
$$P(2, x) = 1 - \exp(-x) - x\exp(-x).$$

(*b*) Use the continued-fraction representation of $P(a, x)$ to verify these results.

D.6 Look up *Stirling's formula* in any good calculus textbook. How does it relate to Lanczos's approximation to the gamma function?

D.7 Use Definitions D.1 and D.2 to prove that

$$B(a, b) = \frac{\Gamma(a)\,\Gamma(b)}{\Gamma(a + b)}.$$

D.8 Prove that $\Gamma(1/2) = \sqrt{\pi}$.

Appendix E. Random-Variate Generators

This appendix summarizes the C functions contained in the Random-Variate GeneratorS library rvgs. The library contains six functions for generating discrete random variates and seven functions for generating continuous random variates. All discrete random variates are returned as type long. All continuous random variates are returned as type double.

Inversion is used to generate all variates in all of the functions. Thus, each function is both synchronized (i.e., one random number produces one variate) and monotone (i.e., larger random numbers produce larger variates). Also, certain variance-reduction techniques [see Chapter 8 of Ross (2002) or Chapter 11 of Law and Kelton (2000)] that exploit inversion can be used to reduce the variability of the estimates of the performance measure of interest. Further reading on random-variate generators is given in Devroye (1986) and Cheng (1998).

Discrete Random-Variate Generators. The six discrete random-variate generators in the library rvgs are based on the discussion and examples in Section 6.2. The discrete parametric random-variables models are summarized in Section 6.4. The functions are the following:

- long Bernoulli(double p) —returns 1 with probability p, 0 otherwise;
- long Binomial(long n, double p) —returns a *Binomial*(n, p) random variate;
- long Equilikely(long a, long b) —returns an *Equilikely*(a, b) random variate;
- long Geometric(double p) —returns a *Geometric*(p) random variate;
- long Pascal(long n, double p) —returns a *Pascal*(n, p) random variate;
- long Poisson(double μ) —returns a *Poisson*(μ) random variate.

Continuous Random-Variate Generators. The seven continuous random-variate generators in the library rvgs are based on the discussion and examples in Section 7.2. The continuous parametric random-variables models are summarized in Section 7.4. The functions are the following:

- double Chisquare(long n) —returns a *Chisquare*(n) random variate;
- double Erlang(long n, double b) —returns an *Erlang*(n, b) random variate;

- `double Exponential(double` μ`)` —returns an *Exponential* (μ) random variate;
- `double Lognormal(double` a`, double` b`)` —returns a *Lognormal* (a, b) random variate;
- `double Normal(double` μ`, double` σ`)` —returns a *Normal* (μ, σ) random variate;
- `double Student(long` n`)` —returns a *Student* (n) random variate;
- `double Uniform(double` a`, double` b`)` —returns a *Uniform* (a, b) random variate.

Appendix F. Correlation and Independence

Given a bivariate data sample (u_i, v_i) for $i = 1, 2, \ldots, n$, the traditional model associated with the statistical analysis of this sample is that the points (u_i, v_i) are the result of repeatedly sampling possible values of a pair of random variables (U, V). As in Section 4.4, the sample correlation coefficient r is then a statistic used to test the hypothesis that U and V are *linearly related*—that is, values of $|r|$ close to 1 are used to argue the existence of constants a, b, c such that $aU + bV + c \cong 0$.

F.1 Bivariate Random Variables

Definition F.1 The random variables U and V (discrete or continuous) are *independent* if and only if

$$\Pr\big((U \leq u) \text{ and } (V \leq v)\big) = \Pr(U \leq u)\,\Pr(V \leq v)$$

for all possible values u and v of U and V. If the random variables are not independent, they are said to be *dependent*.

Example F.1

The usual discrete-event simulation model of a single-server service node makes use of the *assumption* of independence. For example, for each job, the service time and arrival time are assumed to be independent. Also, generally the service time does not depend on the length of the queue, and so, unless the queue discipline is based upon a knowledge of a job's service times, the delay a job experiences in the queue is independent of the job's service time. The interarrival time and the delay in the queue, however, are *not* independent—that is, if the interarrival time is quite large, then any pre-existing queue has time to disappear, and, as a result, the delay in the queue experienced by a job with a large interarrival time is likely to be small. Similarly, a job's delay in the queue and wait in the service node are definitely not independent, particularly if the utilization is close to 1.

Independence is a very strong condition—it means that the value of one random variable does not depend in *any* (linear or nonlinear) way on the value of the other. How does independence relate to correlation?

The answer is that independent random variables are uncorrelated. The converse is *not* necessarily true, however—uncorrelated random variables can be dependent. To understand why this is so, we begin by defining the covariance and correlation of two random variables.

Covariance

Definition F.2   The *covariance* of two random variables U and V is

$$\gamma_{UV} = E\big[(U - \mu_U)(V - \mu_V)\big],$$

where

$$\mu_U = E[U] \qquad \text{and} \qquad \mu_V = E[V]$$

are the means of U and V, respectively.

The sample covariance (Definition 4.4.2) is an estimate of γ_{UV}, which is calculated from data pairs. Moreover, just as there is an alternative formula for the sample covariance, an equivalent expression for the covariance is

$$\gamma_{UV} = E[UV] - \mu_U \mu_V.$$

The derivation of this equation is left as an exercise.

Correlation Coefficient

Definition F.3   The *correlation coefficient* of two random variables U and V is

$$\rho_{UV} = \frac{\gamma_{UV}}{\sigma_U \sigma_V},$$

where

$$\sigma_U = \sqrt{E\big[(U - \mu_U)^2\big]} \qquad \text{and} \qquad \sigma_V = \sqrt{E\big[(V - \mu_V)^2\big]}$$

are the standard deviations of U and V respectively, which are assumed to be nonzero.

The sample correlation coefficient (Definition 4.4.2) is an estimate of ρ_{UV}. Therefore, by analogy, it is reasonable to expect the following:

- $|\rho_{UV}| \leq 1$;
- $|\rho_{UV}| = 1$ if and only if there exists constants (a, b, c) such that $aU + bV + c = 0$.

The proof that these two statements are true is left as an exercise.

Independence Implies Uncorrelated.   One important consequence of independence (presented without proof) is that, if the random variables U and V are independent, then

$$E[UV] = E[U]E[V].$$

From this result, if U and V are independent, then $\gamma_{UV} = 0$. That is,

$$\begin{aligned}
\gamma_{UV} &= E[UV] - \mu_U \mu_V \\
&= E[U]E[V] - \mu_U \mu_V \\
&= \mu_U \mu_V - \mu_U \mu_V \\
&= 0,
\end{aligned}$$

as summarized by the following theorem. This theorem provides the justification for using a nonzero value of the sample correlation coefficient to argue the *non*independence of the two random variables from which the sample was drawn.

Theorem F.1   If the random variables U and V (discrete or continuous) are independent, then they are uncorrelated. That is

- if U and V are independent then U and V are uncorrelated.

Equivalently, in terms of the contrapositive,

- if U and V are correlated (positive or negative) then U and V are dependent.

Example F.2

Consider an urn filled with black balls and white balls. Let p be the proportion of black balls, and suppose that two balls are drawn from the urn, in sequence, with replacement. Let the random variables U and V count the number of black balls (0 or 1) on the first draw and on the second draw respectively. Because the draws are *with* replacement, U and V are independent. Moreover, U and V are *Bernoulli*(p) random variables, and so $E[U] = E[V] = p$. The possible values of UV are 0, 1, with $\Pr(UV = 1) = p^2$. Therefore UV is a *Bernoulli*(p^2) random variable, and so $E[UV] = p^2$. From these calculations, $\gamma_{UV} = p^2 - p^2 = 0$, and so $\rho_{UV} = 0$: As is consistent with Theorem F.1, we see that U and V are uncorrelated.

Example F.3

The converse of Theorem F.1 is not necessarily true. For example, let U be $Uniform(-1, 1)$ and define $V = U^2$. It is intuitive that U and V are *not* independent. They are, however, uncorrelated. The pdf of U is $f(u) = 1/2$ for $-1 < u < 1$, so

$$E[U] = \int_{-1}^{1} \frac{1}{2} u \, du = \cdots = 0$$

$$E[V] = E[U^2] = \int_{-1}^{1} \frac{1}{2} u^2 \, du = \cdots = \frac{1}{3}$$

$$E[UV] = E[U^3] = \int_{-1}^{1} \frac{1}{2} u^3 \, du = \cdots = 0.$$

From these calculations, $\gamma_{UV} = E[UV] - E[U]E[V] = 0 - 0 = 0$, and so $\rho_{UV} = 0$. Thus we see that U and V are uncorrelated but dependent random variables. The dependence of V on U despite ρ_{UV} being zero becomes clear when one views the scatterplot generated by, say, 100 (u_i, v_i) samples of (U, V), along with the associated regression line, as illustrated in Figure F.1.

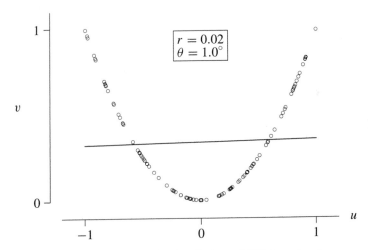

Figure F.1   Monte Carlo simulation of 100 (u_i, v_i) pairs.

The point here is that the correlation that exists between U and V is *nonlinear* and so is not reflected in the computation of the *linear* correlation coefficient.

The primary significance of Theorem F.1 is that the sample correlation coefficient can be used as a test for *dependence*: If r is the sample correlation coefficient of the data (u_i, v_i)

for $i = 1, 2, \ldots, n$ and if $|r|$ is significantly different from 0 then, with great confidence, we can conclude that the (U, V) pair of random variables that generated the data is *not* independent.*

Example F.4

In Example 4.4.2, a modified version of program `ssq2` was used to simulate a FIFO $M/M/1$ service node and generate a steady-state sample of 100 interarrival times, delays, waits, and service times. Scatterplots and the associated correlation coefficients corresponding to four of the six possible pairings were illustrated. Intuition, coupled with the results of this experiment, suggests the following (symmetric) table of ρ_{UV} correlation coefficient values

	interarrival	delay	wait	service
interarrival	1			
delay	−	1		
wait	−	+	1	
service	0	0	+	1

where $+$, $-$, and 0 indicate positive, negative, and no correlation respectively. The 0's in this table are a consequence of Theorem F.1: Because the interarrival times and service times are independent, their correlation is 0. Similarly, the delay and service times are also independent, and so their correlation coefficient is 0 as well. The 1's down the diagonal reflect the fact that each of the four random variables is perfectly correlated with itself. The values of the four remaining correlation coefficients cannot be established "by inspection." It is relatively easy to establish equations for two of them, however. These equations are a consequence of the following two theorems.

Theorem F.2   If U and V are random variables (discrete or continuous), and if $W = U + V$, then the mean of W is

$$\mu_W = \mu_U + \mu_V$$

and the variance of W is

$$\sigma_W^2 = \sigma_U^2 + 2\gamma_{UV} + \sigma_V^2 = \sigma_U^2 + 2\rho_{UV}\sigma_U\sigma_V + \sigma_V^2.$$

The proof of this theorem is left as an exercise. Note that

$$\sigma_W^2 = \sigma_U^2 + \sigma_V^2$$

if and only if U and V are uncorrelated. The independence of U and V is a *sufficient* condition for this equation to be valid.

---
*There is no simple test for independence.

M/M/1 Correlations

> **Theorem F.3**   Let the random variables W, D, and S represent the wait, delay, and service time experienced by a randomly selected job when a FIFO $M/M/1$ service node is in steady state. Define $\rho = \lambda/\nu$, where λ is the arrival rate and ν is the service rate. Then
>
> - the steady-state correlation between W and S is $\rho_{WS} = 1 - \rho$, and
> - the steady-state correlation between W and D is $\rho_{WD} = \sqrt{\rho(2 - \rho)}$.

Proof. Let μ_W, μ_D, and μ_S and σ_W^2, σ_D^2, and σ_S^2 denote the mean and variance of W, D, and S, respectively. To prove this theorem, we begin with the covariance between W and S. Recall that $W = D + S$ and that D and S are independent. Therefore,

$$\mu_W = \mu_D + \mu_S \qquad \text{and} \qquad \sigma_W^2 = \sigma_D^2 + \sigma_S^2.$$

Moreover,

$$E[WS] = E[(D + S)S] = E[DS + S^2] = E[D]\,E[S] + E[S^2] = \mu_D \mu_S + E[S^2];$$

as a consequence,

$$\gamma_{WS} = E[WS] - \mu_W \mu_S = \mu_D \mu_S + E[S^2] - (\mu_D + \mu_S)\mu_S = E[S^2] - \mu_S^2 = \sigma_S^2.$$

Using an analogous argument, we find that the covariance between W and D is $\gamma_{WD} = \sigma_D^2$. Therefore, the two correlation coefficients of interest are

$$\rho_{WS} = \frac{\sigma_S}{\sigma_W} \qquad \text{and} \qquad \rho_{WD} = \frac{\sigma_D}{\sigma_W}.$$

To complete the proof, we need to relate the three standard deviations σ_W, σ_D, and σ_S to the steady-state utilization ρ. Recall (Section 8.5) that, for an $M/M/1$ service node, the service time is *Exponential* (μ_S). Similarly, if the queue discipline is FIFO, the steady-state wait in the node is *Exponential* (μ_W), where $\mu_W = \mu_S/(1 - \rho)$. Because an *Exponential* random variable has its standard deviation equal to its mean, it follows that

$$\rho_{WS} = \frac{\sigma_S}{\sigma_W} = \frac{\mu_S}{\mu_W} = \frac{\mu_W(1 - \rho)}{\mu_W} = 1 - \rho.$$

Similarly,

$$\sigma_D^2 = \sigma_W^2 - \sigma_S^2 = \mu_W^2 - \mu_S^2 = \mu_W^2 - \mu_W^2(1 - \rho)^2 = \mu_W^2 \rho(2 - \rho),$$

and, therefore,

$$\rho_{WD} = \frac{\sigma_D}{\sigma_W} = \frac{\mu_W \sqrt{\rho(2 - \rho)}}{\mu_W} = \sqrt{\rho(2 - \rho)},$$

which establishes the theorem. $\square$

Example F.5

The following table summarizes the FIFO $M/M/1$ service node steady-state correlation coefficients ρ_{WS} and ρ_{WD} for selected values of ρ.

ρ	0.1	0.2	0.3	0.4	0.5	0.6	0.7	0.8	0.9
ρ_{WS}	0.900	0.800	0.700	0.600	0.500	0.400	0.300	0.200	0.100
ρ_{WD}	0.436	0.600	0.714	0.800	0.866	0.917	0.954	0.980	0.995

Both (W, S) and (W, D) have a positive correlation, as expected.

The analytic evaluation of the other two nonzero correlation coefficients, ρ_{RD} and ρ_{RW}, is significantly more difficult. In particular, the delay D_i and the interarrival time R_i of the ith job are related to the delay D_{i-1} and service time S_{i-1} of the previous job (for a FIFO queue, per Section 1.2) by the nonlinear equation

$$D_i = \max\{0,\ D_{i-1} + S_{i-1} - R_i\} \qquad i = 1, 2, 3, \ldots.$$

Not only is this equation nonlinear; to evaluate γ_{RD}, one must know both the paired correlation between S_{i-1} and D_i and the serial correlation between D_{i-1} and D_i.

Example F.6

Fortunately, the two correlation coefficients ρ_{RD} and ρ_{RW} are relatively easy to *estimate* by simulation. Figure F.2 illustrates estimates of these two FIFO $M/M/1$ service

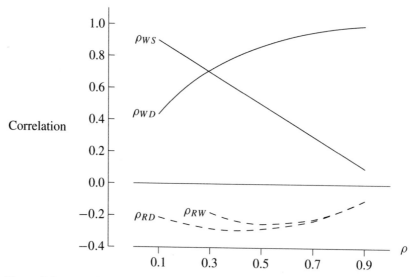

Figure F.2 Correlations between W, D, S, and R as a function of the traffic intensity $\rho = \lambda/\nu$ for an $M/M/1$ queue at steady state.

node steady-state correlation coefficients (as a dashed line). For comparison, the theoretical correlation coefficients from Theorem F.3 are also illustrated (as solid lines). Both (R, D) and (R, W) have a weak negative correlation, as expected.

F.2 Multivariate Random Variables

The high autocorrelation that typically exists in the time-sequenced stochastic data produced by a discrete-event simulation makes the statistical analysis of the data a challenge. Specifically, if we wish to make an *interval* estimate of some steady-state statistic—for example, the average wait in a service node—we must be prepared to deal with the impact of autocorrelation on our ability to make accurate estimates of the standard deviation. The following discussion introduces the notation and assumptions that are typically used in the analysis of discrete-event simulation output.

Let $X_1, X_2, \ldots, X_b$ be a *batch* (sequence) of random variables with common mean μ and common variance σ^2:

$$E[X_i] = \mu \qquad \text{and} \qquad E\left[(X_i - \mu)^2\right] = \sigma^2$$

for $i = 1, 2, \ldots, b$. These random variables are *not* assumed to be independent. Instead, we assume that the correlation between X_i and $X_{i'}$ is

$$\rho(i, i') = \frac{\gamma(i, i')}{\sqrt{\gamma(i, i)}\sqrt{\gamma(i', i')}} = \frac{\gamma(i, i')}{\sigma^2},$$

where the covariance is

$$\gamma(i, i') = E\left[(X_i - \mu)(X_{i'} - \mu)\right]$$

and $\gamma(i, i) = \sigma^2$ for $i = 1, 2, \ldots, b$. In addition, define the *batch mean* as

$$\overline{X} = \frac{1}{b}\sum_{i=1}^{b} X_i$$

and the *batch variance* as

$$S^2 = \frac{1}{b}\sum_{i=1}^{b}(X_i - \overline{X})^2.$$

Both $\overline{X}$ and S^2 are random variables. We will now compute the expected value of each, along with the variance of $\overline{X}$.

Expected Value of the Batch Mean.   The expected value of $\overline{X}$ is particularly easy to calculate. Because the expectation operator $E[\cdot]$ is *linear*, it follows that

$$E[\overline{X}] = \frac{1}{b}\sum_{i=1}^{b} E[X_i] = \frac{1}{b}\sum_{i=1}^{b}\mu = \mu.$$

As was discussed in Section 8.1, this result confirms that the batch (sample) mean $\overline{X}$ is an *unbiased* estimator of μ.

Expected Value of the Batch Variance.   To calculate the expected value of S^2, we begin with the observation that

$$\frac{1}{b}\sum_{i=1}^{b}(X_i - \mu)^2 = \frac{1}{b}\sum_{i=1}^{b}\left((X_i - \overline{X}) + (\overline{X} - \mu)\right)^2$$

$$= \frac{1}{b}\sum_{i=1}^{b}(X_i - \overline{X})^2 + \frac{2}{b}(\overline{X} - \mu)\sum_{i=1}^{b}(X_i - \overline{X}) + \frac{1}{b}\sum_{i=1}^{b}(\overline{X} - \mu)^2$$

$$\vdots$$

$$= S^2 + (\overline{X} - \mu)^2.$$

Therefore,

$$E[S^2] + E[(\overline{X} - \mu)^2] = \frac{1}{b}\sum_{i=1}^{b}E[(X_i - \mu)^2] = \frac{1}{b}\sum_{i=1}^{b}\sigma^2 = \sigma^2,$$

where

$$E[(\overline{X} - \mu)^2] > 0$$

is the variance of $\overline{X}$. Because the variance of $\overline{X}$ is positive, we see that the batch variance S^2 is a *biased* estimator of the common variance σ^2; $E[S^2] < \sigma^2$, because the batch variance *underestimates* σ^2 by the amount $E[(\overline{X} - \mu)^2]$. As will now be shown, the extent of this expected underestimation (bias) is determined by the correlation and can be significant.

Variance of the Batch Mean.   To calculate the expected value of $(\overline{X} - \mu)^2$, recognize that

$$\overline{X} - \mu = \frac{1}{b}\sum_{i=1}^{b}(X_i - \mu),$$

and so

$$(\overline{X} - \mu)^2 = \frac{1}{b^2}\sum_{i=1}^{b}(X_i - \mu)\sum_{i'=1}^{b}(X_{i'} - \mu) = \frac{1}{b^2}\sum_{i=1}^{b}\sum_{i'=1}^{b}(X_i - \mu)(X_{i'} - \mu).$$

Therefore, from the definition of $\gamma(i, i')$ and $\rho(i, i')$,

$$E[(\overline{X} - \mu)^2] = \frac{1}{b^2}\sum_{i=1}^{b}\sum_{i'=1}^{b}\gamma(i, i') = \frac{\sigma^2}{b^2}\sum_{i=1}^{b}\sum_{i'=1}^{b}\rho(i, i').$$

The following theorem summarizes the previous discussion. The details of the proof are left as an exercise.

Theorem F.4   Let $X_1, X_2, \ldots, X_b$ be a batch (sequence) of random variables with common mean μ and common variance σ^2. Let $\rho(i, i')$ be the correlation between X_i and $X_{i'}$ for $i = 1, 2, \ldots, b$ and $i' = 1, 2, \ldots, b$. Then the expected value of the batch mean is

$$E[\overline{X}] = \mu \qquad \text{(the batch mean is an \textit{unbiased} estimator of } \mu),$$

the expected value of the batch variance is

$$E[S^2] = \beta \sigma^2 \qquad \text{(the batch variance is a \textit{biased} estimator of } \sigma^2),$$

and the variance of the batch mean is

$$E\big[(\overline{X} - \mu)^2\big] = (1 - \beta)\sigma^2,$$

where the *bias* in the batch variance is

$$\beta = 1 - \frac{E\big[(\overline{X} - \mu)\big]^2}{\sigma^2} = 1 - \frac{1}{b^2}\sum_{i=1}^{b}\sum_{i'=1}^{b}\rho(i, i').$$

The bias in the batch variance is determined by the extent to which β is different from 1. This is illustrated by two important examples.

Example F.7

If $\rho(i, i') = 0$ for $i \neq i'$ (this is the case if the X_i's are independent, and thus uncorrelated), then, because $\rho(i, i) = 1$ for $i = 1, 2, \ldots, b$,

$$\beta = 1 - \frac{1}{b^2}\sum_{i=1}^{b}\sum_{i'=1}^{b}\rho(i, i') = 1 - \frac{1}{b^2}\sum_{i=1}^{b}\rho(i, i) = 1 - \frac{1}{b} = \frac{b-1}{b}.$$

Therefore, in this case, the expected value of the batch variance is

$$E[S^2] = \left(\frac{b-1}{b}\right)\sigma^2.$$

The result in Example F.7 is the basis for the interval-estimation equation established in Section 8.1: To estimate the variance of an *independent* sample (batch) of size b, the (slight) bias in the batch variance can be removed by using

$$\frac{1}{\beta}S^2 = \frac{1}{b-1}\sum_{i=1}^{b}(X_i - \overline{X})^2 \qquad \text{in place of} \qquad S^2 = \frac{1}{b}\sum_{i=1}^{b}(X_i - \overline{X})^2.$$

Unless b is small, this bias correction is largely irrelevant. As the next example illustrates, however, the bias correction is typically *not* irrelevant for correlated observations—which are the usual case in time-sequenced data generated by a discrete-event simulation.

Example F.8

If $\rho(i, i') = \rho_{|i-i'|}$ for $i \neq i'$, then the correlation matrix has the form

$$\begin{bmatrix} 1 & \rho_1 & \rho_2 & \cdots & \rho_{b-1} \\ \rho_1 & 1 & \rho_1 & \cdots & \rho_{b-2} \\ \rho_2 & \rho_1 & 1 & \cdots & \rho_{b-3} \\ \vdots & \vdots & \vdots & \ddots & \vdots \\ \rho_{b-1} & \rho_{b-2} & \rho_{b-3} & \cdots & 1 \end{bmatrix}.$$

This assumption is known as *weak stationarity*. (See Alexopoulos and Seila, 1998.) In this case,

$$\beta = 1 - \frac{1}{b^2} \sum_{i=1}^{b} \sum_{i'=1}^{b} \rho(i, i')$$

$$= 1 - \frac{1}{b^2} \left(b + 2(b-1)\rho_1 + 2(b-2)\rho_2 + \cdots + 2\rho_{b-1} \right)$$

$$= \frac{b-1}{b} - \frac{2}{b} \sum_{j=1}^{b-1} \left(1 - \frac{j}{b} \right) \rho_j,$$

where ρ_j is the autocorrelation between X's separated by a lag j. Therefore, in this case, the expected value of the batch variance is

$$E[S^2] = \beta\sigma^2 \qquad \text{where} \qquad \beta = \frac{b-1}{b} - \frac{2}{b} \sum_{j=1}^{b-1} \left(1 - \frac{j}{b} \right) \rho_j.$$

We can remove the bias introduced by the autocorrelation by using S^2/β to estimate σ^2. To do so, however, we must know $\rho_1, \rho_2, \ldots, \rho_{b-1}$. In practice, the best we can hope for is to *estimate* these autocorrelations by using Definitions 4.4.5 and 4.4.6—that is, we can estimate β as

$$\hat{\beta} = \frac{b-1}{b} - \frac{2}{b} \sum_{j=1}^{b-1} \left(1 - \frac{j}{b} \right) r_j,$$

where program acs is used to compute r_j for $j = 1, 2, \ldots, k$ and it is assumed that $\rho_j = 0$ for $j > k$, so that, if $b - 1 > k$, then $r_j = 0$ for $j = k, k+1, \ldots, b-1$.

Would the assumptions for Example F.8 hold for the wait times for jobs in a congested $M/M/1$ service node in steady state? First, consider the individual wait times. Does each job's wait time have the same mean and variance, i.e., does $E[X_i] = \mu$ and $V[X_i] = \sigma^2$ for some large index i? Intuitively, the answer is "yes"—all wait times should have identical distributions at steady state. Second, consider the correlation structure. Does the correlation between the wait times for job i and job i' depend only on $|i - i'|$ for the service node at steady state? As a specific instance, is the correlation between the wait times X_{946} and X_{949} (three jobs apart) the same as the correlation between X_{971} and X_{974} (also three jobs apart)? As in the first instance, the answer is probably "yes"—it is reasonable so assume that correlation depends only upon the difference in the indices of the jobs.

Example F.9
For example, the autocorrelation estimates from Example 4.4.3 were used to compute $\hat{\beta}$ for various batch sizes b. The "cut-off" autocorrelation lag was, somewhat arbitrarily, taken to be $k = 100$. The results were

b	16	32	64	128	256	512	1024
$\hat{\beta}$	0.17	0.29	0.43	0.60	0.77	0.88	0.94

and we see, as expected, $\hat{\beta} \to 1$ as $b \to \infty$—that is, for *large* batch sizes, the bias in the variance estimate vanishes. Note, however, that, for this example, positive autocorrelation causes small batches to yield a variance estimate that is (on average) too small by a *significant* factor.

F.3 Exercises

F.1 Prove that the two equations

$$\gamma_{UV} = E[(U - \mu_U)(V - \mu_V)] \qquad \text{and} \qquad \gamma_{UV} = E[UV] - \mu_U\mu_V$$

are equivalent.

F.2 Given that definition of the correlation coefficient of two random variables U and V is

$$\rho_{UV} = \frac{\gamma_{UV}}{\sigma_U\sigma_V},$$

prove that (a) $|\rho_{UV}| \le 1$; (b) $|\rho_{UV}| = 1$ if and only if there exist constants (a, b, c) such that $aU + bV + c = 0$. Hint: Minimize $E\left[(\alpha(U - \mu_U) + \beta(V - \mu_V))^2\right] \ge 0$ subject to the constraint $\alpha^2 + \beta^2 = 1$.

F.3 Relative to Example F.2, suppose the draw is *without* replacement, from an urn with n black and n white balls. (a) What is ρ_{UV}? (b) Note that $\rho_{UV} \to 0$ as $n \to \infty$. Why? (c) Verify the correctness of your ρ_{UV} equation via a Monte Carlo simulation for the case $n = 5$.

F.4 Suppose the relation in Example F.3 is $V = U^3$ (rather than $V = U^2$). (a) What is ρ_{UV}? (b) Comment.

F.5 Prove Theorem F.2.

F.6 How would Examples F.4 and F.6 change if the service discipline is "shortest job first"? Conjecture first, then simulate.

Appendix G. Error in Discrete-Event Simulation

This appendix discusses various sources of error in a discrete-event simulation. We base this discussion on a framework for the simulation-modeling process. We develop this framework by starting with a small subset of the framework and sequentially adding components. An additional benefit to the introduction of this framework is that it allows us to review many of the key concepts presented in the body of the text. We recommend reading this appendix prior to reading Chapter 9, which introduces input modeling.

G.1 Discrete-Event Simulation-Modeling Framework

Discrete-event simulation-model development typically goes through the 11-step process outlined in Section 1.1.2. Lurking behind this process, however, is error that can creep into a discrete-event simulation model from multiple sources. In order to isolate and quantify these various sources of error, we use a discrete-event simulation-modeling framework that we have adapted from Schmeiser (2001) and Nelson (1987). We begin with a small subset of the framework and sequentially add more components.

We will illustrate the framework through a series of examples that increase in complexity, most of which are based on the single-server service node. Fortunately, all of the concepts that are introduced on the simple single-server service-node model generalize to more complex models.

The discrete-event-modeling conceptual framework begins, as did our first model introduced in Section 1.2, with a trace-driven simulation. The letter V denotes an ordered, and possibly partitioned, set of numbers that represent the input to a discrete-event simulation.* The set V is typically referred to as "simulation-input data" and may contain thousands (or even millions) of numbers. The letter Y denotes an ordered, and possibly partitioned, set of numbers that represent

---

*As will be seen later, the choice of the letter V for the input is motivated by the fact that these will later become known as "Random Variates" (once random-number generation and random-variate generation are introduced into the framework).

the output from a discrete-event simulation. The set Y is typically referred to as "simulation-output data," and, like V, may contain thousands (or even millions) of numbers. Since the set V is being transformed to the set Y via a discrete-event simulation algorithm, we connect the two letters with an arrow for our first incarnation of the discrete-event simulation-modeling framework.

$$V \longrightarrow Y$$

Example G.1

In the single-server service-node discrete-event-simulation model, the set V consists of the arrival times $a_1, a_2, \ldots$ and the associated service times $s_1, s_2, \ldots$. In Example 1.2.2, for instance, the set V consists of the following ten arrival and service times:

i	1	2	3	4	5	6	7	8	9	10
a_i	15	47	71	111	123	152	166	226	310	320
s_i	43	36	34	30	38	40	31	29	36	30

In this model, V is an ordered set of 20 numbers, partitioned into two sets of ten numbers each. These numbers are passed through a discrete-event simulation algorithm, such as the one given in Algorithm 1.2.1, yielding the simulation-output data set Y. The set Y could be one or more quantities gathered from the simulation model that might be of interest to the analyst (e.g., wait times, server utilization). For simplicity, we assume that only the *delays* are of interest, so there is just an ordered set with no partitioning. For the 20 values in V (given previously), the ordered set Y follows.

i	1	2	3	4	5	6	7	8	9	10
d_i	0	11	23	17	35	44	70	41	0	26

A group of assumptions has gone into the conversion of the set V to the set Y. These assumptions are collected into what is referred to as a "logic model" $\mathcal{L}$. We maintain the convention that calligraphic letters are always associated with arrows (often representing transformations) in our framework. Thus our framework has now evolved to include the logic model $\mathcal{L}$, which is involved with the transformation from the set of simulation-input data V to the set of simulation-output data Y.

$$V \xrightarrow{\mathcal{L}} Y$$

Example G.2

For the single-server service-node model, examples of the assumptions that compose the logic model $\mathcal{L}$ can be broken into (i) assumptions about the server, (ii) assumptions about the queue, and (iii) assumptions about the jobs being processed. Assumptions about the server include the following:

- there is just one server;
- once a server begins processing a job, there are no interruptions;

- the server is not able to process more than one job at a time;
- the server never takes breaks.

Assumptions about the queue include the following:

- the queue discipline is FIFO (first-in, first-out), so jobs are not allowed to "pass" one another in the queue;
- the capacity (size) of the queue is infinite, so no arriving jobs are rejected.

Assumptions about the jobs include the following:

- jobs do not "feed back" in order to be reprocessed by the server;
- there is no time delay between the completion of the processing of one job and the initiation of service on a subsequent job waiting in the queue.

All of these assumptions must be in place in order to arrive at the ten values in the set Y computed in the previous example.

It is difficult to draw meaningful conclusions about the system from the numbers in the set Y. This is often the case, because there may be thousands or millions of numbers in Y that need to be summarized in a fashion that is helpful to the simulation analyst or decision maker. The simulation-output data in Y can be summarized graphically (such as by a histogram) or a numerically (such as by a sample mean). We will assume that the values in Y are summarized into one or more numerical quantities that will be referred to here as the ordered set $\hat{\theta}$. The decision of *which* numerical quantity or quantities (e.g., means, medians, modes, variances) should be used to summarize the simulation-output data Y is a statistical-inference question that we denote by S. Thus the framework has now expanded:

$$ V \xrightarrow{\ \mathcal{L}\ } Y \xrightarrow{\ S\ } \hat{\theta} $$

Example G.3

For the single-server service node presented in the previous two examples, there are many statistical measures of interest that could be computed from the delays. Examples include the maximum delay, the median delay, and the mean delay. If we are interested only in estimating the population mean delay, then the set $\hat{\theta}$ consists of only a single number, which, in this case, is the average of the ten delays in the set Y, $\bar{d} = 26.7$. The decision to calculate the sample mean delay time to estimate the population mean delay time is embodied in the symbol S, whereas the value of the sample mean delay time, 26.7, is contained in $\hat{\theta}$.

As indicated at the end of Chapter 1, the *trace-driven* approach presented in the examples thus far is rather limited. Simulations performed in the manner of the previous three examples will typically be too short to provide any useful information for decision makers. Efforts to extend the length of a simulation gave rise to the concept of a *random-number generator*, which was introduced in Chapter 2, and the notion of *random-variate generation*, which was introduced in an ad-hoc fashion in Sections 2.3 and 3.1 and then in a more organized manner in Sections 6.2 and 7.2. The addition of random numbers and random variates to the framework expands our framework to include (i) an ordered set of random-number generator seeds X_0, (ii) a random-number generator $\mathcal{G}_r$, (iii) an ordered set of random numbers U, and (iv) an input model $\mathcal{I}$:

$$X_0 \xrightarrow{\mathcal{G}_r} U \xrightarrow{\mathcal{I}} V \xrightarrow{\mathcal{L}} Y \xrightarrow{\mathcal{S}} \hat{\theta}$$

Detailed descriptions of the four new elements (X_0, $\mathcal{G}_r$, U, and $\mathcal{I}$) that have been added to the framework follow.

- X_0 is a set containing a single seed for a random-number generator or an ordered set containing the seeds for a multiple-stream random-number generator. This seed or these seeds are fed into the random-number generator denoted by $\mathcal{G}_r$.
- $\mathcal{G}_r$ denotes a random-number generator that is used to transform the seed or seeds in X_0 in to an ordered set of random numbers U. Careful selection of the random-number generator $\mathcal{G}_r$ will avoid such problems as the hyperplane issue outlined in Example 2.2.7. The subscript r on $\mathcal{G}$ denotes *random*-sampling variability. Two potential sources of random-sampling variability are: (i) the modeler could have the misfortune of choosing a random-number generator seed that leads to an associated set of random numbers U that, for example, has an unusually low or high mean, which biases the simulation output data Y; (ii) the modeler might run a simulation so short that an unusually small number of random numbers produces results that contain large variability.
- U is a set of ordered random numbers. These random numbers are generated by using the seed(s) in the set X_0 and the random-number generator $\mathcal{G}_r$. The number of random numbers in the set is dictated by the length of the simulation run. The purpose of these random numbers is to be converted to random variates in the set V.
- $\mathcal{I}$ is referred to as the "input model," constructed via the methods introduced in Chapter 9. The input model contains the distributions that are used to transform the random numbers in U into the random variates in V. A Monte Carlo or discrete-event "simulation model" consists of the input model $\mathcal{I}$ and the logic model $\mathcal{L}$.

Example G.4

Consider the single-server service-node implementation in program ssq2, described in Section 3.1. The nine symbols used in the framework (from left to right) are interpreted as follows.

- There is only a single stream of random numbers used in the simulation, so there is just a single seed in X_0, namely $x_0 = 123456789$, which is implemented in program ssq2 with the statement PutSeed(123456789).
- The random-number generator, $\mathcal{G}_r$ for this particular simulation, is the Lehmer generator

$$x_{i+1} = 48271 \, x_i \bmod (2^{31} - 1),$$

which was introduced in Chapter 2 and implemented in Random().

- The ordered set U contains 20 000 random numbers that will be transformed to 10 000 interarrival times and 10 000 service times. Since a multiple-stream generator was not used, these 20 000 numbers are not partitioned in any fashion. They are accessed in an as-needed fashion in program ssq2.

- The input model $\mathcal{I}$ consists of an *Exponential*(2.0) interarrival-time distribution and a *Uniform*(1.0, 2.0) service-time distribution.
- The simulation-input data set V is partitioned into 10 000 interarrival times and 10 000 service times.
- The logic model $\mathcal{L}$ contains all of the assumptions associated with a single-server service node listed in Example G.2.
- The simulation-output data set Y contains 10 000 interarrival times, 10 000 service times, 10 000 delays, and 10 000 waits. These 40 000 numbers are part of our conceptual framework, but program ssq2 does not actually store these numbers. The quantities needed to compute statistics of interest are gleaned as the simulation executes, in order to conserve memory.
- The statistical inference $\mathcal{S}$ on the single-server service node requires that sample means on four job-averaged statistics (interarrival time, service time, wait, and delay) and three time-averaged statistics (number of jobs in the service node, number of jobs in the queue, and server utilization) be computed. Program ssq2 does this by summing the appropriate values in the variables sum.delay, sum.wait, sum.service, and sum.interarrival as the simulation executes, then computing the means of the observation-based and time-based statistics upon completion of the simulation.
- There are seven measures of performance associated with the system that are contained in $\hat{\theta}$: average interarrival time, average wait, average delay, average service time, average number of jobs in the service node, average number of jobs in the queue, and server utilization. Program ssq2 prints their values: 2.02, 3.86, 2.36, 1.50, 1.91, 1.17, and 0.74, respectively, after the simulation reaches its terminal condition.

One surprising fact that emerges from this framework is that, once $\mathcal{G}_r$, $\mathcal{I}$, $\mathcal{L}$, and $\mathcal{S}$ are defined, the transformation running from the set X_0 to the set $\hat{\theta}$ is completely deterministic. Unless there is error induced by roundoff, the same seed value(s) in X_0, even on different computers and using different base languages, will result in identical value(s) in $\hat{\theta}$.

We now present four brief examples to introduce the sources of simulation error, each corresponding to one of the four arrows in the framework: $\mathcal{G}_r$, $\mathcal{I}$, $\mathcal{L}$, and $\mathcal{S}$.

Example G.5

Consider a simulation that contains a rare event (e.g., a nuclear meltdown). Say two consecutive unlikely events need to occur for this rare event to occur; the simulation analyst triggers this rare event whenever one random number from $\mathcal{G}_r$ exceeds 0.9999 and the subsequent random number exceeds 0.99999. Probability theory indicates that, for a random-number generator with $m = 2^{31} - 1$, the expected number of occurrences per cycle is

$$m \cdot 0.0001 \cdot 0.00001 \cong 2.15.$$

For the function Random() used in this text, however, there were *no* pairs of consecutive random numbers that achieved this criteria. This will lead the simulation analyst to significantly biased simulation results concerning this rare event (e.g., no meltdowns during the simulation). What happened? If all pairs of consecutive random numbers generated by `Random()` are plotted in the unit square, there is a tiny rectangle in the upper-right-hand corner [to the northeast of $(0.9999, 0.99999)$] that contains no random numbers. This is illustrated in Figure G.1, where no consecutive pairs of random numbers fall in the tiny rectangle. Is this a weakness of `Random()`? Yes, for this particular example, but not in general. Your authors burned many CPU cycles to arrive at this particular failing of `Random()`.

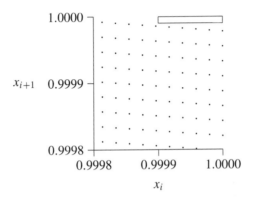

Figure G.1 Adjacent random-number pairs.

Example G.6

Assume that a particular feature of a simulation model is known to be *exactly* normally distributed. Rather than using an exact generator, the simulation analyst unwisely uses the *approximate* generator

$$Z = U_1 + U_2 + \cdots + U_{12} - 6,$$

where $U_1, U_2, \ldots, U_{12}$ is an *independent and identically distributed* sequence of $Uniform(0, 1)$ random variables, which was introduced in Section 7.2. By using this approximate generator, the analyst is introducing error along the $\mathcal{I}$ arrow that will eventually flow into the estimate of the measure(s) of performance $\hat{\theta}$.

Example G.7

It is not a difficult matter to misspecify a logic model $\mathcal{L}$. In the case of the single-server service node, the analyst might assume that the simulation-model queue discipline is FIFO, when in fact the queue discipline in the system happens to be "shortest processing time first." This model misspecification will probably generate an error between the model and system, which will result in error in $\hat{\theta}$.

Example G.8

Incorrect statistical procedures $\mathcal{S}$ can also produce error in the simulation. In the single-server service node, if the interest is in the *population mean delay*, yet we collect the *sample median delay*, we are collecting the wrong quantity, and error will be induced.

It is possible that good fortune will make this error small [e.g., the sample mean and the sample median will typically be nearly the same value if (i) the pdf of the delay is approximately symmetric, and (ii) the run time is long], but this should not be counted on.

The last four examples have discussed *error* in discrete-event simulation modeling, but neglected to indicate what the two quantities are whose error we are discussing. We introduce a new ordered set to the framework, θ, whose elements are analogous to $\hat{\theta}$, but are associated with the *system* rather than with the discrete-event simulation *model*. Only in a few rare, simple systems (e.g., an $M/M/1$ queue) are we able to know the values of the quantities in θ via analytic methods. Indeed, the reason that simulation exists as a discipline is to estimate these (typically unknown) quantities.

The quantities in θ are unknown *constants*, whereas the quantities in $\hat{\theta}$ are *random variables*. [If you contest the second half of this proposition, just change the random-number seed(s) in X_0, and you will see the values of the random variables in $\hat{\theta}$ change accordingly.] The constants in θ could be determined if an observer viewed a stationary system for an infinitely long time period and collected the appropriate statistics on the measures of performance of interest. This costly approach, however, would not allow a modeler to alter the system, as is the case in a discrete-event simulation model. We add the ordered set θ^* to our framework to denote the values of the measures of performance associated with the simulation model for an infinite-length simulation run.

One of the goals of successful discrete-event simulation modeling is to make the error, which we denote by $|\theta - \hat{\theta}|$, as small as possible. Although this is a noble goal, Box (1979) points out that

> "All models are wrong; some models are useful."

Professor Box is clearly a realist when it comes to the error that exists between θ and $\hat{\theta}$. The question that he brings up is whether we can make the error small enough so that the model provides useful information about the system.

Approximating a System with a Model.   The framework is now extended to incorporate the "modeling" portion of the simulation-modeling process—that is, creating an input model $\mathcal{I}$ and a logic model $\mathcal{L}$ that accurately reflect the system under consideration.

If data can be gathered on the system under consideration, which is typically the case, then this data is contained in the ordered, possibly partitioned set D. The process of collecting this data, including the decisions about which aspects of the model data are to be collected, is contained in C_r. The r subscript once again denotes $r$andom-sampling variability.

The process of converting the data set D to an input model $\mathcal{I}$ that captures the probabilistic aspects of the system is embodied in the symbol $\mathcal{P}$. The data can be used directly, forming a nonparametric model where an *Equilikely*$(1, n)$ random variate samples each of the n data values with probability $1/n$. Alternatively, a parametric probability model can be fitted to the data set by using techniques in Sections 9.2 and 9.3. Included in $\mathcal{P}$ is the *type* of distribution selected and the values used for the parameters (if any).

Finally, the process of creating a logic model $\mathcal{L}$ involves the process of making a group of assumptions about the system denoted by $\mathcal{A}$. These assumptions might not capture every nuance

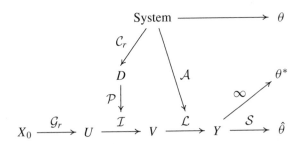

Figure G.2   Discrete-event simulation-modeling framework.

of the system (e.g., weather might be such a minor factor in the system that it can be disregarded with only minimal consequence). Some assumptions are made for implementation reasons; others are made for simplification reasons. The framework, now extended to its completion, is illustrated in Figure G.2.

Example G.9

Although it is a rather idealized system, we assume that the system is an $M/M/1$ queue having the system population mean interarrival time 1.0 minutes and the system population mean service time 0.9 minutes. These values correspond to the traffic intensity 0.9—a rather congested service node.

The first step in the data-collection process C_r is to determine the appropriate data elements to collect. In this case, collecting n consecutive interarrival times, which we denote by $X_1, X_2, \ldots, X_n$, is adequate to create an input model for the arrival process. Assume that we analyze the data set and have the good fortune to recognize that (i) the arrival process is stationary, and (ii) the exponential distribution is a reasonable fit for the interarrival distribution. This would be equivalent to having no error introduced in the process of developing a probability model $\mathcal{P}$. We can use the sample mean $\overline{X}$ as a parameter in our *Exponential*(μ) generator. But will $\overline{X}$ always equal the population value, 1.0? Certainly not. In fact, $\overline{X}$, being a scaled sum of unit exponential variates, will have an Erlang distribution. This *sampling variability* is captured in C_r in the framework. To create an input model for the service-time distribution, we collect m service times $Y_1, Y_2, \ldots, Y_m$. If we again have the good fortune to recognize that the service times are stationary and exponentially distributed, we can again use the sample mean $\overline{Y}$ for the mean service time in the simulation model. Finally, if we make all of the assumptions from Example G.2 concerning a single-server service node, then we have introduced no error into the logic model $\mathcal{L}$.

Assume further that our measure of performance of interest is the expected delay of the third customer that enters the system, given that the system begins in an empty and idle state. We choose this rather obscure measure of performance because Kelton and Law (1985) give an analytic expression for the expected delay of the third customer:

$$E\left[D_3\right] = \rho\mu_S\left[\frac{1+4\rho+2\rho^2}{(\rho+1)^3}\right],$$

where ρ is the traffic intensity and μ_S is the mean service time. For our system parameters—namely, $\mu_A = 1.0$, $\mu_S = 0.9$, and $\rho = \mu_S/\mu_A = 0.9$,

$$E\,[D_3] = 0.81 \left[\frac{1 + 3.6 + 1.62}{6.859}\right] \cong 0.7345.$$

Thus, the set θ consists of this one number: $\theta = 0.7345$. We have purposely designed this example so that the only source of error comes from the random sampling of the data set, C_r (assuming a long simulation run and an ideal random-number generator). Thus for various values of n (the number of interarrival times collected) and m (the number of service times collected), we could draw histograms of the delays of the third customers for multiple replications of the simulation model in order to analyze the effect of this random sampling variability. (See Exercise G.5.)

We close with two comments on the input-modeling process:

- There are simulation models for proposed systems (rather than existing systems) where it is not possible to collect data values in the set D. This situation arose when one of the authors of this text was working at NASA on a simulation model of a proposed space station. The time to construct the space station was needed, but the only way to proceed was to get minimums, maximums, and modes from experts on the time to complete the various construction operations. The *Triangular(a, b, c)* distribution was then used in the discrete-event simulation model.
- It is not necessary to use *all* nonparametric or *all* parametric models in a particular simulation. Mixing the two for a particular model (e.g., trace-driven interarrival times and a parametric model for service times in a single-server service node) is acceptable.

G.2 Summary

This subsection contains a summary of the high-level, abstract framework that has been developed for describing the process and the various sources of error that occur in constructing a discrete-event-simulation model as depicted in Figure G.2.

The uppercase letters X_0, U, V, Y, $\hat{\theta}$, θ^*, θ, and D denote ordered, and possibly partitioned, sets containing one or more numbers. To avoid writing "one or more numbers" in our descriptions of these sets, we assume that there are multiple numbers in the sets. The descriptions of these ordered sets follows.

- X_0 is a set of seeds for a random-number generator, one for each stream used in the implementation of the discrete-event-simulation model.
- U is a set of random numbers created by using the random-number generator G_r to transform the seeds in the set X_0 to random numbers. The random numbers in U are partitioned by the associated stream when multiple streams are employed. The *next-event* approach to discrete-event simulation allows us to generate random numbers one at a time as needed, in order to save memory and CPU time.
- V is a set of simulation-input data ("variates") created by applying the input model $\mathcal{I}$ to the set of random numbers U.

- Y is a set of simulation-output data generated by applying the logic model $\mathcal{L}$ to the set of simulation-input data V. The simulation-output data are typically dependent, although the probability model for each individual observation is often identical for a steady-state analysis once the simulation model warms up.
- $\hat{\theta}$ is a set of point estimators for the unknown measures of performance θ, calculated as a function of the simulation-output data Y. In general, there is some error present (i.e., $\hat{\theta} \neq \theta$).
- θ^* is the corresponding set of measures of performance associated with a simulation run of infinite length.
- θ is the corresponding set of measures of performance associated with the system of interest.
- D is a set of data values collected on appropriate elements of the system of interest in order to build an input model $\mathcal{I}$.

The calligraphic letters $\mathcal{G}_r, \mathcal{I}, \mathcal{L}, \mathcal{S}, \mathcal{C}_r, \mathcal{P}$, and $\mathcal{A}$ in Figure G.2 are all associated with arrows. These isolate seven sources of error associated with the discrete-event-simulation-modeling process. These letters denote transformations, probability models, data-collection methods, assumptions, and so on, as described below.

- $\mathcal{G}_r$ is a random-number generator used to transform the seeds in the set X_0 to the set of random numbers U.
- $\mathcal{I}$ is the input model used to transform the set of random numbers U to the set of simulation-input data V. The process of transforming U to V is known as *random-variate generation*. The input model is typically constructed by analyzing a set of data D, although, in rare cases, an input model is constructed in the absence of data by using expert opinion, bypassing the set D entirely.
- $\mathcal{L}$ is the logic model that captures assumptions made about the system and transformations (often formulated as algorithms) that are used to transform the set of simulation-input data V into the set of simulation-output data Y.
- $\mathcal{S}$ is a statistical-estimation procedure. The $\mathcal{S}$ connecting the set of simulation-output data Y and the set of point estimates of the measures of performance $\hat{\theta}$ involves computation of statistics, which are functions of the set of simulation-output data Y (e.g., sample mean, sample median, or sample variance).
- $\mathcal{C}_r$ denotes the data-collection procedures from the system of interest. It is crucial to collect the appropriate data elements from the system. Also, the data should be collected in an appropriate fashion, via standard sampling techniques.
- $\mathcal{P}$ involves the process of formulating a probabilistic input model that adequately describes the set of data collected in D. The $\mathcal{P}$ connecting the set of system data values D and the input model $\mathcal{I}$ involves either resampling the data (i.e., the nonparametric approach) or fitting a parametric model to the data set.
- $\mathcal{A}$ denotes assumptions made on the system of interest. These assumptions are used to create the logic model $\mathcal{L}$ describing the operation of the system. Incorrect or simplifying assumptions lead to modeling error.

The simulation model consists of the combination of the probabilistic input model $\mathcal{I}$ and the logical model $\mathcal{L}$. Once the simulation model, $\mathcal{I}$ and $\mathcal{L}$, has been determined, the sequence of four arrows leading from X_0 to $\hat{\theta}$ is a sequence of four deterministic functions for a particular random-number generator, $\mathcal{G}_r$, and choice of sample statistics collected, $\mathcal{S}$. All that is needed to arrive at $\hat{\theta}$ are the random-number seeds in the set X_0.

Error can occur in any of the arrows labeled by a calligraphic letter. There is no letter on the arrow attaching the system of interest to the measures of performance θ because there is no error associated with this transition. It is that the values of the measures of performance are unknown that typically necessitates the use of a discrete-event-simulation analysis for a complex system.* Two distinct types of error emerge:

- *Modeling error* captures the difference between the *system* and the *simulation model* ($\mathcal{I} \cup \mathcal{L}$) and is reflected in $|\theta - \theta^*|$ when an ideal random number generator $\mathcal{G}_r$ is used. It has three subcomponents: $\mathcal{C}_r$, $\mathcal{P}$, and $\mathcal{A}$. The *validation* process (see Balci, 1998) involves assessing how well the simulation model approximates the system.
- *Simulation sampling error* accounts for the fact that only a finite-length simulation has been run and is reflected in $|\theta^* - \hat{\theta}|$.

The mean-square error—for example,

$$E\big[(\hat{\theta} - \theta)^2\big] = E\big[\hat{\theta}^2 - 2\hat{\theta}\theta + \theta^2\big] = \cdots = V[\hat{\theta}] + \big(E[\hat{\theta}] - \theta\big)^2$$

—captures the simulation sampling error in the first term $V[\hat{\theta}]$ and the modeling error in the second term $(E[\hat{\theta}] - \theta)^2$. The mean-square error can be computed only on simple "toy" systems, where the values in θ are known.

The discussion here assumes an ideal system that does not change with time. Most real-world systems are changing over time, however, so an infinite sample drawn from the system reflects how the system performed at one particular point in time, not how it will perform in the future.

The r subscript denotes a step in the discrete-event modeling process where error from random-sampling variability is present. Both the random-number generator $\mathcal{G}_r$ and the data-collection procedures $\mathcal{C}_r$ involve random-sampling variability. An "unlucky" single random-number seed on a good generator $\mathcal{G}_r$ could, for example, produce a sequence of unusually small random numbers U, whose mean is significantly less than $1/2$. Likewise, an "unlucky" random sample from a legitimate data-collection procedure $\mathcal{C}_r$ could, for example, produce a sequence of unusually large data values in D. The error induced by random-sampling variability can be minimized by making numerous long simulation replications (in the case of $\mathcal{G}_r$) and by collecting large system data sets (in the case of $\mathcal{C}_r$). Almost universally, the former is cheaper than the latter.

The other sources of error are associated with the calligraphic letters in the diagram are as follows:

- using a poor random-number generator $\mathcal{G}_r$,
- making poor modeling decisions in $\mathcal{P}$ that result in a poor probabilistic input model $\mathcal{I}$,
- using incorrect system data sampling procedures $\mathcal{C}_r$,
- making incorrect or simplifying assumptions about the system in $\mathcal{A}$ that result in a poor logic model $\mathcal{L}$,
- making poor choices in $\mathcal{S}$ when analyzing the set of simulation-output data Y.

---

*Once the model has been developed, there are many techniques available to analyze the model. Simulation, of course, is the emphasis in this text.

Why do we simulate? An "analytic" model is appropriate when mathematics can be used to find the exact values of the measures of performance in θ. For many real-world systems, however, the transformation $U \longrightarrow V \longrightarrow Y \longrightarrow \hat{\theta}$ is so mathematically complex that the axiomatic approach to probability results in mathematically intractable expressions for the elements in the set $\hat{\theta}$. Equivalently, the numbers in the set Y are drawn from an unknown or mathematically intractable probability model.

G.3 Exercises

G.1   For the single-server service node, give the element of the framework given in Figure G.2 that is altered when (*a*) the queue discipline is changed to LIFO, (*b*) the random-number seed is changed to 987654321, (*c*) a larger data set is drawn from the system, (*d*) the choice of service-time distribution is changed from *Exponential* (2.0) to *Gamma* (2.0, 3.0), (*e*) median delays, rather than mean delays, are collected after the simulation run, (*f*) a different multiplier a is used on the random-number generator.

G.2   Consider the single-server service node with and without feedback. Which element(s) of the framework given in Figure G.2 differs for these two models?

G.3   Describe the sets V, Y, and $\hat{\theta}$ for the simple-inventory-system model from Section 1.3.

G.4   Implement a simulation to estimate the variance of the maximum of five independent and identically distributed *Exponential* (2.0) random variables. Use words and symbols to define X_0, $\mathcal{G}_r$, U, $\mathcal{I}$, V, $\mathcal{L}$, Y, $\mathcal{S}$, $\hat{\theta}$, θ^*, and θ. Run the simulation ten times with different seeds, and show that the value in the set $\hat{\theta}$ hovers around the value in the set θ.

G.5   Consider the model described in Example G.9. (*a*) For $n = 12$ and $m = 10$, run 10 000 simulations collecting the delay of the third customer. (Each simulation replication uses different $\overline{X}$ and $\overline{Y}$ values.) Calculate the sample mean and sample standard deviation and plot a histogram of the 10 000 values. (*b*) Read the article by Schruben and Kulkarni (1982) concerning estimating the parameters for the $M/M/1$ queue. Can any analytic conclusions be drawn for this particular simulation?

Bibliography

The page number(s) in brackets after each reference refers to the page number where the reference is cited.

Abramowitz, M., Stegum, I.E. (1964), *Handbook of Mathematical Functions*, National Bureau of Standards, AMS 55. [551]

Ahrens, J.H., Dieter, U. (1974), "Computer Methods for Sampling from Gamma, Beta, Poisson, and Binomial Distributions," *Computing*, **12**, 223–246. [346]

Alexopoulos, C., Seila, A.F. (1998), "Output Data Analysis," from *Handbook of Simulation: Principles, Methodology, Advances, Applications, and Practice*, J. Banks, ed., John Wiley & Sons, New York, 225–272. [386, 564]

Andradóttir, S. (1998), "Simulation Optimization," from *Handbook of Simulation: Principles, Methodology, Advances, Applications, and Practice*, J. Banks, ed., John Wiley & Sons, New York, 307–334. [35]

Andradóttir, S. (2006), "Random Search," from *Elsevier Handbooks in Operations Research and Management Science: Simulation*, B.L. Nelson, S.G. Henderson, eds., Elsevier, New York, forthcoming. [349]

Arkin, B.L., Leemis, L.M. (2000), "Nonparametric Estimation of the Cumulative Intensity Function for a Nonhomogeneous Poisson Process from Overlapping Realizations," *Management Science*, **46**, 989–998. [437]

Balci, O. (1998), "Verification, Validation, and Testing," from *Handbook of Simulation: Principles, Methodology, Advances, Applications, and Practice*, J. Banks, ed., John Wiley & Sons, New York, 335–393. [576]

Banks, J., Carson, J.S., Nelson, B.L., Nicol, D.M. (2005), *Discrete-Event System Simulation*, Fourth Edition, Prentice Hall, Upper Saddle River, NJ. [384, 401, 463, 527]

Barton, R.R., Schruben, L.W. (2001), "Resampling Methods for Input Modeling," from Proceedings of the 2001 Winter Simulation Conference, B.A. Peters, J.S. Smith, D.J. Medeiros, M.W. Rohrer, eds., 372–378. [411]

Barton, R.R. (2006), "Metamodel Simulation Optimization," from *Elsevier Handbooks in Operations Research and Management Science: Simulation*, B.L. Nelson, S.G. Henderson, eds., Elsevier, New York, forthcoming. [349]

Box, G.E.P., Muller, M.J. (1958), "A Note on the Generation of Random Normal Deviates," *Annals of Mathematical Statistics*, **29**, 610–611. [297]

Box, G.E.P. (1979), "Robustness in the Strategy of Scientific Model Building," from *Robustness in Statistics*, R.L. Launer, G.N. Wilkinson, eds., Academic Press, New York, 201–236. [572]

Bratley, P., Fox, B.L., Schrage, L.E. (1987), *A Guide to Simulation*, Second Edition, Springer–Verlag, New York. [8, 59, 401, 403]

Brown, R. (1988), "Calendar Queue," *Communications of the ACM*, **31**, 1220–1243. [221]

Bucklew, J.A. (2004), *Introduction to Rare Event Simulation*, Springer, New York. [108]

Caroni, C. (2002), "The Correct 'Ball Bearings' Data," *Lifetime Data Analysis*, **8**, 395–399. [404]

Carrano, F.M., Prichard, J.J. (2002), *Data Abstraction and Problem Solving with C++: Walls and Mirrors*, Addison–Wesley, Boston. [216]

Casella, G., Berger, R.L. (2002), *Statistical Inference*, Second Edition, Duxbury Press, Pacific Grove, CA. [423, 424, 548]

Chan, T.F., Golub, G.H., LeVeque, R.J. (1983), "Algorithms for Computing the Sample Variance: Analysis and Recommendations," *The American Statistician*, **37**, 242–247. [138, 146]

Chatfield, C. (2004), *The Analysis of Time Series: An Introduction*, Sixth Edition, Chapman & Hall, New York. [183, 406]

Cheng, R.C.H. (1977), "The Generation of Gamma Random Variables with Non-Integral Shape Parameter," *Applied Statistics*, **26**, 71–75. [344]

Cheng, R.C.H. (1998), "Random Variate Generation," from *Handbook of Simulation: Principles, Methodology, Advances, Applications, and Practice*, J. Banks, ed., John Wiley & Sons, New York, 139–172. [552]

Chhikara, R.S., Folks, L.S. (1989), *The Inverse Gaussian Distribution: Theory, Methodology and Applications*, Marcel Dekker, New York. [427]

Chick, S.E. (2001), "Input Distribution Selection for Simulation Experiments: Accounting for Input Modeling Uncertainty," *Operations Research*, **49**, 744–758. [426]

Çinlar, E. (1975), *Introduction to Stochastic Processes*, Prentice–Hall, Englewood Cliffs, NJ. [331]

Cochran, W.G. (1977), *Sampling Techniques*, Third Edition, John Wiley & Sons, New York. [403]

Crane, M.A., Iglehart, D.L. (1975), "Simulating Stable Stochastic Systems, III: Regenerative Processes and Discrete-Event Simulations," *Operations Research*, **23**, 33–45. [386]

D'Agostino, R.B., Stephens, M.A., eds. (1986), *Goodness-of-Fit Techniques*, Marcel Dekker, New York. [426]

David, H.A. (2004), *Order Statistics*, Third Edition, John Wiley & Sons, New York. [409]

Devroye, L. (1986), *Non-Uniform Random Variate Generation*, Springer–Verlag, New York. [348, 552]

Drew, J.H., Glen, A.G., Leemis, L.M. (2000), "Computing the Cumulative Distribution Function of the Kolmogorov–Smirnov Statistic," *Computational Statistics and Data Analysis*, **34**, 1–15. [460]

Efron, B., Tibshirani, R.J. (1993), *An Introduction to the Bootstrap*, Chapman & Hall, New York. [85]

Epp, S.S. (1990), *Discrete Mathematics with Applications*, Wadsworth, Pacific Grove, CA. [540]

Evans, M., Hastings, N., Peacock, B. (2000), *Statistical Distributions*, Third Edition, John Wiley & Sons, New York. [267, 326]

Fishman, G.S. (1971), "Estimating Sample Size in Computer Simulation Experiments," *Management Science*, **18**, 21–38. [386]

Fishman, G.S. (1973), "Statistical Analysis for Queueing Simulations," *Management Science*, **20**, 363–369. [386]

Fishman, G.S. (2001), *Discrete-Event Simulation: Modeling, Programming, and Analysis*, Springer, New York. [58, 59, 60, 205, 215, 308, 401, 463]

Fishman, G.S. (2006), *A First Course in Monte Carlo*, Duxbury, Belmont, CA. [3]

Fu, M. (2006), "Gradient Estimation," from *Elsevier Handbooks in Operations Research and Management Science: Simulation*, B.L. Nelson, S.G. Henderson, eds., Elsevier, New York, forthcoming. [108]

Fujimoto, R.M. (1998), "Parallel and Distributed Simulation," from *Handbook of Simulation: Principles, Methodology, Advances, Applications, and Practice*, J. Banks, ed., John Wiley & Sons, New York, 429–464. [532]

Gallagher, M.A., Bauer, K.W., Jr., Maybeck, K.W. (1996), "Initial Data Truncation for Univariate Output of Discrete-Event Simulations Using the Kalman Filter," *Management Science*, **42**, 559–575. [374]

Gentle, J.E. (2003), *Random Number Generation and Monte Carlo Methods*, Second Edition, Springer–Verlag, New York. [59]

Glynn, P.W. (2006), "Regenerative Method," from *Elsevier Handbooks in Operations Research and Management Science: Simulation*, B.L. Nelson, S.G. Henderson, eds., Elsevier, New York, forthcoming. [386]

Goldsman, D., Nelson, B.L. (1998), "Comparing Systems via Simulation," from *Handbook of Simulation: Principles, Methodology, Advances, Applications, and Practice*, J. Banks, ed., John Wiley & Sons, New York, 273–306. [386]

Goldsman, D., Nelson, B.L. (2006), "Batching Methods," from *Elsevier Handbooks in Operations Research and Management Science: Simulation*, B.L. Nelson, S.G. Henderson, eds., Elsevier, New York, forthcoming. [386]

Graham, R.L., Knuth, D.E., Patashnik, O. (1989), *Concrete Mathematics*, Addison–Wesley, New York. [540]

Gross, D., Harris, C.M. (1985), *Fundamentals of Queueing Theory*, Second Edition, John Wiley & Sons, New York. [102, 103]

Harrell, C., Ghosh, B.K., Bowden, R. (2000), *Simulation Using ProModel*, McGraw–Hill, New York. [527]

Heidelberger, P., Welch, P.D. (1981), "A Spectral Method for Confidence Interval Generation and Run Length Control in Simulations," *Communications of the ACM*, **24**, 233–245. [386]

Henderson, S.G. (2003), "Estimation for Nonhomogeneous Poisson Processes from Aggregated Data," *Operations Research Letters*, **31**, 375–382. [431]

Henderson, S.G., Mason, A.J. (2004), "Ambulance Service Planning: Simulation and Data Visualization," from *Operations Research and Health Care: A Handbook of Methods and Applications*, M.L. Brandeau, F. Sainfort, W.P. Pierskalla, eds., Kluwer, Boston, 77–102. [403]

Henriksen, J.O. (1983), "Event List Management—A Tutorial," from Proceedings of the 1983 Winter Simulation Conference, S. Roberts, J. Banks, B. Schmeiser, eds., 543–551. [193, 205, 207, 219]

Hillier, F.S., Lieberman, G.J. (2005), *Introduction to Operations Research*, Eighth Edition, McGraw–Hill, New York. [85]

Hogg, R.V., McKean, J.W., Craig, A.T. (2005), *Mathematical Statistics*, Sixth Edition, Prentice Hall, Upper Saddle River, NJ. [86, 289, 300, 323, 350, 364, 423]

Hoover, S.V., Perry, R.F. (1989), *Simulation: A Problem-Solving Approach*, Addison–Wesley, Reading, MA. [401]

Horn, R.A., Johnson, C.R. (1990), *Topics in Matrix Analysis*, Cambridge University Press, Cambridge. [75]

Johnson, N.L., Kotz, S., Kemp, A.W. (1993), *Univariate Discrete Distributions*, Second Edition, John Wiley & Sons, New York. [267, 326]

Johnson, N.L., Kotz, S., Balakrishnan, N. (1994), *Continuous Univariate Distributions, Volume I*, Second Edition, John Wiley & Sons, New York. [267, 326]

Johnson, N.L., Kotz, S., Balakrishnan, N. (1995), *Continuous Univariate Distributions, Volume II*, Second Edition, John Wiley & Sons, New York. [267, 326]

Johnson, N.L., Kotz, S., Balakrishnan, N. (1997), *Discrete Multivariate Distributions*, John Wiley & Sons, New York. [267, 326]

Joines, J.A., Roberts, S.D. (1998), "Object-Oriented Simulation," from *Handbook of Simulation: Principles, Methodology, Advances, Applications, and Practice*, J. Banks, ed., John Wiley & Sons, New York, 397–428. [532]

Jones, D.W. (1986), "An Empirical Comparison of Priority-Queue and Event-Set Implementations," *Communications of the ACM*, **29**, 300–310. [219]

Juneja, S., Shahabuddin, P. (2006), "Rare-Event Simulation Techniques: An Introduction and Recent Advances," from *Elsevier Handbooks in Operations Research and Management Science: Simulation*, B.L. Nelson, S.G. Henderson, eds., Elsevier, New York, forthcoming. [108]

Karian, Z.A., Dudewicz, E.J. (1991), *Modern Statistical, Systems, and GPSS Simulation: The First Course*, W.H. Freeman and Company, New York. [527]

Kelton, W.D., Law, A.M. (1985), "The Transient Behavior of the $M/M/s$ Queue, with Implications for Steady-State Simulation," *Operations Research*, **33**, 378–395. [394, 573]

Kelton, W.D., Sadowski, R.P., Sturrock, D.T. (2004), *Simulation with ARENA*, Third Edition, McGraw–Hill, New York. [527]

Kennedy, W.J., Gentle, J.E. (1980), *Statistical Computing*, Marcel Dekker, New York. [551]

Kilgore, R.A. (2002), "Web Services with .Net Technologies," from Proceedings of the 2002 Winter Simulation Conference, E. Yücesan, C.-H. Chen, J.L. Snowden, J.M. Charnes, eds., 841–846. [532]

Kim, S., Nelson, B.L. (2006), "Selecting the Best System," from *Elsevier Handbooks in Operations Research and Management Science: Simulation*, B.L. Nelson, S.G. Henderson, eds., Elsevier, New York, forthcoming. [349]

Klein, R.W., Roberts, S.D. (1984), "A Time-Varying Poisson Arrival Process Generator," *Simulation*, **43**, 193–195. [428, 442]

Kleinrock, L. (1975), *Queueing Systems, Volume I: Theory*, John Wiley & Sons, New York. [102]

Kleinrock, L. (1976), *Queueing Systems, Volume II: Computer Applications*, John Wiley & Sons, New York. [102]

Knuth, D.E. (1997), *The Art of Computer Programming, Volume 1: Fundamental Algorithms*, Third Edition, Addison–Wesley, Reading, MA. [551]

Knuth, D.E. (1998), *The Art of Computer Programming, Volume 2: Seminumerical Algorithms*, Third Edition, Addison–Wesley, Reading, MA. [53, 94, 444, 446, 456]

Kuhl, M.E., Damerdji, H., Wilson, J.R. (1998), "Least Squares Estimation of Nonhomogeneous Poisson Processes," from Proceedings of the 1998 Winter Simulation Conference, D.J. Medeiros, E.F. Watson, J.S. Carson, M.S. Manivannan, eds., 637–645. [442]

Kuhl, M.E., Sumant, S.G., Wilson, J.R. (2004), "An Automated Multiresolution Procedure for Modeling Complex Arrival Processes," *INFORMS Journal on Computing*, forthcoming. [442]

Kutner, M.H., Nachtsheim, C.J., Neter, J., Wasserman, W. (2003), *Applied Linear Regression Models*, Fourth Edition, McGraw–Hill, New York. [404]

Lada, E.K., Wilson, J.R. (2005), "A Wavelet-Based Spectral Procedure for Steady-State Simulation Analysis," *European Journal of Operational Research*, forthcoming. [386]

Lanczos, C. (1964), *Journal of the Society of Industrial and Applied Mathematics*, Series B: Numerical Analysis, **1**, 86–96. [547]

Law, A.M., Kelton, W.D. (2000), *Simulation Modeling and Analysis*, Third Edition, McGraw–Hill, New York. [8, 46, 59, 161, 401, 422, 425, 463, 552]

Lawless, J.F. (2003), *Statistical Models and Methods for Lifetime Data*, Second Edition, John Wiley & Sons, New York. [403, 404, 426]

L'Ecuyer, P., Simard, R., Chen, E.J., Kelton, W.D. (2002), "An Object-Oriented Random-Number Package with Many Long Streams and Substreams," *Operations Research*, **50**, 1073–1075. [59]

L'Ecuyer, P. (2006), "Uniform Random Number Generation," from *Elsevier Handbooks in Operations Research and Management Science: Simulation*, B.L. Nelson, S.G. Henderson, eds., Elsevier, New York, forthcoming. [37]

Lee, S., Wilson, J.R., Crawford, M.M. (1991), "Modeling and Simulation of a Nonhomogeneous Poisson Process Having Cyclic Behavior," *Communications in Statistics—Simulation and Computation*, **20**, 777–809. [441]

Leemis, L.M. (1991), "Nonparametric Estimation of the Cumulative Intensity Function for a Nonhomogeneous Poisson Process," *Management Science*, **37**, 886–900. [436, 437]

Lehmer, D.H. (1951), "Mathematical Methods in Large-Scale Computing Units," Proceedings of the 2nd Symposium on Large-Scale Calculating Machinery, Harvard University Press, 141–146. [39]

Lemieux (2006), "Quasi-Random Number Techniques," from *Elsevier Handbooks in Operations Research and Management Science: Simulation*, B.L. Nelson, S.G. Henderson, eds., Elsevier, New York, forthcoming. [108]

Lewis, P.A.W., Goodman, A.S., Miller, J.M. (1969), "A Pseudo-Random Number Generator for the System/360," *IBM Systems Journal*, **8**, 136–146. [52]

Lewis, P.A.W., Shedler, G.S. (1979), "Simulation of Nonhomogeneous Poisson Processes by Thinning," *Naval Research Logistics Quarterly*, **26**, 403–414. [329]

Lewis, P.A.W., Orav, E.J. (1989), *Simulation Methodology for Statisticians, Operations Analysts, and Engineers*, Volume 1, Wadsworth & Brooks/Cole, Pacific Grove, CA. [59, 401]

Little, J. (1961), "A Simple Proof of $L = \lambda W$," *Operations Research*, **9**, 383–387. [19]

McLeod, A.I., Bellhouse, D.R. (1983), "A Convenient Algorithm for Drawing a Simple Random Sample," *Applied Statistics*, **32**, 182–184. [279]

Marsaglia, G. (1968), "Random Numbers Fall Mainly in the Planes," *National Academy of Sciences Proceedings*, **61**, 25–28. [53]

Marsaglia, G., Bray, T.A. (1964), "A Convenient Method for Generating Normal Variables," *SIAM Review*, **6**, 260–264. [297]

Meeker, W.Q., Escobar, L.A. (1998), *Statistical Methods for Reliability Data*, John Wiley & Sons, New York. [441]

Meketon, M.S., Schmeiser, B.W. (1984), "Overlapping Batch Means: Something for Nothing?" from Proceedings of the 1984 Winter Simulation Conference, S. Sheppard, U.W. Pooch, C.D. Pegden, eds., 227–230. [386]

Nance, R.E. (1993), "A History of Discrete Event Simulation Programming Languages," *ACM SIGPLAN Notices*, **28**, 149–175. [526]

Nelson, B.L. (1987), "A Perspective on Variance Reduction in Dynamic Simulation Experiments," *Communications in Statistics*, **B16**, 385–426. [566]

Nelson, B.L., Yamnitsky, M. (1998), "Input Modeling Tools for Complex Problems," from Proceedings of the 1998 Winter Simulation Conference, D.J. Medeiros, E.F. Watson, J.S. Carson, M.S. Manivannan, eds., 39–46. [401]

Nicol, D.M., Heidelberger, P. (1995), "A Comparative Study of Parallel Algorithms for Simulating Continuous Time Markov Chains," *ACM Transactions on Modeling and Computer Simulation*, **5**, 326–354. [507]

Nijenhuis, A., Wilf, H.S. (1978), *Combinatorial Algorithms for Computers and Calculators*, Second Edition, Academic Press, San Diego. [279]

Niven, I., Zuckerman, H.S., Montgomery, H.L. (1991), *An Introduction to the Theory of Numbers*, Fifth Edition, John Wiley & Sons, New York. [540]

Odeh, R.E., Evans, J.O. (1974), "The Percentage Points of the Normal Distribution," *Applied Statistics*, **23**, 96–97. [296]

Ólafsson, S. (2006), "Metaheuristics," from *Elsevier Handbooks in Operations Research and Management Science: Simulation*, B.L. Nelson, S.G. Henderson, eds., Elsevier, New York, forthcoming. [349]

Park, S.K., Miller, K.W. (1988), "Random Number Generators: Good Ones Are Hard to Find," *Communications of the ACM*, **31**, 1192–1201. [52]

Pegden, C.D., Shannon, R.E., Sadowski, R.P. (1995), *Introduction to Simulation Using SIMAN*, Second Edition, McGraw–Hill, New York. [384]

Press, W.H., Flannery, B.P., Teukolsky, S.A., Vetterling, W.T. (1992), *Numerical Recipes in C*, Second Edition, Cambridge University Press, Cambridge. [551]

Pritsker, A.A.B. (1995), *Introduction to Simulation and SLAM II*, Fourth Edition, John Wiley & Sons, New York. [79, 527]

Qiao, H., Tsokos, C.P. (1994), "Parameter Estimation of the Weibull Probability Distribution," *Mathematics and Computers in Simulation*, **37**, 47–55. [422]

Rigdon, S., Basu, A.P. (2000), *Statistical Methods for the Reliability of Repairable Systems*, John Wiley & Sons, New York. [169, 442]

Ross, S. (2002), *Simulation*, Third Edition, Academic Press, San Diego. [442, 552]

Ross, S. (2006), *A First Course in Probability*, Seventh Edition, Prentice–Hall, Englewood Cliffs, NJ. [77, 460]

Schmeiser, B. (1982), "Batch Size Effects in the Analysis of Simulation Output," *Operations Research*, **30**, 556–568. [384]

Schmeiser, B.W. (2001), "Some Myths and Common Errors in Simulation Experiments," from Proceedings of the 2001 Winter Simulation Conference, B.A. Peters, J.S. Smith, D.J. Medeiros, M.W. Rohrer, eds., 39–46. [566]

Schriber, T.J., Brunner, D.T. (1998), "How Discrete-Event Simulation Software Works," from *Handbook of Simulation: Principles, Methodology, Advances, Applications, and Practice*, J. Banks, ed., John Wiley & Sons, New York, 765–812. [532]

Schruben, L.W. (1982), "Detecting Initialization Bias in Simulation Output," *Operations Research*, **30**, 569–590. [374]

Schruben, L.W. (1983), "Confidence Interval Estimation Using Standardized Time Series," *Operations Research*, **31**, 1090–1108. [386]

Schruben, L. (1992), *SIGMA: A Graphical Simulation Modeling Program*, The Scientific Press, San Francisco. [527]

Schruben, L., Kulkarni, R. (1982), "Some Consequences of Estimating Parameters for the $M/M/1$ Queue," *Operations Research Letters*, **1**, 75–78. [577]

Seila, A.F., Ceric, V., Tadikamalla, P. (2003), *Applied Simulation Modeling*, Brooks/Cole–Thompson Learning, Belmont, CA. [527]

Stephens, M.A. (1974), "EDF Statistics for Goodness of Fit and Some Comparisons," *Journal of the American Statistical Association*, **69**, 730–737. [461]

Summit, S.X. (1995), *C Programming Faqs: Frequently Asked Questions*, Addison–Wesley, Reading, MA. [55]

Swain, J.J. (2005) "Seventh Biennial Survey of Discrete-Event Software Tools," *OR/MS Today*, **32**, forthcoming. [527]

Szechtman, R. (2006), "Conditioning-Based Techniques," from *Elsevier Handbooks in Operations Research and Management Science: Simulation*, B.L. Nelson, S.G. Henderson, eds., Elsevier, New York, forthcoming. [108]

Taaffe, M.R., Clark, G.M. (1988), "Approximating Nonstationary Two-Priority Nonpreemptive Queueing Systems," *Naval Research Logistics*, **35**, 125–145. [442]

Trosset, M. (2005), *An Introduction to Statistical Inference and Its Applications*, The College of William & Mary, Williamsburg, VA, Department of Mathematics. [366]

Tufte, E. (2001), *The Visual Display of Quantitative Information*, Second Edition, Graphics Press, Cheshire, CT. [24, 160]

Vanden Eynden, C. (2001), *Elementary Number Theory*, Second Edition, McGraw–Hill, New York. [540]

Vitter, J.S. (1984), "Faster Methods for Random Sampling," *Communications of the ACM*, **27**, 703–718. [279]

von Neumann, J. (1951), "Various Techniques Used in Connection with Random Digits," *National Bureau of Standards, Applied Mathematics*, Series 12, 36–38. [338]

Wagner, M.A.F., Wilson, J.R. (1996), "Using Univariate Bézier Distributions to Model Simulation Input Processes," *IIE Transactions*, **28**, 699–711. [411]

Wand, M.P. (1997), "Data-Based Choice of Histogram Bin Width," *The American Statistician*, **51**, 59–64. [161]

Welford, B.P. (1962), "Note on a Method for Calculating Corrected Sums of Squares and Products," *Technometrics*, **4**, 419–420. [138]

Whitt, W. (2006), "Analysis for Design," from *Elsevier Handbooks in Operations Research and Management Science: Simulation*, B.L. Nelson, S.G. Henderson, eds., Elsevier, New York, forthcoming. [349]

Wilf, H.S. (1989), *Combinatorial Algorithms: An Update*, SIAM. [279]

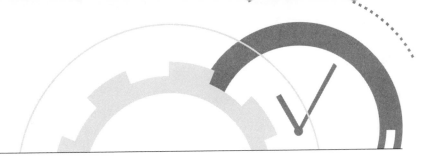

Index

Fractiles of the *Chisquare(n)* Distribution

n	$\alpha = 0.005$	$\alpha = 0.010$	$\alpha = 0.025$	$\alpha = 0.050$	$\alpha = 0.950$	$\alpha = 0.975$	$\alpha = 0.990$	$\alpha = 0.995$
1	0.00004	0.00016	0.00098	0.00393	3.841	5.024	6.635	7.879
2	0.01003	0.02010	0.05064	0.10259	5.991	7.378	9.210	10.597
3	0.07172	0.11483	0.21580	0.35185	7.815	9.348	11.345	12.838
4	0.20699	0.29711	0.48442	0.71072	9.488	11.143	13.277	14.860
5	0.41174	0.55430	0.83121	1.14548	11.070	12.833	15.086	16.750
6	0.67573	0.87209	1.23734	1.63538	12.592	14.449	16.812	18.548
7	0.98926	1.23904	1.68987	2.16735	14.067	16.013	18.475	20.278
8	1.34441	1.64650	2.17973	2.73264	15.507	17.535	20.090	21.955
9	1.73493	2.08790	2.70039	3.32511	16.919	19.023	21.666	23.589
10	2.15586	2.55821	3.24697	3.94030	18.307	20.483	23.209	25.188
15	4.60092	5.22935	6.26214	7.26094	24.996	27.488	30.578	32.801
20	7.43384	8.26040	9.59078	10.85081	31.410	34.170	37.566	39.997
25	10.51965	11.52398	13.11972	14.61141	37.652	40.646	44.314	46.928
30	13.78672	14.95346	16.79077	18.49266	43.773	46.979	50.892	53.672
35	17.19182	18.50893	20.56938	22.46502	49.802	53.203	57.342	60.275
40	20.70651	22.16426	24.43304	26.50930	55.758	59.342	63.691	66.766
45	24.31104	25.90127	28.36615	30.61226	61.656	65.410	69.957	73.166
50	27.99075	29.70668	32.35736	34.76425	67.505	71.420	76.154	79.490

Fractiles of the *Student(n)* Distribution

n	$\alpha = 0.005$	$\alpha = 0.010$	$\alpha = 0.025$	$\alpha = 0.050$	$\alpha = 0.950$	$\alpha = 0.975$	$\alpha = 0.990$	$\alpha = 0.995$
1	−63.657	−31.821	−12.706	−6.314	6.314	12.706	31.821	63.657
2	−9.925	−6.965	−4.303	−2.920	2.920	4.303	6.965	9.925
3	−5.841	−4.541	−3.182	−2.353	2.353	3.182	4.541	5.841
4	−4.604	−3.747	−2.776	−2.132	2.132	2.776	3.747	4.604
5	−4.032	−3.365	−2.571	−2.015	2.015	2.571	3.365	4.032
6	−3.707	−3.143	−2.447	−1.943	1.943	2.447	3.143	3.707
7	−3.499	−2.998	−2.365	−1.895	1.895	2.365	2.998	3.499
8	−3.355	−2.896	−2.306	−1.860	1.860	2.306	2.896	3.355
9	−3.250	−2.821	−2.262	−1.833	1.833	2.262	2.821	3.250
10	−3.169	−2.764	−2.228	−1.812	1.812	2.228	2.764	3.169
15	−2.947	−2.602	−2.131	−1.753	1.753	2.131	2.602	2.947
20	−2.845	−2.528	−2.086	−1.725	1.725	2.086	2.528	2.845
25	−2.787	−2.485	−2.060	−1.708	1.708	2.060	2.485	2.787
30	−2.750	−2.457	−2.042	−1.697	1.697	2.042	2.457	2.750
35	−2.724	−2.438	−2.030	−1.690	1.690	2.030	2.438	2.724
40	−2.704	−2.423	−2.021	−1.684	1.684	2.021	2.423	2.704
45	−2.690	−2.412	−2.014	−1.679	1.679	2.014	2.412	2.690
50	−2.678	−2.403	−2.009	−1.676	1.676	2.009	2.403	2.678
∞	−2.576	−2.326	−1.960	−1.645	1.645	1.960	2.326	2.576